Proceedings of the 4th International Conference

COMPUTER INTEGRATED

MANUFACTURING

21 – 24 October 1997
Singapore

Volume 2

Editors

A. Sen, A.I. Sivakumar and R. Gay

Gintic Institute of Manufacturing Technology
Nanyang Avenue
Singapore

 Springer

A. Sen, A.I. Sivakumar and R. Gay
Gintic Institute of Manufacturing Technology
71 Nanyang Drive
Singapore 638075

Computer Integrated Manufacturing

ISBN 981-3083-66-2 (vol. 1)
ISBN 981-3083-67-0 (vol. 2)
ISBN 981-3083-68-9 (set)

© Springer-Verlag Singapore Pte. Ltd. 1997
Printed in Singapore

Typesetting: Camera-ready by authors
5 4 3 2 1 0

COMPUTER INTEGRATED
MANUFACTURING

INTERNATIONAL CONFERENCE ON COMPUTER INTEGRATED MANUFACTURING (ICCIM)

Proceedings of the ICCIM'91
Ed. B.S. Lim
ISBN 981-02-0684-4

Proceedings of the ICCIM'93
Eds. A. Sen, J. Winsor and R. Gay
ISBN 981-02-1568-1 (set)
ISBN 981-02-1946-6 (vol. 1)
ISBN 981-02-1947-4 (vol. 2)

Proceedings of the ICCIM'95
Eds. J. Winsor, A.I. Sivakumar and R. Gay
ISBN 981-02-2376-5 (set)
ISBN 981-02-2973-9 (vol. 1)
ISBN 981-02-2974-7 (vol. 2)

ICCIM'91 and ICCIM'95 were published by World Scientific Pub. Co. ICCIM'93 was jointly published by World Scientific Pub. Co. and Global Pub.

ORGANISING COMMITTEE

Chairman :   Dr. Robert Gay, Gintic

Members :   Mr. Goh Kiah Mok, Gintic
Mr. Lew Sin Chye, Nanyang Technological University
Dr. Lim Kah Bin, National University of Singapore
Ms. Lim Lai Lee, Gintic
Ms. Jacqueline Ng, Miller Freeman
Dr. Anirudha Sen, Gintic
Dr. AI Sivakumar, Gintic
Dr. Soh Yeng Chai, William, Nanyang Technological University
Dr. V. Subramaniam, National University of Singapore
Dr. Tay Kiang Meng, Gintic
Mr. Tay Meng Leong, Nanyang Technological University
Mr. Stephen Teng, Singapore Industrial Automation Association

TECHNICAL COMMITTEE

Chairman :   Dr. AI Sivakumar, Gintic
Co-Chairman : Dr. Anirudha Sen, Gintic
(Editorial Com)

Members :   Mr. Henry Chang , Productivity and Standards Board
Dr. Opas Chutatape, Nanyang Technological University
Mr. Goh Kiah Mok, Gintic
Mr. Lew Sin Chye, Nanyang Technological University
Dr. Lim Kah Bin, National University of Singapore
Mr. Jasbir Singh, Gintic
Dr. V. Subramaniam, National University of Singapore
Dr. Tay Kiang Meng, Gintic
Mr. Tay Meng Leong, Nanyang Technological University

INTERNATIONAL ADVISORY COMMITTEE

Prof. L. Alting, Denmark
Dr. Jeff Ang, Australia
Dr. Peter Bernus, Australia
Prof. C.B. Besant, UK
Prof Maurice C Bonney, UK
Dr. Jim Browne, Ireland
Prof Hans Jorg Bullinger, Germany
Prof. Luis Camarinha-Matos, Portugal
Prof. G. Doumeingts, France
Dr. Ing I Encarnacao, Germany
Prof. Er Meng Wah, Singapore
Dr. Peter Falster, Denmark
Prof. Lester Gerhardt, USA
Prof. Goh Tong Ngee, Singapore
Dr. Gideon Halevi, Israel
Prof Yukio Hasegawa, Japan
Prof Wolfgang Hoheisel, Germany
Prof. Kazuyoshi Ishii, Japan
Prof. Ji Zhou, China
Prof Hiroshi Katayama, Japan
Prof Fumihiko Kimura, Japan
Prof. Dr. D. Kochan, Germany
Dr. George L. Kovacs, Hungary
Dr. W.P. Lewis, Australia
Mr. Lin Cheng Ton, Singapore
Prof. Lim Mong King, Singapore
Prof. Grier Lin, Australia
Dr. Wout Loeve, The Netherlands

Prof. Robert N.K. Loh, Malaysia
Prof. K.J. MacCallum, UK
Dr. Andras Markus, Hungary
Prof. Nils Martensson, Sweden
Dr. Kenichi Mori, Japan
Dr Klaus-N Muller, Switzerland
Prof. Andrew Nee, Singapore
Dr. Laszlo Nemes, Australia
Prof Paul J Nolan, Ireland
Dr Pierre Pahud, Switzerland
Dr. Dan Patterson, Singapore
Dr. Ingvar Persson, Sweden
Dr. Jean-Marie Proth, France
Mr. Edward Quah, Singapore
Prof. Asbjorn Rolstadas, Norway
Prof. H. Sato, Japan
Prof. Mario T Tabucanon, Thailand
Mr. Tan Ah Sway, Singapore
Prof Mitchell Tseng, Hong Kong
Dr. Varaprasad, Singapore
Prof. N. Viswanadham, India
Prof. K.K. Wang, USA
Prof David J. Williams, UK
Prof Tony Woo, USA
Prof Mike Wozny, USA
Mr. Losu Zabala, Spain
Prof. Zhu Jianying, China

PREFACE

ICCIM '97 is the fourth international conference we have organised in the field of computer integrated manufacturing. It is gratifying to see that since its inception in 1991, ICCIM has established a track record of bringing to the public what latest technology is today and what will be available tomorrow.

This year our theme for ICCIM is "Manufacturing into the next Millennium" - an appropriate title as we approach the end of the twentieth century. In this age of rapid change, it is difficult to see beyond a few years what new markets and technologies will emerge. However, at ICCIM'97 we believe that many of the ideas presented and some of the research work reported will become reality tomorrow. For example, just since the last ICCIM conference in 1995 the growth of Internet-based business has suddenly exploded and made fortunes for many people, very often, unexpectedly. While the scale was beyond expectations, the trend as related to the manufacturing industry was anticipated and discussed in a number of papers from previous ICCIM conferences.

We have divided the proceedings into two volumes. In the first volume, we gathered the papers dealing with systems level management and planning of manufacturing activities, whilst the second volume concentrates more on product related and implementation related issues.

In this volume the following topics are covered :
- Product Design and Concurrent Engineering
- Factory Automation

On behalf of the Organising and Technical Committees of ICCIM '97, we would like to thank all those who have made this conference what it is today - the authors, International Advisory Committee , reviewers and participants.

A special thanks to Ms. Lim Lai Lee for her contribution to the numerous details in organising this event and for collating the articles for these two volumes.

October 1997

A.Sen
A.I. Sivakumar
R.Gay

CONTENTS

Volume 1

Flexibility and Change Management

Business Process Reengineering

CIM Design and Planning

Integrated Databases

Internet and the WWW

CIM Justification

CIM Implementations

MANUFACTURING PLANNING AND CONTROL

Aggregate Planning

Production Scheduling Systems

Performance Evaluation and Simulation

Petri Net Approaches/Soft Modelling Approaches

Logistics Management

Quality Management

Control Issues

ADVANCED MANUFACTURING PROCESSES

PRODUCT DESIGN AND CONCURRENT ENGINEERING

Geometric Modelling

Product Design

Product Analysis

Process Planning

Rapid Prototyping

CE Environments

FACTORY AUTOMATION

Shopfloor Integration

Assembly Automation

Inspection and Diagnostic Systems

Robotics and AGV Systems

NC Programming

Automation Components

ADVANCED MANUFACTURING TECHNOLOGY

PRODUCT DESIGN AND CONCURRENT ENGINEERING

Geometric Modelling

Using Soft Objects to Represent Medical Structures

H.L. Lim, Y.P.Wang, C.K. Chui, Y. Y. Cai

Institute of Systems Science,

National University of Singapore,

Kent Ridge, Singapore 119597

Email: {hllim|ypwang|cheekong|yyycai}@iss.nus.sg

Abstract

Soft object is one of the most effective object representations in 3D computer graphics. Using it, the surface of an object is represented by the summation of the field functions of a set of key points. Soft object has the advantage of being able to represent blobby or tubular objects and their deformations with relatively few primitives and operations. In this paper, we developed an improved rendering method to visualize the surfaces. We compute all pairs of keypoints whose field function interact. For each pair of the points, we compute axial and radial rays from the axis joining them. Newton's method and regula falsi are then used to determine the intersections between the rays and the iso-surface. Based on the intersections, we can subdivide the iso-surface of the soft-object into patches. After computing the normals at the vertices of the patches, the soft objects can be rendered. We explore using soft objects to model blood vessels and catheters for our project for the simulation, visualization and design of catheters.

1. Introduction.

3D graphics has traditionally employed polygonal patches as the basic elements for 3D models. Although the use of polygons has advantages such as speed of rendering and ease of representation, it does have limitations. One of its major drawbacks is that polygons are rigid structures. When they are used to represent organic, soft, or fluid objects, their rigidity makes them difficult to represent the deformation of these objects when they move or deform. An alternative and increasingly popular method to represent these objects is to apply a class of object model called soft objects. Using this representation, each object or its primitives are represented by a set of field functions. Since the field functions within an object and between objects can interactive freely, the surfaces of these objects can be smooth connected. Their deformation can also be realistically displayed.

Because of the advantages of soft objects, we have applied them for representing anatomical models in our project, *daVinci* (Visual Navigation of Catheter

Insertion). The objective of the project is to achieve realistic simulation and visualization of the process of catheter navigation through a network of blood vessels. Through the use of soft objects, we hope to achieve realistic representation of organic tissues in both their original and deformed shapes.

2. The Technology of Soft Objects.

2.1 Representing Surfaces by Field Functions.

Central to the technology of soft objects is its use of field functions to represent iso-surfaces. The idea of using field functions was first proposed by Blinn [1]. He described a technique that approximates atoms by Gaussian distribution functions. Each atom is assigned a Gaussian distribution function:

$$f(r) = ae^{br^2}$$

where the van der Waals potential of the atom is defined as the function of its radius. The potential value at any point is then

$$F = \sum_i f_i(r) = \sum_i a_i e^{b_i r_i^2}$$

At any point where the potential value is equal to or greater than a threshold value, the surface will be visible. If two or more atoms share a common space, their potentials at that point add up, and then this value is compared to the threshold value. Atoms also may have negative potential, creating the effect of repulsion between them.

2.2 Wywill's Representation of Soft Objects.

Wywill et at have improved upon the basic technology of soft objects [2] [3] [4]. They are also the first to use the term *soft objects* for the class of soft, deformable objects. Using their representation, an object contains one or more keypoints. The field function of a key point is of the form

$$f(r) = a\frac{r^2}{R^2} + b\frac{r^4}{R^4} + c\frac{r^6}{R^6} + 1$$

where R is the value at which the field decays to zero. Thus, the isosurface of the field function at that value represents the surface of the object.

In the above equation, a, b and c are constants such that $C(0.0) = 1$, $C'(0.0) = 0$, $C'(R) = 0$, $C'(R) = 0$, and $C(R/2) = 0.5$. Thus, the field value at a point decays from 1 to 0.5 and to 0 when it moves from positions where $R = 0$, $R = 0.5$ and $R =$

1. Compared with Blinn's formulation, Wywill's definition of soft objects can be more quickly computed as there are no exponential functions to be evaluated.

3. The Visualization of Soft Objects.

3.1 Existing Approaches in Rendering Soft Objects.

Blinn displays the soft objects by numerically evaluating the closest surface at each pixel [1]. Each atom in his model is transformed into a standard viewing space, where the axes are parallel or orthogonal to the viewing direction. After the centres and their radii have been transformed to the new coordinate system, the visible field surface of the model at each pixel is computed by tracing a ray from the viewpoint origin through the pixel. If only an atom is present, the ray intersection can be found analytically. However, if more than one atom is present, numerical methods would need to be used. Blinn examined both Newton's method and regula falsi. The former has fast convergence but may not give an answer if the initial guess is not close to the actual solution. The latter is slower but guaranteed to converge. Blinn therefore proposed using a hybrid of Newton's method and regula falsi. Newton's method is first used. If the intermediate result is outside the bracket of solution, regula falsi is then used.

Wyvill instead overlays the space with a 3D grid. All the cubes formed by the grid that are intersected by the surfaces of the soft objects are found. The surfaces within each cube are approximated by polygons. These polygons are then rendered.

To achieve this efficiently, Wywill scans the cubes along one axis to determine some cubes that intersect the surfaces of the soft objects. These cubes are used as seeds, whereby other connected cubes are successively checked to determine whether they intersect the surfaces.

For each intersecting cube, the field values at its eight vertices are found. From these values, the polygons within the cube and part of a surface can be constructed.

Wyvill also describes a method for ray tracing soft objects [6]. He first finds the intersections of each ray with all the bounding spheres of the soft objects. The intersections are ordered in the sequence along the ray. Using these intersections as starting values, the actual intersections between the ray and the surfaces of the soft objects are found numerically.

3.2 A New Method in Rendering Soft Objects.

Amongst the various methods for rendering soft objects, the ray tracing approaches by Blinn and by Wywill can create high quality images but are costly as the surfaces at every pixel need to be found. When the model moves or rotates, the same ray tracing operations need to be carried out from scratch. Wyvill's method of using 3D grid to divide the soft objects into polygons is more general and efficient. However, the 3D grid divides the surfaces into irregular patches. The searching of cubes that intersect the surfaces would also require considerable computations. We therefore propose a new method for computing and rendering the surfaces of soft objects.

3.2.1 Determination of local axes and rays to be traced.

This method takes advantage of the fact that most of the isosurface would usually surround a keypoint if there are no other influencing nearby keypoints. First, we detect all pairs of keypoints whose field functions are likely to interact. For each pair of the points, the axes joining them are first found. This is defined as the local X axis. We also determine another two axes that are non-parallel to each other and the local X axis. They are referred to as the local Y and local Z axes (we will discuss more about the determination of the local Y and Z axes later). The portion of the local X axis within the endpoints is divided into a predefined number of intervals. With each endpoint of the interval as origin, we trace a predefined number of evenly spread-out, axial rays that are parallel to the plane containing the local Y and Z axes. These rays are referred to as *axial rays*. If any of the endpoints is a terminating keypoint (a keypoint that can only be influence by the other keypoint pairs with it), a set of rays would also have to be traced from it and cover the directions that are outside the axial rays. We achieve this by subdividing the hemisphere containing the solid angle not covered by the axial rays and tracing rays in the directions defined by the subdivided angles. These rays are referred to as the *radial rays*. The rays to be traced for a set of terminating keypoints are shown in figure 1.

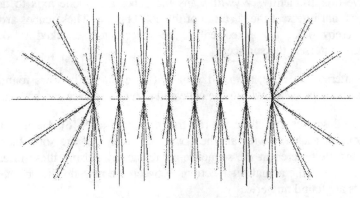

Figure 1: Tracing of rays from a pair of keypoints.

Note that if the field function of a keypoint interacts with the field functions of more than one keypoint, it would belong to more than one keypoint pair. The local-Y and local-Z axes for each pair would have to be selected such that they remain non-parallel to the local-X axis while providing continuity across the pairs.

The ideal situation would be for the same axes to be used for the adjacent pairs. Failing which, we determine the axes that are closest to their counterparts. We then find out if there are solid angles that are not covered by the axial rays of the pairs. Should this be the case, some radial rays would have to be traced to cover these angles.

3.2.2 Computing ray intersections.

After the rays have been defined, each of them is traced. The intersection between each ray and the iso-surface is computed. Like Blinn's method, we first apply Newton's method and then regula falsi to obtain the solution. However, in most situations Newton's iterations would suffice. There may be situations where numerical iterations fail to find a solution. This indicates that there are other nearby keypoints, such that the line is always within the isosurface. In this situation, the point on the line that has the lowest field value is determined and used as its approximation.

The formula of Newton's method is

$$p_{new} = p_{old} - \frac{F(p_{old}) - T}{F'(p_{old})}$$

Where p_{old} is the old guess and p_{new} is the new guess of the iteration, T is the threshold value of the isosurface, $F(p)$ and $F'(p)$ are respectively the field function and the derivative of the field function.

Since we are attempting to find the intersection between a ray and the isosurface, the equation of the ray with direction (ax, ay, az) and origins (x_0, y_0, z_0) can be written as the following parametric equations:

$$x = a_x p + x_0$$

$$y = a_y p + y_0$$

$$z = a_z p + z_0$$

The function of the isosurface can be expressed as a function of r^2:

$$F = \sum_i f_i(r_i^2)$$

As $r_i^2 = (x - x_i)^2 + (y - y_i)^2 + (z - z_i)^2$, where (x_i, y_i, z_i) is the coordinates of the keypoint. By substituting the parametric equation of the ray, r^2 can be expressed as a quadratic equation of p.

Further substituting this quadratic equation to F, we obtain an implicit equation of p. Likewise, the derivative of the iso-surface can also be expressed as a function of r^2, which can be converted to an equation of p:

$$F'(f) = \sum_i f_i'(r_i^2) = \sum_i \frac{df_i(r_i^2)}{d(r_i^2)} \frac{d(r_i^2)}{dp}$$

Using the above substitution, the equation of Newton's method can be expressed totally by p, which can then be evaluated.

3.3.3 Creation and rendering of patches.

After the tracing of rays, patches are defined from vertices that are the surface intersections of adjacent rays. The normals at these vertices are computed. This can be derived from the partial differentiation of the field functions describing the soft objects using the following equations:

$$N = \frac{\partial F}{\partial x} i + \frac{\partial F}{\partial y} j + \frac{\partial F}{\partial z} k$$

Having obtained both the coordinates and normals of the patches, the soft objects can be rendered.

The images of the mesh and shaded surfaces of soft objects generated by the current technique are shown in figures 2 to 9.

3.3.4 Advantages and disadvantages of the current approach.

Compared with Blinn and Wywill's methods, the current approach has the advantage that the computed patches are substantially fewer. It therefore compares favorably with them in terms of speed. Because the patches are directly subdivided from the iso-surface, they tend to have a better fit to the surface topology and their sizes are more uniform. This can improve the quality of image. The disadvantage of the current approach is that an axis needs to be defined for every pair of interacting keypoints. Therefore, the current approach is efficient in objects that are tubular in shape (such as the vascular anatomy), but not in models where the field functions of many keypoints would interact with each other,

4. Using Soft Objects to represent Vascular Anatomy.

4.1 Visualization of Catheter Navigation.

Guidewire and catheter are two major instruments in angiographic vascular catheterization [8][9]. Each of them has multiple variations designed to facilitate a

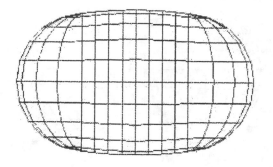

Fig 2: Mesh surface of first sample soft object.

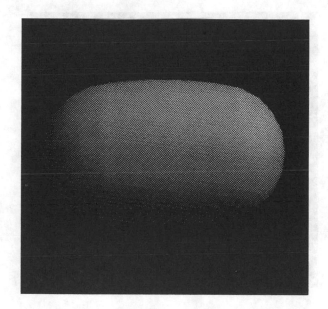

Fig 3: Shaded Image of the first sample soft object.

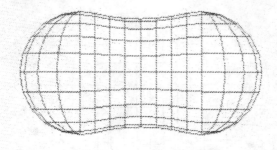

Fig 4 Mesh surface of the second sample soft object.

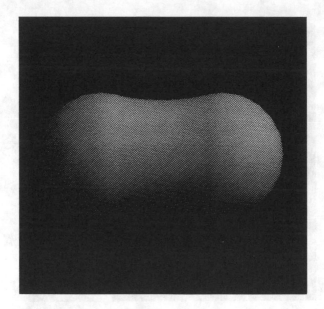

Fig 5 Shaded image of the second sample soft object.

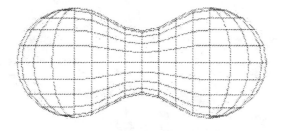

Fig 6: Mesh surface of the third sample soft object.

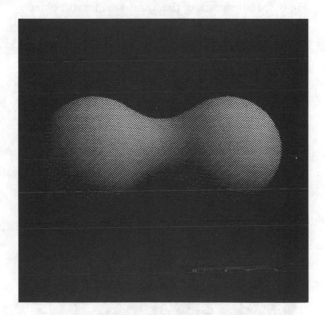

Fig 7: Shaded Image of the third sample soft object.

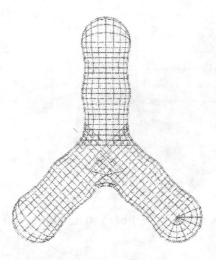

Fig 8: Mesh surface of the fourth sample soft object.

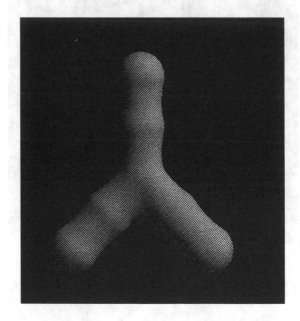

Fig 9: Shaded Image of the fourth sample soft object.

particular function. Catheters are categorized according to their material, shape, length, etc. Some serve multiple purposes in addition to their original intentions. They possess the properties of pliability, torquability, and shape memory. The function of the guidewire is to provide support and guidance for passage of an overlying catheter through the artery and into the lumen of the selected artery.

In the catheterization process, the tip part of pre-shaped catheters usually undergoes various changes. For instance, the insertion of a guidewire into the catheter tip may straighten up the tip shape while the withdrawal of the guidewire would reshape the catheter back to its original configuration. The simulation of the interaction between catheter and guidewire is therefore an important task for the analysis of the deformation in catheterization.

4.2 Basic Information of Vascular Anatomy.

The objective of our system called daVinci (Visual Navigation of Catheter Insertion) [10][11][12][13][14][15] is to achieve realistic simulation and visualization of the process of catheter navigation through a network of blood vessels. In daVinci, the vascular anatomy is represented by the central lines and their corresponding radii of the vessels. Using the marching cube technique, the image data are segmented to obtained coordinates of the central lines and the radii.

This anatomical structure is stored in a file. In the file are blocks which represent the segments of the central line. Each block consists of an index number, a description, a radius and vertices of the segment nodes.

4.3 Representation of Vascular Anatomy Using Soft Objects.

To determine the soft object representation of the vascular anatomy, the anatomical structure file is analysed. Each block of the segment is treated as a soft object, while the vertices of the segment are used as keypoints. Figure 9 and 10 show a model of a bifurcation joint represented by a soft object using different number of keypoints.

5. Conclusion.

In this paper, we have described the use of soft objects to represent medical models such as blood vessels. We have also implemented a new method for displaying these objects. With further development, we hope to improve the efficiency and explore further use of the soft object technology in medical applications.

Acknowledgement:

We would like to thank National Science and Technology Board of Singapore for funding our research.

Fig 10: Shaded image of a blood vessel simulated by
a soft object with sparse keypoints.

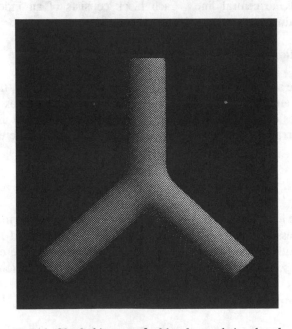

Fig 11: Shaded image of a blood vessel simulated
by a soft object with dense keypoints.

References:

[1] Blinn, J. "A generalization of algebraic surface drawing", ACM Transaction of Graphics, 1982, Vol. 1, No. 1.

[2] Wyvill G., McPheeters C., Wyvill B.", "Soft Objects", Advanced Computer Graphics (Proceedings of Computer Graphics Tokyo '86), Tsiyasu L. Kunii (Editor).

[3] Wyvill B., McPheeters C., Wyvill G., "Animating soft objects", The Visual Computer, Vol. 2, No. 4, Aug, 1986.

[4] Wyvill G., McPheeters C., Wyvill B., "Data structure for soft objects", The Visual Computer, Vol. 2, No. 4, Aug, 1986.

[5] Wyvill B., Wyvill G., "Field functions for implicit surfaces", The Visual Computer, Vol. 5, No. 1/2, Mar 1989.

[6] Wyvill G., Trotman A., "Ray-Tracing Soft Objects", New Trends in Computer Graphics (Proceedings of CG International '90), 1990.

[7] Wyvill G., Wyvill B., McPheeters C., "Solid Texturing of Soft Objects", IEEE Computer Graphics and Applications, Vol. 7, No. 12, 1987.

[8] D. Kim and D. E. Orron, "Peripheral Vascular Imaging and Intervention", Mosby-Year Book, 1992, Inc. Missouri, USA.

[9] CORDIS Corporation Diagnostic and Interventional Radiology, Technical Report, Miami, USA 1992.

[10]C. K. Chui, H. T. Nguyen, Y. P. Wang, R. Mullick, R. Raghavan, James H. Anderson, "Potential Field of Vascular Anatomy for Realtime Computation of Catheter Navigation", First Visible Human Conference, Bethesda, MD, USA, October 1996.

[11]Y. P. Wang, C. Chui, Y. Y. Cai, R. Viswanathan and R. Raghavan, "Potential Field Supported Method for Contact Calculation in FEM Analysis of Catheter Navigation", XIXth International Congress of Theoretical and Applied Mechanics, Kyoto, Japan, August 1996.

[12]Y. Y. Cai, R. Viswanathan, Y. P. Wang, C. K. Chui, R. Raghavan, "Simulation of Catheter-guidewire Interaction for Catheterization by Arc Parameterization", Fourth International Conference on Control, Automation, Robotics and Vision, Singapore, December 1996.

[13]J. Anderson, W. Broky, C. Kris, , R. Viswanathan, and R. Raghavan, daVinci: "A vascular Catheterization and Interventional Radiology-based Training and Patient Pretreatment Planning Simulator", Society of Cardiovascular and Interventional Radiology 21st Annual Meeting, March Seattle, USA, 1996.

[14]Y. P. Wang, C. K. Chui, Y. Y. Cai , H. L. Lim "CathViewer: An Interactive System for Catheter Design and Testing", Fourth International Conference on Computer Integrated Manufacturing, Singapore, Oct 1997.

[15]Y. Y. Cai, Y. P. Wang, C. K. Chui, H. L. Lim, "Integration of Geometric Modeling and Physical Modeling for Catheter Navigation", Fourth International Conference on Computer Integrated Manufacturing, Singapore Oct 1997.

Integration of Geometrical Modeling and Physical Modeling For Catheter Navigation

YY Cai, YP Wang, CK Chui, and HL Lim

Institute of Systems Science
National University of Singapore
Heng Mui Keng Terrace, Kent Ridge 119597
Republic of Singapore
Tel: (65)7726724, Fax: (65)7744990
Email: {yycai I ypwang I cheekong I hllim} @iss.nus.sg

Abstract. Catheter manufacturing is a very lucrative business emerging from high-technology development in the minimal invasive surgery. We reports an integration of physical modeling and geometrical modeling for catheterization application. A potential space model of multiple-dimensional information is developed to describe both physical and geometrical properties of the vascular structure. The physical behaviors of object interactions between vascular walls and catheters can be obtained by applying a Finite Element Method (FEM) for the analysis of force, displacement, deformation, etc. Different from the conventional boundary representation that is usually hollow inside a spatial object, the potential space model is defined in any space points not limited to object boundary. This model can be utilized to extract smooth multiple-branch boundary geometry of the complex vascular network by iso-surfacing technique. The interventional device models are designed as soft objects with various geometrical features and physical properties to provide flexibility of object deformation during the catheter navigation. The approach developed is applied in the design of interventional device, the catheter navigation in the simulation of the catheterization, and visualization of the angiographic vascular network.

1 Introduction

Design and manufacturing automation has been a driving force for the development of CAD systems. Geometric modeling represents one of the central components of CAD systems. Many representation schemes have emerged since mid 1970s. A Constructive Solid Geometry (CSG) object is built from the standard primitives, using regularized Boolean operations and rigid motion. The modeling system PADL by Voelcker and Requicha in the University of Rochester [1] represents the earliest CSG work. Figure 1 (a) shows a typical example of a CSG description of an object: the solid C is made from blocks A and B using a union operation; solid E is then made from the solid C and cylinder D using a difference operation. An object can be unambiguously described by its topological and geometrical descriptions. Briefly, the topological information specifies abstractly vertices, edges, and faces, and indicates their incidence and connectivity, while the geometrical representation specifies the equations of the surfaces and orientations. The oldest formalized schema for representing the boundary of a polyhedron and its topology appears to be the winged-edge representation first proposed by Baumgart [2]. BUILD is the first B-Reps (Boundary Representation Scheme) system developed by the CAD Group at the University of Cambridge [3]. As a typical representation scheme, B-reps, has become a de-facto industrial standard for core representation in a geometric modeler nowadays. Figure 1(b) shows the boundary representation of the same solid mentioned above. Slightly different from the Westerners, the Japanese started their modeling research from the half-space concept. Okino *et al* at the Hokkaido University developed the TIPS (Technical Information Processing System) in 1973 [4]. The half-space scheme is sometimes classified as a case of CSG, however, the geometry meaning and the developed motivation of this scheme are distinct from others.

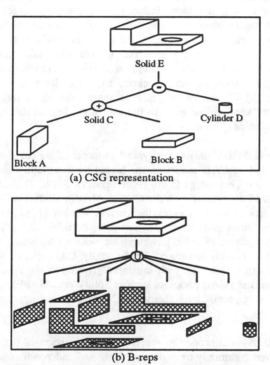

(a) CSG representation

(b) B-reps

Fig. 1. Modeling representations in CSG and B-reps

CSG and B-Reps are two main streams of object modeling. Apart from these, there exist a number of other solid representation schemata which play a peripheral role in solid modeling due to certain limitations. Arbitrary 3D objects can be subdivided into any specified resolution in a hierarchical octal tree structure. Octree was originally proposed by Hunter [5]. Unlike the CSG or B-Reps, the Boolean operation (union, intersection, and difference) using the Octree scheme appears dramatically simple with linear processing time [6]. However, space decomposition is dependent of the coordinate system and this makes it unattractive to rotate objects so represented and also causes inconsistency in congruent objects lying at different orientation. Besides, feature detection and extraction are extremely difficult. Conceptually, Euler operators can be thought of as creating and modeling consistently the topology of manifold object surfaces. By locally deleting or adding vertices, edges, and faces, they can be used to create closed surfaces and to modify these surfaces. As an intermediate language, Euler operators are used in the Geometric WorkBench (GWB), a modeling system developed at the Helsinki University of Technology, incorporating B-Reps [7]. Euler operators can insulate the operations implemented on top of them from details of the data structures used to represent the surface topology. Another advantage is that they ensure topological consistency throughout the modeling process. For quite sometimes, the defined structure of geometry and topology information relating to a solid object, although computationally efficient, was often in terms of low-level details, and was therefore an incomplete definition for product modeling. Ansaldi, Floriani and Falicidieno developed a different graph-based boundary representation called the Face Adjacency Graph (FAG) or Face Adjacency Hypergraph (FAH) in the early 80s [8]. The faces are described as the nodes, whereas the arcs and the hyperarcs represent the relationships among the faces induced by the sets of edges and vertices, respectively. In a feature-based design scheme, the designer can create the part using features directly instead of using lower level geometric entities thus incorporating higher level information into the design. This feature information can then be used for manufacturing planning. The concept of AAG was introduced by Joshi for recognizing features such as slots, pockets, steps for the purpose of automatic process planning [9]. The arc of the graph represents the connectivity between two faces and the attributes are assigned to the arcs according to the geometry of the adjacency relationships.

Romulus (Evans & Sutherland) was a typical example of the early commercialized CAD systems. Modern CAD modelers are becoming more and more capable and versatile. Parametric Technology Corp. (ProEngineer), SDRC (I-DEAS), AutoDesk (AutoCAD), IBM/Dassault (CATIA), CV (CADDS), EDS/UniGraphics (UG), HP(CoCreate) are major players in the current market [17]. Lots of modeling kernels have been developed during the past three decades. Parasolid (EDS), Acis (Spatial) and DesignBase (Ricoh), however, are widely considered as three major geometric modeling kernels nowadays. Traditionally, CAD systems were developed on the Unix platforms. However, with the rapid growing of the Pentium processor, Window platform based CAD packages such as SolidWorks, SolidEdge (Intergraph), PT/Modeler (ProEngineer), Helix (MicroCADAM) etc., are aggressively entering the market.

Geometrical and physical properties are two distinct aspects of real-world objects. They are, however, commonly treated as separate and independent parts in most of the engineering modeling domains. For instance, CAD modeling usually chooses boundary geometry to describe the object shape with little concerns or even complete ignore of the physical behavior of objects. The boundary abstraction of a 3D object is indeed based on a uniformity of the interior of an object. It has been acknowledged that the B-reps is rather suitable for the conventional tasks of design and manufacturing. The hollow assumption inside the object body, however, may be inappropriate for tasks in advanced design, simulation, and concurrent engineering [10]. On the other hand, physically-based modeling has recently become a subject of widespread interest in the graphics and visualization circle due to the advances of medical modeling, virtual reality, etc. Some researchers work on various new methods for representing deformable models while others use physically-based model as a tool to explore the phenomenon of natural motion [11-12]. Computational mechanics techniques such as the finite element method are increasingly used in physically-based modeling [13]. The idea of the FEM is that a continuum can be approximated by dividing it into discrete elements. They make it possible not only to calculate the value of important physical quantities inside a three-dimensional body, but also to analyze the interaction between various objects.

For highly realistic medical/biomedical modeling, simulation, as well as interventional device designing, geometrical and physical properties are equally important. This work reports the integration of physical modeling and geometrical modeling for catheterization application. The multiple-dimensional potential space model and interventional device model are developed. Physical behaviors of object interactions between vascular walls and catheters can be derived based on the potential space model and catheter device model using FEM. Different from the conventional boundary representation that is usually hollow inside an object, the potential space model is defined in any space points not limited to object boundary. The boundary geometry of the vascular object extracted from the potential space model can be very complex with smoothed multiple-branch structure. Besides, physical modeling based on the explicit multiple-dimensional potential space can significantly speed up the analysis of force, displacement, deformation, etc. The interventional devices are designed as soft objects with various geometric features and physical properties. The integration of geometry and material properties within the catheter device provides flexibility of object deformation during the catheter navigation. The approach developed is applied in the design of interventional devices, the catheter insertion in the simulation of the catheterization, and visualization of the angiographic vascular network.

2 Multiple-dimensional Potential Space Model

A central line hierarchical structure is created on the basis of the vascular anatomy. From image data, the cross-sectional vessels segmented are used to obtain the coordinates of the central line assuming that all vessels in 3D space are in circular cylindrical structures with varied radii. With same assumption, it is also possible to calculate the radii of the vessels from the cross-sectional areas of the segmented vessels.

2.1 Vascular Potential Field

The potential space is a multiple-dimensional field where each point defines a scalar quantity and a 3D vector. The scalar factor used is the shortest distance from the space location to the central lines. The 3D vector, however, is the direction of the shortest path from the space location to the central lines (Figure 2). To avoid inconsistency, the potential distances are normalized with respective radii of the vessels. So it is clear that the potential value at a space point tells whether the grid is inside the vessel or on the surface. This is significant to conduct inside/outside testing and boundary testing as well. Similarly, vector unifying is performed.

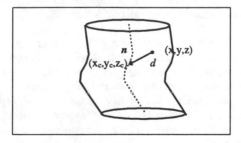

Fig. 2. The potential field consists of a shortest distance (d) and its associated direction (n) from each point (x,y,z) to the central lines (*dash line*).

Computing the shortest distances and directions sometimes can be time-consuming. A discrete model with pre-computed distances and directions can significantly improve the efficiency of the real-time simulation. For vascular network, a sparse grid space with a given thickness embracing the vascular vessels is used to specify the region of the discrete model. A triple-linear interpolation of the distances and vectors at the adjacent grid points is used to compute the normalized distance and unit vector for an arbitrary point in the region. Beside, a volumetric structure derived from the discrete model can be used for visualizing the object geometry.

2.2 Extraction of Vascular Boundary Geometry by Iso-surfacing

The complexity of vascular geometry is one of the reasons for the hesitation to apply the conventional surface/solid modeling techniques for previous researchers. In this work, a "Marching Cubes" algorithm is applied to modeling the vascular geometry [14]. To achieve this, a discrete volume model is created based on the potential space model. That is first to separate the distance information from the potential field and then convert the value into a characteristic value by a 8-connectivity checking. The grid size is equivalent to the voxel size. As a result, a grid near the vessel boundary usually has a value less than eight while a inner grid has a value of eight. This conversion is able to smooth the boundary surface. With the potential volume, the boundary geometry is subsequently extracted using a lookup table based on binary (above/below threshold) values at the corners of a volume voxel to produce high-speed polygon generation. The local gradients are also calculated to produce visually smooth iso-surfaces across neighboring voxels.

Visualization of the three-dimensional vascular networks is one of the necessities for our catheterization simulation. The geometric model constructed from the central line models (with variable radii) together with the boundary geometry extracted from the potential space model provide a viable way for geometric representation. Figure 3 shows the extracted boundary surface of the vascular structure of the Visible Human.

3 Interventional Device Model

Catheter and guidewire are major interventional devices in angiographic vascular catheterization. They have multiple variations designed to facilitate a particular function. Catheters are categorized according to their material, shape, length, etc. Some serve multiple purposes in addition to their original intentions. They possess the properties of pliability, torquability, and shape memory. Guidewire is another type of interventional device with few variation. The function of the guidewire is to provide support and guidance for passage of an overlying catheter through the arterotomy and into lumen of the selected artery.

Catheter manufacturing is a very lucrative business emerging from high-technology development in the medicine field. However, surgeons who practice the interventional procedures often use personnel preference and prior experience in selecting the devices, for example, the angioplasty balloons and stents for cardiovascular procedures, since the devices are not designed as matching systems. This practice can potentially develop inconsistencies in the performing characteristics of those devices. Such inconsistence can produce damage to the vascular wall tissue.

3.1 Deformable Interventional Device

Traditionally, CAD tools are developed for designing and manufacturing of the hard (or rigid) engineering objects. The interventional medical devices, however, often compose of highly soft material which allows the devices to be deformed during the process of operations due to the interaction of the devices and the human angiographic structures. Furthermore, they can very unlikely have an opportunity to validate the design concept themselves at a realistic checking environment. And the validation process is considered as the utmost crucial step in any design procedure. Therefore, conventional CAD methodology and design tools are not appropriate for the tasks of interventional device designing and evaluating: (1) new approaches of soft object design need to be developed and applied to support high-realistic simulation of catheterization; (2) testing and evaluating of catheter, especially, the new designs, require a very friendly integration of the design and evaluation phases.

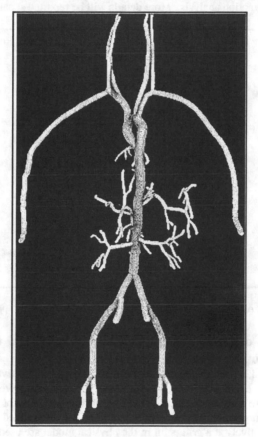

Fig. 3. The boundary surface of the Visible Human vascular structure extracted from the multiple-dimensional potential space model using iso-surfacing technique

3.2 Feature-based Modeling of Interventional Device

A feature-based design approach is applied in this work to model the interventional devices. Typical features of catheter and guidewire include generalized sweeping, tapering, lofting, holing, shelling, threading, filleting, etc. The parametric and variational methods are particularly important to the designing tasks of interventional devices such as catheter and guidewire. We have created varational and parametric based models for catheter, guidewire, balloon and stent, etc. (Fig. 4.). We are currently developing software packages for presenting, designing and validating of interventional devices on the Intel PC-based Window NT platform

781

using the object-oriented paradigm and OpenGL toolkit. It aims to provide high value-added sales tool, customized design tool and realistic validation tool for interventional products. The work features (1) integration of interactive 3D design and real-time navigation based validation; (2) customized design with the realistic vasculatures models; (3) varational and parametric based product designing.

Many interventional devices are soft objects by nature. In the validation procedure, interventional device may have various deformations due to interaction between vasculature and device. We model the device at various layers of backbone, geometry, material and graphics. The FEM is embedded in the design models to support the deformation analysis. More details will be reported separately.

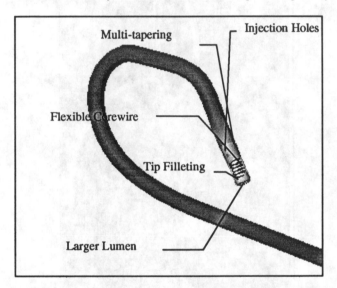

Fig. 4. A catheter designed using the feature-based modeling approach

4 Integration of Geometrical Modeling and Physical Modeling for Catheterization Application

In our integrared model of the vascular potential space and devices, we have a dynamic FEM engine, together with all the material properties of the catheter/guidewire and the vessels. The central line network of the vasculature preserves the anatomical relationship of the vessels. The normalized distance and the unit vector respectively at a grid point in the physical model are used as references to determine the magnitude and direction of the contact force for FEM calculations [15]. The contact force between the vessel wall and catheter/guidewire at a FEM node of the catheter/guidewire is formed when the catheter/guidewire is moved at the grid point outside the vessel. The distance values of the potential space provide explicit information for the inside/outside testing of the catheter/guidewire nodes against the vessels. The central lines play a role similar to gravity centers. This is because when the catheter/guidewire nodes are at the grids outside the original shape of the vessels, these nodes are usually "pulled back" into the vessels by the interaction forces (Figure 5). These forces are pointed to the central lines of vessels model. The FEM engine is applied to solve the deformation of the catheter/guidewire when interaction with the vessels. The model developed provides possible solution for high realistic and real-time catheterization simulation.

An adaptive iterative procedure for the catheter/guidewire FEM analysis is used in this work: (1) inside/outside checking for the catheter/guidewire nodes; (2) generation of the contact forces for the outside nodes; (3) construction of the stiffness matrix; (4) solving the displacement from the stiffness-displace equation; (5) checking the nodes and performing iteration for the optimal displacement update.

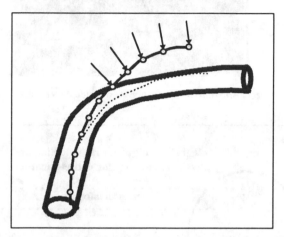

Fig. 5. The contact force pushed the catheter nodes (outside the vessels) back insides the vessels

In order to simulate the vascular catheterization procedures, the research team at ISS has developed a high realistic system for visual catheter navigation [16]. The system consists of hardware device and software system. A mechanical device is designed to manipulate the catheter/guidewire for inserting and moving them inside the vascular tree. The pulse streams generated by the incremental optical encoders are measured and fed to the interface software system.

The navigation is simulated with the visual aides of fluoroscopic and 3D views. The geometry of vascular structure is displayed based on the geometric model. Figure 6 illustrates a three-dimensional view inside the vessel. This internal view displays the geometric structure of the Visible Human simulating a viewpoint at the tip of the catheter during the navigation. The navigation is fully engined by the real time FEM calculation. The integrated model provides both geometrical and physical information for the catheter navigation application (Figure 7).

Conclusions. This work presents an integrated approach of geometric and physical modeling for catheterization application. The geometric and physical properties of the complicated vascular structure is modeled based on the multiple-dimensional potential space. The boundary information extracted is utilized to render the three dimensional structure during the navigation procedure. The moving of the catheter/guidewire inside the 3D vascular networks is determined within the physical model. The integration of the geometrical modeling and physical modeling provides an efficient solution for the high realistic medical simulation.

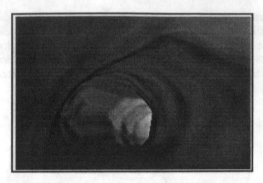

Fig. 6. A fly-through navigation view inside the vascular network

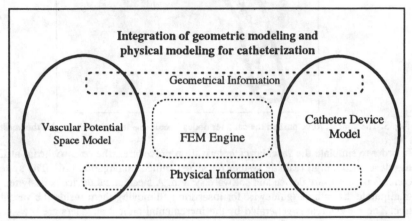

Fig. 7. The integration of geometrical modeling and physical modeling for catheterization.

Acknowledgment. We would like to thank National Science and Technology Board of Singapore for funding this research.

References

1. H. B. Voelcker, An Introduction to PALD: Characteristic, Status, and Rationale, *Technical Memo No. 22*, Production Automation Project, University of Rochester, 1974.
2. B. Baumgart, A Polyhedron Representation for Computer Vision, *Proceedings of the National Computer Conference, AFIPS*, **44** (1975), pp. 589-596.
3. I. C. Braid, *Design with Volumes*, Ph D Thesis, University of Cambridge, 1973.
4. N. Okino, Y. Kakazu, and H. Kubo, TIPS-1, Technical Information Processing System for Computer Aided Design and Manufacturing, in *Computer Languages for Numerical Control* , edited by J. Hatvany, North Holland, Amsterdam, 1973, pp. 141-150.
5. G. M. Hunter, *Efficient Computation and Data Structures for Graphics,* Ph D Dissertation, Princeton University, Princeton, NJ, 1978.
6. D. Meagher, Geometric Modeling Using Octree Encoding, *Computer Vision, Graphics and Image Processing*, **19** (1982), pp. 129-147.
7. M. Mantyla and R. Sulonen, GWB: A Solid Modeler with Euler Operators, *IEEE Computer Graphics Applications*, **2** (1982), pp. 17-31.
8. S. Ansaldi, L. De Floriani, and B. Falicidieno, Geometric Modeling of Solid Objects Using A Face Adjacency Graph Representation, *Computer Graphics*, **19**(3) (1985), pp. 131-139.
9. S. Joshi, *CAD Interface for Automated Process Planning*, Ph D Thesis, Purdue University, 1987.

10. S. C. Lu, A. B. Rebello, D. H. Cui, R. Yagel, R. A. Miller and G. L. Kinzel, A Visualization Tool for Mechanical Design, *Proceedings of Visualization'96*, San Francisco, CA, pp. 401-403.

11. J. Platt, D. Terzopoulos, K. Fleischer and A. Barr, Elastically Deformable Models, *Proceedings of ACM SIGGRAPH'1987*, pp. 205-214.

12. J. Weil, The Synthesis of Cloth Objects, *Proceedings of ACM SIGGRAPH'1986*, pp. 225-232.

13. Y. C. Fung, *Bio-mechanics: Mechanical Properties of Living Tissues*, Second Edition, Springer-Verlag, New York, 1993.

14. W. E. Lorensen and H. E. Cline, Marching Cubes: A High Resolution 3-D Surface Construction Algorithm, *Computer Graphics Vol. 21, No. 4, (Proceedings of SIGGRAPH'87)*, July 1987, pp. 163-169.

15. Y. P. Wang, C. K. Chui, Y. Y. Cai, R. Viswanathan and R. Raghavan, Potential Field Supported Method for Contact Calculation in FEM Analysis of Catheter Navigation, in *Proceedings 19th International Congress of Theoretical and Applied Mechanics*, August 1996, Kyoto, Japan.

16. J. Anderson, W. Brody, C. Kriz, Y. P. Wang, C. K. Chui, Y. Y. Cai, R. Viswanathan and R. Raghavan, daVinci: A Vascular Catheterization and Interventional Radiology-based Training and Patient Pretreatment Planning Simulator, *Society of Cardiovascular and Interventional Radiology 21st Annual Meeting*, March 1996, Seattle, USA.

17. P Marks and K Riley, Aligning Technology For Best Bussiness Results, Design Insight, 1995, USA.

A Mechanism for Coding the Form Feature

Junjun Wu, Tongyang Wang, Minghua Wu and Ji Zhou

CAD Center, HuaZhong University of Science & Technology, P.R.China

Abstract Feature-based modelling is concerned with unambiguous computer representation of the construction and manipulation to the feature-based solid model. This representation should permit us to distinguish different features when manipulating the model interactively and to regenerate the proper model according to the recorded constructive history conveniently. The mechanism of coding the form feature plays an important role in the representation. This paper describes a computational method for fully automating the generation of the feature. In this paper, we propose a coding approach by the nature of the form feature to maintain the design intent while changing the topology of the part, thus to automate feature generation while greatly increasing the flexibility.

1 Introduction

In feature-base modelling, it is greatly required that more flexibilities and convenience should be provided. One of the main embodiments is that the design intents should be maintained while the topology of the part is changed. This is determined by the geometric representation and the maintenance of the feature constructive history. It involves how to record the constructive history and leads to the main purpose of this paper. The constructive history is implemented by referenced relationships among the faces, edges and vertices of the designed object's Breps, that is, the relationships between features.

In reference[1] the author pointed out that the most useful schemes for representing solids have been, and still are, Constructive Solid Geometry(CSG) and Boundary Representations(Breps). Both representations are useful complementary, we believe. Because CSG well supports the parameterized part families and Breps is excellent for graphic user interaction, many feature-based practitioners adopt both CSG and Breps in hybrid form and this trend is likely to continue indefinitely. But the development of a technique for feature generation automatically has turned out to be very difficult when attacked only by pure geometry, especially for arbitrary 3D shapes. The automated feature generation must be implemented at a higher level through attaching some information to the geometric entities.

As we have known, the relationships between features are very crucial in feature-base design. These relations can be expressed explicitly as constraints from which many of the design variants can be derived. It is safe to say that only the relationships between features can be recorded and used to implement the constructive history of the design object. But the relationships are recorded using the codes of the feature. Now the problem is how to code the feature by greatly decreasing the ambiguities simultaneously.

The remainder of this paper is organized as follows. The related works are discussed, then the role of coding feature face, then representations for coding mechanism for the form feature, at last drawing the conclusion.

2 Related Works

In a commercial system named SolidWorks 96, a problem exists. Consider the example shown in Figure 1. It illustrates a design object with SolidWorks 96. It has been designed in the following three steps:

1. create a block by drawing a rectangular profile on the sketch and extrude it by a specified depth.

2. add a torus on the top of the box by revolving a circular profile around a axis by 360 degree;

3. chamfer the circular edge on the right by specifying a constant radius.

Change the definition of the torus after defining the design object on the left. After regeneration, the result is shown on the right. We can see that the chamfered edge is the one on the left.

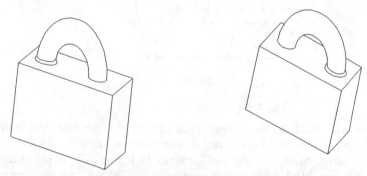

Fig. 1. An example created by SolidWorks'96

Why does it happen? It is because SolidWorks 96 cannot distinguish the 2 edges. The common character of these two edges is that they all are the intersections between the top surface of the block and the torus' surface. This example created by SolidWorks 96 demonstrates that the coding mechanism used by SolidWorks 96 has a limitation in some cases.

In the prototype of feature-based design system called GHCAD, developed by Huzhong Univ. of Sci & Tech and Tsinghua Univ., all the faces are coded according to the sequence of searching the topology of the design object's Breps. The code which includes three identifications is taken as an attribute attached to the face. The three identifications are the feature identification, pattern identification and face identification decided by the searching sequence. Although this is a valid method, there exist some drawbacks. One of the main drawbacks is the existence of ambiguity when executing the same steps as what have been done to SolidWorks 96 above.

Another is that it is assumed the topology of the design object can not be variable, so it cannot derive many design variants, thus, its flexibility is greatly decreased.

To overcome the drawbacks discussed above, it is necessary to find a new method to code the feature. Anyway, the new method for coding feature should enhance the capability of decreasing ambiguity while increasing the flexibility.

In reference[3], Vasilis Capoyleas tried to solve this problem by assigned a name to all the entities and identify it from its representation. Although his naming schema worked well to some extent, there existed a main drawback. To name an element he had to name many relative elements and he had to handle a graph structure to distinguish the element needed from other elements.

3 The Role Of Coding Feature Face

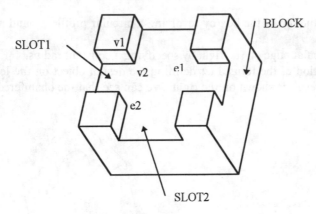

Fig. 2. A part with two slots

In a feature-based model, the basic topological entities are face, edge and vertices. It is obvious that new faces will not be created after boolean operation. As to edges and vertices it is not the same. See the part depicted in figure 2. This part includes 3 features, that is, one block and two intersecting slots. The constructing sequence is from block to slot1, then to slot2. After boolean operations some new edges(eg. e1,e2) and new vertices(eg. v1,v2) are generated, but no new faces are created.

From the above example, it can be concluded that it is the feature face, and only the feature face that can be acted as the role to record the relationships between features. That is to say the parent_son relationships are maintained by the information on feature face. It is difficult to treat the edge and vertex as the fundamental entities to record the relationships between features, for they are dynamically changed. Although edge and vertex are not taken as the fundamental entities, they themselves can be recorded by the feature face. One edge can be recorded by two intersecting feature faces and one vertex by three intersecting feature faces. It is one of the key points in feature-based modelling. So it is pertinent to say that only the feature faces can be taken as the basis of coding.

When designing a part using the feature-based modelling system, usually begin with a base feature. Then add the other features to this base feature. The parent-son relation is created during the constructing process. The parent-son relations are recorded by recording the feature entities such as faces, edges and vertices. All these feature entities can be recognized by decoding the code attached to the feature faces in advance and the relationships between features can be acquired through the recognized entities. Hence the constructing history of the part can be replayed.

It should be made certain that all the features be coded and the referenced faces, edges and vertices be recorded during the constructing process of the part in a sound feature-based, constraint-based design system.

4 Coding Mechanism For Form Feature

4.1 Feature Classification

Not all the features' forms are generated with the same method. It is useful to classify various features by their natures because every type of features has its special character. In reference[2], C. M. Hoffmann classified feaures into three types. The first type is generated features which are formed by extruding, revolving or sweeping a profile along a trajectory. The second type is datum feature which includes datum point, datum axis and datum plane and the third is modifying feature, that is, fillet, chamfer, shell and so on. This method of classifying form features is quite useful for coding feature.

It should be noted that most of the features belong to the first type and the first feature in a design body is sure to fall into this type. The feature volume of this type can be gotten easily. Datum feature is used to construct the other features easily during the design process and this kind of feature has no feature volume. Modifying feature operates directly on the Breps of the design object. The character of Modifying feature is that the feature volume cannot be gotten easily. We do not consider datum feature and modifying feature here only to limit the scope of this paper.

4.2 Coding Principle

Before presenting the method of coding feature, the coding principle must be provided.

As discussed above, the feature face is the basic element in feature-based design. So we code the feature face to identify it uniquely. We can identify edge with the codes of the faces between which the intersection is the edge and some extra information. The extra information is added here to decrease the ambiguity. As to vertex, it can be identified by 3 faces whose intersection is the vertex itself. The extra information is added to decrease the ambiguity.

4.3 Coding Method For 2d Sketch

Because all the generated features are formed by sweeping a profile along a trajectory, it is natural to consider that the profile is the fundament of the generated features. Every face on the generated feature has a parallel on the profile. We can directly perceive through the sense that the edge on the profile is the parallel. So it is necessary to establish a relation between the elements of the profile and the generated feature's faces.

Let's first see the sketch entities in the 2d sketch. The sketch entities include straight-line, arc, circle and curve. Except for the circle marked by only one point, all the sketch entities can be marked by two points. What we concern about is the identification of the entity, not the location and size. For instance, we can mark a spline curve by its start point and end point. The shape is not the thing that we should concentrate on. So point is the basic entity in 2d sketch. And The other entities such as straight-line, arc and so on can be recorded by points that related to them.

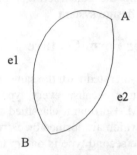

Fig. 3. A 2D profile consisted of 2 arcs

We assign an identification which can be defined by a constant to mark the point. It must be satisfied that the constant is greater than zero or equal to zero. So straight-line, arc, circle and curve can be encoded by the following triples

$$E = [c1, c2, rn]$$

In the above triples, E means entity in 2d sketch. c1 and c2 stand for the codes assigned to the points related to E. When E is a circle, the value of c2 is zero. Now pay attention to rn. It means referenced number. The reason it is introduced into the above description is that we should consider the case depicted in figure 3.

In most of the cases, the number of the entities that references the same two points is one. But in figure 3, the profile is closed and the 2 entities both references point A and point B. In order to identify the 2 entities, rn is included in the triples. It records the generating sequence of the entity. We assume that the value of c1 is 1 and c2 is 2. If e1 is sketched first, the representation of e1 and e2 are as the following descriptions

$$e1 = [1, 2, 1]$$
$$e2 = [1, 2, 2]$$

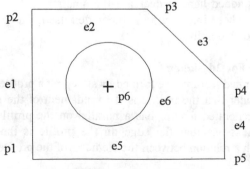

Fig. 4. A 2D profile

A constant is attached to the point while sketching it. The value of the constant is increased by 1 according to the sketching sequence and the value can be reused only if the sketch point to which it attaches does not exist because of deleting it from the sketch. With this mechanism the topology of the profile can have great flexibility. Consider the example depicted in figure 4. All the sketch points are assigned the value varied from 1 to 6. The straight line e1 is recorded by points p1 and p2 and straight line e2 is recorded by the point p2 and point p3, and so on. The circle is recorded by point p6. Every straight line is parallel to a new plane on the feature F1 formed by extruding this profile. As to the circle, a conical face. If another feature F2 referenced the face which is parallel to the entity e1. It can be assured that the profile's shape can vary arbitrarily if the two points related e1 are not deleted, while satisfying the relation between F1 and F2 at the same time.

4.4 Coding Method For Form Feature

The coding method in 2d is the basis of coding form feature. The identification of face has the following description

$$ffi = [\, fi, Esp, Est, pi, op \,]$$

In the above description, ffi means feature face identification. The following section will discuss every domain in it.

fi stands for feature identification. In order to maintain the parent_son relation and to manipulate the model interactively, it is useful to introduce the feature identification into the structure.

Esp means entity in the sketch in which the profile is sketched and Est stands for entity in the sketch in which the trajectory is sketched. In feature-based design, most of the features are generated features which are formed by extruding, revolve and sweeping a profile along a path. Every side face is parallel to one of the entity in the sketch. This is the reason why the above description includes Esp. It is the same to Est. As to the front face, the value of Esp is -2 and to the back the value of Esp is -1. In the case of revolusion with 360 degree rotating, there are no front and back face and it is unnecessary to code them.

pi means part identification. In assembly environment, more than one part will be inserted into the assembly document. To every feature in a part, the feature identification is unique, but with more than one feature, it does exist that more than one features have the same feature identification. In order to handle this case, add the part identification into the face identification. But it also should be noted that in the process of designing a part, it is unmeaningful to assign a part identification to the part. The part identification is generated when it is inserted into the assembly document. So the part identification is dynamically changed and it is determined by the sequence inserting the assembly document.

Even with the above domains, there exist some special cases that cannot be handled. See the example depicted in figure 5. The Brep is formed by extruding a profile and a draft angle. There exist two methods to handle the corners if the profile is expanded. If the EXTEND method is adopted, the edges in the expanded profile should be extended to their intersection point. The extensions are then swept back along the path and intersected to produce two new faces and a new edge. Or the ROUND method will be adopted. For the edge extensions do not intersect, an arc tangent to the two edges is created and swept to create a closed face in this case. The example belongs to the later. Considering the four cone surfaces in the four corners, they differ from the other faces on the Brep. So the coding mechanism is invalid to this case. This case can be solved by given an additional domain named op in the description. And for the necessity to extend the system, it is useful to append an optional domain to the feature face identification.

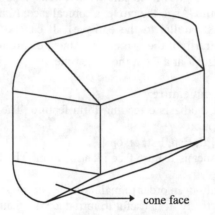

cone face

Fig. 5. Extrude with round draft

4.5 Record Edge
Edge can be recorded by the following triples
$$Ed = [ffi1, ffi2, op]$$
In it, ffi1 and ffi2 are the two faces with which the edge is formed. And op expressed by duple [u, v] which is used to record the value of the face's parameter space.
Consider the example depicted in Figure 1. Both of the 2 circular edges are formed by intersecting face f1 and f2. We can select a point on the edge respectively, It can be made certain that the duples of the 2 points are different from each other. So with the triples we can distinguish the 2 circular edges easily.

4.6 Record Vertex
Vertex can be recorded by the following description
$$V = [Ed1, Ed2, Ed3]$$
In the above triple, Ed1, Ed2 and Ed3 are the codes of the 3 intersected edges. With it, every vertex has a unique expression.

5 Conclusion
With the above coding mechanism, every feature element such as vertex, edge and face can be identified uniquely. It is helpful to decrease ambiguities greatly and automate feature's generation properly. Also, the flexibility of the feature-based model is enhanced.

References

1. Wesley M A, Lozano-Perez T, Lieberman L I, Lavin M A and Grossman D D, A geometric modelling system for automated mechnical assembly, IBM J. Res. DEV., vol.24, No.1(1980), pp64-74
2. C. M. Hoffmann, and R. Juan, Erep, an editable, high level representation for geometric design and analysis, in P. Wilson, M. Wozny, and M Pratt, (Eds.) Geometric and Product Modelling, North Holland(1993), pp129-164.
3. Vasilis CapoYleas, Xiangping Chen and Christoph M Hoffmann, Generic naming in generative constraint-based design, CAD, Vol.28, No.1(1996). pp17-26.

Computation of the Skeleton of 3D Polyhedral Solid

Yujin Hu, M.M.F. Yuen* and Ji Zhou

CAD Center
Huazhong University of Science and Technology
Wuhan, Hubei, P. R. China
*Hong Kong University of Science and Technology
Clear Water Bay, Kowloon, Hong Kong

Abstract. The skeleton of an object, also known as the medial axes transform, is an alternative geometric representation that has great potential in engineering. In this paper, a new method constructing the skeleton of 3D polyhedral solid is present. The basic principle of the method is the abstract Delaunay triangulation of a set of active topological entities(ATE), which are a new classification of the object boundary. In order to conveniently construct Delaunary triangle, the conception of the adjacent connecting graph(ACG) of the ATE is introduced. The proposition and algorithm for determining the topology and geometry of the object skeleton are given. Finally, some examples illustrate the computation properties of the method.

1. Introduction

The skeleton of an object, also known as the medial axes transform, is an alternative geometric representation that has great potential value in engineering. It was first proposed by Blum[1] as a technique for biological shape description. It is defined as the locus of the centers of all inscribed spheres of an object, where an inscribed sphere is a sphere that is not contained by any other interior sphere.

The skeleton is closely related to the Voronoi diagram of an object. Lee[2] says that the skeleton is a subset of the Voronoi diagram. Lavender[3] also says that if the object boundary is suitably defined the object skeleton may be recovered from the Voronoi diagram. So the key constructing the object skeleton is to construct its Voronoi diagram.

Methods generating skeletons can be classified into two main types, namely discrete boundary method and exact tracing method. The discrete boundary method[4][5][6][8] is based on the properties of the Delaunay triangulation of point sets. The object boundary is firstly discretized into a point set, then Voronoi diagram of a point set is computed via its dual Delaunay triangulation. Due to the discrete point sets can't completely represent the continuous object boundary, so this method only gets the approximate skeleton of an object except the boundary of an object is infinitely subdivided. The exact tracing method[7][9][10][11] doesn't need to discritize the object boundary, and it directly solve a system of non-linear equations of the object skeleton. It begins from the intersection of the object boundary, then traces the equidistant between the object boundaries so that all equidistant of the object boundaries are found. However, because it is difficult to recover non-linear equations of the object skeleton into explicit equations, much computation is required to solve the system of non-linear equations.

In this paper, a new method constructing the object skeleton is proposed. The object boundary is represented into a new set of topological entities. At the same time, by means of the relationship of the skeleton and Voronoi diagram, the definition of the active topological entity sets (ATES) of the skeleton, which is a set of the topological entities making a contribution to the skeleton, is introduced. Thus the computation of the skeleton is transferred into the computation of Voronoi diagram of ATES. Because the computation of Voronoi diagram of ATES is very complex, we only compute its dual --- the abstract Delaunay triangle/tetrahedron. Then based on the abstract Delaunay trianglation and some skeleton properties, the topology and geometry of the skeleton entities are separately constructed. In order to compute conveniently abstract Delaunay triangle/tetrahedron a conception of adjacent connecting graph(ACG) is introduced.

2. Basic Definitions

2.1. Voronoi Diagram and Skeleton Entity

Definition 1 Let $P = \{p_1, p_2, \ldots\ldots p_n\} \subset R^m$ be a set, where $2 \le n \le \infty$ and $p_i \ne p_j$ for $i \ne j$, $i, j \in I_n = \{1, 2, \ldots\ldots n\}$. We call the region given by

$$V(p_i) = \bigcap_{j \in I_n} H(p_i, p_j) \quad i \ne j$$

the Voronoi domain associated with entity p_i, and the set given by

$$Vd(P) = \{V(p_i), \ldots, V(p_n)\}$$

the Voronoi diagram of P. Where

$$H(p_i, p_j) = \{p \in R^m | d(p_i, p) = d(p_j, p)\}$$

$d(p_i, p)$ is the minimum Euclidean distance between entity p and entity p_i. $H(p_i, p_j)$ is equidistant from p_i and p_j. If we connect all the entities adjacent Voronoi domains, Delaunay triangle sets $DT(P)$, a dual of Voronoi diagram $Vd(P)$, would be formed. Fig. 1a shows $Vd(P)$ and $DT(P)$ of a point set on a plane. If P not only includes points but also other entities such as 3D line, 3D plane, then $Vd(P)$ is Voronoi diagram of mixed entity sets, $DT(P)$ is the abstract Delaunay tetrahedron sets of mixed entity sets. Fig. 1b shows Voronoi diagram of a block. It can be classified into two parts, interior Voronoi diagram(fine lines) and exterior Voronoi diagram(dot lines). Since points on the Voronoi diagram are equidistant from two or more boundary entities of the block, it is easy to see that the interior Voronoi diagram of the block is actually the skeleton of the block. The shade part in Fig. 1b shows the equidistant between top face and bottom face of the block, and the complete skeleton of the block is shown as Fig. 1c.

The skeleton of an object is still represented the structure of vertex, edge and face. We generally call them skeleton vertex, skeleton edge and skeleton face.

Definition 2 Skeleton face is a bounded portion of the equidistant surface of two boundary entities. It is bounded by skeleton edges and skeleton vertices. It is also defined the locus of the center of the sphere tangent to two boundary entities. As shown in Fig. 1c.

Definition 3 Skeleton edge is a bounded portion of the equidistant curve of three or more boundary entities. It is bounded by two skeleton vertices. It is also defined the locus of the center of the sphere tangent to three or more boundary entities. It connects two or more adjoining skeleton faces. See Fig. 1c.

Definition 4 Skeleton vertex is an equidistant point of four or more boundary entities. It is also defined the center of the sphere tangent to four or more boundary entities. See Fig. 1c.

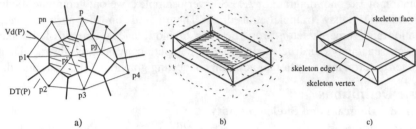

Fig. 1 Voronoi diagram and the skeleton of a block

2.2 Active Topological Entity

Due to the object skeleton is contained in Voronoi diagram of the object boundary topological entities, it motivates us to think if the topological entity creating the object skeleton can directly be separated from the boundary topological entities of the object. Thus we introduce definitions as follow.

Definition 5 Let p_i , p_j be two topological entities of an object O. If it satisfies the following condition

$$\bigcap_{j \in I_n} H(p_i, p_j) \cap O \neq \phi \quad i \neq j$$

then p_i is called active topological entity(ATE), where I_n represent all boundary topological entities.

Definition 6 Let $P = (p_1, p_2, \ldots\ldots p_n)$ be a set of boundary topological entities of an object, then all ATEs in P, written as $\tilde{P} = (\tilde{p}_1, \tilde{p}_2, \ldots\ldots \tilde{p}_m)$, $m \leq n$, is called as active topological entity sets(ATES).

3. Construction of ATES and its Adjacent Connecting Graph

3.1 Classification of the topological entity

In order to construct conveniently the active topological entity and the skeleton of an object, we will further classify the object topological entities, vertices, edges and faces, into new topological entities.

Definition 7 If edge angle $\alpha_e(t)$ between two adjoining faces of an edge E is all greater than 180 degree, then the edge E is called concave edge E_a. Similarly, if edge angle $\alpha_e(t)$ is all less than or equal 180 degree, then the edge E is called convex edge E_v.

Definition 8 If vertex is a intersection of concave edges, then the vertex is called concave vertex V_a. Similarly, if vertex is a intersection of convex edges, then the vertex is called convex vertex V_v. If vertex is a intersection of only one concave edge and two or more than two convex edges, then the vertex is called mixed_I vertex MV_1.

If vertex is a intersection of only one convex edge and two or more than two concave edges, then the vertex is called mix_II vertex MV_2.

3.2 Construction of Active Topological Entity

Proposition 1 Let $P = (p_1, p_2, \ldots\ldots p_n)$ be a set of boundary topological entities of an object, then all faces, concave edges, concave vertex as well as mix_II vertex are ATEs, and others are not ATEs.

This property is not difficult to proof. Only need to proof the intersection of Voronoi region of no ATE p and object O is empty, i.e. $\bigcap\limits_{j \in I_n} H(p, p_j) \cap O = \phi$.

3.3 Construction of Adjacent Connecting Graph

Because ATE is a alternative represention of the boundaries topological entity of an object, so among ATEs they still preserve the relationships similar to the topological connections of object boundaries. We call a graph consisting of these connecting relationships as adjacent connecting graph(ACG).

ACG is mainly used to implement the abstract Delaunay triangulation of ATES. Fig. 2a shows a simple 2D L-shaped polygon. ATES consists of one concave vertex and six boundary edges. The adjoining relationship between two topological entities can be represented as a graph with two nodes, one for each ATE, and one edge connecting the nodes. Fig. 2b shows the complete ACG of ATES for the L-shaped polygon. Fig. 2c and Fig. 2d show the example in 3D.

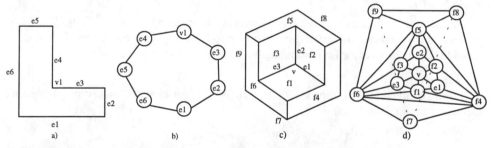

Fig. 2 Adjacent connecting graph of objects

From above construction of the ACG, we discover that the ACG of ATES exists the features as follow:

1). ACG is closed.

2). Each node of ACG connects at least with two nodes (2D) or three nodes (3D).

3). If the ACG is conceived into a domain (2D) or solid (3D), then its boundary generally is segment/triangle. So we often call the boundary face of ACG as boundary triangle.

4. Abstract Delaunay Triangulation

Although the definition of ATES reduces the searching space computing Voronoi diagram, constructing exact Voronoi diagram of ATES is still complex because computing Voronoi diagram of ATES has to involve the computation of surface intersection. Consequently, we don't directly compute the Voronoi diagram of ATES, but we compute its dual abstract Delaunay triangle or tetrahedron by mean of ACG. Once abstract Delaunay triangle or tetrahedron is exactly determined, the topology of the skeleton and the geometry of the skeleton vertex is actually determined.

4.1 The Criterion of Abstract Delaunay Triangulation

Proposition 2 If the radius of inscribed sphere tangent to any four ATEs in ATES is minimum, then the tetrahedron constructed by the four ATEs is an abstract Delaunay tetrahedron.

4.2 The Unit Normal Vector of Abstract Delaunay Triangle

The ACG is an abstract topological graph. It reveals the topology information between the ATEs, but it does not inform any geometry information between the ATEs. Actually, the inscribed sphere corresponding to abstract Delaunay tetrahedron in the ACG is closely related to the geometry of ATEs, thus we define a unit normal vector for each abstract Delaunay triangle face.

Definition 9 Let $B(c,r)$ be the inscribed sphere corresponding the abstract Delaunay tetrahedron with ATEs p_i, p_j, p_k, p_l, where c, r are the center and radius of inscribed sphere respectively. Let q_i, q_j, q_k, q_l be the corresponding four tangent points, $\hat{N}(q_i, q_j, q_k)$ be a unit normal vector of the plane patch constructed by three tangent points q_i, q_j, q_k, which directs the center of $B(c,r)$, then $\hat{N}(q_i, q_j, q_k)$ is called the unit normal vector corresponding to abstract Delaunay triangle $\triangle_{p_i p_j p_k}$, it denotes $\hat{n}(p_i, p_j, p_k)$.

The unit normal vector of abstract Delaunay triangle has two functions. One is to provide the searching direction for solving non-linear system equations. Other is to provide the tangent vector for computing the geometry of skeleton edge.

4.3 Algorithm

Based on above criterion and definition of ATES, we give the algorithm of abstract Delaunay triangulation as follow.

Input: ATES $\tilde{P} = (\tilde{p}_1, \tilde{p}_2, \ldots \tilde{p}_m)$.

Output: tetrahedron sets $Tetr = (t_1, t_2, \ldots t_s)$.

Step 1 Initializing tetrahedron list $Tetr$.

Step 2 Construct cyclical list CG to store the ACG of $\tilde{P}$. Each node in CG records the complete information of each boundary segment/triangle in the ACG, which contains the unit normal vector and neighborhoods of each boundary segment/triangle.

Step 3 If CG is not empty, then select a node $T(p_i, p_j, p_k, pt_i, pt_j, pt_k, \hat{n})$ from cyclical list CG. Where p_i, p_j, p_k is three ATEs constructing boundary triangle $\triangle_{p_i p_j p_k}$, pt_i, pt_j, pt_k are three pointers pointing to adjacent boundary triangle, $\hat{n}$ is the unit normal vector of $\triangle_{p_i p_j p_k}$. If CG is empty, then output list $Tetr$.

Step 4 Search fourth ATE p_l from $\tilde{P}$ so that the inscribed sphere of four ATEs corresponding to abstract Delaunay tetrahedron $t(p_i, p_j, p_k, p_l)$ is minimum.

Step 5 Compute exactly the unit normal vector of each triangle of abstract Delaunay tetrahedron $t(p_i, p_j, p_k, p_l)$.

Step 6 Record abstract Delaunay tetrahedron $t(p_i, p_j, p_k, p_l)$ and correspond-ing inscribed sphere $B(c, r)$.

Step 7 Delete the node $T(p_{i}, p_j, p_k, pt_i, pt_j, pt_k, \hat{n})$ from CG. Then check if abstract Delaunay triangles $\triangle_{p_i p_j p_l}$, $\triangle_{p_j p_k p_l}$, $\triangle_{p_k p_i p_l}$ are nodes of CG respectively. If some of them are nodes in CG, then delete them from CG. If some of them are not nodes of CG, fill them into CG, goto Step 3.

Finding fourth entity above algorithm is very key and important. It requires much more times to solve non-linear system equations which govern the center and radius of the inscribed sphere of any four entities. In general, the construction of the non-linear system equations is not difficult and there is a general method to formulate them[3][11][6]. However, to solve non-linear equations has to consume more times. Newton-Raphson method is adopted to solve non-linear equations in this paper. Initial value $\vec{X}$ of the solution is determined with $\vec{X} = k\hat{n}$, where k is an experienced coefficient. Moreover, because the solution of non-linear equations is not unique, i.e. the inscribed sphere of any four entities may be not unique, so the solved solution has further to be checked. The algorithm computing minimum inscribed sphere describes in detail as follow.

Input: Node $T(p_{i}, p_j, p_k, pt_i, pt_j, pt_k, \hat{n})$ and ATES $\tilde{P} = (\tilde{p}_1, \tilde{p}_2, \ldots\ldots \tilde{p}_m)$.

Output: Inscribed sphere $B(c, r)$ or null.

Step 1 Let three entities corresponding to triangle nodes be p_i, p_j, p_k, initial minimum radius min_r be 1.0e+10. Moreover, initializing $B(c, r)$ is empty.

Step 2 For all ATEs $l = 1, m$.

2.1 Select an entity p_l from P.

2.2 Determine initial value of the solution and solve non-linear system equations to obtain the inscribed sphere $B(c, r)$ corresponding to p_i, p_j, p_k, p_l.

2.3 If $B(c, r)$ is empty, goto Step 2.1

2.4 Check if $B(c, r)$ is inside object, if false, goto Step 2.1

2.5 Check if the projects of the center of $B(c, r)$ on the entities p_i, p_j, p_k, p_l are all inside corresponding entity respectively, if false, goto Step 2.1.

2.6 Compare the radius of $B(c, r)$ with min_r, and record corresponding tangent points(i.e., the coordinates of the projects in Step 2.5) as well as inscribed sphere $B(c, r)$.

Step 3 output $B(c, r)$.

5. Construction of Topology for Object Skeleton

The constructions of topology for the object skeleton mainly determine the connection among skeleton vertices or skeleton edges. In order to define the topology of object skeleton, two definitions related to the object skeleton are introduced as below.

Definition 10 The trunk of a skeleton is a portion of the skeleton which is not incidence with the object boundary.

Definition 11 The branch of a skeleton is a portion of the skeleton which is incidence with the object boundary.

Abstract Delaunay triangulation exactly determines the geometry of the skeleton vertex. Moreover, it also determines the topology between skeleton vertices or skeleton edges exactly because abstract Delaunay tetrahedron set is dual to Voronoi diagram.

Proposition 3 In abstract Delaunay tetrahedron set $Tetr = (t_1, t_2, \ldots \ldots t_s)$, if any two tetrahedrons share same Delaunay triangle Δ_{ijk}, then the inscribed sphere centers of the two tetrahedrons define a skeleton edge of the trunk of object skeleton.

Proposition 4 In abstract Delaunay tetrahedron set $Tetr = (t_1, t_2, \ldots \ldots t_s)$, if any several tetrahedrons share same Delaunay edge l_{kl}, then the inscribed sphere centers of the several tetrahedrons define a skeleton face of the trunk of object skeleton.

Proposition 5 In abstract Delaunay tetrahedron set $Tetr = (t_1, t_2, \ldots \ldots t_s)$, if the Delaunay triangle of a tetrahedron is boundary triangle Δ_{ijk}, then the inscribed sphere center of the tetrahedron and the convex vertex of the object beside Δ_{ijk} define a skeleton edge of the branch of object skeleton.

Proposition 6 In abstract Delaunay tetrahedron set $Tetr = (t_1, t_2, \ldots \ldots t_s)$, if one boundary Delaunay edge l_{kl} is shared by several tetrahedron, then the inscribed sphere centers of the several tetrahedrons and the convex edges of the object beside l_{kl} define a skeleton face of the branch of object skeleton.

For above properties, we explained as below with the doubly tapered block shown as in Fig. 3a. Fig. 3b shows its ACG with six ATEs, which are described as nodes a, b, c, d, e, f respectively. Fig. 3c shows five abstract Delaunay tetrehadrals. Their inscribed sphere centers are exactly corresponding the skeleton vertices v_1, v_2, v_3, v_4, v_5 of the object skeleton respectively. Due to tetrahedron $abde, abfd, dbef$ share db edge, tetrahedron $abde$ and $abdf$ share triangle Δabd face, tetrahedron $abde, dbef$ share triangle face Δdbe, tetrahedron $abdf, dbef$ share Δdbf, so the inscribed sphere centers of three tetrahedrons $abde, abfd, dbef$ construct three skeleton edges $v_1 v_2, v_1 v_3, v_2 v_3$ and one skeleton face $v_1 v_2 v_3$. Moreover, due to triangle Δabe is a boundary triangle, vertex B is the convex vertex beside triangle Δabe, so v_2 and B define a branch edge $v_2 B$; due to tetrehadrals $abde$ and $abfd$

share the boundary edge ab, so v_1, v_2 and the convex edge BC beside ab define a branch face $v_1 v_2 BC$ of the skeleton. Similarly, another three tetrahedrons will construct the skeleton edges $v_3 v_4, v_3 v_5, v_4 v_5$ skeleton face $v_3 v_4 v_5$ etc.

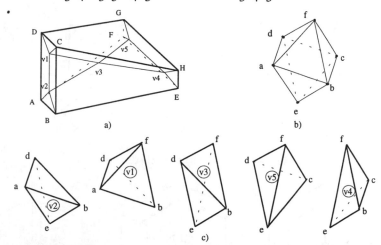

Fig. 3 Topology construction of skeleton for doubly tapered block

6. Computation of Geometry for Object Skeleton

According to the definition of the skeleton edge and skeleton face, we know the geometry of skeleton entities is closely dependent on the geometry of an object boundary. In general, the geometry of skeleton entities of polyhedral solids can describe with simple algebraic equation. Because the object boundaries consist of lines/planes, then the skeleton curve/surface corresponding to skeleton edge/face is a portion of quadrics of the following types: a) point/point (plane), b) plane/plane (plane), c) line/line (hyperparabolic), d) point/line (cylinder parabolic), e) point/plane (circle parabolic), f) line/plane (parabolic cylinder).

Based on the topological construction and geometrical approximate mentioned above , the algorithm generating the trunk and branch of the object skeleton is given as follow.

Input tetrahedron sets $Tetr = (t_1, t_2, \ldots \ldots t_s)$

Output the trunk and branch of the object skeleton
Step 1 For all tetrahedrons do
      For all tetrahedrons edge which is interior CG do
          1.1.1 searching all tetrahedrons sharing same common edge.
          1.1.2 if common edge is interior, then compute exact skeleton trunk.
          1.1.3 else (i.e., common edge is boundary edge), then compute exact skeleton
                    branch.
Step 2 Display the skeleton of an object

7. Example and Conclusion
7.1 Example
In this section, we show a few examples of our method. Fig 4a shows a 2D example with curve boundaries. It consists of 29 ATEs, i.e., 21 edges and 8 concave vertices. The

trunk of skeleton has 27 skeleton vertices (including 7 pairs degeneracy skeleton vertices) and 26 skeleton edges (including degeneracy skeleton edges). The branch of skeleton has 13 skeleton edges. All the skeleton edges are described as Bezier curves and line. Fig. 4b shows another 2D example with multiple-connection polygon.

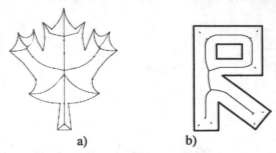

a)   b)

Fig. 4 Skeletons of two 2D domain

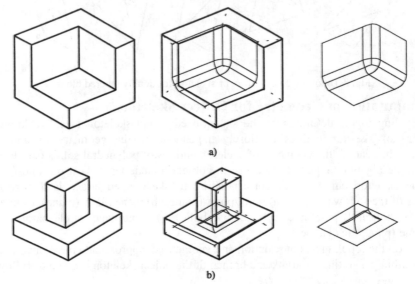

a)

b)

Fig. 5 Two polyhedrals and its skeletons

Fig. 5 shows two 3D polyhedral examples. The total ATEs shown in Fig. 5a are 9 faces, 3 concave edges and one concave vertex. The skeleton in this case consists of 12 trunk faces and 15 branch faces, and on the skeleton trunk there are 19 skeleton vertices and 32 edges. In this example there is no degeneracy skeleton vertex. Fig 5b shows other examples with 19 ATEs, and its skeleton consists of 6 trunk faces and 14 branch faces. In this example there are more degeneracy cases related to concave edge and mixed vertex. In this paper, the tolerance we check coincided point (or degeneracy) is same level tolerance as to solve non-linear system equations.

7.2 Conclusion

In this paper we have presented a method for skeleton construction of an object in 3D. Because the method is based on Voronoi diagram, the topology correctness of the skeleton can be ensured. In addition, we have not to divide the object boundary infinitely and not to trace the object skeleton, so there are less computation complex than discrete boundary method and exact tracing method. Although this method can not directly construct exact skeleton of the object with curve shaped boundary, we believe the same conception and classification introduced in this paper can be lead to an algorithm for skeleton construction of the object with non-intersection curve shaped boundary. Actually we have done for 2D domain with curve shaped boundary as shown in Fig. 4a, but the calculation of non-linear equations for curve shaped boundary is more complicated.

References

1. H. A. Blum Transformation for extracting new descriptors of shape. In W. Wathen-Dunn (Ed.) Models for Perception of Speech and Visual Form MIT Press, USA(1967)pp.362-380
2. Lee, DT. and Drysdale,RL, Generalization of Voronoi Diagrams in the Plane, SIAM J. Comput. 10(1981)pp.73-87
3. David Lavender and Adrian Bowyer et al. Voronoi Diagrams of Set-Theoretic Solid Models, IEEE Computer Graphics and Application 16(1992) pp.69-77
4. T. K. H. Tam and C. G. Armstrong, 2D Finite Element Mesh Generation by Medial Axis Subdivision, AES, 13(1991)pp.313-324
5. J. A. Goldak X. Yu, Constructing Discrete Medial Axis of 3D Objects, International Journal of Computational Geometry & Applications, 1(1991) pp.327- 339.
6. J. M. Reddy Computation of 3D Skeletons Using a Generalized Delaunay Triangulation Technique, Computer Aided Design, 27(1995)pp.677-694
7. E. Sherbrooke and N. Patrikalakis Compution of the medial axis transform of 3D polyhedra, Proc ACM/IEEE Symposium on Solid Modeling and Application, Salt Lake City,Uath, (1995) pp.187-200
8. DJ Sheey,CG Armstrong and DJ Robinson, Computing the Medial Surface of a Solid from a Domain Delaunay Triangulation, Proc ACM/IEEE Symposium on Solid Modeling and Application, Salt Lake City,Uath, (1995) pp.201-212.
9. Niranjan Mayya and V. T. Rajan, An efficient Shape representation Scheme Using Voronoi Skeletons, Pattern Recognition Letters 16(1995) pp.147-160.
10. D. Dutta and C. M. Hoffmann, On the Skeleton of Simple CSG Objects, Journal of Mechanical Design, 115(1993) pp.87-94
11. C. M. Hoffmann, How to Construct the Skeleton of CSG Objects, Technical Report CSD-TR-1014, Computer Sciences Department, Purdue University, Sept. 1990

Creating Blending Surface and N-sided Region Surface with Rational Bezier Patches

Luo Hongzhi,   Yi Wensen,   Li Shiqi,   Zhou Ji

CAD Center, Mechanical Institute at Huazhong University of Science and Technology, Wuhan, P. R. China

Abstract.   This paper presents mathematical techniques for constructing G^1 continuous blending surfaces and fill arbitrary N sided regions using rectangular rational Bezier patches. In this paper the derivation of the filling of an N-sided region with the rational Bezier surface is described. An algorithm for generating an N-sided region surface of G^1 continuity with the rational Bezier surface and the test results of the algorithm are presented. This algorithm can be used for any order of rational Bezier patches and its shape can be interactively adjusted by changing the fullness of the surface. Rectangular rational Bezier patches are preferred for the surface because nearly all CAD/CAM system can handle this type of patch quite effectively. In addition, they are supported by STEP.

1 Introduction

The generating Blending and N-sided surface is important in product geometric modeling and automatic manufacturing systems. Many papers published have discussed using non-standard surfaces to building G^1 or nearly G^1 blending surface or N-sided surface[1-3]. It is harmful for application and data exchange to bring non-standard surfaces which do not supported by STEP into CAD/CAM system. They are only applicable in an island of system or in particular circumstance. For the unified treatment of product geometric model, system integration and data exchange of product, we are hoping to build blending surface and N-sided surface with standard surface which is supported by STEP, and can be treaded for the life cycle of the product geometric model in a unified way. There are a few algorithms published to fill a N-sided hole with standard surface, such as bezier patches, but N is limited to three, four and five[4].

This paper puts forward a new method to create blending surface and N-sided region surface with rectangular rational Bezier patches. In this paper, the derivation of the filling of an arbitrary N-sided region with the rational Bezier surface is described. An algorithm for generating an N-sided region surface which has G^1 continuity with the adjacent rational Bezier surface and the test results of the algorithm are presented. The algorithm can be used for any order of rational Bezier patches and its shape can be interactively adjusted by changing the fullness of the surface. The form of the N-sided region surface is in conformity with STEP.

2 Basic description of N-sided region

In an affine coordinate system of space E^3, a rectangular rational Bezier patch $R(u,v)$ has the representation

$$R(u,v) = \frac{S(u,v)}{W(u,v)}, \qquad 0 \le u,v \le 1, \tag{1}$$

where

$$S(u,v) = \sum_{i=0}^{n} \sum_{j=0}^{m} r_{ij} w_{ij} B_i^n(u) B_j^m(v); \tag{2}$$

$$W(u,v) = \sum_{i=0}^{n} \sum_{j=0}^{m} w_{ij} B_i^n(u) B_j^m(v); \tag{3}$$

in which $\{\vec{r}_{ij}\}$ is net vertices of patch $R(u,v)$, and w_{ij} is corresponding weights of vertex $\vec{r}_{ij}$. $B_i^n(u) = c_k^l t^l (1-t)^{k-1}$ is basic function of Bernstein.

Theorem 1. *Let rectangular rational Bezier patches $R_1(u,v)$ and $R_2(u,v)$ be C^0 continuous along a common boundary curve u=0. Then $R_1(u,v)$ and $R_2(u,v)$ are G^1 continuous if the following conditions hold*

$$R_{2u}(0,t) = \lambda(t) R_{1u}(0,t) + \mu(t) R_{1v}(0,t) \qquad 0 \le t \le 1 \tag{4}$$

where $\lambda(t) < 0$ for $0 \le t \le 1$.

2.1 Boundary condition for N-sided region

Let $\{P_j(u,v)\}, j = 1,2,\cdots,n$ are rational patches surrounding an isolate N-sided region D (see Figure 1). Assume that there are four boundary Bezier curves connect to corner vertex of the region, and the tangent vectors at the corner vertices are co-linear. We request that to construct a N-sided rational Bezier surface S in the region, S is G^1 continuous with $\{P_j(u,v)\}, j = 1,2,\cdots,n$. and the patches of $S : \{\vec{S}_j(u,v)\}, j = 1,2,\cdots,n$ are G^1 continuous along their common boundary curves.

2.2 Consistency condition

When a patch must be smoothly joined with two patches across two adjacent edges at a corner vertex, the inconsistent problem might be encountered.

Fig. 1. Boundary condition at a N-sided region

Let $\vec{S}_j(u,v), j = 1,2,\cdots,n$ be blending patches, then following condition must be hold at four corner vertices of $\vec{S}_j(u,v)$

$$\frac{\partial^2 \vec{S}_j(0,0)}{\partial u \partial v} = \frac{\partial^2 \vec{S}_j(0,0)}{\partial v \partial u}, \quad \frac{\partial^2 \vec{S}_j(1,0)}{\partial u \partial v} = \frac{\partial^2 \vec{S}_j(1,0)}{\partial v \partial u}$$

$$\frac{\partial^2 \vec{S}_j(0,1)}{\partial u \partial v} = \frac{\partial^2 \vec{S}_j(0,1)}{\partial v \partial u}, \quad \frac{\partial^2 \vec{S}_j(1,1)}{\partial u \partial v} = \frac{\partial^2 \vec{S}_j(1,1)}{\partial v \partial u} \tag{5}$$

3. G^1 Blending at the corner vertex of N-sided region

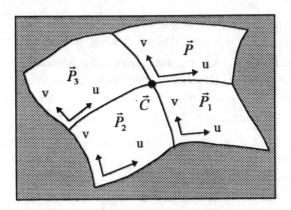

Fig. 2. Blending at corner vertex $\vec{C}$ of N-sided region

As shown in Figure 2, assume that there exist three basic patches $\vec{P}_1(u,v), \vec{P}_2(u,v)$

and $\vec{P}_3(u,v)$ at common vertex $\vec{C}$, and they are satisfied with boundary condition of N-sided region. Now we want to insert a patch $\vec{P}(u,v)$ which has G^1 continuity with $\vec{P}_1(u,v)$ and $\vec{P}_2(u,v)$ simultaneity. Without losing generality, we can assume that at the corner vertex $\vec{C}$, the across boundary tangent vectors of $\vec{P}_1(u,v), \vec{P}_2(u,v)$ and $\vec{P}_3(u,v)$ are satisfied with following conditions:

$$\vec{P}_{1u}(0,1) = \frac{1}{k_1}\vec{P}_{2u}(1,1) , \quad \vec{P}_{3v}(1,0) = \frac{1}{k_2}\vec{P}_{2v}(1,1) \tag{6}$$

Due to that there are at least G^1 continuity between $\vec{P}_1(u,v)$, $\vec{P}_2(u,v)$ and $\vec{P}_3(u,v)$, and according to theorem 1, following condition must be hold

$$\vec{P}_{1u}(0,1) = \lambda_{12}(v)\vec{P}_{2u}(1,v) + \mu_{12}(v)\vec{P}_{2v}(1,v)$$
$$\vec{P}_{3v}(u,0) = \lambda_{32}(u)\vec{P}_{2v}(u,1) + \mu_{32}(u)\vec{P}_{2u}(u,1) \tag{7}$$

where, $\lambda_{12}(v), \mu_{12}(v)$ are the blending function between $\vec{P}_1(u,v)$ and $\vec{P}_2(u,v)$, and $\lambda_{32}(v), \mu_{32}(v)$ are the one between $\vec{P}_2(u,v)$ and $\vec{P}_2(u,v)$. Further, we can force that the endpoint condition of them are as follows

$$\lambda_{12}(1) = \frac{1}{k_1}, \lambda'_{12}(1) = 0, \mu_{12}(1) = 0, \mu'_{12}(1) = 0,$$
$$\lambda_{32}(1) = \frac{1}{k_2}, \lambda'_{32}(1) = 0, \mu_{32}(1) = 0, \mu'_{32}(1) = 0, \tag{8}$$

from (6),(7) and (8), we have

$$\vec{P}_{1uv}(0,1) = \vec{P}_{2uv}(1,1) / k_1, \quad \vec{P}_{3vu}(1,0) = \vec{P}_{2vu}(1,1) / k_2 \tag{9}$$

because the patches $\vec{P}_1(u,v)$, $\vec{P}_2(u,v)$ and $\vec{P}_3(u,v)$ exist, there must be

$$\vec{P}_{1uv}(u,v) = \vec{P}_{1vu}(u,v)$$
$$\vec{P}_{2uv}(u,v) = \vec{P}_{2vu}(u,v)$$
$$\vec{P}_{3uv}(u,v) = \vec{P}_{3vu}(u,v) \tag{10}$$

from (9) and (10), we can derive

$$k_1\vec{P}_{1uv}(0,1) = k_2\vec{P}_{3vu}(1,0) \tag{11}$$

let $\vec{P}(u,v)$ be G^1 continuity with $\vec{P}_1(u,v)$ and $\vec{P}_3(u,v)$ along their common boundary curve, and the blending function are $\lambda_1(u), \mu_1(u)$ and $\lambda_3(u), \mu_3(u)$ respectively, from theorem 1, we have

$$\vec{P}_v(u,0) = \lambda_1(u)\vec{P}_{1v}(u,1) + \mu_1(u)\vec{P}_{1u}(u,1),$$
$$\vec{P}_u(0,v) = \lambda_3(v)\vec{P}_{3u}(1,v) + \mu_3(v)\vec{P}_{3v}(1,v) \tag{12}$$

then, the twist vector of the patch $\vec{P}(u,v)$ at $\vec{P}(0,0)$ is

$$\vec{P}_{vu}(0,0) = \lambda_1'(0)\vec{P}_{1v}(0,1) + \lambda_1(0)\vec{P}_{1vu}(0,1) + \mu_1'(0)\vec{P}_{1u}(0,1) + \mu_1(0)\vec{P}_{1uu}(0,1)$$
$$\vec{P}_{uv}(0,0) = \lambda_3'(0)\vec{P}_{3u}(1,0) + \lambda_3(0)\vec{P}_{3uv}(1,0) + \mu_3'(0)\vec{P}_{3v}(2,0) + \mu_3(0)\vec{P}_{3vv}(1,0) \tag{13}$$

let
$$\lambda_1(0) = k_1; \quad \lambda_1'(0) = 0; \quad \mu_1(0) = 0; \quad \mu_1'(0) = 0$$
$$\lambda_3(0) = k_2; \quad \lambda_3'(0) = 0; \quad \mu_3(0) = 0; \quad \mu_3'(0) = 0 \tag{14}$$

from (11)-(14), we can easily derive that

$$\vec{P}_{uv}(u,v) = \vec{P}_{vu}(u,v)$$

this means that at the vertex $\vec{C}$, the inserted patch $\vec{P}(u,v)$ is satisfied with consistency condition. So patch $\vec{P}(u,v)$ does exist. Now we have proved following theorem.

Theorem 2 *Suppose $\vec{C}$ is common corner vertex of four rectangular rational Bezier patches $\vec{P}_1, \vec{P}_2, \vec{P}_3$ and $\vec{P}$, at vertex $\vec{C}$ their across boundary tangent vectors are co-linear. If $\vec{P}_1, \vec{P}_2, \vec{P}_3$ does exist, and they are C^0 and G^1 continuity along their common boundary curves, then $\vec{P}$ is satisfied consistency condition if end-point condition of blending function between $\vec{P}$ and its adjacent patches is set as (14).*

4 G^1 **Blending at the center vertex of N-sided region**

As in Figure 3, $\vec{O}$ is common vertex of N patches in the N-sided region. Given vertex $\vec{O}$ and rational Bezier curves $\{\vec{C}_j(t)\}, j = 1, 2, \cdots, n$ which run between center points of each boundary curves and *vertex* $\vec{O}$. We can construct following G^1 blending equations along each $\vec{C}_j(t)$ for correspondent patches according to theorem 1:

$$\begin{cases} \vec{S}_{2u}(0,t) = \lambda_1(t)\vec{S}_{1v}(t,0) + \mu_1(t)\vec{S}_{1u}(t,0) \\ \vec{S}_{3u}(0,t) = \lambda_2(t)\vec{S}_{2v}(t,0) + \mu_2(t)\vec{S}_{2u}(t,0) \\ \dots \\ \vec{S}_{1u}(0,t) = \lambda_n(t)\vec{S}_{nv}(t,0) + \mu_n(t)\vec{S}_{nu}(t,0) \end{cases} \tag{15}$$

The status of two patches blended along their common boundary curve $\vec{C}_j(t)$ is: the first point of $\vec{C}_j(t)$ is $\vec{O}$, at which the across tangent vector is not co-linear in general; while the end point is the common vertex of four patches, the across tangent vector is co-linear. It is easy to know from the theorem 2 and the relationship of tangent vectors at the first point of $\vec{C}_j(t)$ that the consistence condition will be satisfied at the first and end point of $\vec{C}_j(t)$, if end condition of blending function be set as follows:

$$\begin{cases} \lambda_j(0) = \alpha_j; \lambda'_j(0) = 0; \mu_j(0) = \beta_j; \mu'_j(0) = 0; \\ \lambda_j(1) = \gamma_j; \lambda'_j(1) = 0; \mu_j(1) = 0; \mu'_j(1) = 0; \end{cases} \tag{16}$$

when t=0, from (15) and (16), we can obtain the equations of tangent vectors

$$\begin{cases} \vec{S}_{2u}(0,0) = \alpha_1\vec{S}_{1v}(0,0) + \beta_1\vec{S}_{1u}(0,0) \\ \vec{S}_{3u}(0,0) = \alpha_2\vec{S}_{2v}(0,0) + \beta_2\vec{S}_{2u}(0,0) \\ \dots \\ \vec{S}_{1u}(0,0) = \alpha_n\vec{S}_{nv}(0,0) + \beta_n\vec{S}_{nu}(0,0) \end{cases} \tag{17}$$

and the equations of twist vectors

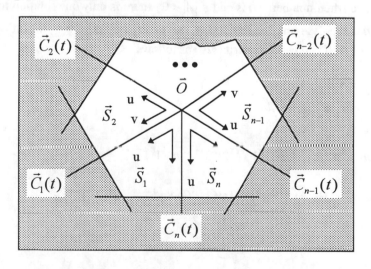

Fig. 3. Blending at center vertex $\vec{O}$ of N-sided region

$$\begin{cases} \vec{S}_{2uv}(0,0) = \alpha_1 \vec{S}_{1vu}(0,0) + \beta_1 \vec{S}_{1uu}(0,0) \\ \vec{S}_{3uv}(0,0) = \alpha_2 \vec{S}_{2vu}(0,0) + \beta_2 \vec{S}_{2uu}(0,0) \\ \cdots \\ \vec{S}_{1uv}(0,0) = \alpha_n \vec{S}_{nvu}(0,0) + \beta_n \vec{S}_{nuu}(0,0) \end{cases} \tag{18}$$

by solving equation (17), all $\{\alpha_j, \beta_j\}, j = 1,2,\cdots,n$ could be obtained, let

$$\begin{cases} \vec{X}_j = \vec{S}_{juv}(0,0) = \vec{S}_{jvu}(0,0) \\ \vec{b}_j = \beta_j \vec{S}_{juv}(0,0) \end{cases} \tag{19}$$

in (18), then (18) can be rewrite as

$$AX = B \tag{20}$$

by solving equation (12), we can get twist vectors $X = [\vec{x}_1, \vec{x}_2, \cdots, \vec{x}_n]^T$. And using of equation (21)

$$\vec{x}_j = m_j n_j \left[\frac{w_{j11}}{w_{j00}} (\vec{r}_{j11} - \vec{r}_{j00}) + \frac{w_{j01} w_{j10}}{w_{j00}^2} (2\vec{r}_{j00} - \vec{r}_{j01} - \vec{r}_{j10}) \right] \quad (j = 1,\cdots,n) \tag{21}$$

we can obtain net vertex $\vec{r}_{j11}$ corresponding to $\vec{x}_j$.

5 Solution of the twist vectors equations

In equation (20), $|A| = \alpha_1 \alpha_2 \cdots \alpha_n - 1$. We know that $\alpha_j < 0, j = 1,2,\cdots,n$. from theorem 1, so when number n is odd, $|A| \neq 0$, there is only one solution for (20); while as n is even, A may be singularity, due to $\alpha_1 \alpha_2 \cdots \alpha_{n-1} \neq 0$, the order of A is at least $n-1$. Then we can rewrite (20) as follows

$$\begin{cases} \vec{x}_2 = \alpha_1 \vec{x}_1 - b_1 \\ \vec{x}_3 = \alpha_1 \alpha_2 \vec{x}_1 - \alpha_2 b_1 - b_2 \\ \cdots \\ \vec{x}_n = \alpha_1 \alpha_2 \cdots \alpha_{n-1} \vec{x}_1 - \alpha_2 \alpha_3 \cdots \alpha_{n-1} b_1 - \cdots - \alpha_{n-1} b_{n-2} - b_{n-1} \end{cases} \tag{22}$$

where

$$\vec{x}_1 = \frac{\alpha_2 \alpha_3 \cdots \alpha_n \vec{b}_1 + \alpha_3 \alpha_4 \cdots \alpha_n \vec{b}_2 + \cdots + \alpha_n \vec{b}_{n-1} + \vec{b}_n}{\alpha_1 + \alpha_2 + \cdots + \alpha_n} \tag{23}$$

There are three cases to solve equation (20) according to the value of the numerator and denominator of $\vec{x}_1$:

    a. the denominator of $\vec{x}_1$ is not equal to zero, then $\vec{x}_1$ exist, equation (20) has sole solution.

    b. the numerator and denominator of $\vec{x}_1$ are all equal to zero, then $\vec{x}_1$ is not sole, equation (20) has infinite solution. In this case, we can find a suitable $\vec{x}_1$, and get a suitable solution for equation (20).

    c. the denominator of $\vec{x}_1$ is equal to zero, but the numerator of $\vec{x}_1$ is not, then $\vec{x}_1$ does not exist, equation (20) has no solution. In this case, we should adjust some value of items in equation (20), and translate case c to case b.

Following is one of the methods to solve the equation (20) in case c: Let the numerator of $\vec{x}_1$ is equal to zero, e.g.

$$\alpha_2\alpha_3\cdots\alpha_n\vec{b}_1 + \alpha_3\alpha_4\cdots\alpha_n\vec{b}_2 + \cdots + \alpha_n\vec{b}_{n-1} + \vec{b}_n = 0 \tag{24}$$

$$\vec{b}_j = -\beta_j\left\{\left[2m_jw_{j1}/w_{j0} + m_j(m_j - 1)w_{j2}/w_{j0} - 2m_j^2(w_{j2}/w_{j0})^2\right]\right.$$
$$(\vec{r}_{j1} - \vec{r}_{j0}) + m_j(m_j - 1)(\vec{r}_{j2} - \vec{r}_{j1})w_{j2}/w_{j0}\Big\}$$
$$j = 1,2,\cdots,n \tag{25}$$

where m_j is the order of curve $\vec{C}_j(t)$, $\vec{r}_{jk}$ is control vertex of curve $\vec{C}_j(t)$, w_{jk} is weight of $\vec{r}_{jk}$

Rising the order of curve $\vec{C}_j(t)$ by one, we can get an adjustable vector $\vec{r}_{j2}$ on $\vec{C}_j(t)$. Changing $\vec{r}_{j2}$ will change the value of $\vec{b}_j$ while not affect other value, the result is very important. Solving equation (25) for new $\vec{b}'_j, j = 1,2,\cdots,n$, we can get $\vec{r}'_{j2}, j = 1,2,\cdots,n$, then find $\min\{|\vec{r}_{j2} - \vec{r}'_{j2}|, j = 1,2,\cdots,n\}$ to make change. Now case c has been changed to case b.

By applying all the results mentioned above, we can easily generate a rational Bezier blending surface $\bar{S}:\bar{S}_j(u,v), j = 1,2,\cdots,n$ on a N-sided region which is of G^1 continuity with all adjacent patches, also the consistency condition is easy to be satisfied and the inconsistency is easy to be corrected. The algorithm is very powerful and flexible.

6. Examples

Several examples are shown as follow.

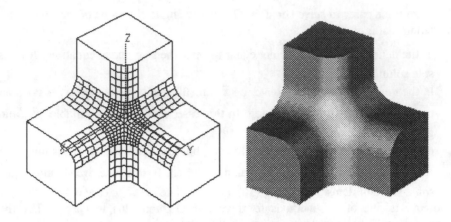

Fig. 4. Wireframe and shaded display of Blending surfaces
and a six-sided region surface

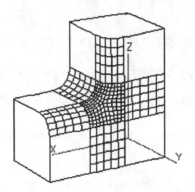

Fig. 5. Blending surfaces and a five-sided region surface

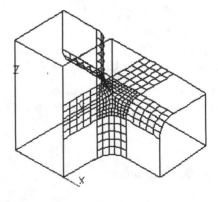

Fig. 6. Blending surfaces and a six-sided region surface

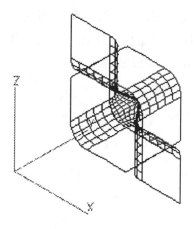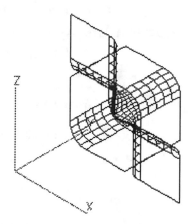

Fig. 7. Adjust fullness of a N-sided surface

7. Conclusion

The preceding sections have shown in details the key techniques that use rectangular rational Bezier patches to construct an arbitrary N-sided region surface to fill a N-sided hole and guarantee G^1 continuity along their common boundary edges. The main contribution of the new method is to build the blending surface and N-sided surface with standard surface form which is widely used and is supported by STEP. So the method will play an important role in geometric modeling and manufacturing field. Due to the length limit, we could not describe how to create center vertex and how to construct $\vec{C}_j(t)$ in N-sided region in detail.

References

1. Calmull E E, Clark J. Recursively Generated B-Spline Surfaces on Arbitrary Topological Meshes, Computer Aided Design, 10(1978) 350-355
2. Chiyokura H, Kimura F. Design of solid with free-form surfaces, in: ACM Computer Graphics, 17(1983) 289-298
3. Wang Lazhu, Zhu Xinxiong. Constructing N-sided patches with NURBS, CADDM, 5(1995) 26-32.
4. Sarrage R F. G^1 interpolation of generally unrestricted cubic bezier curves, Computer Aided Geometric Design., 4(1987) 23-39
5. Luo Hongzhi. Study and Practice for Integral Geometric Modeling System, Doctoral dissertation. Huazhong University of Science and Technology, 1(1992), pp. 78-88.

Optimal Expected-Time Algorithm For Finding Largest Empty Circle And Its Application On PCB Stamping Mould Design

Wang Jia ye, Boey Seng Heng, Lee Eng Wah

Gintic Institute of Manufacturing Technology
71 Nanyang Drive
Singapore 638075

Abstract. The design of PCB stamping mould presents us with a problem of finding the largest empty circle (LEC) between polygons. This paper presents a new algorithm for finding the LEC with the expected time to find a local LEC of $O(1)$ and preparation time of $O(N)$, where N is the total number of holes and sides of profiles.

1 Introduction

A major step in PCB manufacturing process is to cut the profiles and small holes after etching. Stamping is an efficient approach to complete the step for single and double layers PCBs. However, an adequate amount of ejectors pins must be provided on a stamping mould (Fig. 1) to push the PCBs out of the punch in order to prevent the PCBs from breaking. A difficult task in the 2D design of stamping mould is to determine the locations and sizes of the ejectors. The sizes of the ejectors are specified by their diameters, with their values chosen as large as possible. As shown in Fig. 2, the empty space between the holes and profiles must be wide enough to place an ejector. This leaves us with the problem of finding the *largest empty circle* (LEC) in the space between the holes and profile for placing an ejector. This problem could be solved by creating Voronoi diagram for holes and profiles.

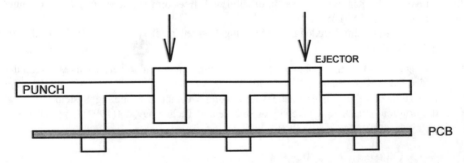

Fig. 1 The PCB Stamping Mould.

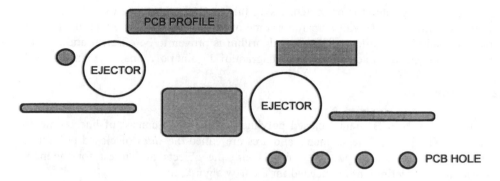

Fig. 2 Design of PCB Stamping Mould.

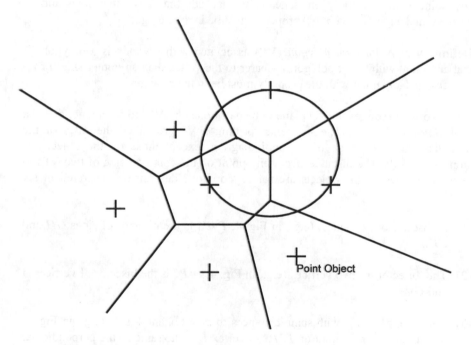

Fig. 3 Voronoi Diagram for a Group of Point Objects.

The Voronoi diagram for a group of point objects is shown in Fig. 3. It is noteworthy that the center of the LEC for a group of points is located at one of the vertices of the Voronoi diagram. It is shown in [1] that these tasks of creating Voronoi diagram and finding the center of LEC can be completed in a total time of $O(NlogN)$, where N is the number of the points. It is further proven in [2] that the center of the LEC for a group of polygons is located at one of the vertices of their Voronoi diagram. The time complexity of creating the Voronoi diagram of polygons is also $O(NlogN)$. In this paper, a new algorithm for finding the LEC of a group of holes and profiles for the design of ejectors for PCB stamping mould is presented. Using this new algorithm, the expected time to find a local LEC is $O(1)$ with preparation time of $O(N)$, where N is the total number of holes and sides of profiles. This new algorithm is proven to be efficient and robust, and it can be extended to create Voronoi diagram of disjoint polygonal object as in [3].

2   Algorithm

The profiles on a PCB contain closed polylines, which often consist of line segments and arcs. The circles, line segments and arcs are called the *sides* of closed polylines. These circles and closed polylines, in 2D, are the objects of interest for the new algorithm. Two definitions are needed for the new algorithm:

> **Definition 1:** A *Voronoi polygon* **(VP)** associated with a side is a polygon that any point within that polygon is nearer to that side than any other sides and vertices, and any point with the property must be in the polygon.

> **Definition 2:** A *Voronoi polygon* **(VP)** associated with a vertex is a polygon that any point within that polygon is nearer to the vertex than any other sides or vertices, and any point with the property must be in the polygon.

The new algorithm requires the 2D plane to be partitioned by VPs to form a net called a *Voronoi diagram* (Fig. 4). The segments forming a VP are called the **sides** of the Voronoi diagram. Every side of Voronoi diagram, except those on the objects, is common to two VPs. The distances from any point on the common side of two VPs to the vertices or sides on the objects are same. Voronoi sides can be generated in six cases.

(1) A bisector of two sides (e.g., in Fig. 4, *P4P5* is the bisector of sides *GH* and *EF*).

(2) The bisector of two vertices (e.g., in Fig. 4, *P1P2* is the bisector of vertices *A* and *G*).

(3) The trace of points with same distances to a vertex and a side (e.g., in Fig. 4, the distance of any point on *P2P3* to vertex *G* is the same as the perpendicular distance of side *AB* to that point).

(4) The trace of points with same distances to an arc and a line segment.

(5) The trace of points with same distances to an arc and a vertex.

(6) The trace of points with same distances to two arcs.

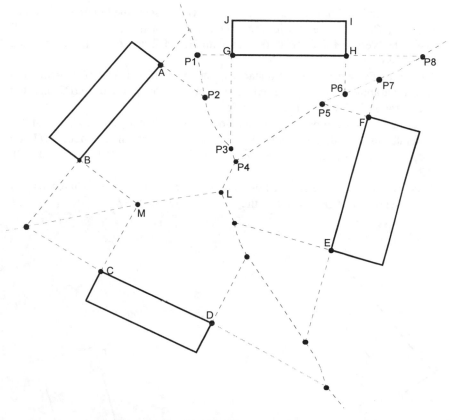

Fig. 4   A Voronoi Diagram.

The traces in cases (3) and (4) are parabolas. The traces in cases (5) and (6) can be parabolas or arcs. A VP associated with a side is bounded by two rays perpendicular to the side from its two end points (e.g., **GP3** and **HP6** for VP associated with side GH in Fig. 4).   A VP associated with a vertex is bounded by two rays perpendicular to the neighbouring sides of that vertex   (**GP1** and **GP3** for VP associated with vertex G in Fig. 4). The region bounded by the two rays is called the **bounded region**.

Our new algorithm is based on an incremental process. For example, when a new object, a circle or a polyline (**GHIJ** in Fig. 6), is added to an existing Voronoi diagram as shown in Fig. 5.   The existing Voronoi diagram is amended to create a new Voronoi diagram (Fig. 4) starting from any vertex of the new object, say **G** in Fig. 6, as illustrated below:

- Since **G** is in the VP associated with vertex **A** and **A** is in the bounded region of **G**, the Voronoi side of the two vertices **A** and **G** is generated by intersecting their bisector with their bounded region.

- The bisector *P1P2* of the two vertices *A* and *G* intersects the bounded region of *G* at *P1*; *P1P2* intersects with *AK* at *P2*. *AK* is in the common side of VP associated vertex *A* and VP associated with side *AB*; and *P2* is in the bounded region of *G*.
- The trace of points with same distances to a vertex *G* and side *AB* is generated as the parabola *P2P3*, where *P3* is the intersection point of *P2P3* and the boundary of the bounded region of *GH*.
- After *P3*, the trace of points with same distances to line segment *GH* and *AB* is generated as the side *P3P4*, which is the bisector of the line segments *GH* and *AB*. This side is in the bounded region of *GH* and in the VP associated with *AB*.
- This tracing process generated polygon *GP1P2P3* as the VP associated with *G* and polygon *ABMLP4P3P2A* as the VP associated with *AB*.

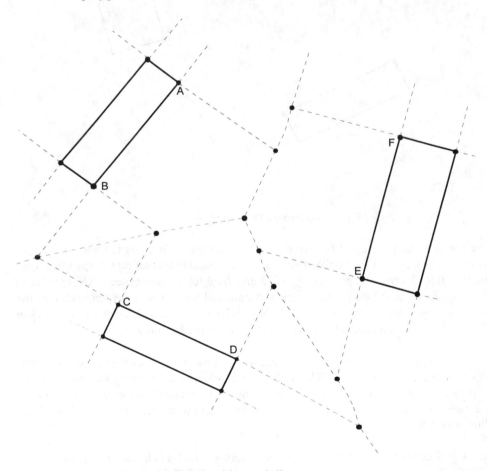

Fig. 5 An Existing Voronoi diagram.

The similar approach can be applied to generate or amend the VPs associated with *GH*, *EF*, *H* and *F*. After deleting all the unwanted line segments created during the process of generating new VPs, the new Voronoi diagram as shown in Fig. 4 is formed.

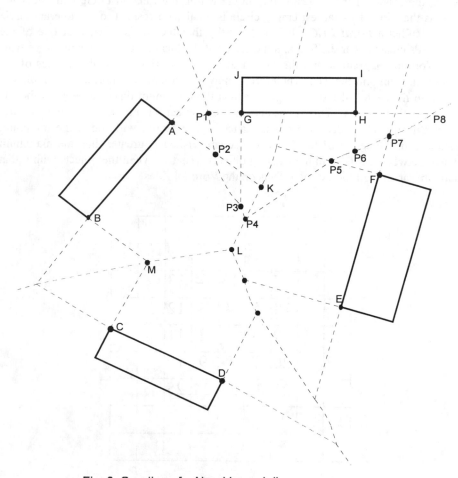

Fig. 6 Creation of a New Voronoi diagram.

From the algorithm described above, it is obvious that a VP is determined by its neighbouring objects.

In order to apply the new algorithm efficiently for ejector design, an estimated **starting point (SP)** must be specified by the designer. This SP is the estimated location of the ejector and it is in the area where the LEC could be found. The incremental process can be carried out by finding objects surrounding the SP using a spiral search method as shown in Fig. 7. This spiral search method partitions the 2D plane into small squares (or cells) with the search sequence as indicated by the numbers in the squares of Fig. 7.

In order to speed up the search for new objects, a *tolerant circle (TC)* centered at SP is created in which the LEC is expected. Since the center of the ejector is expected to be within the tolerant circle, the problem of designing an ejector becames the process of locating the center of the largest empty circle within the tolerant circle with the same radius as the LEC. This largest empty circle is local with respect to the tolerant circle and it is called a *Local LEC*. It is noteworthy that the SP is situated at one of the several VPs created near the SP; and the center of the local LEC is at one of the vertices of the Voronoi diagram within the TC, or at the intersection point of the sides of the Voronoi diagram and the TC. Therefore, a new object can be ignored if its minimum distance to the center of the TC is greater than the maximum distance between the TC and the existing objects. This is because the new object has no effects on the Voronoi sides within the TC under such circumstance. The search will end if the minimum distance from the center of the TC to the square visited is greater than the maximum distance between the TC and the existing objects. That is, when the objects within that square are far enough to have no effects on the Voronoi sides.

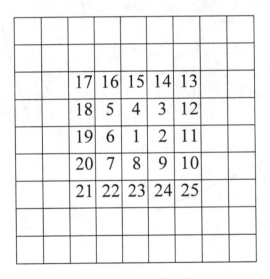

Fig. 7 The Spiral Search Method.

3   Conclusion

Since the local LEC is determined by the local objects (sides and vertices) surrounding the SP and the incremental process only searches for objects surrounding the SP as provided by the designer, there is no need to create the Voronoi diagram for the whole PCB. This makes the new algorithm more efficient for locating the Local LECs for ejector design. Although the time complexity of the new algorithm may be $O(N*N)$ for the worst case, the total optimum expected time to find a local LEC will be $O(N)$ due to the hugh number of objects with uniform distribution in a single PCB layout. It is therefore efficient to apply the algorithm for the design of ejectors in a PCB stamping mould.

Reference

1. M.I.Shamos and D.Hoey. Closest-Point Problems. In Proc. 16th Annual IEEE Symposium on Foundations of Computer Science, October 1975.
2. Martin Held. "*On the Computational Geometry of Pocket Machining*", Springer-Verlag 1991.
3. Wang Jia Ye, Wang Wenping, Liu Ding Yuin. Optimal Expected-Time Algorithm for Voronoi Diagram of Disjoint Polygonal Object. Proceedings of ICMA'97,1997.

Product Design

The Research On Design Principles To The CAD/CAM Geometric Platform　Based On Shared Standard Semantic Models Of Step/Ap203

Ren, Aihua and Wei, juxia

Dept. of Computer Science and Engineering, Beijing University of Aeronautics and
Astronautics, Beijing, China 100083
email: renah@buaa.edu.cn

Abstract.　 This paper expounds the approaches and the design principles to the CAD/CAM geometric platform based on shared standard semantic models of STEP/AP203[3] in the CIMS environment. general purpose geometric platform frame which is independent of specific graphic systems is given in this paper. The frame is fit for different kinds of specific graphic systems. The advantages for the approaches show that the software design of CAD/CAM is independent from specific graphics systems. Within the scope of some powerful graphics system, the CAD software can be run on these graphic systems by plugging the systems' APIs in the correspondent API(Application Program Interface) sockets.

Key words　　STEP tools　　Object-oriented design　　Geometric modular

1 Introduction

Today, the focus on CAD/CAM developing is the integration of CAD/CAM with many subsystems which have different constructions and various technical functions, based on "Data Base of Product Definitions".

Among the majority of present commercial CAD/CAM systems, the graphic information is transmitted mainly by IGES(Initial Graphic Exchange Specification) or by neutral file between CAPP and CAD or different CAD systems. In the CAD/CAM system based on STEP, product data are exchanged mainly by sharing the STEP product model data base or by the STEP neutral file (ISO10303-21). In STEP based system, not only the graphic information can be transmitted but also the whole product message can be transformed. STEP is more satisfied for information integration of modern manufacturing systems than IGES. But for the developers and programmers of STEP based CAD/CAM，they have to understand both data model definitions of STEP and data model definitions of the specific commercial graphic system. In order to make STEP based CAD programmers can concentrate their jobs on STEP only and make the developed CAD/CAM system be independent from the specific commercial graphic system, it is necessary to develop a generic STEP based object-oriented geometric platform. It makes the CAD/CAM system can use various graphic system as the base support. This platform is more opening. Therefore the flexibility is very important. Based on this framework structure, STEP based data model can be automatically transformed into any data form of the specific commercial graphic system. And the framework can supply many operations to STEP/AP203 model data. In fact, these operations are supported by the presented supported graphic systems . On the aspect of CAD programmers, they only need to know how to use STEP data form to develop the software, and no need to know how to use the supported graphic system. Through this framework the CAD/CAM developer can deal with their design in an uniform model without considering particular graphics system underneath.

2 Basic concept and design idea

STEP documents   consist of three sets (see Figure 1),(1) Representation methods and   implement   methods;(2)   Integrated   resources;(3)   Application   protocols representation methods providing the data model description language[2], EXPRESS. EXPRESS is used to defined all STEP data.   Implementation methods direct the developing work on Express data model.   Integrated resources is composed of generic resources and application resources. They are used to describe product information.   Generic resources are independent from any specific application and can be used by any applications. Application protocols add some necessary restrictions and modifications on generic resources to make them fit for particular application needs. Application protocols are defined for the sake of making STEP be suited to some particular environment. AP203 is one of the STEP application protocols. It is about 3-dimensional product shape representation and configuration controlled data definitions within the product design phase. Product shape representation is a kernel part in the product design. AP203 refers to lots of data definitions of ISO 10303-42[4] to describe 3D product shape.

In a CAD/CAM integrated system, the graphics information between CAD and CAPP or among various application systems can be transferred by STEP neutral file[1] or be transferred by sharing product model data. It is necessary for the developing of CAD/CAM integrated system to keep the unique representation of STEP semantic model and the complete property of this model. Also, it is convenient to display the semantic model in a visual graphics model during the developing of CAD/CAM systems. The graphic operation can be done directly in the visual mode. Therefore, it is very important for developing CAD/CAM integrated system directly to use the geometric representation graphics package which is identical to STEP semantic model or to adapt any graphics package (X) as the visual display and as the graphics

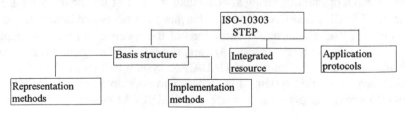

Fig. 1 The structure of STEP

operation for STEP geometric representation. Therefore the problems on graphics representation of CAD/CAM integrated system based on STEP is described as (take an example of AP203[3]):

How to make the STEP semantic sharing on geometric representation and how to give the graphics operation and visualization based on the sharing. In other words, this is the problem on graphic operation and visualization of STEP/AP203 semantic model or the problem on transformations between the two semantic models of STEP/AP203 and X graphics package.

Problem solving:

By using framework mechanism 203-FM to solve the problem. (see Fig. 2)

Objectives:

- CAD/CAM programs can do the sharing of STEP/AP203 semantic model.
- CAD/CAM developers need only concentrate their attentions on STEP semantic model system structure design.
- Through supported graphic system, graphic operations and displays for STEP/AP203 semantic models are provided.

3 System model

Let the framework model of geometry platform with heterodoxy architecture be 203-FM, then 203-FM consists of quintuple at least:

$$203\text{-}FM=\{Q, M, GS, I, R\} \tag{1}$$

where,

Q: a Set of all the data models and objects in 203-FM;

M: a Set of operations of 203-FM;

GS: a Set of architectures of various graphic systems;

I: an Interface model of users or on Interface of software components, providing various functions and I/O forms which meet component technology;

R: Integrated relation models, representing various inclusion and collection relation among Q, M. GS and I.

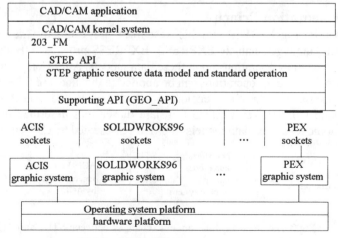

Fig.2 System structure

203-FM defines the semantic models which meet the STEP/AP203 requirements. It can be used as a geometric description platform for various graphic packages, providing an unified operating interface and being independent from any specific graphic systems. The inner structure of 203-FM is with the cutable feature and it provides reusable C++ library by using O-O technique.

4 System Architecture

As Fig. 2, the architecture shows hierarchical relationship among CAD application software, generic geometry platforms, particular graphic systems and operation systems.

The 203-FM provides the selection choice among the various commercial graphic package for CAD/CAM system developer (open property, flexibility and quality). The developing work is mainly as follows:

(1) To create a generic framework 203-FM with STEP model data definitions and geometric operations;

(2) To add standard operations upon STEP data model represented by C++ classes.

(3) To develop the application interface GEO-API which can recognize specific graphic system and can make it compatible with STEP standard data models.

Since 203-FM is built upon different graphic packages, therefore it has different selections on these graphics packages. GEO-API provides the interface sockets to recognize and automatically match different graphic system.

203-FM provides a group of standard logical operations for graphic application programmers. The objects operated by 203-FM are the data models of STEP geometric represented graphics resources. Since GEO-API is developed, these standard operations and the entities represented by STEP data model can be shown by using different graphic systems . (see Fig.3).

From the system structure (see Fig.2) it is shown that CAD application users work upon CAD/CAM system. The CAD/CAM system designers use 203-FM through STEP-API to operate particular graphic package. Considering the CAD/CAM system, all the defined product data conform with STEP requires.

5 Implementation Principles

As shown in Fig.4, STEP graphic resources are originally described by data description language , namely EXPRESS. EXPRESS provides only data description entities without any operations upon these data models. After Compiling   EXPRESS model into C++ codes , operations can be applied to the model of C++ codes. C++ has the characteristic of a class enclosure which meets the STEP object-oriented requirement. It is the key work to translate the entities described in EXPRESS of STEP graphic resources data models into that represented by C++ classes .

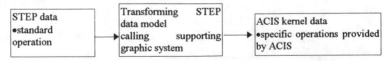

Fig.3 The steps for data model exchanges between STEP and ACIS

203-FM implementing steps are as following:
(1) To map and transform STEP graphic resources data models to the data models of any specific graphic system; First the supported graphic system must have object-oriented graphic data description, than the transformation from STEP graphic resources data models to the data models of the supported graphic system can be done. There are two key transforming methods(or operations): one is the attribute transformation from a STEP graphic resource object to a corresponding object of the supported graphic system. The other method does the opposite thing, which feedbacks the changing of the object made by the supported graphic system to STEP graphic resources, and the same change to STEP corresponding object will be made, i.e.

$$203\_FM\{Os\} \Longleftarrow \Rightarrow GS\{Og\}.$$

(2) To add operations to graphic resources entities.

As shown in Fig.5, methods adding relies on the supported graphic system because STEP doesn't provide operations on data model. In the implementation of 203_FM, an operator(STEP_GEO_OPERATOR) is designed to call supported operations, such that

$$GS\{M\} \Rightarrow 203\_FM\{M\}$$

By designing a data transformation algorithm including specific graph calling algorithm and generic algorithm, 203_FM graphic operations can be realized. The security and enclosure of the original supported system is still kept. This operator can be represented as:

$$STEP\_GEO\_OPERATOR(DMT,INC)$$

Where,

DMT: Data model transformation algorithm;

INC: Algorithm of methods adding to 203_FM and methods calling to supported graphic system.

(3) To code C++ application programs, test the correctness of the expanded classes and verify the correctness and consistency of the transformation. This work is mainly to design the application program for the expanded classes and make the specific instances described by STEP graphic resources visible on screen so that the correctness of the expanded classes developing can be verified. (4)To build generic geometric platform framework named 203_FM to provide higher abstract level of C++ classes description.

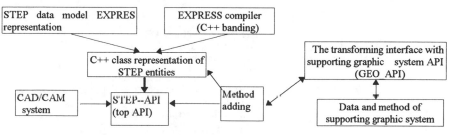

Fig. 4   The principle of 203_FM geometry platform developing

After developing a STEP transforming upon a solid modeling graphics system(e.g. ACIS), we can give a complete operation set on STEP data models, and make the abstract level further. Then we provide the operation set to users as a standard operation interface. This make the interface become one framework form and independent from any specific graphic systems. On the framework the operations is called STEP_API. When users select a specific graphic system, the interface software which transformates STEP operations to that specific graphic system is loaded. The software for these selection and allocation is called a socket of bottom API(GEO_API). For instance: the transformation for ACIS[5] is named ACIS sockets, the transformation for SolidWorks96[6] is named SolidWorks96 sockets, etc., namely 203_FM={O,M,GS,I,R};

Where   GS={Sockets(GS1), Sockets(GS2), …Sockets(GSn),}

(5) To integrate the sockets developed for various graphic subsystems to the generic geometric platform framework(203_FM). This kind of integration means that the management on STEP models transformation can be achieved during the executing of a socket selection on graphic packages.   Then the different C++ libraries can be created and selected according to particular needs.

(6) To provide an opening and reusable C++ library. Because the framework is given in a C++ class library form, users can expand the classes directly on the library.

6 Conclusions

(1) In this paper a design project of generic STEP geometric semantic model platform has been given. In fact, it is a implementation plan for graphic resources of STEP/AP203[3]. It makes CAD/CAM programmers develop an application by using STEP data models directly. And in this way the different application programs can share STEP semantic model directly by reference.

(2) The object-oriented technique and basic software component methodology are used during the designing and implementing of the platform defined according to 203_FM structure and its API. This not only ensures to fit with STEP, but also enlarges the open property during the designing and developing of CAD/CAM systems.

(3)   According to the conceptions, architectures and principles in this paper, the developed platform has been verified in a STEP based CAD/CAM integrated system.[8]

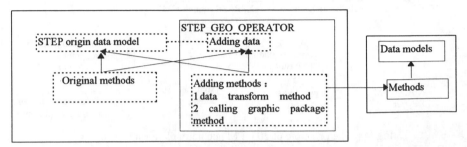

Fig. 5 The concept of method extending

References

1. ISO 10303--21 "Clear Test Encoding of the Exchange Structure"
2. ISO/DIS 10303--11 "Product Data Representation and Exchange --Part 11: The Express Language Reference Manual", 1992,8.
3. ISO/DIS 10303--203 "Product Data Representation and Exchange--Part 203: Application Protocol: Configuration Controlled Design"
4. ISO 10303--42 "Integrated Resources: Geometry and Topology Representation", 1992..9
5. "ACIS Geometric Modular Programmers Reference" SPATIAL TECHNOLOGY INC, 1995.
6. Alan Smith, "A Near look for Solid Modeling" Computer_Aided Engineering, 1996.4 P19--P24
7. "HP--PHIGS Graphics Techniques" Herlett--Packard Co., Vol.1, 1991.1
8. "LONICERA system" , 720 institute of Beijing University of Aeronautics and Astronautics, Beijing, China

A CAD-System for Dressmaking

Karl-Heinz Roediger

University of Bremen, Department of Mathematics & Computer Science
P.O. Box 33 04 40, D-28334 Bremen, Germany
e-mail: roediger@informatik.uni-bremen.de

Abstract. This paper reports about an R&D-project in the dressmakers' trade in Germany. A system for all tasks in small dressmakers' workshops is nearly completed. It covers office tasks as well as CAD of basic and model patterns. The special situation of the target group raised some problems: The final system should be a low cost one; it should also be handled very easily; and special attention should be paid to work organization. These constraints have been solved in a low cost system under Windows 95, which is also suitable as a system for those countries that will no longer be the workbench of the so-called first world.

1 Introduction

In Germany as well as in other - also developing - countries one hardly may find any information technology in small dressmakers' workshops. The only technical instrument apart from a - in the developing countries not everywhere electrical - sewing machine might somtimes be a rotary iron. Nevertheless there are a lot of repetitive tasks to be done in these workshops: Correspondence, bookkeeping, billing, maintaining of customer and supplier data as well as the construction of basic, model and production patterns. Especially the last task, often consisting of variant construction, is time-consuming. The fairly old production methods together with high wage costs have brought the dressmakers' trade in Germany into difficulties: They cannot realize the prices, they should ask for their hand-made clothes, because only few people are able to pay them. On the other hand, the dressmakers cannot pay the wages for their employees because of the prices they actually realize

They now become acquainted with the same problems Middle Europe's clothing industry had suffered from years ago: The production of ready-to wear clothing went over to Eastern Europe or - much more - to Asia; if they could not survive with design and pattern construction they went bankrupt. Now Germany's and also Middle Europe's dressmaking is in a dilemma, which is characterized by a fairly high demand for new clothing on one hand and production costs that are much too high on the other. The demand results from people wishing something different to the ready-to-wear clothes or having a so-called problem figure. The amount of the production costs is mainly resulting from the production methods. Nearly every new dress is constructed »from scratch«, that means, despite the fact that most of the new things are only variants of formerly made ones, the dressmakers start the process by designing the basic patterns followed by the model and the production patterns. This repetitive activities take a long time and are not the dressmakers' favorite ones. Therefore the dressmakers are looking for ways to modernize their old fashioned production methods.

In this situation the dressmakers' guild of Bremen, a county of Germany where - as all-over the country - the trades are organized in specific guilds, asked for help from the University of Bremen in 1993. Together with the guild we have defined a research and development project on »CAD in dressmaking«. In October 1994 this project sponsored by the program »Arbeit und Technik« (»Work and Technology«) of Bremen's county government was started. The objectives of the project are to look into the dressmakers' work organization and to propose new organizational forms if necessary, to investigate the requirements for a computer system and to compile a system that meets these requirements out of standard products or to develop a new one, and to examine the needs for training IT skills and to arrange adequate courses.

2 Work Organization

The project was started with a long phase of work analysis. Every dressmaker was observed and interviewed several hours, during which we got to know big differences in the work process even between the individual dressmakers. They all learned certain methods of pattern construction, mainly »Hohenstein« and »Mueller & Son«; during their working life they have modified these basic methods to their very own one. We - computer scientists - again learned what industrial science is postulating against F.W. Taylor's »Principles of Scientific Management« (1911): There does not exist only one best way for all dressmakers, there are many individual best ones. It was our first lecture from work organization.

The second lecture was: Whatever the individual design style looks like, all clothes can be traced back to basic patterns as they exist for dresses, skirts, suits, jackets, trousers, blouses, shirts etc. The way of constructing a new model pattern is one of varying a basic one by size, by appliqués, by odds and sods like collars, revers, pockets etc.

The third observation was one of a very specific division of labour: Because of the employees' education pattern construction is the privilege of the female masters in every workshop; everthing else is also done by the employees. Because all masters have their own specific method of creating new patterns, because it shows their hand and is often some sort of a trade mark for their workshop, they do not believe their employees being able to do it the same way. Nevertheless the resulting problem is that in cases of the master's absence or illness nobody is able to prepare the sewing of new clothes. If the journeywomen ever had this special skills they have lost it over the years they did not practice it. All participants in our project therefore decided that pattern construction should be no longer a privilege of a single person, because of the possible problems in work organization.

A fourth lecture when looking into work organization was: Whatever somebody is trying to do with a computer in a workshop like those under investigation, work with a computer is only a part of the whole working day of dressmakers. They are doing complete work along the production process of clothing. Except pattern construction they never divide this process neither horizontally nor vertically. Normally the whole process lies in one hand. The uniqueness of every piece might be the reason for such an organization of work.

These observations have several implications for the future work organization and for the requirement definition: First, a new software system must be open for all different methods of pattern construction as well as be dedicated to every individual process in order to give the right support. Second, a new system should contain a database with all basic patterns in order to support variant construction. Third, work with and at the computer must become an integral part of the whole production process. It must be easy to use and exactly match the work routine. Easy switches between pattern construction and office work like maintaining a customer's measures should be possible. These implications might be fullfilled by a new software system. Fourth, to enable the journeywomen for the construction of well fitting patterns is a matter of training pattern construction and IT skills. That means as a result to qualify the journeywomen as well as the masters for the use of computers.

3   COAT - A System for the Dressmakers' Workshops

After the investigation of work organization the existing CAD systems have been studied, especially those ones used in the textile industry, in order to look what features they are offering and how far they probably meet the requirements resulting from the specific work organization in small workshops. Pattern construction, grading and cutting cloth have been automated in the textile industry years ago. The proprietary systems used there are rather expensive, more than small workshops can pay, and they are not easy to use. They require a professional user being familiar with all features of a complete CAD system. The working situation of a dressmaker however is characterized by a high variety of totally different tasks during a day. Moreover, lot size in the clothing industry is always more than one; in dressmaking it is generally one; that means, having constructed the patterns for one piece of clothing a dressmaker will use the computer at the earliest when having finished the sewing for that piece. Because of these circumstances dressmakers are in terms of human-computer interaction research novices and occasional users by sometimes using a computer only a few minutes per day.

The industrial solutions to these problems are designed for a work place, where one specialist is practicing computer aided construction the whole day. This manageress is more or less familiar with the functionality and the user interface of the system. Nevertheless the industrially used systems often are so complicated that even the specialists cannot explain the functionality and the tools: They often do things without knowing their exact meaning, and they must keep in mind a lot of commands and actions in order to do the right things at the right time. Additionally, as in the small workshops most of the clothing enterprises use their own very special way of pattern construction and dressmaking and this method is sometimes implemented in the software. Some dressmakers involved in our project have tried to work with these systems. The only thing they could tell at first was: »Not that way!«.

In order to achieve all possible effects of rationalization by introducing computers we looked for a complete and simple software solution for all tasks in a dressmaker's workshop. That includes all office tasks as word processing, bookkeeping, billing, maintaining of customer and supplier data as well as a specific system for the construction and grading of new clothing. Because of the specific target group this project has some specific constraints.

First, the complete solution consisting of hard- and software must be extremely cheap because most of those workshops do not have any budget for information technology. Second, the solution must be extremely user friendly because nobody of the target group had any experience with computers and - even in the future - will work only a very short time at or with a visual display unit. The system also shall become part of the vocational training of the future dressmakers. Third, the solution must be work-oriented and should not affect their normal work organization, because the main task of a dressmaker is consulting, designing, cutting and sewing. Usually construction is only a small portion of the whole dressmaking process; for the expert women it is some kind of routine matter. Therefore the whole system must be embedded in the normal process of work organization.

Because there is no standard solution for these constraints, the project participants, computer scientists, dressmakers and teachers in dressmaking, have decided to develop a completely new system for a commercial PC environment, called COAT (CAD and Office Applications for the Taylors' Trade). A new training course in the use of a personal computer for administrative as well as for constructive tasks in dressmakers' workshops was developed. An evolutionary and participatory way of system development was chosen in order to fullfill the requirements of work organization and usability.

Concerning the first constraint, the upper limit for hard- and software investment in our project was about 5.500 USD. The hardware platform we use in our project is a Pentium PC with 16 MB RAM, 1 GB hard drive, a CD-ROM drive, a PCI graphics accelerator and a 17" or better 21" monitor; the operating system is Windows 95®. All together this may cost less than 3.000 USD today (Halfhill 1996). The difference between both sums should be more than enough for the purchase of standard software and the development of new dedicated CAD software, because the development is sponsored by the county government of Bremen.

3.1 Software Development

Looking around one may find a lot of standard software. But a complete and unified solution for all tasks in a dressmaker's workshop as mentioned above cannot be found. That means either to solve only parts of the problems or to force casual users to learn more about computers, operating systems and interfaces than necessary and opportune. Therefore it was at first decided to create a new unified system with a uniform graphical user interface. This decision is a great opportunity to create an adequate solution for the target group and a big challenge at the same time.

The main objectives of all software development are usability and flexibility. Concerning the frequency in which a PC tool is needed in a dressmaker's workshop an easy-to-use interface is of great importance. To create such an interface the whole system is implemented in an uniform way under Windows 95®. Most of the necessary interactions are of the types »drag and drop«, »cut and paste« or »point and click«. No commands must be learnt by heart; what can be done with the system is offered in menus, lists, and buttons. Whenever possible defaults are underlayed. All tools are self-explanatory - because of the terminology - for professional dressmakers; additionally they are explained in a help system. To enable the use of the system in other countries it has a language switch: any given language may be easily read in and used; until now it is realized in English, German and Italian. A Portuguese and a Spanish version are in progress. In one word: we have tried to realize the state of the art in human-computer interaction (Preece et al. 1994, Shneiderman 1992).

According to the working situation with alternating activities a very flexible system, which allows immediate changes between the different tasks, is necessary. Figure 1 presents the overall structure of COAT. It consists of two parts: an office and a workshop one. Correspondence, customer file, bookkeeping, stock, and supplier files are included in the office part; it is mainly designed out of standard packages like a word processing system, a database management system, spread sheets etc. together with connecting parts implemented via MS Access®. The central module of the total system is the database management system which is also serving the workshop part, the CAD module. The database consists of data about customers (adresses, measures, individual pecularities etc.), suppliers (materials, prices, discounts, lead time etc.), materials (stock, prices etc.), patterns etc.

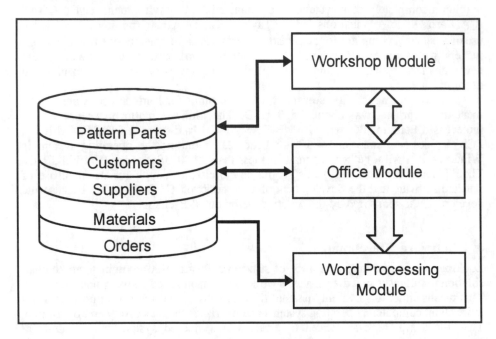

Fig. 1. Structure of COAT
The broad lines and arrows denote procedure calls between components of the overall system, the slim ones are showing data interfaces.

The workshop part consists of the CAD module together with the different data- bases for basic, model, and production patterns as well as for the bits and pieces. The database is the connecting link between workshop and office. Pattern construction can be supported automatically (size), partly automated by presenting databases for the small odds and ends, and by special construction tools with which the patterns can be modified just as dressmakers are doing it manually now. Nobody is forced to use the basic patterns, all can be done with the construction tools as well. The system also supports exploration and experimental action; because all construction steps are recorded in the background every state may be reconstructed by a sophisticated undo function.

Moreover, the dressmakers may store their newly created model patterns in a further database in order to use them for other customers or some years later for the same ones according to the first theorem of fashion »It has all been done before«. As well as in, for example, mechanical engineering, the basic idea of such a software design is that of variant construction. Most of what dressmakers are doing is in that way; designers however may additionally use the construction tools. As well as we know from the dressmakers examining the prototypes, the system is exactly fitting their problems. In most cases it reduces the time spent for pattern construction from hours to seconds.

Flexibility has a twofold meaning: The system must be flexible according to different styles of working as well as flexible in itself. The first kind of flexibility has been explained before. The second one is realized by multiple entries and multiple transitions offered by the system. Whatever one wants to do, she or he may start with or change to it. If during the construction process a dressmaker notices that material is missing, she may directly write an order; if a special basic pattern is missing in the pattern database, the dressmaker may construct it from scratch by using the construction tools.

3.2 Design Knowledge

The experiences with CAD systems in manufacturing have shown that geometric information about the final product design is not sufficient, especially in situations in which someone is reusing an old design, for example in variant construction. As mentioned before, variant construction is also widely spread in fashion design. In order to correctly use older designs or patterns dressmakers as well as manufacturing engineers need information about it at a semantically higher level than the pure geometric elements typically offered by CAD systems. Moreover, because dressmakers are not familiar with hyperbolic functions and splines, geometric data about patterns are of no use for them. Several attempts were made in CAD research to preserve the design knowledge in terms of knowledge-based systems (Kratz 1991), in terms of features of the product design (Mäntylä, Dana, Shah 1996) or in terms of macros, scripts and meta languages in which the design process can be documented.

In COAT we have chosen scripts, a meta language which can be read and understood by the dressmakers. The scripts have been especially designed and evaluated for this application. A script is automatically generated during every pattern construction process; it represents the process as an ASCII text. At the same time the scripts realize the undo function as they support the dressmakers by giving them that information about a pattern which they need to create a variant. With these scripts we also solve the grading problem: Instead of calculating the nonlinear process of grading with numerous tables of jump in values we take these scripts, feed them with the measures of the new size as parameters and reconstruct the model by interpreting the script. Dressmakers may complete or modify their constructions by using the script language, the elements of which are offered in drop down lists at every construction step. The only input they are forced to are parameters (length, width, height in cm).

3.3 Participatory Design

The whole process of software development is organized in a participatory, prototype-oriented and iterative way (Schuler and Namioka 1993). After an extensive phase of work and requirements analysis the description of work and the requirements definition were written by computer scientists and evaluated by dressmakers. The allover design of the new system was done in cooperation. The system specification was again elaborated by computer scientists with an object-oriented tool. This has formed the basis for the first prototype presented to the dressmakers. After having finished the first course in using computers they have evaluated this prototype in their workshops and criticized it in joint meetings.

The best results of these critique meetings are achieved when the discussions take place in small groups. This critique together with the scheme for iterative development again form the plan of action for the next phase. A new version of the total system is offered to the dressmakers approximately every two month. The office part is now stable; there is no further development. The part under iteration is now only the CAD system, which is implemented in Microsoft Visual C++®.

Participatory development is without doubt the best way to develop efficient and usable systems. But also without doubt, it is time-consuming for both sides, developers and users. Especially in small and medium-sized enterprises time is money: If one person is absent sometimes work stands still. Regarding this, the demands for time e.g. for project management, introductory courses, or participatory development should be handled very carefully.

4   Qualifying for the Forthcoming Challenges

A bigger problem was that no dressmaker in the project had ever worked with a computer before. They are familiar with sewing machines and rotary irons; personal computers however have been a totally new experience. All participants in the project are however convinced that the only possibility for the survival of the taylors' trade in Middle Europe is to renew the old fashioned production methods by introducing computers and software systems into the workshops as quick as possible. It is often argued that the introduction of information technology destroys workplaces. As it is done in this project no workplace is at risk; on the contrary we hope to create new workplaces because the situation concerning orders is very good. If the workshops can produce the clothes in a shorter time and therefore more economically some new workplaces may be realized. The central aspect of these expectations is that all dressmakers, masters as well as journeywomen, achieve as much IT skills as they need in order to rationalize their production methods. This in mind the first training was the most important step during the project. If it had failed we could have stopped the project.

So a special introductory course in using a computer had to be designed very carefully. In this training course a lot of time was spent with practicing the handling of mouse, cursor, and graphical user interface. Although the dressmakers are accustomed to handling needles and threads some of them have had problems with their micro motor avtivity. We succeeded in arousing and preserving the dressmakers' enthusiasm for the new media by working out the course strictly problem-oriented. We did not teach the use of an operating system, of a graphical interface or of a word processor. We took problems from their working life as designing sheets of paper with individual logos, maintaining the customer file, billing, advertising for new products, ordering of material, constructing simple patterns etc. The first course covered twelve evenings three hours each; presently we offer two evenings every second month in order to refresh the skills, to discuss problems and to teach new features of the incrementally developed CAD system. Regarding the capabilities of the dressmakers it seems to be sufficient.

Because pattern construction shall no longer be a privilege of the masters we also integrated the journeywomen of the workshops involved in the project into these training courses. They have passed them with the same, sometimes with more enthusiasm than their employers. Together with teachers we are now elaborating new elements for the primary vocational training of dressmakers concerning the use of a PC in administrative and constructive tasks of a workshop. The vocational school where the first training will be offered is well equipped with personal computers. The school will use COAT for primary training.

5 Outlook

As mentioned before, we are developing COAT in an evolutionary way: The dressmakers are using preliminary versions of the system during their normal work. They criticize these versions during our regular meetings and we, academic researchers, improve these versions in a spiral approach (Boehm 1988). The project ends in spring 1998; then the system will be completed and evaluated.

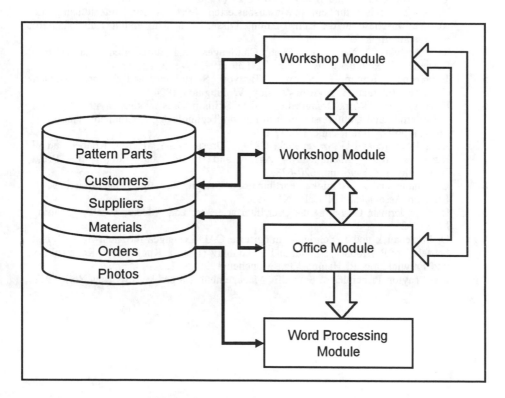

Fig. 2. Structure of COAT and CADwalk

In the meantime, we have started a new project in the same domain called CADwalk, a term which is derived from Computer Aided Design and catwalk. With this project we intend to support the consulting process between a dressmaker and her customer. We take photos of a customer with a digital camera, project them onto a monitor, and clothe the person with the new dress which she has just selected. In doing so, we provide a fast impression of the new garment before the first fitting. So the customer can see whether the new dress meets her requirements in shape, material, colour, and design. Because the computation or simulation of the material properties as for example the fall of the folds or of the material at all in reasonable time so far is an unsolved problem on PC platforms, this development will take a longer time. Figure 2 shows the structure of both systems, COAT and CADwalk. CADwalk is only a module operating on the same database which now is enlarged by photos of customers. The dressing is done by converting the two-dimensional patterns of COAT into three-dimensional clothes in CADwalk.

6   References

1.   B.W. Boehm, A Spiral Model of Software Development and Enhancement, *IEEE Computer* **21** (1988) No. 5, pp. 61-72.
2.   T.R. Halfhill, Inside the Web PC, *BYTE* **21** (1996) No. 3, pp. 44-56.
3.   N. Kratz, Architektur eines wissensbasierten Systems zur Unterstützung der Konzeptionsphase in der Konstruktion, Dissertation, Universität Kaiserslautern, Kaiserslautern 1991.
4.   M. Mänttylä, D. Nau, and J. Shah, Challenges in Feature-Based Manufacturing Research, *Communications of the ACM* **39** (1996) No. 2, pp. 77-85.
5.   J. Preece, Y. Rogers, H. Sharp, D. Benyon, S. Holland, and T. Carey, Human-Computer Interaction, Addison-Wesley, Wokingham 1994.
6.   K.-H. Roediger, Work Organization and IT Skills in Dressmaking, invited Paper for the International Conference on Information Technology for Competiti-viness, Florianopolis, Brazil, June 19-21, 1997.
7.   K.-H. Roediger and U. Szczepanek, CAD in Dressmaking, in: R.J. Koubek and W. Karwowski (eds.), Manufacturing Agility and Hybrid Automation I, IEA Press, Louisville, KY 1996, pp. 620-623.
8.   D. Schuler and A. Namioka, Participatory Design: Principles and Practices, L. Erlbaum Associates, Hillsdale, NJ 1993.
9.   B. Shneiderman, Designing the User Interface, 2nd. ed., Addison-Wesley, Reading, MA 1992.
10.   U. Szczepanek und T. Schriefer, Integrierte CAD-Lösungen für Klein- und Kleinstbetriebe, in: P. Brödner, H. Paul und I. Hamburg (Hrsg.), Kooperative Konstruktion und Entwicklung, R. Hampp Vlg., München 1996, S. 137-143.
11.   F.W. Taylor, Principles of Scientific Management, Harper & Row, New York 1911.

A New Generation CAD System for Railway Weighing Apparatus and Its Implementation

Meng Xiangxu   Li Xueqing   Gong Bin

CAD Laboratory,
Institute of Computing Technology, Academia Sinica, P. R. China
Department of Computer Science, Shandong University, P. R. China

Abstract. Railway weighing apparatus (RWA) can be widely used in various plants, coal mines and railroads. The main dimensions of RWA often need to be changed to adapt to the application circumstance, but all of RWA has the similar topological structures. In this paper, we describe a new generation CAD system model for RWA based on knowledge representation and parametric geometry technology. The parametric geometry modeling can support complex drawing representations. The implementation of the system is also described.

1 Introduction

Railway weighing apparatus (RWA) can be widely used in various plants, coal mines and railroads. The main dimensions of RWA often need to be changed to adapt to the application circumstance, but all of RWA has the similar topological structures. The product design is expensive and error-prone tasks because of complex calculations and engineering drawings. In this paper, we describe a new generation CAD system model for RWA based on knowledge representation and parametric geometry technology. The new parametric geometry modeling based on directed hypergraph is a major component of the system. It can support the representation of dimensional constraints and geometric constraints. Moreover, the geometric structure changes and design constraints can also be represented. The parametric geometry modeling is finished while interactively constructing geometric objects. The constraint consistency is ensured by construction steps. Overconstrained or underconstrained sketches can not take place. The parametric geometry modeling can support complex drawing representations.

The CAD system for RWA is composed of four parts: *user interface, specification analysis and expert system, graph deduction and dimension parameter value transfer, product drawings output subsystem.*

User interface is used to accept user input (such as: product specifications and select design constraints) and provide a friendly interactive environment.

Specification analysis and expert system(SAES) is used to analyze product specifications and make a reasoning to calculate the main dimensions and basic structures of the RWA. A rule-based method is adopted to select product types and structural parameters. Experiential knowledge and experiential formulae can be easily incorporated into the subsystem. In the process of reasoning, user can also change and select the key value of product parameters interactively. So, it is fast and convenient to make a new product design in the system.

Graph deduction and dimension parameter value transfer(GDPT) is used to create correct engineering drawing instances from parametric drawings. It accepts initial main dimensions from the SAES and transfer to parametric assembly drawing. Then solve the parametric assembly drawing by graph deduction to make an instance of the parametric drawing, and transfer the new dimension derived from the drawing to the subassembly drawings. Finally all of the parametric drawings are calculated by the subsystem and create the engineering drawing for the product.

Product drawings output subsystem is to adjust the drawings and output it to plotter or display. User can also transfer the design results to other CAD software system (for example, AUTOCAD).

2 The Design Flowchart and System Architecture

2.1 The Traditional Design Flowchart

In the railway weighing apparatus (RWA) design fields, the engineering drawings of a family of products are similar and can be determined by several key parameters. Considering the structures of RWA, the structures of 100 tons RWA are similar to 30 ton's. The differences between them are the changes in specification for rails, steel plates, angle steel and I-steel, in the number of spandrel girders and rails. Figure 1 shows traditional design process.

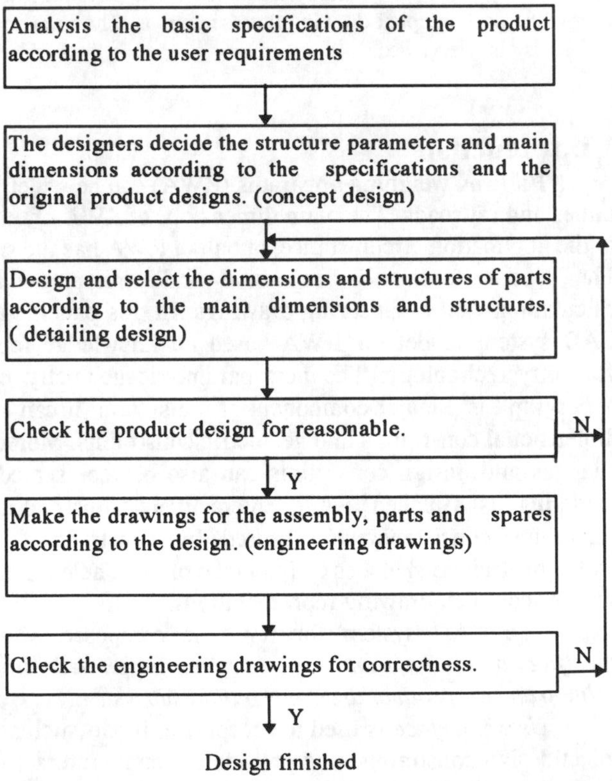

According with the design flowchart, the parametric CAD system should satisfy with the following conditions:

Analysis the basic specifications of the product according to the user requirements

↓

The designers decide the structure parameters and main dimensions according to the specifications and the original product designs. (concept design)

↓

Design and select the dimensions and structures of parts according to the main dimensions and structures. (detailing design)

↓

Check the product design for reasonable.　　N

↓ Y

Make the drawings for the assembly, parts and spares according to the design. (engineering drawings)

↓

Check the engineering drawings for correctness.　　N

↓ Y

Design finished

Figure 1

(a) The modeling should combine the traditional design calculation, optimization design and engineering drawings together.

(b) The parameterization of drawings can support the topological changes for the structures of products.

(c) The engineering drawing instance can be created automatically according to the given parameters.

(d) The modeling should ensure that the relative dimensions and parameters of spares, parts and assembly are consistent.

2.2 The Parametric CAD System Architecture For RWA

In view of above-mentioned reasons, the new generation parametric CAD modeling for RWA is as follows (as show in Figure 2):

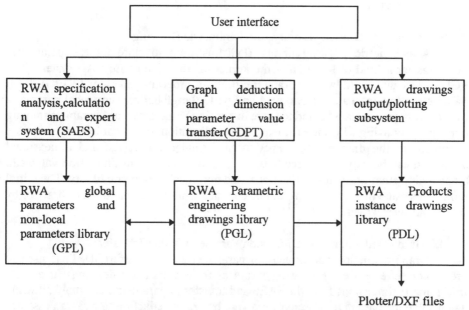

Figure 2

2.2.1 Specification Analysis, Calculation And Expert System

. Specification analysis, calculation and expert system(SAES) is used to analyze product specifications and make a reasoning to calculate the main dimensions and basic structures of the RWA. A rule-based method is adopted to select product types and structural parameters. Experiential knowledge and experiential formulae can be easily incorporated into the subsystem. In the process of reasoning, user can also change and select the key value of product parameters interactively. So, it is fast and convenient to make a new product design in the system.

There are two kinds of design processes. One is certainty process that is basis on the numerical calculation, experiential formulae, FEM analysis and optimization. The other is concept design process that is basis on the design knowledge and logical reasoning. The design can be considered as a recursive process between synthesis and analysis.

The subsystem was written by PROLOG. The input data source can be specifications, dimensions and design constraint. By the execution of the system, the basic structures, main dimensions and other parameters can be gotten. The results were written to the RWA global parameter and non-local parameter library.

There are three kinds of knowledge in the knowledge library.

(a) RWA construction and protocol knowledge: to describe the RWA characteristic and the selection of basic structures.

(b) Numerical knowledge: the numerical and dimensional constraint among the parts.

(c) Design rules: the standards and rules for the RWA and the design experience.

2.2.2   Parametric Drawing Library And Parametric Drawings

A parametric drawing library (PGL) is used to store all the parametric drawings for a kind of RWA. It is the kernel for the parametric CAD system. The PGL library was build by the parametric drafting program generator (PDPG). Each parametric drawing can retrieve, read and write the global and non-local parameter library. It is easy to transfer and to share dimensional and structural parameter among parametric drawings. Because the expression calculation and variable assignment are supported in the drawing, the experienced formulae for design and dimensional deduction can be expressed directly in the drawing. The result dimension value can be passed to other drawings by the global and non-local parametric library. The initial parameter value of each drawing can be obtained from global parameter library or by user input interactively.

2.2.3   Graph Deduction And Dimension Parameter Value Transfer

Graph deduction and dimension parameter value transfer(GDPT) is used to create correct engineering drawing instances from parametric drawings. It accepts initial main dimensions from the SAES and transfer to parametric assembly drawing. Then solve the parametric assembly drawing by graph deduction to make an instance of the parametric drawing, and transfer the new dimension derived from the drawing to the subassembly drawings. Finally all of the parametric drawings are calculated by the subsystem and create the engineering drawing for the product.

2.2.4   Product Drawings Output/Plotting Subsystem

Product drawings output/plotting subsystem is to adjust the drawings and output it to plotter or display. User can also transfer the design results to other CAD software system (for example, AUTOCAD).

3.   RWA Parameter Modeling and Dimensional Deduction

The aims to the RWA CAD system are:

For a group of given specifications and main dimensions, all the specified dimensional value of each drawing can be obtained automatically by the design calculation and constraint expression deductions based on the dependence relationship among parametric drawings.

For the RWA products design, the dependence relationship can be expressed by a tree structure. Figure 3 shows a typical structure for a RWA.

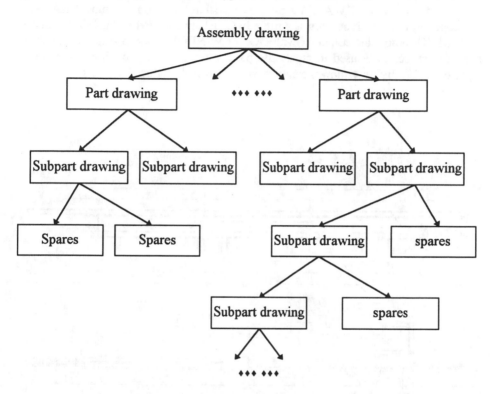

Figure 3 Typical structure for a RWA

Because the PDPG system supports expression deduction automatically, dimensional deduction can be realized as follow:

(a) Scanning and processing each drawing on the node of dependence relationship tree by breadth-first search.

(b) For each drawing, read each parameter initial value from the GPL and obtain freedom parameter from user interactively. Calculate the constraint expressions and make the graph deduction to create the instance of parameter drawing. Write new design parameters to the GPL and shared by other sub-drawings.

(c) Store the result instance drawing to the PDL for plotting. The drawing can also be converted to other system data format for further processing (Such as AUTOCAD, Pro-engineer).

(d) Repeat scanning from (b) to (d) until all the node was handled.

By the end of processing, the whole set of specific detailing design drawings for RWA was obtained. The whole calculation course was finished in several minutes.

4 The Implementation and Examples

A parametric RWA CAD prototype system based on the model has been developed by the researcher. It was written in C++ language and PROLOG. It can run on IBM PC compatible computer with DOS or WINDOWS 3.X operating system. It has been successfully used to create the parametric engineer drawings for various types of RWA. Figure 4 show the engineering drawings produced by the system.

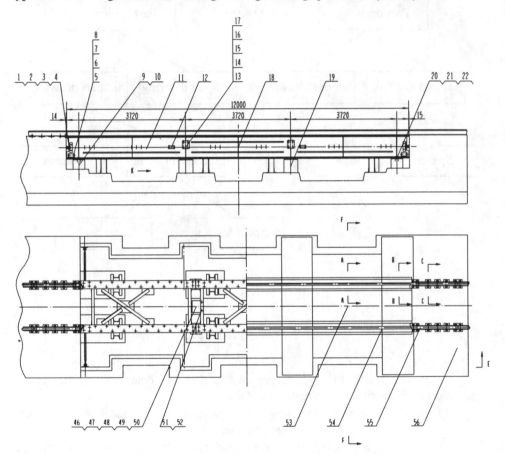

Figure 4   The assembly drawing of RWA produced by the system. When the main dimensions are changed by users, all of the drawings in a RWA will be regenerated to satisfy with the constraints.

5 Conclusion

In this paper, we describe a new generation CAD system model for RWA based on knowledge representation and parametric geometry technology. The new parametric geometry modeling based on directed hypergraph is a major component of the system. It can support the representation of dimensional constrains and geometric constrains. Moreover, the geometric structure changes and design constraints can also be represented. The parametric geometry modeling is finished while interactively constructing geometric objects. The constraint consistency is ensured by construction steps. Overconstrained or underconstrained sketches can not take place. The parametric geometry modeling can support complex drawing representations. The implementation suggests that the modeling is practical and convenient for parametric geometry design. It can be used to realize various products oriented parametric designs and drafting. The model will be extended to 3D Geometry to support parametric solid CAD modeling.

6 Acknowledgments

The project was supported by Shandong Natural Science Foundation. We gratefully acknowledge their support. The author also thanks all research fellows for many helpful discussions and suggestions.

7 References

1. V. C. Lin, D. C. Gossard, R. A. Light, Variational geometry in computer aided design, computer graphics, 15, 3 (1981) 171-177
2. R. A. Light, D. C. Gossard, Modification of geometric models through variational geometry, CAD, 14, 4, (1982) 209-214
3. B. Aldefeld, Variational geometries based on a geometric-reasoning method, CAD, 20, 3, (1988) 117-126
4. A. Verroust, F. Schonek, and D. Roller, Rule-Oriented Method For Parameterized Computer-Aided Design, CAD, 24, 3, (1992)
5. X. Meng, J. Wang, A New Generator For Parametric Drafting Program, CADDM, 2, 1, (1992) 38-45
6. D. Roller, An approach to computer-aided parametric design, CAD, 23, 5,(1991) 385-391
7. H. Suzuki, H.Ando, F. Kimura, Geometric constraints and reasoning for geometrical CAD system, Comput. & Graph, 14, 2, (1990) 211-224
8. K. Martini, Hierarchical geometric constraints for building design, CAD, 27, 3, (1995) 181-191

Designing a Computer-Aided Product Design Advisor

Tay Eng Hock

Department of Mechanical Engineering
National University of Singapore
10 Kent Ridge Crescent
Singapore 119260

Abstract. In this paper, the specifications and the customer requirements of a computer-aided product design advisor are identified and described. These specifications are derived from a survey and experiential interviews conducted in a design firm.

After identifying the needs of potential users, a proposed integrated architecture for design capture and analysis is described. The architecture will provide the development tools and methods to build and integrate design tools. A designer will be able to easily make enquiries to the common knowledge sources containing design rules and guidelines, manufacturing capabilities, existing products and parts, and a classification scheme. A common user interface will provide communication between the user and the underlying tools. The architecture will evolve over time and development will start with a common representation of the data and interfacing mechanisms.

1 Introduction

Two major goals of firms in the nineties are to significantly reduce product cost and time to market (TTM). To meet these goals, quality product designs will have to be developed that will minimize the impact on the manufacturing facilities allowing for rapid product introduction. The design community has recognized that today's design tools and environments are insufficient in meeting their future needs.

The typical design evaluation tool can only deal with single parts and do not consider the system aspects of the design. Further, much of the data to provide answers is not readily available from design capture systems, such as UG-II^{TM} or Pro/Engineer ^{TM}, and therefore must be entered again into these evaluation tools. In many cases, this occurs too late into the design cycle to impact the design positively. If these questions and subsequent answers were asked earlier during the actual design process, then manufacturing considerations would be easier to incorporate into the product's design.

From the survey of user needs and subsequent reports, there exists a very large demand to implement a tool that will allow for quick location of standardized parts being able to locate using non-geometric attributes. Further, there is no single product/part design process among those surveyed. Different designers use Computer-Aided Design (CAD) tools in different ways: some to capture the design beginning with the conceptual design, others do the conceptual design on paper, and transfer it to a CAD system as the design becomes more detailed, while a third group does the conceptual and detailed design work on paper, and loads it to a CAD system only when the design has been detailed.

Therefore, the computer-aided product design advisor should have access to Product/Part Information stored in the CAD systems, to detailed manufacturing process information, and to a battery of design evaluation tools such as tolerancing and costing. A system for determining overall design "goodness" based on predetermined criteria will be included.

2 Design Tools

Design tools fall into the following categories:
1. Design capture - where the design is input into some system
2. Evaluation/analysis - where the design is measured for some variable(s)
3. Design rule checkers - where the design is checked for violations of design practices and standards
4. Advisory tools - where the design is analysed and suggestions are provided to produce a better design
5. Synthesis tools - where the design is generated by the tool, given a set of design specifications.

In the mechanical design space the technology is in the infancy stages of design capture and evaluation/analysis tools. In the electrical arena, where the problem is more constrained and perhaps better defined, tools have advanced to the point where design synthesis has been partially realised. This is a component of the future design system that will allow the designer to perform a sustained high level of performance creating a design and have the system perform the more mundane tasks automatically.

3 Design Survey Results

The survey was targeted at three groups within the company: product/part designers, manufacturing process designers, and design tool developers. Distribution of the survey was by hardcopy, electronic mail, and posting to the Design for Assembly (DFA) and Design For Manufacturing (DFM) notes files. Responses were collected by written response, electronic mail, and many personal interviews. Due to the complexity and length of the survey, and the workloads of the respondents, some difficulty was encountered in completing the surveys.

Some of the findings are categorised and summarised below:

The Product/Part Design Process

There is no single Product/Part Design Process within the company. Different designers use Computer-Aided Design (CAD) tools in different ways. Some to capture the design beginning with the conceptual design, others do the conceptual design on paper, and transfer it to a CAD system as the design becomes more detailed, while a third group does the conceptual and detailed design work on paper, and loads it to a CAD system only when the design has been detailed.

In many cases, the difficulty in finding parts to perform a function, cause designers to design a new part. Since most designs are targeted for a specific plant, even more proliferation can occur. Likewise, the difficulty in locating a standard design for a part feature, causes these to redesigned also. These duplications generate unnecessary costs.

The design function that an organisation performs varies widely within the company. Few perform only a single function such as new designs only or even Engineering Change Orders (ECOs) only.

Manufacturing capabilities are factored into designs by interaction between the designers and manufacturing engineering. This is a manual interaction. This works well for products that are introducing a new manufacturing process, but those using an existing process often receive attention too late in the design process to influence the design.

Design tradeoffs should be tracked, including the reason for a decision made. A design advisor tool could help the designer to evaluate the different tradeoffs.

Product/Part Design Parameters

The majority of those surveyed indicated that there were at least nine different design parameters. When asked to indicate which were critical design parameters, the majority could agree on only two parameters: cost and product introduction time.

The emphasis on different design parameters changes from design to design, and frequently changes during a design project. Designers are frequently trying to hit a moving target of design parameters.

Design Advisor Requirements

A design advisor could improve the design process in a number of ways. By reducing Time to First Revenue Shipment (FRS), reducing cost, increasing product quality, and improving the design, among other things.

The design advisor should have access to product/part information stored in the CAD systems, to the detailed manufacturing process information, and to a battery of design evaluation tools (EMI, Tolerancing, Costing, and others), including a system for determining overall design "goodness". It should include both a common parts library with detailed part information about existing parts (geometry, function, requirements, specifications, etc.), and a common feature library with details of common features that could be incorporated directly into designs in a CAD system.

The design advisor should be flexible enough to be used by a designer, no matter what their designing style, the type of design function they are performing, or the kind of product or part they are designing.

The interface to the design advisor should be flexible. The users should be able to indicate how much help and design assistance they want, and when their designs should be evaluated (in an ongoing mode, when saved, on demand, etc.).

Summary

the survey results clearly pointed to the need for a design advisor system that will check designs early for various considerations such as DFM. A design advisor system could improve the availability of manufacturing process information to design suggestions based on industry and corporate standards. This would result in a number of improvements in the design process: increased product quality, reduced time to FRS, reduced design time, reduced numbers of common mistakes, reduced design duplication at both the part and feature levels.

4 Proposed Design Advisor

A design advisor is required that offers design suggestions to the product designer that assists in attaining cost and other product goals/metrics. The design shell will integrate and utilise current evaluation tools to perform product analysis. Furthermore, tools will be developed that perform evaluation and estimation on incomplete designs, such as conceptual designs, and suggest design alternatives thereby assisting to produce a better final product design. In its final form the design advisor will greatly enhance the manufacturing to engineering relationship by providing the design engineer with specific manufacturing information at the time of design. It will provide design engineering with estimates of manufacturing costs, recommended manufacturing process, and provide manufacturing process engineers with product information. It will advise in areas of expertise such as safety, plastics, and sheet metal as well as methods to reduce process steps or improve manufacturing yield. Rules included will be analytic, comparative and deductive in nature. Included with each rule will be an impact assessment, both positive and negative, so that the designer can make an informed decision whether to apply the rule or not.

Furthermore, the design advisor should be an overall architecture which integrates CAD tools, point out design solutions, workbenches, and part databases. It provides and automates data translations/extractions, manages the search for data, and manages remote processing. It does this focused at the design attribute level and not at the tool or file level.

5 Design Methods

There are several approaches or methods that one could take for the design advisor. There re three methods chosen to allow for incremental development of the total design advisory system. The methods are: post-design, palette-design, and embedded-design. The post-design method utilises feature extraction from CAD files to generate input to evaluation modules. This is similar to the traditional CAD post processing methods. The pallette-design method utilises a context based menu it guide the designer in the selection of desired components. The embedded-design method utilises tight integration between the modeling tool(s) and evaluation tools by focusing at the design attribute level. The embedded-design method is the long term vision for implementation of design advisors.

6 Design Structure

There are several options for structuring advisors and rule bases. One option is to build a single advisor inference engine with a single rule base providing just one set of applicable rules. Another is to build several smaller domain specific advisor inference engines with their own separate rule base providing multiple sets of applicable rules. A hybrid approach is to develop the smaller domain specific advisors and couple them into a domain independent advisor each with their own rule base. Each of these advisory implementation options would operate on the current product model continued within the blackboard.

The first option single advisor with a single rule base has the following advantages and disadvantages:

- advantages
    1. one advisor to support
    2. one rule base to support
    3. able to deal with multiple domain tradeoffs through conflict resolution
    4. incremental knowledge
    5. intermediary between the product model and the rule base
- disadvantages
    1. potential size problem due to the number of different domains required
    2. must validate new knowledge using the complete advisor
    3. all domains may not be best represented in a single representation
paradigm

4.   may have to extract the rules from the many existing advisors, some of which may   be unknown or embedded.

The second option, multiple advisors with independent rule base has the following:
- advantages
    1.   independent advisors, one per domain area
    2.   modular
    3.   independent testing and validation
    4.   able to utilise each advisor independently
    5.   able to support multiple types of advisors, such as externally developed advisors that   use a foreign structure
- disadvantages
    1.   may have many duplicate rules
    2.   no mechanism for resolving conflicts between competing domains
    3.   no mechanism for supporting interaction between domains.

The third option or the hybrid approach:
- advantages
    1.   one rationalisation advisor with several independent domain specific advisors
    2.   all of the positive points of the second option
    3.   the rationalisation advisor can be added later when conflict resolution and interactions   are understood better, this gives potential for applying learning to a rationalisation   advisors
    4.   ease of expansion by adding in domain specific advisors
    5.   could potentially build a hierarchy of advisors that fully segment the rule sources.
- disadvantage
    1.   may be difficult to obtain the methodology for resolving conflicts between domains.

The third option has been selected as the structure for developing the advisor.

7 Conclusion

The survey has clearly indicated the need for a design advisor that will impact the market place in terms of product that are cost effective and timely. While the characteristics of such a design advisor are being worked out, the design structure and design method has been adopted to be the hybrid structure and the embedded-design method.

References

1.	Finger, S. and Dixon, J.R., A Review of Research in Mechanical Engineering Design, Part 1: Descriptive, Prescriptive, and Computer-Based Models of Design Processes, Research in Engineering Design, Vol. 1, pp. 51-67, 1989.

2.	Pugh, S., Total Design: Integrated Methods For Successful Product Engineering, Addison-Wesley Publishing Company, 1991.

3.	Sriram, D. et al, An Object-Oriented Framework for Collaborative Engineering Design in Computer-aided Cooperative Product Development, MIT, Cambridge, USA, Springer-Verlag, 1989.

4.	Tomiyama, T., Kiriyama, T. and Umeda, Y., Toward Knowledge Intensive Engineering in Proceedings of the 1994 Lancaster International Workshop on Engineering Design CACD '94, Lancaster, U.K., 1994.

Development of
An Intelligent Grating CAD System----GRTCAD

Ma YongSheng

Gintic Institute of Manufacturing Technology, 71 Nanyang Drive, Singapore 638075

and

Gao HuiXing

SOE Dept., Ngee Ann Polytechnic, 535 Clementi Road, Singapore 599489

Abstract. This paper reports a CAD product developed for grating design. In this system, methods for designing different types of grating panels are modelled using an object oriented approach. Design rules can be defined interactively by using friendly dialogue boxes. Advanced generic 2D dividing algorithms with the consideration of grating area boundaries, openings, gap size, supporting structure and panel properties are coded. Drawing geometry, layout and bill of material are generated automatically. Industrial feedback has indicated improvement on design quality and productivity. **Key words: CAD, Grating.**

1. Terminology

Grating panel: A completed piece of grating, cut to size banded and notched as per cutting drawing.

Bearing bars: Load carrying bars of uniform section in rolled steel spanning between supporting steel members. It is also known as *Loading Bars*.

Crossing bars: Rods of uniform section in square twisted steel, electro-forged at right angle to the bearing bars.

Length of panel (L): The overall measurement of a panel when measured parallel with the bearing bars.

Width of panel (W): The overall measurement of a panel when measured parallel with the crossing bars.

Bearing bar pitch: The standard distance between two neighbouring bearing bars.

Crossing bar pitch: The standard distance between two neighbouring crossing bars.

Openings: The spaces cut out in grating areas to accommodate some obstacles, such as pipes, columns, tanks, etc.

Nominal circle: A circle defined by the user and used for circular platform to measure panel material size. Particularly, the panel width is measured along this circle.

Standard width: The width of a standard panel (W). For polar bearing bar direction, this parameter is measured along the circumference of the nominal circle.

Minimum width: The minimum size of a separated panel in the width direction. Similarly, for polar bearing bar direction, this parameter is measured along the circumference of the nominal circle.

Gross area of panel: The total area of flooring including openings, notches, etc; as used in material estimation (Area = L x W).

Net area of panel: The real geometrical area of a panel excluding openings, notches, etc, as used in weight calculation and certain accurate quotation.

Bill Of Material (BOM): A list of grating panels with series number, group number, gross length and width, gross or net area, cutting drawing number, etc. This document is used in production planning and invoicing.

Cutting list: A set of production assignment documents with cutting drawings for all grating panels, planned time schedule, and other technical requirements.

Packing list: Document contains information for a set of panel groups for delivery purpose. Panels are grouped according to panel size, total weight of each group, and other delivery instructions.

Start cutting point: A point indicating the distance between the first crossing bar in a panel and the parallel panel cutting edge measured along the bearing bar direction.

2. Introduction

Grating is necessary in shipbuilding, refinery construction, steel plant construction, etc. In this process, a platform with different layouts is covered by a set of panels so that the platform can be used for servicing or loading purpose. However, so far, there is no any specific CAD system for grating design. Grating design requires experienced engineers to calculate panel layout from structural layout.

Grating design includes selecting grating category according to the load range, and calculating gratings layout for a given area, which is constrained by the supporting structure and obstacles according to certain distributing patterns. Grating layout drawings are then produced. The total gross grating area is calculated so that quotation for estimating and invoicing can be obtained. Each panel's geometry is later used to produce the panel cutting drawing. This cutting drawing is used to cut the panel shape from standard grating panel. The net area of each panel is used to calculate the weight. This is used for planning and shipment packing. Throughout the process, the Bill Of Material (BOM) is required.

Grating areas can mainly have rectangular or circular shapes. Certain openings are necessary in order to accommodate pipe lines or any other obstacle. Loading bar alignment among panels is important in order to achieve consistent appearance. Because the panel material comes with standard width and length, dividing methods should be optimised to reduce material waste. Traditionally, the grating design and material estimation were done manually.

3. System description

The system is essentially a specially developed, automatic design and drawing generation system for the grating engineering industry. PC and AutoCAD were selected as the hardware platform and the basic graphic tool respectively. This CAD system can automatically produce design drawings for a grating area with a given supporting structure. The profiles of the supporting structures and the openings for obstacles consist of straight lines and circular arcs. After users click in the drawing contours of the outer support structure, openings, and inner supports in the grating area, grating layout drawings are automatically generated with the consideration of the technical rules for dividing the grating area into panels. Rules are selected by the user. It can also calculate the grating area, BOM, start cutting point, cutting list, packing list while panels are created. All operations are menu-driven. After the system generates and displays the drawing of panels, users can still modify these panels interactively.

3.1 Specifications

- The system can deal with an area with inner support lines, even when they have cross openings.
- Standard panel width is used whenever possible.
- Gap generation is automatic for a fixed gap value and floating gap values optimised in a range.
- If the panel width is smaller than the minimum width specified, the small panel is then merged with its neighbouring panel.
- Panels can be modified easily.
- Text in panel labelling and other automatically generated documents, such as bill of material, cutting points, can be modified easily.

3.2 Data Structure and Repository

For creating the relationship between a geometrical block and a panel data structure, a special user defined object type is used. Each grating piece is drawn as an AutoCAD block which has many attributes, such as a unique name and a grouping number. A separate data file is created when panels are generated. All panel information including segment definition, position and block name are written to this file. During an interactive operation process, such as modifying the position of a panel grouping number, when a panel is clicked, the system can return the panel's name. Thus, a relationship between the panel and the data file is established. If there is any modification on any panel, its data records are updated accordingly. The panels with the same dimensions, in the same project, can be assigned with the same cutting number if the user selects this option. The user can take a batch or several batches to process the layout in a design session.

4. Panel Generation Algorithm and General Rules

Two types of grating areas are considered, i.e., rectangular and circular platforms. Three panel distribution patterns are designed according to the bearing bar direction, i.e., horizontal, vertical, and polar patterns. Many intelligent judgements can be performed, such as standardising the width of the panels, selecting the suitable dividing line within an internal opening, setting the gap between panels, and labelling panels automatically. Certain technical requirements, such as maximum and minimum grating panel dimensions are observed.

4.1 Standard Panel Width and Sub-Standard Width

In order to minimise material waste, it should be appreciated that generated panels should have standard widths whenever possible. In Fig.1, the grating area is a simple rectangle. The bearing bar direction is vertical, so the panels created are in a row. In Fig.1(a), the gap value is set as a fixed value, 10 mm, and the standard panel width is 905mm. When dividing the grating area from right to left, the first eight identical panels are created with the standard width 905mm, while the last one has a width of 740mm.

For grating panel cutting, it is always preferred to cut next to a bearing bar so that the side edge of the panel is neat with a solid bar. Therefore, if the standard panel width cannot be achieved, a width that can accommodate certain number of bearing bars is preferred. There could be many width values that satisfy this condition. These values are referred as sub-standard panel widths. Sub-standard widths are calculated by this formula: *Bearing_bar_pitch X (No. of bearing bars - 1) + Bearing_ bar_width.*

In the examples shown in this paper, the bearing bar thickness is 5mm and the pitch is 30mm. If the minimum and the maximum numbers of bearing bars are 4 and 45 respectively, the sub-standard values are the integer series from 95mm to 1325mm with an increment of 30mm.

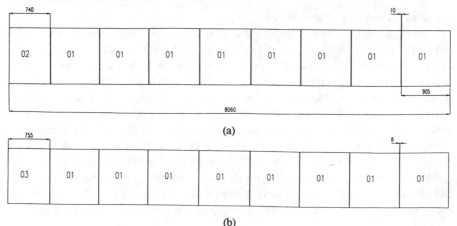

(a)

(b)

Fig.1 Simple rectangular platforms

To achieve the sub-standard, the gap value has to be adjustable. In Fig.1(b), the gap value is adjusted to 8mm. The width of the last panel becomes 755mm which is one of the sub-standard widths. When the user selected either horizontal or vertical loading bar direction, the rules in this section are applied.

When the user activates the "Auto-adjust gap" option, panel gaps will be dynamically adjusted in a range, between the minimum and the maximum values allowable, to ensure that the panel widths are equal to the standard width or sub-standard widths. Otherwise, gap values will be kept constant throughout the panel generation process.

4.2 Rules to Deal with Openings

Most of the openings are either circles or rectangles. There must be a dividing line passing through each opening. Otherwise, grating panels have to be installed from the top of the obstacle, and doing this is often not possible.

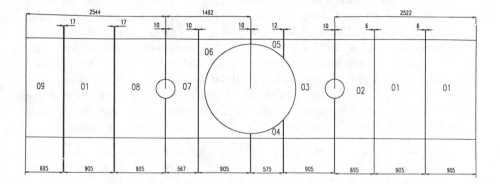

Fig.2 Grating area with circular openings

For a circular opening, the dividing line must go through its centre point. An example is shown in Fig.2. There are three circular openings. With the gap value varying from 6mm to 17mm, the panels created have either standard or sub-standard widths.

However, when the opening is not a circle, e.g. a rectangle, the dividing line can be at any position within the range in which the opening has the maximum height. This range is defined as the limit zone of the opening.

For example, when using the vertical bearing bar direction, opening (1) in Fig.3 can be divided at any horizontal position between A and B, while opening (2) can be between P and Q. Again, in order to generate panels with widths either equal to the standard width or sub-standard widths, dividing line positions are optimised with great effort. In this system, the limits of each opening in the horizontal and vertical (or in radial and angular for polar pattern) directions and their corresponding zones in which the opening reaches limits are calculated.

It should be noted that if the bearing bar direction is vertical, dividing lines to generate panels are also vertical. Then, the dividing line through an opening must fall in the common horizontal range of the corresponding zones for its upper and lower limits. Similarly, limits zones can be found if the bearing bar direction is horizontal. If the bearing bar direction is polar, dividing lines are in radial directions, the dividing line through an opening, must fall in the common range of the corresponding angular zones for the outer and inner limits related to the reference centre of the platform.

GRTCAD provides an option for the user to decide whether the dividing line positions can be away from the centre of their limit zones. This is controlled by the "Dividing at square centre" toggle button.

In Fig.3, all panels' widths are adjusted by changing the neighbouring gap values which vary from 10 to 14. Note that opening (1) and opening (2) are overlapping, and there is some overlapping limit zone. The system automatically identifies this type of overlapping and decides a dividing position within the common limit zone, hence positively avoids small panels being generated. It should also be noted that the first dividing line from right to left results in two panels on the right side, i.e., panel "01" and "02". Similar rules are applicable to a polar pattern although the restrictions for a polar pattern is much more stringent because the dividing line must be in the radial direction in order to have reasonable load distribution.

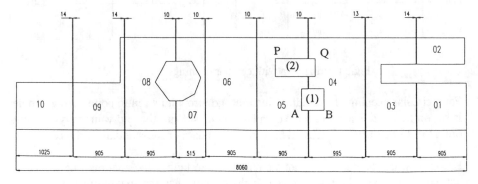

Fig.3 Rectangular grating area with irregular openings

4.3 Panel Group Numbering
The user can group identical panels into a group so that the machining of these panels can be carried out without repeating set-up processes. This system mainly identify those panels with simple shapes, such as rectangles and typical fan shapes. This is done automatically when the user toggle the "Group Numbering" switch on. Otherwise, the

panel number is assigned in the sequence of their generation. The user is also allowed to insert, modify, or confirm the starting panel number for the current platform before starting to generate panels. For some special cases, where an opening is small rectangle, if gaps can be adjusted, the opening is divided at one end so that the panel manufacturer only needs to cut the opening on one side panel.

4.4 Rules to Deal with Inner Support Lines

Quite often, the grating area spans in a large space, inner support structure are therefore necessary to support grating panels to achieve satisfactory rigidity. The support structure design is out of the scope of this system. However they are considered as inner support lines in grating panel generation. For rectangular platforms, supporting lines are perpendicular to the bearing bar direction, while for polar platforms, the support lines are in radial directions.

In GRTCAD, the grating area is divided into sub-areas considering the gaps along the supporting lines, and the panels in each sub-area are then generated in order automatically. In Fig.4, it can be observed that there are two inner support lines in the horizontal direction (the bearing bar direction is vertical). Note the lower support line crosses the big opening and the system has created grating panels properly.

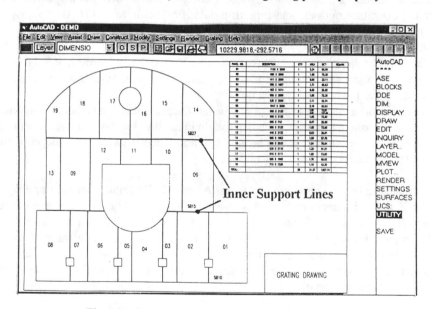

Fig.4 Grating area with inner support lines

4.5 Avoiding Small Panels

Panels generated cannot be too small because small panels are inconvenient to cut and install. In GRTCAD, in order to avoid small panels, the next generated panel size is predicted. If the panel width is smaller than the minimum width, the next small panel will not be generated. Instead, the current panel will include this small area. Hence, the current panel becomes bigger. For polar bearing bar direction, this minimum allowable width is measured along the circumference of the nominal circle, which is also used to measure the standard width in the dividing process.

5. Editing Capability

GRTCAD provides many user-friendly interactive modification tools, such as changing the shape and name of a panel, dividing one panel into two or merging two into one, etc. When panels are created or modified under the normal AutoCAD environment, they can be re-integrated into the grating database by running a "read-in" program. These modification functions make this system more flexible to meet customers' requirements and to cope with complicated geometry. A dialogue box showing panel modification options is shown in Fig.5.

6. User Interfaces

A customised menu group under "Grating" is inserted into AutoCAD's top menu bar. If the user clicks the grating command item, the GRTCAD is activated, and the first menu for selecting operations is then displayed. After clicking a grating operation button, a few AutoCAD-like dialogue boxes will be displayed for selecting operation options, and inputting data. For example, GRTCAD displays a dialogue box to allow the user to set initial parameters for panel generation. These parameters include:

- Bearing bar direction, the choice can be either Horizontal, or Vertical, or Polar.
- Standard panel width, minimum width, bearing bar pitch, thickness, and height; crossing bar pitch, starting panel number, etc.
- Dividing principles applied, such as whether the gap value can be floated, and the range for auto-adjustment; whether the dividing line for a rectangular opening must be at the centre or it can be adjusted; and whether panels are grouped with one panel number if they are identical, etc.

These dialogue boxes are defined by using AutoCAD Dialogue Control Language (DCL). A map with important menus is shown in Fig.5. Used items in dialogue boxes include *Radio buttons, Editing boxes*, and *Toggle switches*. The user's settings for each dialogue box are backed up in a data file and reloaded when the system is started. By doing so, the user will save a lot of time for inputting these values. This data file is updated instantly when the settings are modified. Hence, the system can "learn" from the user.

7. System Output

Grating drawings are the most important output. Fig.4 shows an example for a rectangular platform. In Fig.6, a polar platform is shown. All panels can be output individually to create cutting drawings. The system can automatically point out the position of cutting start point for each panel in order to align all the crossing bars of the neighbouring panels. These cutting start points are displayed on the

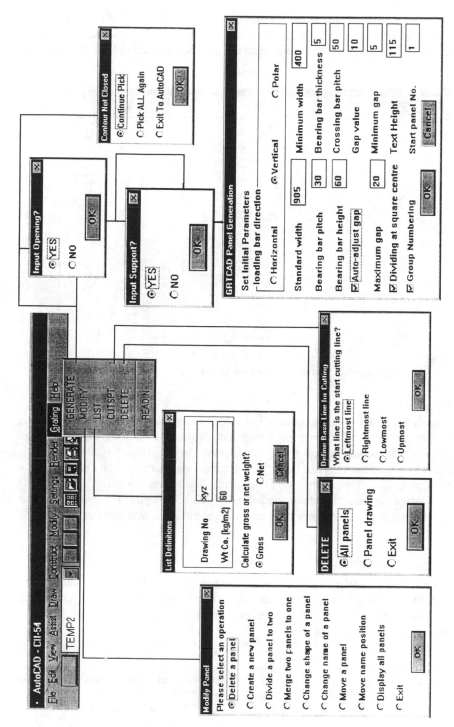

Fig.5 Menu Map in GRTCAD

863

manufacture drawings. All the output data documents, such as bill of material, cutting list and packing list, are generated automatically using the integrated data base. Hence, it eliminates human errors. The output data can also be exported to Lotus 1-2-3 spread sheets.

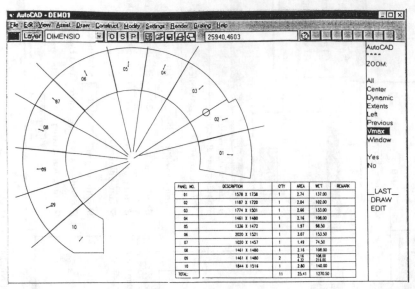

Fig.6 Circular grating area with openings

8. Conclusion

The intelligent features in this system allow the user to cut down operation and training time. The easy-to-use data entering format has shorten the learning curve. The user needs less man power to deal with grating panel design. Higher quality and more accurate grating drawings can be produced with high efficiency. Human errors are eliminated in the different output data documents. The approach used in this paper has been adopted by several other specialised CAD development projects such as generating scaffolding structures and design 3D metal form structures in construction engineering.

9. Acknowledgement

The authors would like to acknowledge Sangjin Dahe Industries Pte Ltd (S) and the EDB of Singapore government for their sponsorship, and Ngee Ann Polytechnic for permitting the relevant projects being carried out. AutoCAD ® is a trade mark of AutoDesk Inc.

References

1. Sangjin Dahe Industries Pte Ltd (S), Company Document, *Specifications for Grating Panel Cutting and Operational Documentation* (1994).
2. Y. Ma and H. Gao, *Grating CAD System --- GRTCAD User's Guide*, (1996).

Simultaneous Consideration of ecological & economical Parameters by using a feature-based design system

Dieter H. Müller, Kai Oestermann

Department of Mechanical Engineering, Bremen University
Badgasteiner Straße 1, 28359 Bremen, Germany
E-mail: bik@bik.uni-bremen.de

Abstract. This paper deals with a new method for the isochronous consideration of ecological and economical parameters in early stages of mechanical design. The method was transformed in a computer aided device for the ecological and economical optimisation of the product definition phase.

1. Introduction

Today's products do not only have to work satisfactorily. In addition to the functional requirements they have to comply with other - partly contradictory - demands /1/, /2/.

Due to the increasing responsibility of companies for the ecology and the increasing demand for environmentally conscious products, a sustainable industrial production is one of the challenges facing the industrialised world. Hence there is the need for an equal consideration of ecological and economical parameters within the process of product design.

One decision model for this is a complete eco balance consisting of the elements life cycle inventory and environmental impact assessment. Due to the complexity of a complete eco balance, this method is not practicable within the product development in small and medium sized enterprises (SME) which represent the backbone of German economic strength.

At present methods for product optimisation like „Design for Cost" and „Design for Environment" are only applicable in a sequential manner. In consequence of that the possibilities to obtain a comprehensive optimisation of products are strongly reduced (Fig. 1).

Therefore it is necessary to develop auxiliary means and methods for the simultaneous optimisation of functional, ecological and economic parameters, especially in the context of product development in SME.

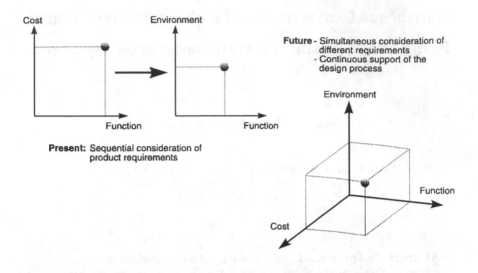

Fig 1: Differences between sequential and simultaneous consideration of different requirements during the process of product optimisation

2. Evaluation Method

In this paper a method will be presented which leads to a more transparent and more feasible evaluation concept for an ecologically oriented production. This method was developed within a project at the Bremen University.

Using this method it will be possible to assess the ecological effects of a certain way of product realisation at an early stage of product planning. This point is an important feature since approx. 70% of the product costs and all product features are pre-determined by the first steps of product development (Fig. 2).

On the other hand it is important that the solution obtained fits equally-balanced to ecological and economical requirements.

For reasons of simplicity, our method is restricted to an eco balance at the level of a life cycle inventory of the production process (Fig. 3).

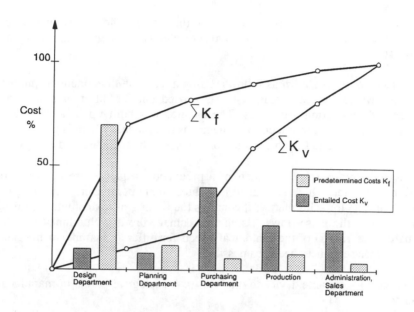

Fig. 2: Fixed and entailed costs during product realisation /3/.

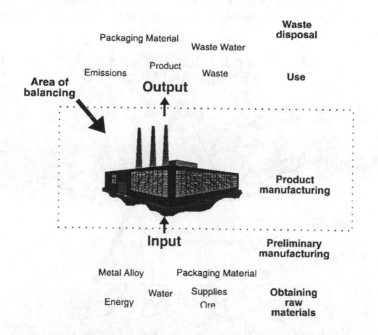

Fig. 3: Area of balancing for SME´s

The evaluation method used is the method of the eco-toxicological equivalent coefficient as used by the Department of Environmental Technology at the Bremen University /4/.

The project process is shown as follows: Using a machine manufactured product the life cycle inventory shown in Fig. 4 was carried out for different manufacturing methods and production sequences. The balance was built up in a modular way, where the smallest unit (production sequence) is one module which is connected with other modules linked to the border lines of the system by the mass and energy flows.

The environmentally relevant effects of a production sequence are described by an emission parameter, which is divided into three single parameters, one for gas, one for fluid and one for solid matter. It contains the finding of toxicological equivalents connected with the mass flows. Using this damage-profit-evaluation a simple and transparent description of the ecological effects for different manufacturing methods and production sequences becomes possible.

This method is a suitable device to uncover weak ecological points in manufacturing sequences.

In a second step, a computer based design system was developed for the ecological and economical assessment of technical products. Within this system a connection was established between manufacturing methods, product cost and functional requirements.

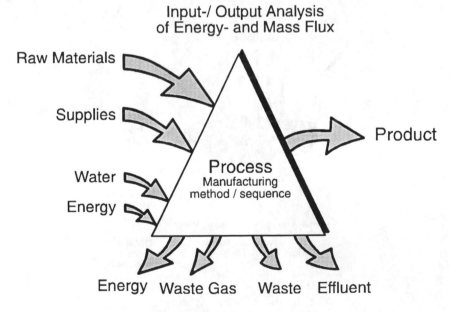

Fig. 4: Life cycle inventory /4/

3. Main results

The result of the project is a feature-based design system (Fig. 5) including ecological and economical decision parameters for the design of machine-manufactured, rotationally symmetrical products (e.g. shafts, bolts etc.). The term „feature" is defined as:

Feature = Form-Feature & Semantical Information /5/.

In addition to the geometrical information, features contain further information about the product, for example, possible manufacturing methods, production costs and eco-toxicological data.

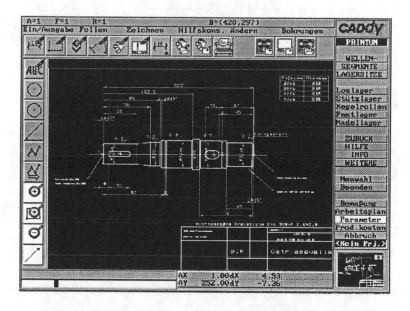

Fig. 5: Feature-based design system

Taking into account the required product functionality, alternative manufacturing methods (e.g. hard machining instead of grinding) or production sequences (e.g. dry machining) are suggested by the system during the design process. Also design actions like the renunciation of a definied surface quality, or the use of different materials, are implemented into the system.

At the same time the production costs are determined by the system. The basic evaluation method for this is the VDI-guideline 2225 /6/. Thus the opportunity of a more complete product optimisation is given to the designer by the simultaneous consideration of the product parameters function, cost and environment.

4. Example

The following example is meant to illustrate the feasibility of the evaluation method in general and the functionality of the developed feature-based design system. As an example we used a machine manufactured product, the shaft of an electromotor. The development task was the consideration of ecological and economical parameters during the design process.

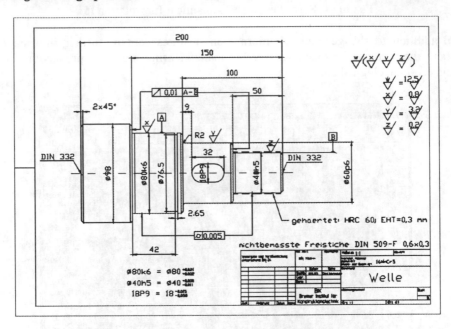

Fig. 6: Shaft of an electromotor

As a first step the preliminary design concept of the shaft was worked out by using the design system. After receiving the input data the process of product optimisation started. The decision of the final design concept was given by the comparison of the energy and ecological requirements of resources and the product costs.

The solution suggested by the system was the replacement of the manufacturing sequence „grinding of roller bearing seat" by „hard machining of roller bearing seat".

The application of the evaluation method lead to the conclusion that the demand for energy per piece could be reduced fr om 10.43 kWh to 9.24 kWh (Fig. 7).

As a result of the application of the evaluation method it was possible to illustrate the optimisation in money values. In this practical case a production cost reduction of about 6% has been achieved (Fig. 7).

Demand of energy		
Variant 1 [kWh]	Manufacturing sequence	Variant 2 [kWh]
0,030	Sawing	0,030
0,985	Roughturning	0,9 85
0,046	Smoothturning	0,046
8,168	Direct hardenening after cementation	8,168
1,203	Grinding of roller bearing sets	-
-	Hard machining of roller bearing sets	0,006
10,432	Total demand [kWh]	9,235

Production cost		
Variant 1 [DM]	Manufacturing sequence	Variant 2 [DM]
30,12	Material cost	30,12
2,56	Sawing	2,56
15,06	Roughturning and smoothturning	15,06
26,95	Direct hardenening after cementation	26,95
8,20	Grinding of roller bearing sets	-
-	Hard machining of roller bearing sets	3,20
82,91	Total Cost [DM]	78,41

Fig. 7: Demand of energy and Production cost

5. Conclusions

The sequential consideration of different requirements like function, cost and environment in early stages of mechanical design is necessary for today's products. We think that the method described in this paper is one step in this direction.

Mainly enterprises using metal-cutting manufacturing should be interested in the results. The basic idea of our simplified ecological balancing method and isochronous consideration of function, cost and environment is also applicable to other manufacturing technologies.

References

1. Müller, D.H and Oestermann, K. (1994). Wie können Marketing und Konstruktion gemeinsam das Produkt definieren ?. *Tagungsband des VDI-EKVIP Zirkel Führungskräftetreffens 9/94*, pp. 2.1-2.15. Düsseldorf: VDI-Verlag.

2. Müller, D.H and Oestermann, K. (1993). Integrative Zusammenarbeit von Konstruktion, Marketing und Design. *Proceedings of 9th International Conference on Engineering Design, Den Haag*, pp 1197-1201.

3. Vayna, S. (1991). Strategien zur Integration von Entwicklung, Konstruktion und Arbeitsvorbereitung. In *Die Konstruktion als entscheidender Wettbewerbsfaktor: Ziele, Strategien, Maßnahmen*; Tagung Bad Soden 7. und 8. März 1991, pp. 57-66. VDI-Gesellschaft Entwicklung, Konstruktion, Vertrieb. Düsseldorf: VDI-Verlag.

4. Räbiger, N. and Haase, C. (1995). Ein Bewertungsmaßstab für die ökotoxikologische Schwachstellenanalyse von Produktionsprozessen. *Tagungsband Colloqium Produktionsintegrierter Umweltschutz "Abwässer der Pharmazeutischen Industrie und Biozidherstellung"* 9/95, pp. 281-292. GVC - VDI-Gesellschaft Verfahrenstechnik und Chemieingenieurwesen und IUV-Institut für Umweltverfahrenstechnik, Universität Bremen (Hrsg.)

5. Krause, F.L., Ciesla, M. Reiger, E., Stephan, M., Ulbrich, A. (1994): Featureverarbeitung -Kernkomponente integrierter CAE-Systeme. In *Produktdatenmodellierung und Prozessmodellierung als Grundlage neuer CAD-Systeme*. Fachtagung der Gesellschaft für Informatik e.V. Paderborn CAD 94. Jürgen Gausemeier (Hrsg.). Müchen, Wien: Hanser.

6. VDI-Richtlinie 2225 (1996). Technisch-wirtschaftliches Konstruieren. VDI-Gesellschaft Entwicklung Konstruktion Vertrieb. Düsseldorf: VDI-Verlag

A Concurrent Integrated CAD (CICAD) Method for Ship and Sailing Yacht Design

Kok Thong Tan and Thomas P Bligh

Department of Engineering, Cambridge University, United Kingdom

Abstract. The objective of this paper is to introduce a new approach that eliminates the sequential nature of the ship design process. This new approach is called the Concurrent Integrated CAD (CICAD) System. It establishes a concurrent computer-aided design procedure, instead of following the traditional sequential design approach. It is modelled and analysed by utilising I-CAM Definition (IDEF) techniques. The paper starts with a brief review of the evolution of ship and yacht design processes from craftsmanship to computer-aided design methods. The authors point out the problems of the ship design process, and investigate the impacts and trends of applying CAD methods, in order to justify the new approach to an integrated CAD method. The paper demonstrates how the sailing yacht design process could be executed concurrently, and estimates its significance. Some preliminary findings of this new model are reported.

1 INTRODUCTION

The shipbuilding industry is actively involved in applying the computer technology since the 1960s [Hays et al, 1994]. Therefore, in the present information age, ship and yacht designers are able to enjoy the benefits of it, especially in reducing tedious calculations, and improving the design accuracy and quality. The industry perceived this transformation as a revolutionary one, for it had digitised traditional 2-D ship design drawings, carried out complex design calculations, and enhanced the productivity through computer numerical control machines [Ross, 1995]. In principle, computer technology does contribute to the ease of making changes throughout the ship design process. Thus designers may have relatively more time to explore many potential designs. Current innovation is concerned with the development of an Integrated Computer-aided Design (CAD) method for ship designs, which combines all stages of ship design programs into a single system, and even embraces the elements of the product's life cycle from conception to disposal [Mistree et al, 1990]. All of these exciting developments in the shipbuilding industry consolidate the belief that computer technology has revolutionised the ship design process. However, the principle of existing CAD methods capture the features of the Ship Design Spiral[1] concept, which is sequential, and therefore suffers from a number of problems, such as designers adversely alter other design characteristics while concentrating on a particular design aspect. For a design that employs the Ship Design Spiral concept, the sequential process is

---

[1] The Ship Design Spiral concept was introduced by Evans in 1959, and enables ship design problems to be solved systematically [Evans. 1959].

inevitable as each phase of design process requires the output from the preceding one. For example, a designer has to conduct a weight estimation, in order to determine the position of the centre of gravity, then stability studies can be carried out by using this value to obtain the righting arm, $\overline{GZ}$. As a result, this computer technology does not actually reform the ship design process. We, however, believe that it is possible to remodel the design process to avoid the ineffective sequential nature of the Ship Design Spiral. The objective of this paper, therefore, is to illustrate this new approach to an integrated CAD method for ship and sailing yacht design.

2 PROBLEMS OF THE SHIP DESIGN PROCESS

Ships, like aircraft, are examples of extremely complex systems. The design and building of ships was wholly a craft until the middle of the eighteenth century, before science affected ships design appreciably [Rawson et al, 1966]. For the next two centuries, the designer started with a number of assumptions and worked through the rules, which only imperfectly modelled the real situation, to see if the design satisfied the requirements, specified beforehand [Evans, 1959]. In 1959, Evans made a significant contribution to visualising and modelling the process of ship design, which is now known as the 'Ship Design Spiral' [Mistree et al, 1990]. The purpose of this attempt was to assist designers in organising the thought process, so as to enable ship design problems to be solved more effectively [Evans, 1959]. While some refinements have been made over time, the principle remains unchanged. Nowadays, this spiral model has become a widely accepted approach for both ship and sailing yacht design [Hassan et al, 1992, Larsson et al, 1994]. A typical design spiral for sailing yacht designs is shown in Figure 1. This spiral consists of eleven stages, where each stage corresponds to a design task. The design process keeps on iterating all these stages until it converges towards the centre, the final solution.

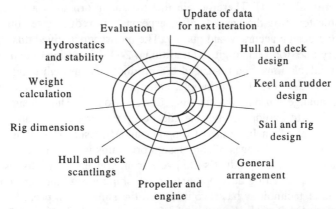

Fig. 1. The design spiral for sailing yacht designs [Larsson et al, 1994].

The major characteristic of the Ship Design Spiral concept is that the design process is sequential and iterative. Designers must satisfy a limited set of criteria at each design stage, however, having satisfied them, goes on to the next stage with little idea of how good the design actually is. Therefore the design is continuously altered by means of consecutive iterations until it reasonably fulfils all the requirements. As each design stage depends on the output of the preceding one, the process is arranged in a sequential manner. This leads to a design philosophy that based on a straightforward sequential and iterative procedure [Kroemker et al, 1996]. Moreover, for certain design stages, internal iterations are required, and this makes the design process even more time consuming.

Surprisingly, it is not easy for practising designers and naval architects to justify the effectiveness of the Ship Design Spiral approach, because by and large, it has been exercised for decades, and has become the routine procedure. Experienced designers may occasionally skip stages, as they are able to make a good guess, based on a rule of thumb [Condylis, 1997, Simpson, 1997]. These designers, however, have to comply with the design spiral, when subtle modification to the design takes place.

Despite the general acceptance of the spiral model, a few academics and operations researchers began to question its effectiveness. A number of general surveys were taken, and the results show that this method, by its very nature, may obstruct the exploration of optimum design [Kroemker, 1997, Kroemker et al, 1996, Mistree et al, 1990, Lyon et al, 1985]. Due to this sequential, iterative, tedious and time-consuming design process, most of the designers are still unable to achieve near optimal designs. This is because the sequential process requires a great deal of design time and thus designers have limited time to explore many potential designs [Kroemker et al, 1996]. Moreover, the longer the design time, the less competitive are designers in bidding the contract [Keane et al, 1996].

In addition to that, the spiral approach is unable to provide the designers with a global view of the designs, especially at an early stage, because the design information emerges at each stage as they work through the spiral. Consequently, they are incapable to recognise the effects of modifications on the design quickly. One drawback to this is that designers may adversely alter other design characteristics while concentrating on a particular design aspect [Condylis, 1997, Simpson, 1997].

3 THE EXISTING TOOLS FOR SHIP DESIGN

The advent of computer technology lead to the development of Computer-aided Design and Manufacturing (CAD/CAM) programs. The shipbuilding industry, likes others, extensively used this new technology from the 1960s [Hays et al, 1994]. Today, CAD/CAM is a key technology for ship design and construction [Hays et al, 1994]. Typical CAD programs for ship design are built around a geometry generation capability which allows the designer to create on the computer screen a mathematically defined hull surface. Furthermore, a number of distinct programs allow the post-processing of these surfaces in the areas of analyses and predictions, such as hydrostatics and stability calculation, resistant and powering, Velocity Prediction Program (VPP), Computational Fluid Dynamics (CFD) etc. Compared with the pre-computer era, these computer-based tools undoubtedly offer a number of advantages, which include replacing time-consuming laborious tasks, improving design accuracy, making changes easy [Ross, 1995].

The application of these tools have had great impact on ship designs. First, many of the design tasks are carried out at great speed by the computer, therefore, in an earlier stage more information about the design is available [Excell, 1997]. As the laborious tasks have been removed, it opens up greater opportunities for intellectual exploration, and eases the process of making changes. Second, as today's powerful hardware and software can effectively model and fair a hullform precisely, the manual lofting process can be eliminated [Hays et al, 1994, Larsson et al, 1994]. Third, it is possible to view a perspective 3-D plot of the hull prior to the construction. Compared with the manual approach, where only three standard views are employed, this visualisation is considerably important, because hulls that look good in lines plan may look quite ugly in reality [Larsson et al, 1994]. Fourth, the computer networking allows a multi-user environment. This forms a huge database that may provide new ideas or opportunities for designers, as well as ease the selection of components such as engines, deck equipment and fittings. Fifth, the data generated during the design process can be tailored in format and content so that they can support the ship production process [Ross, 1995]. From an integrated CAD/CAM feedback program, potentially costly mistakes can be ironed out before manufacture. Thus, this results in a production-oriented design, which may reduce the cost and time of manufacturing [Excell, 1997]. Sixth, the digitisation of design information, instead of traditional engineering drawings, promotes interdisciplinary co-operation [Mulligan, 1997].

Many believe that reusing data in other computerised tasks will maximise the benefits of computerisation, and this is called *Integration* [Hays et al, 1994]. Thus, they regard the integrated ship design program as a compelling concept, and one whose time has come [Ross, 1995, Hays et al, 1994]. This notion is driven by economic aspects. As it involves enormous sum of cost to computerise any given task in the design and manufacturing process, it is difficult to justify the computerisation if that task is taken as a standalone item. As a result, there is a definite trend that to integrate the computer-aided ship design programs [Ross, 1995]. In this case, the ship design process may be enhanced through individual programs sharing their results with each other, preferably from a common database, which may be extended to computer-aid manufacturing (CAM). A few state-of-the-art integrated ship design programs have been developed, the representative ones include HULLTECH, AutoSHIP System, FORAN, HICADEC, IMSA, TRIBON, NAPA and NAVSEA CAD-2. In general, these programs can be categorised into two groups, where it is either integrated among modules, or by means of a product model [Ross, 1995]. Integration among the modules of the ship design process is the most common one; it means that the various programs of design tasks are designed to communicate data with one another to at least some extent. The HULLTECH, AutoSHIP and IMSA are examples of this group. The second one, a product model, shares an integrated database of various ship design programs, which means there is no need for data conversion among the modules. This is a more advanced level of integration. However, it suffers from inflexibility of future growth [Ross, 1995]; recent developments involve object-oriented system design which exhibits the concept of abstraction, encapsulation and inheritance. This results in simpler structures which are easier to maintain [Wu, 1994]. Either group, however, captures the features of the Ship Design Spiral concept.

This trend towards integration seems to be a revolutionary concept for the shipbuilding industry, as it may reduce the lead time and cost, as well as improve the quality of the products in general. Computervision Corporation even anticipates that there should be saving of around 25% on shipbuilding costs [Pullin, 1997]. Hence, there cannot be serious

876

opposition to the notion that an integrated system may enhance the shipbuilding industry. However, it would be interesting to find to what extend this new approach resolves the problems of the ship design process, and whether the remaining problems are of great significance.

The above integrated CAD approach, with its powerful computing capability, may minimise the problems of the ship design process, in terms of removing laborious tasks and making change easy. However, designers literally experience the same design process, as they did before the computer era, except that now they are provided with 3-D images and high speed calculations. In principle, they still adhere to the sequential nature of the Ship Design Spiral. Thus, this approach suffers from the consequences of the spiral approach, and therefore, still does not resolve the problems of the ship design process.

The significance of these problems is evidenced by various observations of designers. For a given specification, designers will require approximately the same number of iterations to converge to an acceptable design, no matter whether the programs are integrated or not. This is because in either case, the designers are lacking a global view of the design, thus the design is adjusted, on a trial and error basis, at respective design stages; this may well adversely alter other design characteristics, which is very annoying [Condylis, 1997, Simpson, 1997]. In fact, this can be avoided if they were provided with the global view of the design throughout the design process. Compared with previous computer-based tools, the above integrated one is unable to assist designers to attain at a solution much more rapidly, because the rate of converging towards a solution remains constant.

4 TOWARDS A NEW DESIGN APPROACH

As the existing approach does not resolve the problems of the ship design process, there is a need to establish a new approach to an integrated CAD system. The new approach must be able to provide the right information to the designers at the right time, in order to assist the designers to converge towards a solution quickly.

This approach is called the Concurrent Integrated CAD (CICAD) System. In general, it will be modelled and analysed by utilising I-CAM Definition (IDEF) techniques, developed by the US Air Force to perform modelling activities in support of enterprise integration [KBSI, 1995]. For the analysis of requirements, the IDEF∅ technique, which is a structured approach, will be employed to capture a static view of the functions performed by designers. This is to explore the possibility of executing the design process concurrently, and to establish the function model. Besides, it helps focus on a specific aspect within the system. Subsequently, the IDEF4 technique, which is an object-oriented approach, will be used in the system development process in order to incorporate desirable life cycle qualities such as modularity, maintainability and reusability. This paper deals with the function model only.

In essence, CICAD is developed by remodelling the sequential ship design process in a concurrent manner. Since each task is highly dependent upon the output of the preceding one, it is crucial, at this point, to examine to what extend the process can be overlapped. This section demonstrates how the sailing yacht design process could be overlapped, and estimates its significance. The yacht design process is selected because it captures the basic tenets of the Ship Design Spiral, and has a moderate complexity.

To begin with, each design stage is modelled in a top-down fashion by adopting the IDEFØ technique, which minimises the need for elaborate descriptive text, and its graphical presentation provides a clear representation of a complex aspect of organisation. The model is supported by the inputs, controls, outputs and mechanisms (ICOMs). As this technique is designated to perform the process planning, it must be modified in order to represent the actual stages of the Ship Design Spiral. Two crucial modifications were made. First, a closed loop structure was incorporated to account for the nature of the iterations in the Ship Design Spiral. Second, the controls and mechanisms were omitted, because they do not play an important role in determining the interdependency, otherwise, they might complicate the entire model.

The overview of the design model is shown in Figure 2. The top level is known as the 'environment', and is given a node number L0 (Level 0), this is to identify the inputs and outputs that cross the boundary of the model. To clearly identify the subject and extent of the model, a 'context' diagram node number L n_i (Level n_i) is established. One level down is the 'viewpoint', node number L n_im_i (Level n_im_i); this is used to establish each aspect of the 'context'.

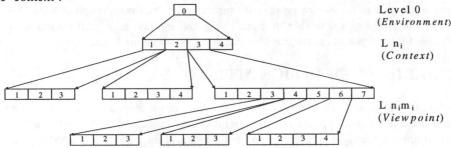

Fig. 2. Overview of the design model

The core diagram, L0 (see Figure 3) illustrates the generic description of sailing yacht design stages, which comprise of requirement development (L1), conceptual design (L2), preliminary design (L3) and detail design (L4). During the first stage, the design requirements are developed; it usually involves clients and designers, where the clients will try to elucidate their demands or wishes for the product. Those challenging requirements will then inspire the designers to generate many sketches in the conceptual design stage, where these initial solutions will be evaluated intuitively, in order to select suitable combinations. If the clients are satisfied with the combination of these potentially most promising solutions, then the preliminary design will be carried out to embody the ideas in a design model. The design is continuously adjusted to converge into a best possible solution. This is the part where the Ship Design Spiral concept is employed extensively, and many computer-based tools have been developed to support it. The result of the preliminary design will be used to bid the contract; therefore, it appears to be the most crucial part of the design process. The preliminary design is the key issue for a designer to be successful. If he does succeed in receiving the contact, then the detail design will be conducted. This involves the refinement of the preliminary design, and converts it into manufacturing instructions. For the requirement development, conceptual design and preliminary design stages, there are three closed loops respectively. The first two loops are intended to determine the viability of each corresponding stage, while the third loop decides the acquisition of the design contract.

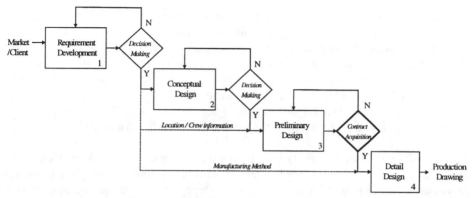

Fig. 3. Diagram Level 0, the sailing yacht design stages

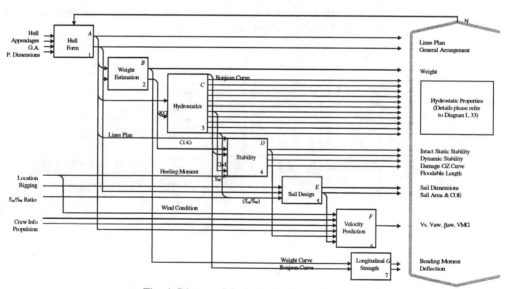

Fig. 4. Diagram L3, the preliminary design

The development of the new approach focus on the preliminary design stage, and its schematic diagram, L3 is shown in Figure 4. The reasons are twofold. First, this is the design stage that suffers most from the problems of the ship design spiral, whereas the conceptual and detail design stages do not. Second, the greatest chance of arriving at an optimum solution occurs during the preliminary design stage. This is because for sailing yacht designs, the conceptual designs do not have any significant meaning in term of performance, as they are usually in the form of sketches, which are approximate and immeasurable. Therefore, to achieve a best possible design, it requires the repetitive and subtle modifications throughout the preliminary design stage. We have concentrated on the preliminary design stage, because this will allow the designer to achieve an optimal design, and, this is the most effective way to improve the quality of a product, rather than through improving the organisational and production efficiency.

Diagram L3 indicates that the existing preliminary design is inevitable sequential, and it's order has been arranged in a fairly effective way. This can be ascertained by conducting the network analysis, which is known as the design structure matrix (DSM) [Smith et al, 1994]. The result shows that all the tasks can be completed sequentially without having to make any guess or assumption throughout the design process; thus, it appears to be a reasonably effective order. To a certain extend, this has justified the spiral concept of Evans, where he was attempting to assist designers in organising the thought process, so as to enable ship design problems to be solved more efficiently [Evans, 1959].

Tasks C, D and F (see Figure 4) have been expanded into Level $n_i m_i$, and are L33, L34 and L36 respectively (see Figures 5, 6 and 7). This is to provide an insight into elementary processes, in order to investigate the interdependency of each design task at a lower level. The diagrams L33, L34 and L36 show that certain design tasks do not actually require the output of the preceding one, but they do require some input from elsewhere. Moreover, a small numbers of tasks may even be carried out at once after the conceptual design. These two findings hint at the possibility of overlapping the whole design process.

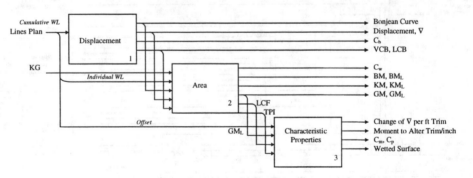

Fig. 5. Diagram L33 : Hydrostatics

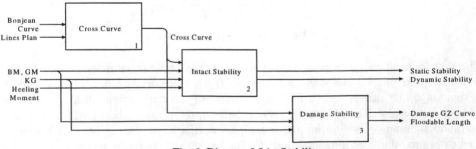

Fig. 6. Diagram L34 : Stability

880

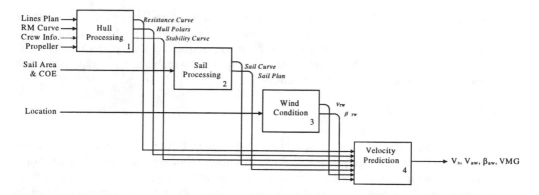

Fig. 7. Diagram L36 : Velocity Prediction

Assuming that the duration of task *A* is 6; task *B* is 2; task *C* is 2; task *D* is 3; task *E* is 2; task *F* is 3 and task *G* is 2 units of time, and the intervals to produce each output within a specified task are spread uniformly, an overlapped design process can be constructed (see Figure 8).

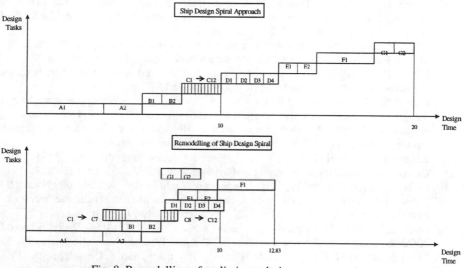

Fig. 8. Remodelling of preliminary design process

Figure 8 shows that the design process could be overlapped by up to 40%. Moreover, the right information becomes available to the designers at an average 30% earlier than in the spiral approach.

5 FUTURE WORK

Two aspects of the function model need to be investigated further prior to system development. First, the model requires refinement to ensure the acquisition of an accurate description of the problem situation. Second, the duration of each design task needs to be considered in terms of CPU and interactive times. This is to clarify where the significant improvements of the concurrent approach arise. Subsequently, the system development process will employ the IDEF4 technique to incorporate desirable life cycle qualities such as modularity, maintainability and reusability

6 CONCLUSION

This paper deduces that existing integrated CAD systems do not profoundly resolve the problems of the ship design process. Therefore, a new design approach, the CICAD, which overcomes the sequential nature of the Ship Design Spiral, is proposed. CICAD resolves the problems by integrating the design process concurrently; this advances the right information to the designers, although not exactly at the right time, it nevertheless would provide a better global view of the design.

REFERENCES

1. Condylis, T., 1997, Industrial collaboration : C & S Yacht Design, Southampton, UK.
2. Evans, J. H., 1959, Basic Design Concepts, Naval Engineers Journal, November, pp. 671 - 678.
3. Excell, J., 1997, Has CAD killed the designer ?, Design Engineering, UK, Vol. March, pp. 25.
4. Hassan, N. E. A., and Thoben, K., 1992, Practical Design of Ships and Mobile Units, Elsevier Applied Science, London, Vol. II, pp. 2.1284 - 2.1297.
5. Hays, B., and McNatt, T., 1994, The Integrated, Computer-based Ship Design Environment, Marine Engineering, Computational Mechanics Publications, Southampton, pp. 75 - 84.
6. KBSI, 1995, IDEF4 Object-oriented Design Method Report, Human Resources Directorate, Logistics Research Division, KBSI (Knowledge Based System Inc.).
7. Keane, R. G., and Tibbitts, B. F., 1996, A Revolution in Warship Design : Navy-Industry Integrated Product Teams, Journal of Ship Production, Vol. 12, No. 4, pp. 254 - 268.
8. Kroemker, M., 1997, Personal communication : Bremen Institute of Industrial Technology and Applied Work Science (BIBA), Germany, mk@biba.uni-bremen.de.
9. Kroemker, M., and Thoben, K. D., 1996, Re-engineering the Ship Pre-design Process, Computers In Industry, Elsevier Applied Science, London, Vol. 31, pp. 143 - 153.
10. Larsson, L., and Eliasson, R., 1994, Principles of Yacht Design, Adlard Coles Nautical, UK.
11. Lyon, T. M., and Mistree, F., 1985, A Computer-Based Method for the Preliminary Design of Ships, Journal of Ship Research, Vol. 29, No. 4, pp. 251 - 269.
12. Mistree, F., Smith, W., Bras, B. A., Allen, J. K., and Muster, D., 1990, Decision-based Design : A Contemporary Paradigm for Ship Design, SNAME Transactions, Vol. 98, pp. 565 - 597.
13. Mulligan, H., 1997, Computed Aided Drawing, Engineers Journal, Ireland, pp. 25 - 29.
14. Pullin, J., 1997, Shipshape and Computer Fashion, CIM Focus, London, Vol. Feb., pp. 15 - 16.
15. Rawson, K. J., and Tupper, E. C., 1966, Basic Ship Theory, Longman, London.
16. Ross, J. M., 1995, Integrated Ship Design and Its Role in Enhancing Ship Production, Journal of Ship Production, Vol. 11, No. 1, pp. 56 - 62.
17. Simpson, A., 1997, Industrial collaboration : Andrew Simpson & Associates, Poole, UK.
18. Smith, R., and Eppinger, S., 1994, A Predictive Model of Sequential Iteration in Engineering Design, MIT's Schools of Engineering and Management, Massachusetts.
19. Wu, B., 1994, Manufacturing Systems Design and Analysis, Chapman & Hall, London.

Product Analysis

Optimisation of Mould Design in Die Casting of Pewter Parts through Numerical Simulation

B.H. Hu, S.W. Hao, X.P. Niu, K.K. Tong and F.C. Yee

Gintic Institute of Manufacturing Technology
71 Nanyang Drive, Singapore 638075

Abstract. Mould design, especially the gating system design, plays a very important role in the die casting of high quality products. Numerical simulation was used for the evaluation and design of the gating and venting systems for the hot chamber die casting of a "merlion", a typical pewter part widely used for souvenirs in Singapore. It was found that the improper geometry and location of the gating and venting systems in the old design resulted in air being entrapped during both the earlier and later stages of mould filling which caused blisters in the casting. In addition, improper cavity numbers and plunger speed led to an excessive cavity filling time, and subsequently resulted in cold shut problems. The casting mould was re-designed with a new gating system, with the aid of CAE techniques. The mould filling process was simulated on the screen before physically cutting the die. The new design gave a more homogenous mould filling and an efficient air venting system was designed following the results of the simulation. Together with the reduction of the cavity filling time, the problems of blisters and cold shuts were solved and high quality pewter parts were achieved.

1 Introduction

Die casting is a high productivity process, in which molten metal is injected into a precisely dimensioned steel mould to produce a part by reproducing, with high accuracy, the finest detail of the impression of the mould cavity. Mould design, especially the gating system design, plays a very important role in the die casting of high quality products[1-4]. A gating system is a composite of the various channels cut in the die that guides the melt from the furnace to the cavities to form the final components. It dictates from which portion the part starts to be filled, the filling direction and speed, whether the flow is turbulent or non-turbulent, and where the overflows and vents should be cut. Finally, it contributes to the quality of the casting in areas such as gas porosity, shrinkage porosity, flow lines, cold shuts and surface finish [2,4,5].

Pewter, a tin based alloy, is typically suitable for die casting due to its low melting point and good fluidity. However, die casters still face problems in areas such as gas porosity and cold shuts during production.

2 Numerical Evaluation Based on an Old Mould

The "merlion", shown in Fig. 1, is a typical pewter product widely used for souvenirs in Singapore. Traditionally, it has been made by spin casting or cold pressing. However, due to the increasing demand for the products, a high productivity process is required. One such process that has been tried in the past was hot chamber die casting. However, it was found that air was entrapped and cold shuts existed within the castings produced(Fig. 2). Table 1 gives the general information for such a casting. The equipment used for this study was an old one phase hot chamber die casting machine. Some key parameters of the machine are shown in Table 2, and Table 3 shows the alloy and die information used for simulation.

The principle of the hot chamber die casting process is briefly illustrated in Fig. 3. The molten metal fills the shot sleeve chamber through the melt inlet. The plunger then moves downward to inject the melt through the gooseneck, nozzle and gating system into the casting cavity. The symmetrical half configuration and geometry of the old gating design and the venting locations of the mould, together with the castings, are shown in Fig. 4. The gate thickness is 0.8mm, about half the thickness of the "merlion" part. The total gate area is 48mm^2 for 10 "merlion" parts.

Table 1 General information of the casting

Alloy	Casting weight	Mould Material	Mean wall thickness	Overall Dimensions
Pewter (2%wtCu, 1.6%wtSb)	8.5g	Hot work steel	1.5mm	53x24x1.6

Table 2 The machine information

Tonnage	Plunger diameter	Hydraulic pressure	Plunger velocity
50 ton	50	100 bar	0.2m/s

Table 3 The alloy and die information

Density	Specific heat	Latent heat	Liquidus	Solidus	Die conduct.	Pouring temp.	Die temp.
7300 kg/m^3	222 J/kg K	7080 J/kg	220°C	180°C	25.6 w/m °C	300 °C	101 °C

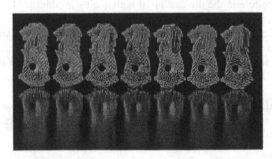

Fig. 1 "Merlion": A typical pewter part

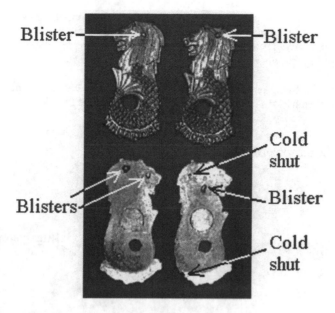

Fig. 2 Blisters in castings

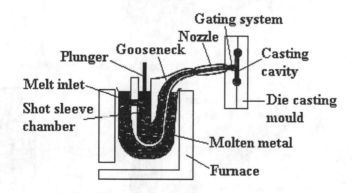

Fig.3 Principle of hot chamber die casting process

A commercial CAE package (MAGMAsoft) was used for numerical analysis. Figures 5 and 6 are selected simulation results (temperature distributions) of the mould filling at two different filling stages. The temperature of the melt is indicated by the grey scales. The grey colour at the bottom of the scale indicates the lowest temperature, and the white colour on the top of the scale the highest. The unfilled portion of the casting cavity is shown in light grey colour.

Fig. 5 shows the filling results at 30% cavity fill. It can be seen that as the liquid metal hits the basin of the sprue, air is entrapped in both the straight and the basin portions of the sprue. This is due to the increasing cross section of the sprue and the basin along the flow direction. Also, the fact that the air is entrapped inside the metal stream, and no possibility exists for it to be vented out of the cavity. Therefore, it will be brought further into the casting cavities during the later stage of filling. This will subsequently result in gas porosity or blisters in the casting.

Fig. 6 represents the filling pattern at 90% cavity fill. It can be seen that the melt fills the tail portion first and then back-fills the head of the "merlion". Since the overflows and air-vents in

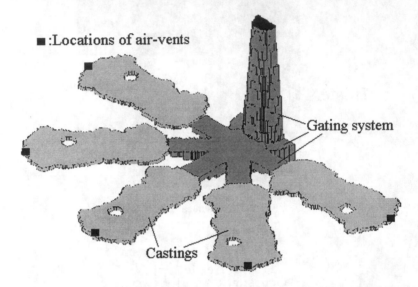

Fig. 4 Three dimensional layout of the casting gating system

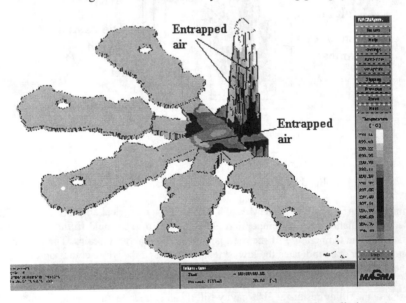

Fig. 5 a) Old die design: Mould filling (30% cavity fill) - Overall view

the old design are located at the tip of the tail instead of at the head of the "merlion", it is impossible to remove the entrapped air. This explains why blisters were often found at the head potion of the "merlion" (Fig. 2). The cavity filling time was calculated as 29.6ms, which seems to be excessive for such a small pewter part. Defects, such as cold shuts, were also found in the castings (Fig. 2).

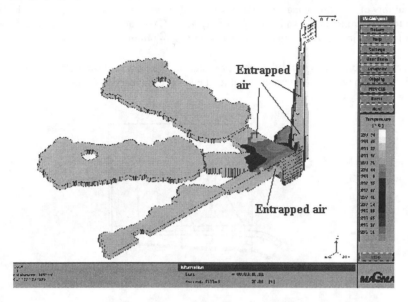

Fig. 5 b)Old die design: Mould filling (30% cavity fill) - Section view

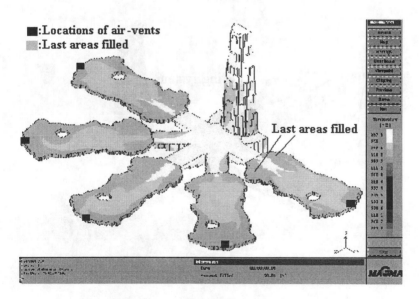

Fig. 6 Old die design: Mould filling (90% cavity fill)

3 New Design

Based on the above evaluation, the mould was re-designed with a new gating system with the aid of CASTFLOW, a computer aided gating design software. In order to get a more homogenous mould filling, a tapered runner system was introduced for this part. Considering the capability of the casting machine, the cavity numbers were reduced from ten to eight. A typical tapered runner is shown in Fig. 7. The layout (3-D) of the new symmetrical half gating system is shown in Fig. 8. The total gate area is 25.6mm^2 (4mm x 0.8mm x 8).

In order to demonstrate visibly the melt flow during the various filling stages and the melt solidification behaviour, to determine whether the new design needed further optimisation, a

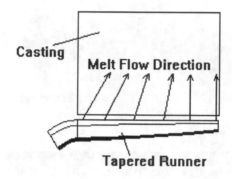

Fig. 7 Tapered runner

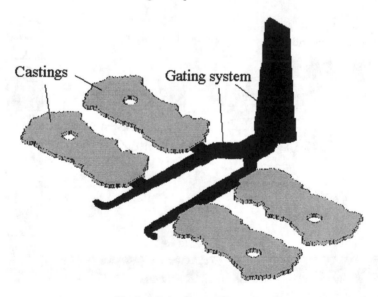

Fig 8. 3-D layout of the new runner
and gating system for "merlion" parts

simulated "shot test" was conducted using MAGMAsoft. The simulation used the same alloy and process parameters as those used for the old design, except that the plunger speed was increased to 0.3m/s.

Figures 9 and 10 show the simulation results (temperature distributions) of the mould filling at 30% and 90% cavity fill respectively.

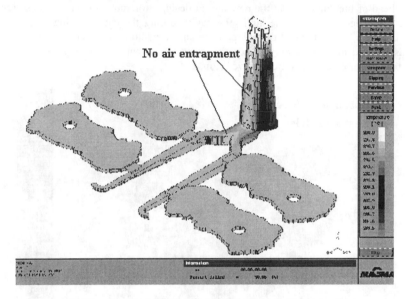

Fig. 9 a) New design: Mould filling (30% cavity full) - Overall view

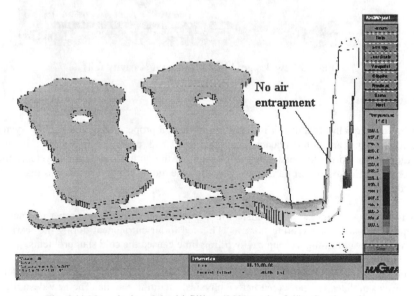

Fig. 9 b) New design: Mould filling (30% cavity full) - Section view

It can be seen from Figures 9 and 10 that, by using the new gating system, the melt steadily fills the sprue and runner without any gas entrapment.

With the new design, the melt also fills the tail-end of the "merlion" first, before back-filling the head portion. This is easily understood, from the geometry of the "merlion", irrespective of where the gate is located. Therefore, new air-venting channels are added to the last filled areas near the head of the "merlion" in the new mould design, to remove the air in these regions. However, in order to remove the air in the cavity during the earlier filling stages, the air venting channels at the tail end of each part remain. In addition, the calculated cavity filling time was reduced to 15.8ms. Further experiments verified that high quality castings without any blisters and/or cold shuts were achieved (Fig. 11).

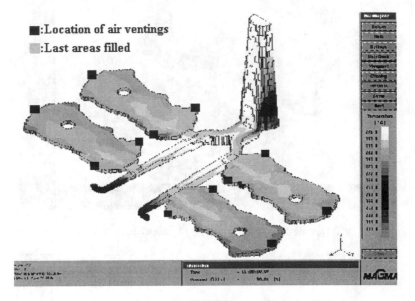

Fig. 10 New design: Mould filling (90% cavity full)

4 Conclusions

To secure good quality castings, it is important to have a proper gating and venting system in the mould to guarantee homogenous mould filling and efficient air venting. Numerical simulation was used for the evaluation and design of the gating and venting systems for hot chamber die casting of pewter parts, through which the mould filling process was made visible and predictable.

It was found that the cause of the blisters in the old design was the improper geometry and location of the gating and venting systems. This led to air entrapment during the earlier and later stages of mould filling. A long cavity filling time caused the cold shut problems.

With the aid of the CAE technique, a new gating system was designed. The mould filling process was simulated on the screen before physically cutting the die. The new design gave a more homogenous mould filling and an efficient air venting system was designed following the

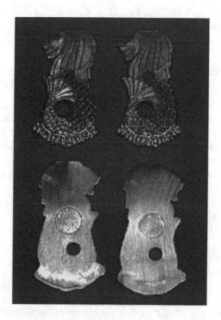

Fig. 11 Hot chamber die cast "Merlion" parts

results of the simulation. Together with the reduction of the cavity filling time, the problems of blisters and cold shuts were solved and high quality pewter parts were achieved.

Acknowledgements

This study is part of the work for the first phase of a project in collaboration with Phoenix Pewter-Phoenix Marketing Service, Singapore. The project was jointly funded by Phoenix Pewter and the Economic Development Board through the "Local Enterprise Technical Assistance Scheme". Special thanks are given to Mr. Kelly Tang, General Manager and Mr Tang Hon Wah, Technical Manager of Phoenix Pewter for their support and invaluable contributions to this project.

References

1. A. C. Street, *The Diecasting Book*, 2nd Edition, Portcullis Press Ltd (1986), 1986, pp.3-17
2. E.A. Herman, *Die Casting Dies: Design*, North American Die Casting Association (1992), pp.15-24
3. B.H.Hu, K.K.S.Tong, F.C. Yee, D. Sudheeran, and K.T. Tan, *Process Diagnosis and Optimisation in Pressure Die Casting*, Proceedings of the 3rd International Conference on Die & Mould Technology, Taipei, Taiwan, A (1995) 137-142
4. K.K.S.Tong, B.H.Hu and F.C. Yee, *Industrial Application of Computer Simulation In Mould Design for Pressure Die Casting*, Proceedings of International Conference on Mechanics of Solids and Materials Engineering, Singapore, A (1995) 235-240
5. T.H. Siauw and S. Tartaglia, *A Design Procedure for Feed Systems Incorporating Tapered Tangent Runners Used in Aluminium Pressure Die Casting*, Proceedings of the 2nd South Pacific Diecasting Congress, Sydney, (1983) 105-109

Fatigue Analysis based on FEA

Wolf-U. Zammert

Dept. Mechanical Engineering, Fachhochschule Technik Esslingen, Germany

Abstract. A PC software package is presented that enhances the performance of FEA-Systems to enable complete fatigue analysis. The program MAKE-FATIGUE is a software-system for general fatigue analysis of structural components, taking into account statistical effects in material behavior and load history. It incorporates the evaluation of fatigue life subject to deterministically or randomly distributed thermal and/or mechanical loads and the calculation of factors of safety due to periodically alternating loads with respect to the behavior in cases of overload. The several methods of fatigue analysis implemented in MAKE-FATIGUE are described and discussed by examples.

1   Introduction

It is the desire of every engineer to obtain a realistic view of the distribution of stresses in a component and to evaluate the consequences of fatigue life and factors of safety. The solution of this problem is the combination of FEA with numeric fatigue analysis.

FEA offers a complete view of the distribution of strains and stresses in a mechanical component. Few FEA-systems also offer a fatigue module for the calculation of usage factors. These fatigue modules however are not applicable for general analysis in the wide range of fatigue calculations. This task can be solved by using supplementary software like MAKE-FATIGUE. MAKE-FATIGUE is a PC software system for general fatigue analysis of structural components, taking into account statistical effects in material behavior and load history:

- evaluation of fatigue life subject to deterministically or randomly distributed mechanical loads,
- evaluation of fatigue life subject to deterministically or randomly distributed thermal loads,
- evaluation of fatigue life subject to a combination of deterministically or randomly distributed mechanical and thermal loads,
- calculation of factors of safety with respect to the behavior in cases of overload.

The input of stresses and load history can either be done manually or by using FEA results files. Structural elements like 2D-, 3D-solids and 3D-shells can be processed. Material properties are handled by files which are compatible with dBASE. The results of fatigue calculations due to an FEM calculation can be completely reviewed in the corresponding postprocessor. All features of the postprocessor can be applied. The MAKE-FATIGUE system consists of 2 modules: LIFETIME for evaluation of fatigue life and usage factors and DURABILITY for calculating factors of safety with respect to Goodman diagrams.

2   General Fatigue Analysis

2.1   Creep Fatigue Analysis

If structural components are being operated subject to elevated temperatures, the dependence of material properties upon temperatures has to be considered. With rising temperatures the strength decreases in general while the ductility increases. If the temperature stays below a limit T_K, which may be different from material to material, the temperature does not affect the microstructure of the material being deformed by the load. Temperatures exceeding the recrystallization temperature T_K induce a decrease of the load-dependent hardening of the microstructure: the material starts to creep. Material properties which have been measured in short-time tests at elevated temperatures e.g.

rupture strength and yield strength are only reliable, if the service temperatures are below the recristallization temperature of the corresponding material. Above the recristallization temperature the material strength not only depends upon temperatures it also depends on time. For fatigue design creep data are used, which show the dependence between temperature T_i, stress σ_i and time to rupture t_{Bi}, **Fig. 1**.

If a machine is operated at elevated temperatures subject to constant loads in a stationary working point, a constant creep fatigue acts on its components (e.g. thermal power plants). In general however, machines are working in different constant operating modes, e.g. full throttle - idle. In that case the components are exposed to a changing creep fatigue on different load levels. On every level a part of the initial creep fatigue strength is "consumed". **Fig. 2** shows as an example the load and temperature history of a pipe intersection that is stressed by a constant interior pressure. The superposition of the load and temperature history levels results in stress levels of the time t_i in which a constant temperature and a constant load act on the component. The collection of all stress levels of a component is called a stress spectrum. Stress spectra with different creep portions can occur in different ways. In general they result from a superposition of load and temperature history, but they can also occur by constant load and changing temperatures or constant temperatures and changing loads. In theses cases the stress spectrum consists of a load or temperature history only.

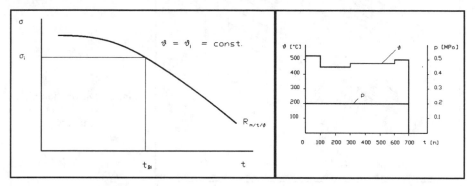

Fig. 1   Determination of time to rupture t_{Bi}
of the component on stress level i
due to service temperature ϑ_i

Fig. 2   Load and temperature history
of a pipe intersection

An assumption for creep fatigue is however, that the hold time on any level lasts sufficiently long. If the stress levels change too quickly, a superposition of creep fatigue and fatigue due to cyclic stresses should be taken into consideration.

Robinson and Taira [1, 2] assume for their creep fatigue damage rule that the creep strength of a component subjected to varying loads and temperature is exhausted, if the accumulation of the creep fatigue ratio has reached its limit. The creep fatigue ratio $D_{\vartheta i}$ is calculated for every stress level as the ratio of hold time at the level t_i to time to rupture t_{Bi}, see Fig. 1. This model leads to the linear damage accumulation rule of Robinson and Taira:

$$\sum_{i=1}^{n} \frac{t_i}{t_{Bi}} = D_\vartheta$$

D_ϑ	creep fatigue ratio
t_i	hold time on stress level i
t_{Bi}	time to rupture on level i
t_i/t_{Bi}	creep fatigue ratio on level i. (1)

If the creep fatigue ratio D_ϑ reaches the limit $D_\vartheta = 1$, the creep fatigue strength of the component is exhausted. Creep fatigue ratios which do not reach this limit can be endured $1/D_\vartheta$-times. The estimated time t_B until the component breaks down is therefore

$$t_B = \frac{\sum_{i=1}^{n} t_i}{D_{\vartheta}} = \frac{t_{ges}}{D_{\vartheta}}$$

Fig. 3 and **Fig. 4** show the results of the creep fatigue analysis of a pipe intersection subjected to varying temperatures and constant internal pressure according to Fig. 2. The distribution of usage factors is shown in Fig. 3. Fig. 4 includes the distribution of the estimated lifetimes in the structure. For reasons of better representation the contour values have been set to specific data by the /cval-command.

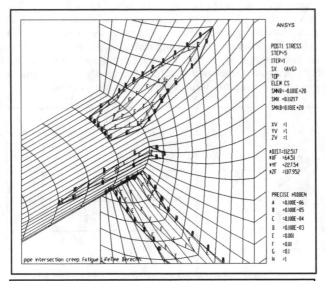

Fig. 3 Distribution of usage factors in the structure

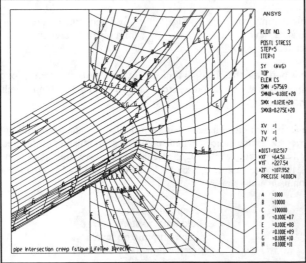

Fig. 4 Distribution of estimated lifetimes in the structure

2.2 Fatigue Analysis for Cyclic Loads

It is an experimental experience that σ-N-curves for steel and cast iron do not show any failure below a specific stress amplitude σ_A, even if the number of cycles is increased, **Fig. 5**. The stress amplitude $\sigma_{A(DFS)}$ at the failure limit of N_G cycles is the endurance stress amplitude. Plotting the endurance stress amplitude against the corresponding mean stress in the test leads to a Goodman diagram. In contrast to cubic volume-centered metals like steel or cast iron, face-centered metals like aluminum or magnesium alloys exhibit curves, which are apparently asymptotic to the N axis. For such materials, the stress $\sigma_{A(ZFS)}$ corresponding to some arbitrary number of cycles N_G is taken as the endurance limit.

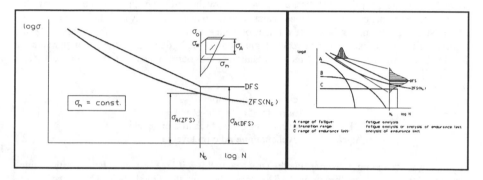

Fig. 5 σ-N-curves and Goodman diagram to σ-N-curve

Fig. 6 Position of stress spectra

Subdividing the σ-N-curves into the range of fatigue and endurance strength, the position of the cyclic stresses to the σ-N-curve determines the method of fatigue analysis, **Fig. 6**. Structural stresses according to spectrum A, which are mainly in the range of fatigue, demand a fatigue analysis on the basis of σ-N-curves whereas components, mainly stressed periodically with constant amplitude in the range of the endurance limit (spectrum C) are dimensioned against the endurance limit with Goodman diagrams. At first glance it is not evident which is the more appropriate method of fatigue analysis for components with stresses in the transition range between fatigue and endurance, because both methods seem to be applicable. Due to the relatively extensive amount of stresses in the range of the endurance limit the use of Goodman diagrams seems to be more plausible. The actual stress spectrum must either be transferred into an equivalent stress spectrum with constant amplitude or a statistical mean stress amplitude has to be determined for fatigue. analysis [3-5]. Because the precise experimental investigation of Goodman diagrams needs a large amount of σ-N-curves and therefore is too expensive and time consuming, it is usual, to replace the Goodman diagram by straight lines, **Fig. 7**, which render an approximation by using only few characteristic material data. The approximation of the Goodman diagram can easily be transferred to the Haigh or Goodman diagram.

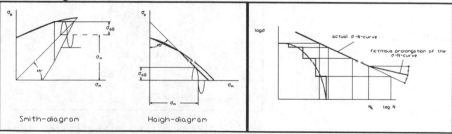

Fig. 7 Approximation of Goodman or Haigh diagrams by straight lines

Fig. 8 Stepped stress spectrum and σ-N-curve for life estimation

897

2.2.1 Fatigue Analysis for Randomly Distributed Cyclic Loads

The basis of life estimation in a fatigue analysis is the precise investigation of the load history to be expected in service. The load history causes a stress spectrum, which usually consists of a mixture of randomly distributed high and low stresses. The range of time, in which the load history is recorded by a sample, must include every relevant service load condition in its characteristic frequencies. Numerous counting methods have been developed to transform the recorded load or stress history into a load or stress spectrum. The usually continuous spectrum is then split into approximating steps. This stepped spectrum and the σ-N-curve supplies the starting point for a numeric life estimation. In the range of the endurance limit the σ-N-curve is modified for life estimation according to [6], so that stress amplitudes below the endurance limit can be included, **Fig. 8**. The component has an initial resistance against fatigue, which is partially consumed on every step of the stress spectrum. Let N_i be the number of cycles to failure on a stress level i and n_i be the number of cycles of that level, the cyclic fatigue ratio $D_i = n_i/N_i$ determines the partial damage on level i. The linear accumulation of the cyclic fatigue ratios corresponds with the Palmgren-Miner rule [7, 8].

$$\sum_{i=1}^{k} \frac{n_i}{N_i} = D_\sigma$$

D_σ cyclic fatigue ratio
n_i number of cycles on stress level i
N_i number of cycles to failure on level i
n_i/N_i cyclic fatigue ratio on level i.   (2)

According to the Palmgren-Miner rule the structure will fail, if $D_\sigma = 1$. Cyclic fatigue ratios which remain below the limit can be endured $1/D_\sigma$-times. The estimated time to failure N will be

$$N = \frac{\sum_{i=1}^{k} n_i}{D_\sigma} = \frac{N_{ges}}{D_\sigma}$$

Fig. 9 represents the stress amplitude spectrum of a pipe intersection. This load history can be due to deterministically or randomly distributed internal pressure. The resulting distribution of usage factors is shown in **Fig. 10**.

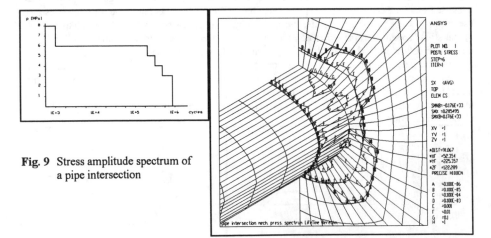

Fig. 9 Stress amplitude spectrum of a pipe intersection

Fig. 10 Distribution of usage factors in a pipe intersection subject to varying pressure amplitudes

898

From a statistical point of view, fatigue data show large scatter and therefore fatigue material data usually are mean values (50%), because the mean tendency can already be determined with a relatively small number of specimen. As a rule the failure probabilities needed in life estimations of structural components are clearly lower than 50%, so that the life estimations calculated on a 50% basis have to be converted into values corresponding to the required failure probability. Conversion will be done with a safety factor S, which includes the statistical consideration of scatter and number of specimen representing the material data. If in Eq. (2) N_i is regarded as the mean value N_{i50}, the appropriate value for the cyclic fatigue ratio considering the required failure probability will be $N_i = N_{i50}$ / S. Life estimations only make sense, if the statistical assumptions above are clearly quoted.

2.2.2 Analysis for Creep Fatigue Interaction

In process engineering, power plants or combustion engines etc., structural components are often operated in temperature ranges, in which creep fatigue has to be expected. If the resulting stresses include mean and cyclic stresses, the resulting damage will be produced by a combination of creep and cyclic fatigue. Both mechanisms of damage have to be considered in dimensioning the component. So the creep and cyclic fatigue stresses are not constant but vary according to the service point, they have to be recorded in appropriate spectra. The simplest method of accumulating creep and fatigue damage is the superposition of the Robinson-Taira and Palmgren-Miner rule. **Fig. 11** shows the scheme of linear damage accumulation using creep fatigue and σ-N-curves.

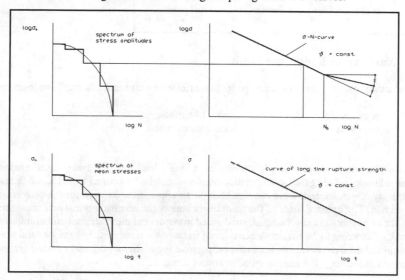

Fig. 11   Linear damage accumulation for creep fatigue interaction

The superposition of both linear accumulation rules leads to

$$\sum_{i=1}^{n} \frac{t_i}{T_{Bi}} + \sum_{i=1}^{k} \frac{n_i}{N_i} = D_{ges} \qquad (3)$$

Statistical aspects are now applied on the creep and fatigue data as mentioned above [9]. Fatigue design will be sufficient, if $D_{ges} < 1$. By using the method of linear superposition of creep and fatigue spectra it has to be kept in mind that the creep and fatigue ratios can only be compared with each other, if they are related to the same period of time. If the periods of time do not match, one fatigue ratio has to be converted to the period of time of the other. The creep fatigue interaction can then be endured $1/D_{ges}$-times.

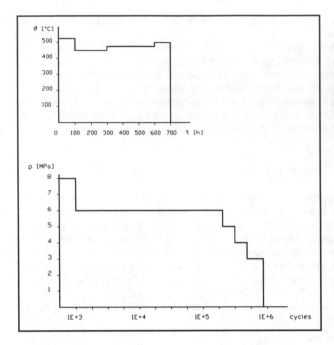

Fig. 12 represents the stress spectra for a combined creep and fatigue stress of the exemplary pipe intersection. The appropriate σ-N-curve will be selected for a mean temperature.

Fig. 12   Load spectra causing thermal and mechanical stresses

2.2.3 Analysis for Endurance Limit

The endurance limit of specimen and components is affected by numerous factors. The main factors are:

- type of load,   - quality of surface,
- mean stress,   - stress concentration,
- size.

Components, which are periodically stressed on a single load step are dimensioned against the endurance limit by reducing the Goodman diagram for unnotched specimen by reduction factors for size, quality of surface and stress concentration to an appropriate diagram for the corresponding cross section. The external loads of the component supply the nominal normal and shear stresses within the corresponding cross section, subdivided into nominal mean stresses and nominal stress amplitudes. According to the material's behavior of fracture the nominal mean stresses and nominal stress amplitudes are converted into nominal comparative mean stresses σ_{vm} and nominal comparative stress amplitude σ_{va}. The maximum comparative stress σ_{vo} is

$$\sigma_{vo} = \sigma_{vm} + \sigma_{va} \qquad (4)$$

In FEA stress concentrations have already been calculated, therefore the consideration of a stress concentration factor will not be applied in the Goodman diagram. In FEA the comparative stresses σ_{vm}, σ_{va} and σ_{vo} correspond with the maximum stresses at the edge of a notch or discontinuity. The upper maximum comparative stresses σ_{vo} characterize the operating stresses of a component in the reduced Goodman diagram. Due to an overloading of the structure these operating stresses will be changed approaching the limitation of the Goodman diagram. The distance between σ_{vo} and the limitation constitutes a certain safety margin against failure. Overloading the structure can be performed by several ways. The different ways of overloading are shown in **Fig. 13**. They have to be selected by the designing engineer in correspondence with the operating conditions and the design of the component. If overloading of the structure results in an increase of the stress amplitudes while the mean stresses remain unaffected, case A will be applied. The safety margin will be exhausted, if the comparative stress σ_{va} reaches the limitation of the reduced Goodman diagram, which is called the upper endurance limit σ_{oG}. If the comparative mean stress σ_{vm} and the comparative stress amplitude σ_{va} increase, case B has to be selected. σ_{oG} will be determined by drawing a line from the origin through the upper maximum comparative stress σ_{vo}. If the lower maximum comparative stress σ_{vu} remains constant and σ_{vo} increases, case C will be applied. C is similar to B. Only the starting point of the line is not the origin but the intersection between the 45°-line and a horizontal line at the level of σ_{vu}. The last case D will be selected, if the stress amplitude remains constant while the mean stress increases. Fig. 13 shows the dependency between the endurance limit and the case of overload.

The upper endurance limit σ_{oG} consists of an endurance strength amplitude σ_{GA} and a corresponding mean stress. σ_{oG} and σ_{GA} are used to calculate factors of safety. S_o is the factor of safety against the upper endurance limit σ_{oG}. S_a is the factor of safety against the endurance strength amplitude σ_{GA}. The are calculated according to Eq. 5:

$$S_o = \frac{\sigma_{oG}}{\sigma_{vo}} \quad \text{and} \quad S_a = \frac{\sigma_{GA}}{\dot{\sigma}_{va}} \tag{5}$$

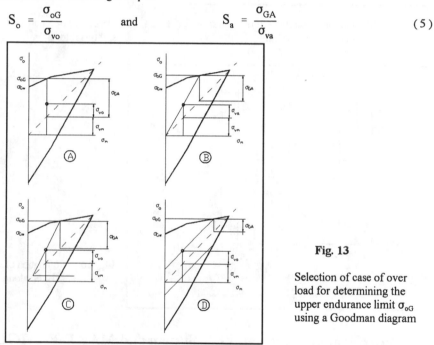

Fig. 13

Selection of case of over load for determining the upper endurance limit σ_{oG} using a Goodman diagram

Fig. 14 represents the distribution of safety factors against the upper endurance limit for a pipe intersection subjected to periodically alternating pressure between 3 and 9 MPa. The mean pressure therefore is 6 MPa and the amplitude pressure 3 MPa. The safety factors are shown at the top surface of the structure being modelled with shell elements.

Fig. 15 shows the definition of a path, along which the tendency of the safety factor against the upper endurance limit can be observed. The resulting values along the paths for the top- and bottom-position of the shell elements are represented in **Fig. 16**.

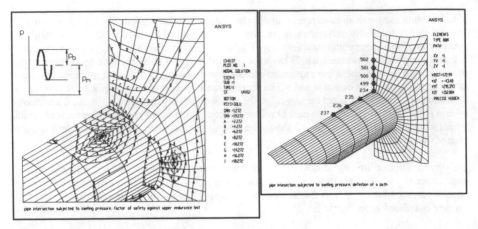

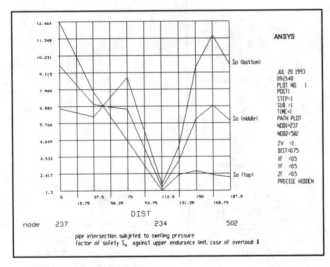

Fig. 14 Distribution of safety factors against the upper endurance limit

Fig. 15 Definition of a path for safety factors

Fig. 16 Variation of safety factors along a path

3 Interfacing between an FEA-System and MAKE-FATIGUE

The stress spectra have to be defined as load steps in the FEA-system. The following remarks illustrate the interfacing between the FEA-system ANSYS® and MAKE-FATIGUE.

Interfacing between ANSYS Rev. 5.X and MAKE-FATIGUE uses temporary files, as shown in **Fig. 17** for an analysis of endurance limits. The structure will be modelled as usual and the load steps being defined in the solve-module. Starting the MAKE-FATIGUE system first initiates a preprocessor via a batch job which reads the stress tensor from file.rst and calculates the von Mises stresses at every node of the elements. These results are written to a temporary file from which MAKE-FATIGUE reads its input. MAKE-FATIGUE performs a fatigue analysis and writes the results to another temporary file from which a postprocessor makes the ANSYS Rev. 5.X pseudo-results file lifetime.rst for fatigue analysis or dauer.rst for the analysis of endurance limits.

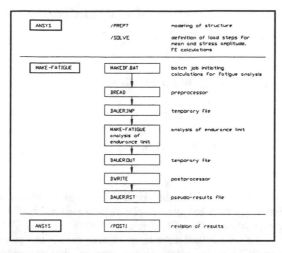

Fig. 17 Interfacing between ANSYS Rev. 5 and MA-KE-FATIGUE

The results of the fatigue analysis are fatigue ratios (usage factors) and estimated lifetimes. They are treated as "stress-values" SX and SY in the pseudo-results file. The factors of safety S_o against the upper endurance limit σ_{oG} and S_a against the endurance strength amplitude σ_{GA} are processed in the same way in an analysis of endurance limits for further reviewing in the ANSYS® postprocessor POST1.

4 Literature

[1]   Robinson, C.L., Effect of Temperature Variations on the long-time Rupture Strength of Steels. Trans. ASME, 74 (1952), No. 5, p. 777-781.

[2]   Taira, S., Lifetime of Structures Subjected to Varying Load and Temperature. Colloquium on Creep in Structures, Stanford University (California), Int. Union of Theoretical and Applied Mechanics, 1960.

[3]   Kececioglu, D.; Haugen, E.B., A unified Look at Design Safety Factors, Margins and Measures of Reliability. Proceed. of 7th Reliability and Maintainability Conference 1968, p. 520 - 531.

[4]   My Dao Thien, N.; Massoud, M., On the Relation between the Factor of Safety and Reliability. Trans. ASME, Series B, J. of Engng. for Industry 96 (1974) 3, S. 853 - 857.

[5]   Haugen, E.B., Probabilistic Approaches to Design. London, New York, Sidney: J. Wiley & Sons 1968.

[6]   Haibach, E., Beurteilung der Zuverlässigkeit schwingbeanspruchter Bauteile. Luft-fahrttechnik-Raumfahrttechnik 13 (1967), No. 8, p. 188 - 193.

[7]   Palmgren, A., Die Lebensdauer von Kugellagern. VDI-Z. 58 (1924), p. 339 - 341.

[8]   Miner, M.A., Cumulative Damage in Fatigue. J. of Appl. Mech., Trans. ASME 12 (1945), p. A159 - A164.

[9]   Zammert, W.-U., Betriebsfestigkeitsberechnung. Braunschweig, Wiesbaden: Vieweg 1985.

The Design and Implementation of a Visualization Tool for Structural Analysis

Yujin Hu, Jun Wu and Ji Zhou

CAD Center
Huazhong University of Science and Technology
Wuhan, Hubei, P. R. China

Abstract. This paper presents a finite element visualization facility, FEVS(Finite Element Visualization System), which has been implemented based upon PCs with Window95, NT and a graphic library --OpenGL. The main features of the FEVS are interactively easy operation, efficiency and robustness of algorithms, and powerful graphic processing functions. The design ideas, system structure, data management, interactive manner and various methods of finite element results processing and displaying are discussed in detail in this paper. Hierarchical data structure and algorithms for fast generating and displaying the contours, isosurface and cross sections using volume rendering are also proposed.

1. Introduction

The finite element(FE) analysis has become a very important tool in structural analysis as a result of the increase in computer power and the progress in numerical computation technologies. However, it produces a large amount of 3D unstructured data. To be able to interpret the data and understand the simulated process the scientists and engineers require visualization tools to be capable of extracting the interesting information from the data and displaying it in an easy comprehensive form.

FEVS is developed through analyzing the practical needs in engineering applications and adopting many new algorithms in visualization field. Since many advanced techniques in computer technology such as OpenGL and MFC programming are used, the system provides a user-friendly interface and many effective ways to display the analysis results. User can first use directly interface to analysis program to input the necessary data(including file input), then inquire the geometrical or topological or physical information, and can process static and dynamic data such as stress, strain and displacement etc., and can analyze or display interesting results in many easy comprehensive form. In this paper we will concentrate to discuss the design ideas, system structure, data management and main algorithms of FEVS. Figures shown in this paper were originally generated in color but are displayed in the paper as shades of gray.

2. System Function and Structure

FEVS is based upon Wondow95, NT and OpenGL environment, it provides a user-friendly graphical user interface and extensive graphical and numerical tools for gaining an understanding of analysis results. Users can be easy not only to operate by menu or commandlines but also to operate by multil-windows. Fig. 1 shows the interface of FEVS. In order to be maintained and further expanded in the future the system is designed into a modular structure. According to the functionality the system is divided into five main modules shown in Fig. 2. Each main module is also subdivided into many small modules.

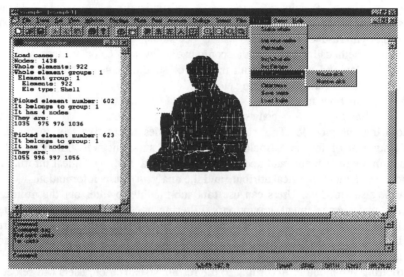

Fig. 1 The interface of FEVS

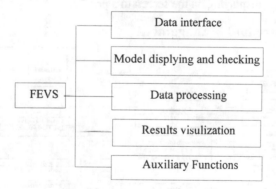

Fig. 2 System Structure for FEVS

A. Data interface module

Data interface module is data exchange module with popular FE analysis program. FEVS provides the interface to ADINA, ALGOR and ANSYS etc.. It supports many element types from FE analysis program. such as 2 to 3 nodes truss, 2 to 3 nodes beam, 3 nodes plate, 3 to 16 nodes shell, 4 to 27 nodes solid element and all plane element etc.

B. Model displaying and checking module

This module is used to display or check FE model in many easy comprehensive form , such as mesh, hidden, shrink, light shading and transparent model etc. users can rotate, pan and zoom model to make it at the best position and the best size.

C. Data processing module

This module is mainly used to extract the data that users interested from the basic output results of the FE analysis program. Also, it can compute new data types from the basic ones such as Von Mises stress and bending stress etc.

D.   Results   visualization   module

This module is the core of the whole system. In order for user to easily understand the simulation results or make design decisions, this module provides many visualization methods for user, such as contour display with color or meshes, deformed geometry display and animation with meshes or shading colors, criterion display for given value and 2D function graphing etc..

E. Auxiliary functions module

It   mainly   consists   of   three   sub-modules:

a. inquiring sub-module: This sub-module provides interactive tool to inquire the model's summary information(including number of sub-structures, number of element groups, number of elements and nodes etc.), geometrical and topological information of each node or element, physical attribute and the analysis results information.

b. filtering sub-module: Users can use this sub-module to filter out the uninterested elements, nodes or element groups etc. Filtering methods include color filtering, geometrical filtering, topological filtering, physical attribute filtering, analysis results range filtering etc.. After the filtering, users can concentrate on the most interested part of the model.

c. outputting sub-module: According to the results visualization, users can select some interested information to generate a report file and output it. Users can also output every virtual images displaying on the screen to a printer.

3. Data Structure and Management

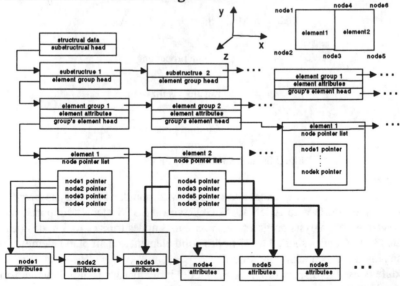

Fig. 3 Data Structure

In FEVS, the input information from FE analysis program can be classified into geometrical data(node coordinates etc.), topological data(element and element group etc.), physical attribute data(material, plate thickness etc.), load and constraint data, analysis results(node displacement, stress, temperature etc.) For storing and accessing these data more efficiently, FEVS uses Object-Oriented Data Base to manage them. In the computer the data relations are presented in Fig. 3(assume every element in the first element group have only four nodes like element1 and element2).

The table 1 shows the information included in the element group structure and node structure.

Table 1 structures for the element group and node

element group structure	node structure
element type	world coordinates x,y,z
number of elements	node displacement $\Delta x, \Delta y, \Delta z, \Delta\theta x, \Delta\theta y, \Delta\theta z$
node number in a element	node stress $\sigma x, \sigma y, \sigma z, \tau x, \tau y, \tau z$
material	node load Fx,Fy,Fz,Mx,My,Mz
layer	node constraints Cx,Cy,Cz,Cθx,Cθy,Cθz
displaying color	node temperature t

4. Visualization Methods

4.1. Visualization of Displacement and Modes of Vibration

The method of displaying model's displacement is putting the undeformed model together with the deformed one in different colors. So users can see the model's deformed shape and each node's displacement clearly as shown Fig.4(left-window shows the hidden model, right-window shows deformed and undeformed meshes in different colors)

Fig.4 Deformed and undeformed meshes in different colors

FEVS can also simulate the model's deforming procedure in two ways according the result data provided by the FE analysis program. If the result data give only nodes' undeformed and deformed positions without intermediate positions, FEVS will select a suitable(For example sin(x)) interpolation function and a number of interpolation, then calculate out a series of nodes' intermediate positions from the undeformed position to deformed position, finally display the model according these intermediate positions in a sequence. Thus a simulation of the model's deforming procedure is achieved. In this method, all the intermediate positions are computed out according to the interpolation function. If the interpolation function does not fit the actual deforming procedure, the simulation is inappropriate. To get a more accurate simulation of the deforming procedure, it is necessary for analysis program to provide the intermediate positions in every given time interval instead of interpolation. This method is called deforming history animation, and it need more extra data.

There is vibration in the period of deforming. Vibration has many frequencies, and every frequency reflects a mode of vibration. FEVS uses static model or simulation by interpolation to view a mode of vibration as the same way described before.

4.2 Data Visualization

A.   data   processing

The FE analysis programs output basic data types such as displacement components $(\Delta x, \Delta y, \Delta z, \Delta \theta x, \Delta \theta y, \Delta \theta z)$, stress components $(\sigma x, \sigma y, \sigma z, \tau x, \tau y, \tau z)$, node temperature (t) etc. data processing is to extract one of them or compute a new data type from them shown as Fig. 5

processed data field

extracting   processing

stress   displacement temperature Von Mises Tresca bending accuracy
component component   stress   stress stress   stress

Fig. 5 Data Processing

Table 2. Methods of visualization

methods	implementation	part of data field described
2D function graphing	various ways of selecting nodes	node selected
Deformed display and animation	deformed geometry display and animation of displacement or of time varying results	part or whole geometry model
Criterion display		
isolines	number of isolines can be set	surface of the data field
contour	grade shading smooth shading	
isosurface	shown by colored mesh light shading on the model surface shown in transparency	whole data field
cross section through model	light shading in the model surface shown in transparency	part of interior data field
isosurface in volume rending	light integration	whole data field

After the data processing, every node has a correspondent scalar value, so the model become a unstructured scalar data field.

908

B. Visualization of the processed data field

As shown on table 2, FEVS provides many methods to visualize the data field from the surface of the model to the interior of the model, from the local data field to the whole data field.

X-Y plotting is used to show the data value variation over a series of nodes selected as shown in Fig. 6.

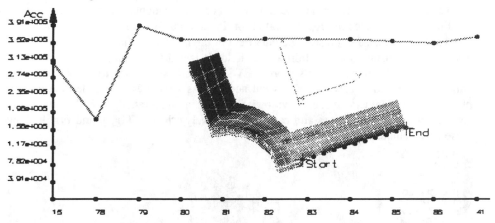

Fig. 6 X-Y plotting of nodes' Von Mises stress variation

In the same way, FEVS can use X-Y plotting to represent how a variable changes over time.

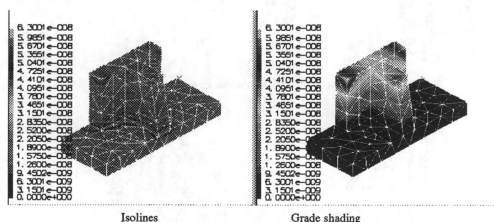

Isolines   Grade shading

Fig. 7 Isolines and grade shading

Typically, FEVS uses isoline and grade shading color to present the data value on model's surface. The method to display the isoline and shading color has following steps: first find the data value range, then divide the range according to the isoline number set by the user and get many small value section $\Delta v_i = v_i - v_{i-1}$. We assign a uniform color to each Δv_i, the principle is the smaller the value, the darker the color, the bigger the value, the brighter the color. Then use the algorithm described in Section 4, we get a clear isoline or grade shading color representation of the data value on model surface. The area that has the brightest isoline or the color block is the location of the biggest data value. The relation between the color and the data value is also shown at the side of the window as shown in Fig. 7.

If we set the color number to a big one, for example, bigger than 100, the color blocks are so small that grade shading color become a smooth shading. The smooth shading is a more accurate representation of the continuous data field. Isosurfaces and cross section are used to present the data field in the interior of the model. Isosurface is a surface of constant value within a data field. It is a three-dimensional extension of two-dimensional isoline. What's more, FEVS can also use a block to represent a data value range which is between two isosurfaces. Cross section is a user defined cutting plane on which data value are shown by isoline or shading method. In FEVS, the plane can be defined in many ways and can be interactively oriented. Fig. 8 and Fig. 9 show these two methods.

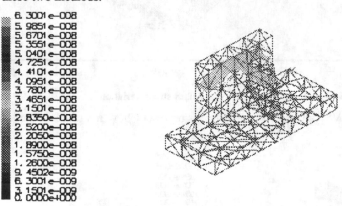

Fig. 8 Isosurface show in surface mesh(value=0.5*maxvalue)

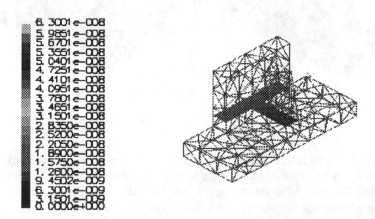

Fig. 9 Cross section shown in surface mesh

910

There are three ways of showing the isosurfaces(cross section). a. model and isosurface(or cross section) shown in different mesh colors. b. shading the isosurface(or cross section) in the model's surface shown in Fig.8(Fig.9). c. regard the model as a transparent object and isosurface(or cross section) a opaque one, show the isosurface in transparency.

Method of isosurfaces showing can only present the data value in the isosurface and the rest of the data field is still unknown. By using volume rendering technique we can get the whole data field's information at the same time. In this method, the isosurface is regard opaque, but the rest of the data field is semi-transparent, which emits light with a color reflect its value. By integrating the light emitted by them, we get a whole understanding of the data field. Fig. 10 is a example of this method. Readers can compare Fig.8 and Fig.10 to see the advantage of this method.

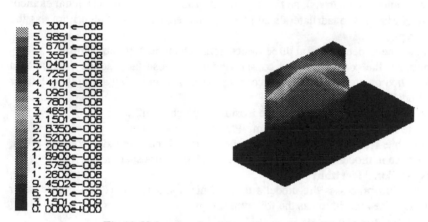

Fig.10 Volume rendered isosurface(value=0.5*maxvalue)

5. Main Algorithms of FEVS

In this section, we will discuss four main algorithms in the system. They are isoline generation algorithm, isosurface generation algorithm, cross section algorithm, isosurface generation using volume rendering method. In these algorithms, we assume the data value are linear in the edges of a element.

5.1　　　　　　Isoline　　　　　　Generation　　　　　　Algorithm

This algorithm can be applied to the model consisting of solid elements(hexahedron, prism, tetrahedron), plate elements and shell elements. The algorithm has the following steps:

a. find the model's visible surfaces and translate these surfaces to triangles or quadrilaterals.

b. set the number of isolines n, find the data value range v_{min} and v_{max}. calculate the corresponding data value of isoline $v_i = v_{min} + (v_{max} - v_{min}) * i / n$, ($i = 1,2,3,...n$) and assign each value a color according to the colormap defined before.

c. divide each quadrilateral into four triangles by interpolate a point in its shape center.

d. assume the current isoline value is v, current triangle is $\triangle ABC$ and the values of

A,B,C have following relations: $v(A) \le v(B) \le v(C)$.Then if $v(A) \le v \le v(C)$, we can be certain that the v-valued isoline lies between the edge AC and one of the other two edges AB or BC. Similarly, if $v(A) \le v \le v(B)$, the other end of the isoline lies in edge AB. if $v(B) \le v \le v(C)$, it lies in edge BC. Calculate out the positions of the isoline ends by linear interpolation and connect them in the color correspondent to v.

e. for every isoline value v_i, do step d and draw the isolines in the triangle with correspondent colors.

f. for every visible surface triangle do step d and e, we get the model's all isolines .

5.2. Isosurface Generation Algorithm

Since isosurfaces exist only in three-dimensional solid elements(tetrahedrons, pentahedrons and hexahedrons), so this algorithm ignores the two-dimensional element types such as triangles, quadrilaterals and beam, truss elements. The algorithm has the following steps:

a. divide all the nonetetrahedral three-dimensional elements into tetrahedrons.

b. to maintain the correctness and consistency of the isosurface, adjust the diagonal orientation on ambiguous faces according to the isovalue using asymptotic decider method[1].

c. process each tetrahedron and get the isosurface patch in it[2].

d. calculate the light shading effect using Phone lighting method.

e. display the isosurface according to one of the three methods: colored mesh, isosurface in model's surface mesh and isosurface in transparency.

5.3 Cross Section Algorithm

This algorithm processes the model's three-dimensional elements and ignore other element types. The algorithm has the following steps:

a. get a plane defined by users.

b. get the surface of the model using hidden method.

c. construct the topological relations among the surface faces.

d. calculate the intersection points of the given plane and the surface faces.

e. connect the intersection points in a loop according to the topology of the surface faces and decide if the loop is a outer loop or inner loop.

f. calculate the intersection points of the given plane and the elements in the model.

g. triangularize the cross section plane using Delaunary method[3].

h. draw the isolines or shading colors in the cross section according the values in the intersection points.

5.4. Isosurface Generated by Volume Rendering

Similarly, this algorithm processes the model's three-dimensional elements and ignore other element types. Our algorithm is similar with the ray casting method [4] in which a ray is traced through the cells, the intersection of the ray with each cell is computed. The algorithm has following steps:

a. get a isosurface value input by users.

b. calculate the box of the model projection in the screen.

c. for every pixel in the box, we do[5]

1. a ray is cast to the model and if the ray intersect with a element a line section is obtained. Assume we get N line sections in the ray path.

2. calculate the color and opacity of every line section. If in the line section there is a point whose value equal to isosurface value, set its pacity=1.0;

3. sort the N ray line sections from near to far from the viewpoint.

4. integrate the N ray line section according to formulation below:

$$C_{acc} = (1 - O_{acc)}) * C_{new} + C_{acc}$$

$$O_{acc} = (1 - O_{acc}) * O_{new} + O_{acc}$$

if $O_{acc} \geq 1.0$, stop integration.

5. calculate the shading effect on the pixel whose ray intersect with the isosurface.

6. output the color in the pixel.

5. Conclusion

In the present work, the visualization tool, FEVS, has been developed based on PC with window95, NT and OpenGL environment using MFC. FEVS provides a user-friendly interface and extensive graphical and numerical tools for gaining an understanding of analysis results. Data processing and data inquiring provides powerful data manipulation capabilities for user. Direct interfaces to more analysis program are available.

In the future, the visualization for vector and tensor fields as well as streamline will be added into FEVS.

References

1.   Yu Hongqian, Hu Yujin, Zhou Ji, Isosurface Generation in Finite Element Models. in: Proc. of the First International Conference on Engineering Computation and Computer Simulation. Changsha.1995

2.   Doi A, Koide A. An Efficient Method of Triangulation Equi-Valued Surface by Using Tetrahedral Cells. IEICE TRANSACTION, 1991,E74(1)

3.   George P L.Automatic Mesh Generation: Applications to Finite Element Methods. New York: Willy, 1991

4.   Levoy. M. Display of Surfaces from Volume Data, IEEE CG&A, 8(5),1988

5.   Wen sili, Tang weiqing, Liu shenquan. Direct Integration in Volume Advances in Computer Aided Design & Computer Graphics, Vol(2),1993

Process Planning

CathViewer: An Interactive System
For Catheter Design and Testing

Wang Yaoping, Chui Cheekong, Cai Yiyu, Lim Honglip

Institute of Systems Science
National University of Singapore
Heng Mui Keng Terrace, Kent Ridge 119597
Republic of Singapore
Tel: (65)7726724, Fax: (65)7744990
Email: {ypwang l cheekong l yycai l hllim} @iss.nus.sg

Abstract. This paper reports on a comprehensive, interactive simulation system of design and testing for catheter designers and manufacturers. The system features a physical computation model of finite element method built with specific biomedical knowledge. A 3D vascular structure which is segmented mostly from the Visible Human Data is used in the system as a realistic testing environment. In additional to conventional tools available in CAD systems, an easy-to-use catheter shape editor is also provided for manipulating catheter geometry. Design Engineers can freely design new catheters or modify the existed ones and then insert them into any position of the vascular structures for various purposes of testing. The navigation and deformation of the catheter and vascular structures in their interaction are computed real time using a customised Finite Element Modelling and Analysis module. The elastic property of catheter can be pre-defined or modified interactively during the navigation procedure.

1 Introduction

Minimally invasive therapeutic procedures, including surgery and interventional radiology, reduce patient discomfort, hospital stay, and medical costs. The socioeconomic impact of compensation for lost work time is also reduced. But interventional radiologists often face critical uncertainties in predicting the outcome of the procedures they plan to perform. More objective assessment of the equipment used in the procedures and better understanding of the treatment mechanism are also needed[1,2]. However, the evaluation of equipment relating to interventional procedures often depends on opinions of various operators. These may be accompanied by personal bias and experience, leading to differing views. In addition, the mechanism of treatment of vascular disease also remains unclear. Although several hypotheses have been put forward, majority of these studies were based on post-mortem examination. Different plaque morphology may respond differently. Moreover, patients also respond to balloon injury or various other devices differently. Generating a model to reproduce the varied scenarios will facilitate greater appreciation of these processes, thereby providing new information for risk assessment of patients and for the development of new devices. Developing a system to simulate the process not only provides a better understanding of the behaviour of these equipment but also allow assessment to be more objective. These factors will enhance modification and further development of the equipment.

However, conventional design systems/tools for designing catheter and other associated interventional devices are unable to address these issues properly. Engineers in catheter manufactures usually apply the existed CAD systems with no or very little direct medical knowledge for their catheter designs. They select the shapes for the catheter hooks and choose the material properties for the head, transition joint and body based on their past experience and some feedback from physicians practising in catheterization. Furthermore, they can very unlikely have an opportunity to validate the design concept themselves at a realistic checking environment. And the validation process is considered as the utmost crucial step in any design procedure[3].

Cathviewer is a joint R&D project involved with medical professionals, mechanical engineers and computer scientists to develop an interactive system for designing and testing catheter and other interventional devices in a medical knowledge based environment. An intuitive drawing tool is provided for engineers to draw, display and modify catheters very easily. A catheter library with all Cordis catheters [4] is also available for designer's reference. A comprehensive vasculature of human blood vessel from Visible Human data is employed in the system for providing the realistic testing bed for the newly designed catheters. An accurate biophysical model allow designers to realistically simulate, on the computer, the interaction between the vascular system and the interventional devices (the catheter, guidewire and various devices such as balloons, stents, etc.) through navigation manipulations, such as pushing, pulling and twisting the devices. The ultimate goal of this project is to develop an accurate and realistic Interventional Simulator [5] to simulate interventional procedures. **CathViewer** for Catheterising Viewer, as a part of this Interventional Simulator, emphasises on the development of new catheters. We will incorporate this model into the Interventional Simulator and deliver an invaluable tool for objective evaluation of the equipment, for development of new equipment, and for pretreatment planning in complex patient cases.

2 CathViewer Architecture

The design and implementation of CathViewer have been influenced by the following considerations.

- Incorporation of models. The model-based approach is followed. Physical models are used to represent the catheter and vascular systems.
- Real-time performance. The run-time system should avoid using time-consuming search and memory-management techniques and should provide predictable response times.
- Accurate vascular system. The credibility of the system relies heavily on the accuracy of the modelled vascular structure. Vascular structures have to be re-constructed from real human data, or professional model.
- An user-friendly drawing tool, good file management and comprehensive catheter library and material library.

The architecture of CathViewer is divided into several components, the Vascular Model Preparation Shell, Drawing Shell, Material Properties Manipulation Shell, and Finite Element Computation Shell. Fig. 1 shows the interaction between the various components.

Users can interact with the system through Vascular Model Preparation Shell and Equipment Manipulation Interface. The Vascular Model Preparation Shell represents a series of techniques we have exploited to develop accurate digital models to be used in the system. The digital model is refined in an iterative manner to improve its accuracy in modelling the required structure. The Equipment Manipulation Interface contains Drawing Shell, Material Properties Manipulation Shell and Navigation Shell. Drawing Shell is used to create and modify the shape of the interventional equipment through a graphic user interface. Material Properties Manipulation Shell deals with physical properties of the interventional equipment. Navigation allows the interventional equipment created/modified from the Drawing Shell and Material Properties Manipulation Shell to be tested on the vascular model from the Vascular Model Preparation Model. The 3D Rendering and Visualisation Shell provides the graphical and visualisation routines required to view the modelled interventional equipment and vascular structures. The Finite Element Computation Shell provides the numerical computation engine to analyse effect of forces on interventional equipment and simulate its navigation in the vascular structures.

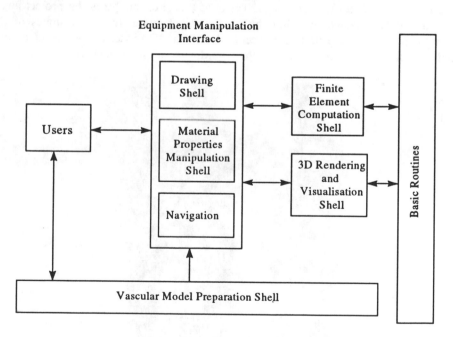

Fig. 1 Architecture of CathViewer.

919

3 Vascular Model Preparation Shell

Aiming to provide a realistic checking and validating environment for catheter design, we use human primary vasculature, cardiac and other secondary tertiary networks in Cathviewer as a virtual patient for the testing purpose. The primary arterial vasculature is from the VHD™ data, the secondary tertiary is from other scanned human vasculature data while the cardiac vessel network is from a professional steel model[6]. The vascular anatomy is represented by the central lines and their corresponding radii of the vessels. These vessels are assumed in the existing version of Cathviewer as cylindrical structures with varied radii. And these structures are treated as rigid.

The primary arterial vasculature and some associated secondary tertiary networks are shown in Figure 2. The extraction of the co-ordinates of the central lines and associated radii of the vessels is achived by applying semi-automatic and manual volume segmentation methods to the VH cyro-section data and the registered CT data[7]. The VHD™ data is stored in numerous 2D slices of uniform thickness of 1mm. We first manually segment the contours of the blood vessels concerned at each slice by using Photoshop, a commercial program. Then the centre of the projected cross-section of a vessel is calculated and the resultant co-ordinates of this centre is taken as the co-ordinates of the centre of the vessel at this slice. The corresponding radius of the vessel is also obtained by the equation: $R=A\cos\alpha/\pi$, where R denotes the radius, A denotes the area of the vessel at this slice, and α is the project angle which can be obtained by using the relationship between the co-ordinates of the central line at this and its two adjacent slices. The 3D model of the vessel is then built for the vessel by connecting all the co-ordinates at involved slices.

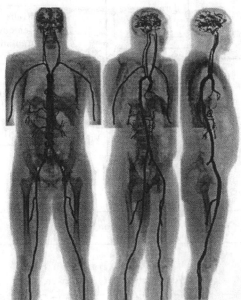

Fig. 2 Human primary vasculature and some secondary tertiary networks

Alternatively, we construct a 3-D model for the cardiac vessels from a professional steel model by using Coordinate Measuring Machine (CMM) Probing. A CMM equipped with an analogue probe or touch probe is a prominent example of the contact inspection method used to determine the 3D co-ordinates of the model. The determined co-ordinates is then fed into a process to generate the computational 3D model used in CathViewer. The related radii of the vessels are also found by using a mechanical measuring technique. The shape of the vascular system as shown in Fig. 3 has been examined and refined further by the medical professional using the Editor Tool.

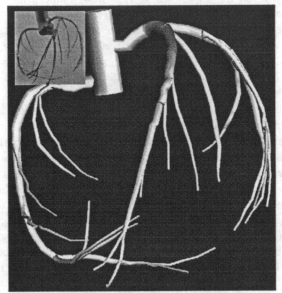

Fig. 3 Cardic networks used in CathViewer.

4 Drawing Shell

The shape of catheter determines how it is being used in the interventional procedure. The design of the catheter shape can be done using standard CAD tools. The Drawing Shell can import files created using these tools and convert them into the format used in CathViewer.

We have also develop a specific shape editor to design catheter. Assume that a catheter designer will start off with a straight rod in designing his/her new catheter. First consider creation and adjustment of the unstressed catheter shape. The UI begins by providing a straight starting form to work on, all the interactions except *new catheter* come under *adjustment*. The shape is built of rods, glued end-to-end without corners, starting from a fixed *proximal* end (where the design is to attach to the part of the catheter that joints the designed head to the end that never enters the patient), from which the parametrization by arc length *s* begins, and a free *distal* end which moves when, for instance, a point in the middle is bent. We will also call rod ends *proximal* and *distal*. There should be a user choice of marking the joints, or leaving them invisible.

Rods can be straight, flat arcs, or pieces of helix. The user does not need to select actively between them, since an arc and a straight line are special cases of a helix, and a straight line is a special case of an arc. Since computation is fastest in the straight case, and faster for arcs than helices, it makes sense to replace a *nearly straight* rod by a straight one, and a *nearly flat* helix by an arc, where *nearly* means that some number is less than a pre-set tolerance, say 0.01.

Adjusting the shape by changing one of the rods is like working with pliers on a wire, rather than adjusting control points on a curve whose ends stay fixed. It matches the way a doctor experiments with new shapes, by bending existing catheters. A rod can be lengthened, along its current straight/circular/helical path; it can be bent, by changing the arc or helix parameters; or it can be twisted around its joint with the previous rod. A rod r_i can be broken into two rods, or merged with the next rod r_{i+1} (which then disappears, while the length of r_i increases by the length of r_{i+1}).

Each rod is defined in its own coordinates, and mapped into the catheter coordinates by its position matrix. But it is convenient to note here that in each rod's internal coordinates it starts from (0,0,0) with tangent in the positive x-direction. In any interaction with the shape, the user needs to select a rod, a point on it, or a general point on the catheter. This selection will typically involve input of (x_{in}, y_{in}, z_{in}) values from a 3D stylus, or screen (x_{in}, y_{in}) values from a 2D mouse. As the input changes, different segments (at most one at any time: test the rods in order from proximal to distal) use highlighting to single that (x_{in}, y_{in}, z_{in}) is within their *touched* zone or that the line of $(x_{in}, y_{in}, z_{abitrary})$ meets it. Clicking the stylus or mouse button when a particular rod R is highlighted will start the interaction with R (such as twisting it around its start, or changing its length) that was picked by the most recent tool selection.

Many tool actions require widgets for user control. All the control functions of the classes involve changing one parameter at a time, so the 2D mouse interactions can all use a slider: for instance, in changing a length, display a slider with the current length, and let the user drag it to a different value.

The shape editor contains a set of tools to vary the length and cross sectional radius of this straight catheter; select reference point(s) along the catheter, bend and twist with respect to these points to create the shape of the catheter hook. We have yet to determine how well will this concept of shape editing be received by the industry. Not only do we have the ability to develop shape, we also have sophisticated rendering techniques to allow the catheter to be viewed in the material envisaged by the designer.

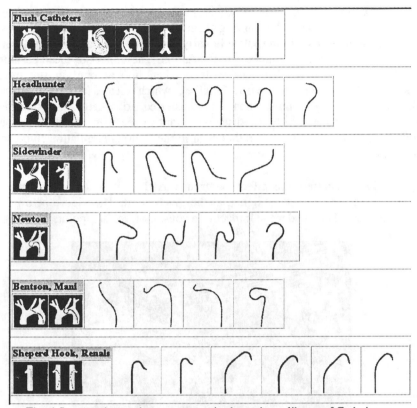

Fig. 4 Some catheters that are present in the catheter library of Cathviewer.

5 Manipulation of Material Properties

We are building a library of all the physical information for each material used in the manufacturing of catheters. Design engineer can assign varying physical properties to the elements defining the shape of catheter. The material models include rubber elasticity and plasticity for metals and foam, equation of state, and a graphical user interface for user material definition. The Young's modulus E and shear modulus G of *polyurethane*, which is commonly used to make various kinds of catheters, are used in cathviewer as the default material values for the newly designed catheter.

Along the length of the arc, there are distinct *physical bits* with assignable values of bendingstiffness (E), twiststiffness (G), and diameter ($2R$). Since these need not share endpoints with the geometric rod subdivisions of the shape, they are managed by a separate linked list. Each physicalbit in the list records the mechanical parameter values appropriate to it. The user is able to select an arbitrary pair of points p0 and p1 on the shape, create a new physicalbit and adjust the parameters by using sliders. The selection of the points is done by the geometric touch detection above; The rod(s) touched report the corresponding values of arc length (from distal end of the entire design), and the linked list of physicalbits is revised accordingly.

In the first version, the physical parameters of a physicalbit are to be treated as constants; later we can consider varying them along its length. One of the obvious schemes, mathematically, is interpolate linearly between values at p_0 and p_1, but this need not be the most easily manufactured variation. The impact of these parameters on the bending and twist stiffnesses will be taken into account in the function elemesh() (to be created by the FEMengine), which creates a model usable in simulations. This model must then be transferred to where an experimental simulation can use it: typically, the user will want to see it in 3D close up, going through the task it is designed for, rather than in the training simulation fluoroscopic shadow play.

Design engineer can test the effect of each variation of material properties by examining the effect of forces assigned at the given point. The Figure below shows how the catheter is bend under the impact of a constant force at a point.

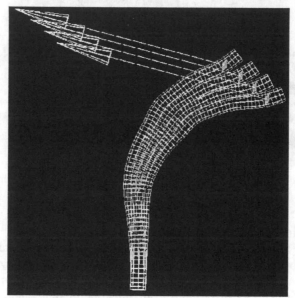

Fig. 5 Catheter bends under the impact of a constant force.

6 Finite Element Modelling and Analysis

With the goal of creating a realistic real time interactive environment for CathViewer, an incremental Finite Element Modelling (FEM) module has been developed for the analysis of catheters navigation in the vascular structure[8]. This FEM analysis uses a new algorithm to speed up the determination of interaction between the walls of vessels and the catheter, which is normally the most expensive portion in FEM calculation of contact problem. Note that the FEM nodes need not correspond to a point defining the hook shape of catheter.

In this FEM analysis, a catheter as a slender structure is discretized into a flexible multibody system with a finite number of 3D FEM beam elements. The blood vessels are assumed to be rigid circular tube-structure (a majority of the great vessels are of

924

circular cross sections and they are well stretched by the surrounding muscles). The catheter navigation is considered as a sequence of movements such as push, pull and/or twist of the flexible multibody system inside the rigid tube-structure. The move of an element of the multibody system at each step is assumed to consist of a rigid-body displacement and a relative deformation. We first move the elements of the system as rigid bodies by applying multibody dynamics method at each step and then find the deformations at their equilibrium position by applying finite element method. Contact forces between the catheter and the walls of the vessels are introduced at the catheter nodes, which are outside the vessels during the rigid moves, to push the catheter back to the tubes. Those contact forces, however, are functions of the position of the catheter at each time step and their determination is very time consuming.

We have developed an efficient semi-implicit algorithm by using the hierarchy structure of the vasculature of the vessels for computing the contact forces involved. The computation of contact force is involved with two steps:
- checking whether catheter nodes are outside the vessels;
- calculate the contact force at the nodes if they are outside.

In order to achieve an efficient in/outside checking, all the branch segments of the vasculature are linked through a topology relation. As shown in Figure 6, a current vessel segment is connected with its parent, brothers and children segments. Each catheter node is associated with one branch segment and the in/outside checking for this node is conducted only with respect to its segment and parent-brother-child segment group, rather than to compare with all the branch segments of the vasculature. It therefore speeds up the checking process by a very good percentage. The contact force at the node outside is defined as $F=cdN$, where d is the distance between the node and the surface of the vessel, N is the unit vector pointing from the node to the surface along the distance direction, while c is the coefficient representing some physical characterisitc parameters of the contact. It is currently set up with an arbitrary number since the flexibility of the vessels has not been taken into consideration in the present study.

7 The Future

We plan to develop a hemodynamics model which will compute deformations of the associated arteries and fluid dynamics (pressure, velocity and temperature) of the blood flow in the arteries. In addition, the morphology, geometry and material properties of the plaque and diseased arteries will be taken into consideration. Careful computation of finite element analysis will be performed to take into account the interaction of the interventional devices with the diseased arterial wall and with the blood flow. The analysis of plaque cutting and removal will also be implemented.

Beyond the immediate goals to complete and refine the simulation and design system, there is a need to create additional patient models and a library of pathologies. Given the degree of variability of the sinus anatomy in the general population, it is important to have a variety of models representative of a diverse population. The models will help developers to test the newly designed catheters within a wide range of simulated conditions and could serve as an evaluation of the new catheters.

To facilitate and reduce the expense of creating patient models for simulation, better software tools are necessary. These same software tools would also facilitate creating patient-specific models from routine medical imaging examinations. This would enable clinical uses of CathViewer, allowing the design engineer to develop customised catheters; and the surgeon to preoperatively plan and rehearse a surgical procedure using these customised catheters. The design engineer and surgeon could experiment with competing design and surgical approaches respectively. The advent of medical digital imaging standards and the increasing appearance of medical image data networks have made it clinically practical to introduce new applications that use the patient's imaging data. In so doing, CathViewer can improve the efficiency and effectiveness as the catheter development system for interventional radiology procedures.

Acknowledgement: Support of this research by National Science and Technology Board of Singapore is gratefully acknowledgement. The generous help of Dr. T. Poston and B. T. Nguyen on the drawing shell of this research is sincerely appreciated.

References

[1] J.L. Berry, V.S. Newman, C.M. Ferrario, W.D. Routh, and R.H. Dean, A method to evaluate the elastic behaviour of vascular stents, 22nd *Annual Scientific Meeting, Society of Cardiovasculatur & Interventional Radiology*, 1996, 7: 381-385.

[2] United States Food and Drug Administration guidance for the submission of research and marketing applications for interventional cardiology devices: PTCA catheters, atherectomy catheters, lasers, intravascular stents. Rockville, Md: FDA, May 1993.

[3] J. Thrner, J. Pegna, M. Wozny, *Product Modeling for Computer-aided Design and Manufacturing*, North-Holland, Amsterdam, 1991.

[4] CORDIS Corporation, Diagnostic and Interventional Radiology, *Technical Report*, Miami, USA 1992.

[5] J.H. Anderson, R. Raghavan, Y.P. Wang and C.K. Chui, daVinci – A vascular catheterization simulator, *Journal of Vascular and Interventional Radiology*, 1997, Vol.8, No.1, Part2, P261.

[6] Kyoto Scientific Specimen Co., Ltd, *Coronary Artery Model/key chart*, Kyoto, Japan.

[7] C.K. Chui, H.T. Nguyen, Y.P. Wang, R. Mullick, R. Raghavan, J.H. Anderson, Potential field of vascular anatomy for real-time computation of catheter navigation *First Visible Human Conference*, Bethesda, MD, USA, October, 1996.

[8] Y.P. Wang, C.K. Chui, Y.Y. Cai, R. Viswanathan and R. Raghavan, Potential field supported method for contact calculation in FEM analysis of catheter navigation, *XIXth International Congress of Theoretical and Applied Mechanics*, Kyoto, Japan, August, 1996.

A CAPP System Based on
Concurrent Engineering System

Ying Sihong, Wang Rong, Ma Dengzhe

CIM Research Institute of Shanghai Jiao Tong University, Shanghai, 200030, China

ABSTRACT: This paper analyzes some PDM software, as PDM system is based on Concurrent Engineering and supports concurrent operation. The author points out that PDM is useful to manage CAD data, but is weekly effective to CAPP data. Now, researching on CAPP is emphasized on generation of process routings, not management of process routing, on part, not whole product, and most important thing is that how to share data resource is never considered.

The author puts forward thought of developing CAPP system based on Concurrent Engineering, and that single database, control mode of design process is the key to carry out concurrent operation in CAPP system. A functional system named SIPM has been developed after researching on methods of process planning, process routing data model and technical document management. It makes that all process planning engineers can share all data resource of their plant, and that is possible to established the interface between CAPP and MRPII. SIPM system has been used in many factories in China.

KEY WORDS: Concurrent Engineering, CAPP, CIM

1 Introduction

From 1995, most companies have many CAD seats to support product design, 41% drawings are in CAD format, and 60% of companies have two or more CAD systems. How to manage the design results and make design change becomes the most important problem. There are 42% companies have some kind of PDM system. PDM is an essential strategic technology underling a successful product development and support environment, and that provides a concurrent operating environment to design engineers. Now most companies are reported MRP (or MRPII or ERP) systems in use. Again, most of these are on different platforms than the design/document creation systems. MRP systems manage parts and materials through the scheduling and production process and many are beginning to offer elements of document management. Usually, PDM system outputs the BOM of a product in neutral format, and MRP system reads, that finishes the information integration, see figure 1.

Fig.1 Information integration between PDM and MRP

PDM is useful to manage CAD data, but is weekly effective to CAPP data since PDM system is based on how to managing engineering documents such as drawing, specification, definition. So a total information integrated system should like figure 2. Process planning is that function within a manufacturing facility that establishes which processes and parameters are to be used to convert materials from an initial form to predetermined final form. Process planning may be defined as the act of preparing detailed work instructions to produce a part. For example, in the machined part domain, using a raw material shape as a starting point, the process planning engineer prepares a list of processes needed to convert this material into predetermined final shape. In addition to operation sequencing and operation selection, the selection of tooling jigs and fixtures is also a major part of the process planning function. And also a CAPP system should provide a concurrent environment for process planning and scheduling engineers.

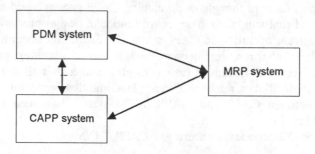

Fig.2 A full information integrated system

2 PDM System

The common problems are focused on PDM system are:
- ensuring everyone works to correct revision level
- distribution of drawings
- time to implement an engineering change
- time to access drawings
- control of variants/configurations
- exchange of drawings with suppliers, customers
- revising paper-based drawings

The author has learned some PDM software such as Iman, Metaphase, Pro/PDM and AutoManage as well. The main functions are similar, they are:
- request for engineering action
- engineering change order or engineering change
- document management including support for multiple versions and multiple formats
- product structure definition
- effectivity of products and occurrences by date, by product unit number, by customizable parameters

928

- configuration management
- test, maintenance and diagnostic information

PDM system is essentially document control and management, whose basic unit consists of controller and browser. The controller is used to coordinate operations among multiple users and provide a concurrent processing environment for them. The controller actually controls attributes of document. The browser is applied to identify document format and reinstate details of document. The more the browser supports document categories, the more its function is perfect. In general, electronic document is made up of one or a group files, and therefore file(s) and its(their) attributes constitute the substance of PDM management. On the other hand, process routing is a flow or a kind of graph. Especially in the case of non-linear process routing, it is a kind of data structure due to its various representation methods and depends on specific application system. Thus management attributes of process routings should be specific data structure and its attributes and can not be represented with one or a group files and then can not be joined into document structure managed by PDM system. Hence, how to provide a concurrent process planning environment must be accomplished by CAPP system.

3 Theory Model of Concurrent CAPP System

The aim of CAPP system is to solve problems of product process planning automation. Traditional CAPP system mostly is a kind of design mode based on serial works and design works at individual design stage are relatively independent, while prior design knows little about demand of subsequent design and subsequent design only has to accept passively results of prior design. Once problems are found appeared in prior design during subsequent design, the prior design has to restart. This causes the efficiency of product design to be reduced and the costs of product development to be increased. The more serious problem must be mentioned is that traditional CAPP system seldom takes into account constraints of production planning management resource and manufacturing resource from MRPII such as production orders, lot size and etc.. For this reason, it could hardly realize the optimization of manufacturing cycle and manufacturing costs of product.

Under concurrent environment, a CAPP system must be guided by theories on manufacturing- oriented, cost- oriented and product- oriented design. The procedure of process planing in CAPP system must be controlled by feedback information from MRPII. From these, it will optimize reasonability of process planing, economics of manufacturing process routings and availability of manufacturing resource, and integrate process planing with practical production and market requirement to the greatest extent.

The theory model of concurrent CAPP system oriented integration with MRPII is a multiple input/output closed-loop feedback system, which should have approximate stability in a wide range of balanced point and rapid time response. This can be best illustrated by Figure 3.

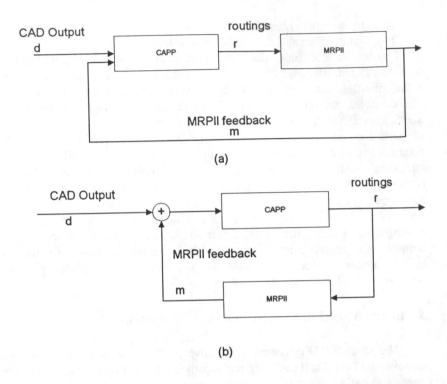

(a)

(b)

Fig.3 A closed-loop feedback model integrating concurrent CAPP system with MRPII system

In figure 3, figure 3(b) is transformed from figure 3(a), where d can be BOM, parts information and other design output information; r is results of process planning which can be expressed as $r = f_1(d, m)$; m is feedback information from MRPII including production lot size, historical production situations and so forth, and it is historical feedback of process routings and belongs to integral feedback in closed-loop system that can be expressed as follows:

$$m = \int_{t=0}^{T} f(r,t)$$

where $t = 0$ is start time of certain process routing and $t = T$ is feedback time of certain process routing.

From the above analysis, when the system input is stepped function, the system time response performance can be shown as curve 1 in figure 5, which keeps stable by means of adjusting feedback continuously in a long time. To settle the stable problem of time response, a differential feedback may be inserted into the system for forecasting. Now the construction of system can be illustrated by figure 4, where p = f_2 (r') is forecasting feedback. The system time response performance(curve 2 in figure 5) is obviously more perfect than the former and therefore that makes CAPP system response more rapidly to changes of design and production demand, and realizes concurrent engineering demands on the system. So, the author takes the concurrent CAPP system with forecasting as theory prototype.

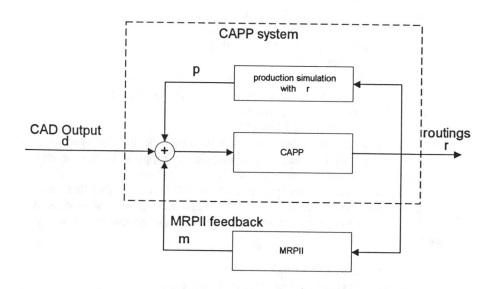

Fig.4 Theory model of concurrent CAPP system with forecasting

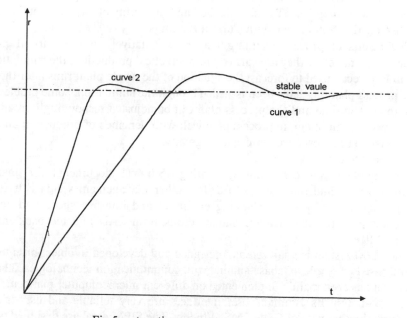

Fig.5 system time response curves

4 A Concurrent CAPP System with Forecasting - SIPM System

The Architecture of SIPM system, as illustrated in figure 6, consists of following functional modules:

. Manufacturing Resource Modeling

This functional module describes base data required for process planning including settings of plants, departments, workshops or sub-plants, situations of technical equipment, machines or workcenters, states of raw materials, standard terms of operations or sub-operations and their relations, user-defined GT rules and etc.. Whether the arrangement and entries of base data is scientific and intact will effect directly efficiency of process planning.

. Task of Process Planning Management

In SIPM system, this module is quite important because the whole process planning is task-driven. In this module, various function is supplied to handle tasks such as assignment, modification, completion

status statistics, graphic analysis and so on. With the help of this module, SIPM system ensures data coordination and workflow management of process planning in the case of concurrent operations, that is, when the system is oriented to multiple users.

. Product Structure Modeling

Product structure is the base of process planning.

. Process Planning

In SIPM system, this module is the main module, which can finish process routings' design, process procedures' design and various specialization process routings. In the meantime, it supplies GT design method and the ability of   typical routing reuse.

. Production Simulation With Current Routings

The results of process planning may be simulatively run in virtual production environment before they are carried out in practical production, the simulation results can be directly used to forecast the situation of the whole plant runs after these results of process planning have been issued. Through the simulation, some problems can be found out in time and the process plan can be adjusted respondingly in advance. In this way, it can integrate process plan well with demands of production and improve the system time response performance greatly.

. Input/Output

This module can carry out generating technical documents for management automatically and interface to MRPII or other management systems. It can reduce repetitive jobs of process planning engineers and improve speed and accuracy of process planning and achieve output results from CAD and feedback information from MRPII.

The SIPM system is analyzed and designed and developed in object-oriented method and based on single database and network communication technologies. The network system has been mainly implemented on different microcomputer platforms by means of task-driven. Its graphical user interfaces are very reliable and user-friendly and inherit the character of GUI of MS-Window. The SIPM database designed on the base of single database theory is central to all SIPM modules and used to exchange and synchronize information and avoid all unnecessary redundancies and ensure high flexibility for user specific customization.

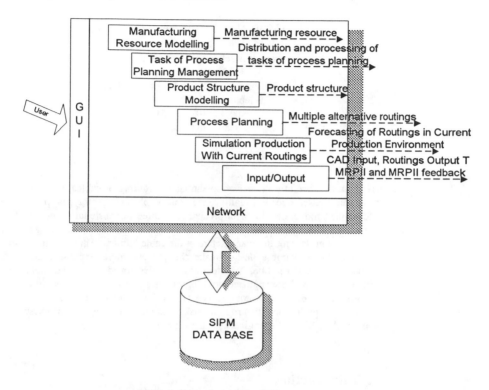

Fig.6 Architecture of the SIPM system

5 Conclusion

Due to commercial PDM system itself, PDM system has no way to concurrently manage process planning. While the SIPM system is analyzed and designed and developed on the base of practice and research in many factories or enterprises, this CAPP system gets over defects of traditional CAPP system, that is, traditional CAPP system can not be put into practice widely. And it is also shown that a CAPP system can be powerful and effective only by putting it into concurrent engineering integration environment.

Reference
1. B.Wu, Manufacturing System Design and Analysis, Chapman and Hall, 1992
2. S.Evans, The Implementation of CAPP in a Rapidly Developing Manufacturing Environment, Proc. 4th European Conf., Automated Manufacturing, 309-318, May 1987
3. Leo Alting and Hongchao Zhang, Computer Aided Process Planning: the state-of-the-art survey, Int. J. Prod. Res., 1989, Vol. 27, No. 4, 553-585

Designing and Planning Cable Harnesses using Immersive Virtual Reality

Foo Meng Ng, J.M. Ritchie and R.G. Dewar

Department of Mechanical and Chemical Engineering,
Heriot-Watt University,
Riccarton, Edinburgh EH14 4AS, UK.
Telephone: +44 (0)131 449 5111
Fax: +44 (0)131 451 3129
e-mail: f.m.ng@hw.ac.uk

Abstract. Currently the design, planning and routing of cables within electro-mechanical assemblies is usually carried out on physical prototypes towards the end of the product development process. The three dimensional harness is then transposed into two dimensions and drawn on to a board. This drawing is then used as a template for the manufacture of the cable harness. These characteristics of conventional harness design make the process more expensive and time consuming than it need be. A system has been developed which uses immersive virtual reality to allow engineers to design harnesses in a virtual chassis without the need for a physical prototype. The system incorporates novel tools to assist the user in performing the routing task. Design and planning information is generated on-line as the engineer plans the route.

1. Introduction

The design and planning of cable harnesses that are use to connect electrical modules within an electro-mechanical device is often considered a tail-end design activity. This process is, in fact, a complex design problem which is time-consuming and costly. The end connectors used will be dependent on the type of socket provided by individual modules, which in turn determines the types of wires to be used. These individual wires vary in size, and together, determine the stiffness and mass distribution of the entire cable harness. In turn, the stiffness, bend radii and mass of the cable harness then dictates the position and number of fasteners required.

Current practise often requires a physical prototype of an electro-mechanical device to be built to allow engineers to manually determine the cable lengths, paths and positions of fasteners. Once a suitable cable path is found, the components involved are selected from tables and the results entered into a database. This information is then converted into two dimensional drawings, part lists and instructions. Figure 1 shows a typical cable harness laid out in two dimensions.

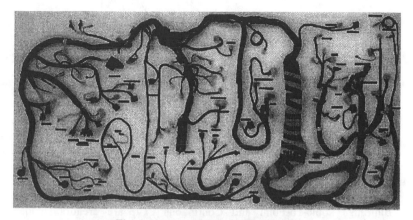

Fig. 1. An assembled cable harness

The routing problem is further complicated by the vulnerability of the cable harness to decisions made upstream. The cable harness may have to be reconfigured after only minor changes that affect the chassis and the individual modules within the prototype machine. The routing process can even result in the late and expensive re-design of the machine chassis to allow the cables to reach their terminal points.

2. Current State-of-the Art in cable harness routing systems

Proprietary software packages do exist to assist in the design of cable layouts, however these tend to be difficult to use and do not always come up with a suitable solution. A significant amount of manual 'tailoring' is often required. The authors are aware of at least one industrial application where this is the case.

There is little published literature on the subject of cable routing with the noticeable exception of the work being done at Stanford University's Centre for Design Research [1, 2, 3]. They have developed a system called First-Link which incorporates artificial intelligence to try and semi-automate the routing process.

However, the authors believe that these CAD-based cable routing systems are unable to entirely integrate the intuitive and creative nature of the expert human operator. By using immersive VR the user is presented with a richer environment. This environment excludes the need for any routing automation and allows users to experiment with alternative solutions.

3. Cable Layout using Immersive Virtual Reality

The virtual design and planning cable routing system presented here is implemented on Hewlett-Packard workstation with additional VR hardware and software from Division Ltd. CAD models of a prototype assembly can be imported directly into the system which negates the need for any extra component modelling.

The virtual environment can be displayed on the computer monitor and the user can interact with virtual objects using a conventional mouse. This is know as "Desktop VR". However, the authors contend that it is necessary to use "Immersive VR" (that is, using an Head Mounted Display (HMD) which gives a stereo image of the virtual environment) to enable users to comprehend complex three dimensional assemblies. A three dimensional mouse is used to interact with the immersive virtual world.

This virtual cable routing process shown in Fig. 2 focuses on the expert user's creativity and intuition to complete the task rather than trying to automate the procedure. The system developed has three modes of operation - namely, Point-To-Point, Continuous Path and Rubber Banding. Point-To-Point and Continuous Path are creation modes, whereas Rubber Banding is an editing mode. Figure 3 shows the floating toolbox in a virtual world. This is used to select the required mode. This figure also shows a section of cable laid around some of the virtual objects in the environment. Notice the straight sections and nodes of the cable.

When the user is creating or editing sections of cable, if there is a collision between one of the sections and an object (or even another section of cable), visual and audio cues are used to highlight the occurrence and warn the user of the clash.

A Size Management VR tool has also been developed, which provides the user with the ability to alter their size with respect to the virtual objects. This allows the user to shrink to a manageable size to perform detail work or to enlarge themselves to create rough cable routes.

Internally, the system represents the network of nodes and cables as a multi-linked structure [4]. This provides a compact and flexible portrayal of the harness. As the cables are routed in real time, behind the scene, the system is generating a text file detailing the bill-of-materials and process planning information that is associated with the physical cable harness (for example, estimated length of the cables, types of fasteners and end connectors used).

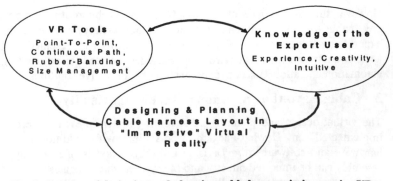

Fig. 2. Concept of designing and planning cable harness in immersive VR

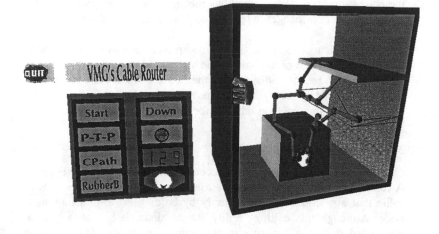

Fig. 3. A floating toolbox, some components and a cable harness in the virtual world

The following sections describe the three modes of operation of the system in more detail. They also detail how collisions are dealt with in the virtual world as well as mentioning the size management facility in the system.

3.1 Point-To-Point

The Point-To-Point technique of routing cables provides the capability to generate rough cable routes rapidly by picking positions or nodes in the virtual environment. The user simple probes a point in space and a section of cable appears between this point and the last node that was created. If the user picks an existing node, they can move this node about in three dimensions. This editing facility within point-to-point is effectively rubber banding which is described later. Additionally, this picking of another node makes that node active and any subsequent point chosen in space will create a section of cable between it and the active node. By choosing existing nodes, multiple branches can emanate from a single node. Some examples of this can be seen in Figure 3.

3.2 Continuous Path

The other technique developed for routing cables is known as the Continuous Path method. This technique routes a cable by extruding a new section from a user selected node. Thus, by picking an existing node, a new node is created and this is then attached to the virtual hand until the node is dropped. This method has rubber banding implicit in it as far as the new section changes in length and pose as the virtual hand moves.

This method is more suitable for detailed routes such as routing a cable around the curvature of the hole in the virtual object as shown in Figure 4. The user can see the collisions that the new section of cable is involved in and immediately take action to move the section and so avoid the clash. Again, this method allows for creating nodes with multiple branches.

3.3 Rubber Banding

Once the entire cable layout is produced, modifications may be required. A rubber banding facility allows the user to re-position entire sections of cable that are knitted together simply by holding on to and moving a node. Although this editing facility stands alone, it has already been mentioned that it is available in the point-to-point tool and, to some extent, in the continuous path tool.

3.4 Collision Detection

The ability to touch and feel objects in the real world is one that is taken for granted. However, this haptic feedback is still in its infancy in virtual environments [5, 6, 7]. For this reason, the system developed here has used alternative visual and audio cues to highlight collisions. A full polygonal collision detection algorithm is available in the Division software. Thus, when a collision occurs, the system utilises the messages sent from the algorithm to make objects in the virtual world turn to a wireframe representation to highlight to the user that something is amiss. This visual cue is accompanied by an audio cue which takes the form of a simple 'beep'.

When a cable section clashes with a virtual component in the chassis the chassis component turns to wireframe. When one cable touches another, both cables turn to wireframe. A section of cable is in collision with a component of the virtual chassis, in Figure 5. Here the chassis component has turned to wireframe. Using the rubber banding tool, the user is able to alter the position of the cable, thus removing the clash.

3.5 Size Management

Another VR tool that assists the human user in routing is the Size Management tool. This tool provides the user with the ability to enlarge or shrink themselves in the virtual environment. By increasing the size of the user with respect to the objects within the virtual environment, cables can be quickly routed using the point-to-point method. If detailed cable routes are required, users can shrink themselves. This provides the user with more information about valid paths and collisions and may be more suited to the continuous path method.

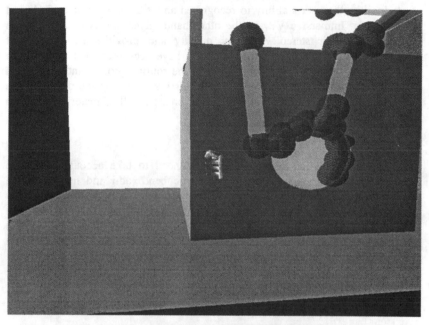

Fig. 4. Detail route using Continuous Path VR tool

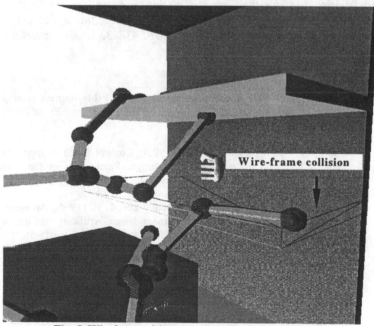

Fig. 5. Wireframe object warning of a clash with a cable

4. Conclusions

Current CAD cable harness routing systems are only able to provide users with two or three degrees of freedom of movement on a flat screen. This inhibits the user's ability to recognised and edit flaws in a cable layout design. Immersive VR, on the other hand, gives the user the ability to change their stereo viewpoint quickly and provides six degrees of freedom. In this paper novel VR tools have been presented to assist the user of virtual environments perform the routing process intuitively and efficiently. The ability also exists in this system to log the details of the routing process which can be used to create a bill-of-materials and two dimensional schematics of the harness.

5. Future Work

In the future the system will be extended to take account of bundle stiffness and mass to facilitate realistic bend radii and help with the positioning of harness fixtures. Also it would be advantageous to add new nodes bisecting existing straight lengths of cable to allow further path editing. Adding a catalogue database to the system will allow designers to choose suitable connectors, bundles and fasteners on-line. Another feature which may prove useful is an undo/redo option when creating nodes. Finally, it will be important to conduct industrial trials to evaluate the ergonomics and usability of the system.

Acknowledgements

The authors are grateful to NCR (Scotland) Ltd., Kinloch Electronics Ltd. and Division Ltd. for their support of this work and access to their expertise and knowledge. The support of the EPSRC, through access to the equipment provided under grant GR/K 41823, is also gratefully acknowledged.

References

1. Conru, A.B (1994), A genetic approach to the cable harness routing problem, Proceedings of the IEEE Conference on Evolutionary Computation, vol 1, pp 200-205.

2. Conru, A.B and Cutkosky, M.R. (1993), Computational support for interactive cable harness routing and design, Proceedings of the 19th Annual ASME Design Automation Conference, vol 65, pp 551-558.

3. Park, H., Cutkosky, M.R., Conru, A.B. and Lee, S. H. (1994), An agent-based approach to concurrent cable harness design, Artificial Intelligence for Engineering Design, Analysis and Manufacturing, vol 8, pp 45-61.

4. Langsam, Y., Augenstein, M.J. and Tenenbaum, A.M. (1996), Data structures using C and C++, 2nd Ed., Prentice Hall, New Jersey.

5. Taylor, P (1995), Tactile and kinaesthetic feedback in virtual environments, Transactions of the Institute of Measurement and Control, vol.17, no.5, pp.225-233.

6.   Caldwell, D.G., Lawther, S., and Wardle, A. (1996), Tactile perception and its application to the design of multi-modal cutaneous feedback systems, Proceedings of the IEEE International Conference on Robotics and Automation, vol.4, pp.3215-3221.

7.   Tateno, T, and Igoshi, M (1996), Hierarchical processing for presenting reaction forces in virtual assembly environments, Journal of the Japan Society for Precision Engineering, vol.62, no.8, pp.1182-1186.

A New Solution to the Integration of CAPP and PPC

Wang Liya
CIM Research Institute, Shanghai Jiao Tong University, P.R.China
Jiang Zhiping
CIM Research Group, Automation Department, Shanghai Jiao Tong University, P.R.China
Yan Junqi
CIM Research Institute, Shanghai Jiao Tong University, P.R.China
Wu Dezhong
CIM Research Institute, Shanghai Jiao Tong University, P.R.China

Abstract: The paper proposes a new architecture of multiple layer, concurrent and distributing integration of CAPP and PPC which supports the function integration and the overall optimization of their process plans and production plans. The architecture is further solidified with both the implementation of a system which is based on an integration platform and the realization for a prototype of a products production planning layer .

Keywords: Computer-aided process planning, production planning and control, functional integration, architecture, system realization

1 Introduction

The integration method for computer-aided process planning(CAPP) and production plan and control (PPC) has long been concerned in both research field and engineering field as well. Currently the mostly used, however, is an information integration method by CAPPs interfacing with PPCs. One of the widely known defects of such a method is that the CAPP and the PPC are still two functionally independent sub-systems[1,2]. With this architecture, the dynamic feature of manufacture resource and the feedback message from PPC is difficult to be fully considered in the stage of CAPP, and the PPC has to invariantly organize production based on whatever results come from CAPP. Therefore, it would be common that, in a real manufacture environment, a detailed process plan is infeasible or hard to be implemented. In order to solve these problems, we propose a new solution to support *function integration* (FI) of CAPP and PPC which takes into account, instead of merely information integration, the influence of the environment to a CAPP and a PPC, the mutual restriction to each other and the joint optimal objectives. In accordance with this criterion, we suggest in this paper a new architecture of multiple layer, concurrent and distributing integration, which support FI of CAPP and PPC. To demonstrate the instructive meaning of such an architecture to a class of job-shop enterprises featured with multiple products and small lot sizes, we put forward an implementation to a system based on an integration platform as well as an universal logic adaptable to the decision making of every layer of the system. A realization detail for the development of a prototype for a products layer is also given.

This work was supported by Natural Science Foundation of China

2 The new integration architecture

By analyzing their individual function and mutual relationship, we find that PPC is a layered procedure from planning down to real manufacturing, whereas the CAPP is only a technically preparatory procedure in advance to real manufacturing. Therefore, there is an obvious boundary line between these two procedures whether in terms of the time they apply or in terms of functions they take. We can thus be hinted that, in order to realize FI, the traditional function division among CAPP and PPC should be broken, and a new combination to their functions is necessary.

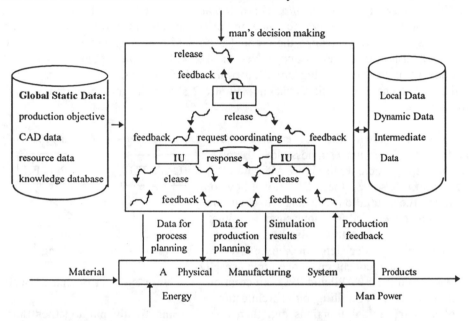

Figure 1   A structure for multiple-layer, concurrent and distributing integration

Figure 1 represents a new architecture of multiple layer, concurrent and distributing integration, in which multiple relationships among layers as well as among components of the same layer are expressed. It is easy to see that the hierarchy is such that all the components in the architecture are   distributed in horizon and layered in vertical. Also depicted are the control relation from upper layers to lower layers, the feedback relation form lower layers up to upper layers, and the distributing relation among components of same layers, respectively. It is noted that such an architecture has an outstanding feature that the complexity for decision making is decomposed. For example, the people can make decision on different layers according to their diverse frequencies and/or diverse   responses, which can be deemed as a heuristic process from macroscopic to microscopic. In this way, the decision-making process for an object on a layer can be done at its own block (unit) in distribution. Moreover, even a larger problem can also be dissolved to be processed in many small related blocks. It is seen that such an architecture is not only in accordance with the mode most enterprises organize their production but also the commonly used control structure in computer integrated manufacturing systems(CIMSs)[3].

From Figure 1, we can see that every node in the multiple layer distributing structure is an integration unit (IU). It is also the basic function unit for CAPP and PPC integration, composed of three fundamental components. One is a *CAPP function unit* (CAPPFU) which ensures to produce process plans which can manufacture required parts, components and products according to the original design; the second is a *PPC function unit* (PPCFU) which is responsible for working out the reasonable quantities of the parts, components and products of a specific period of time from the requirements; and the third is called an *integration decision function unit* (IDFU) which is separated from the original function used for time and/or cost control in traditional CAPP and PPC sub-systems, which realize the optimal control of process plans and production plans on a basis of timing or cost objectives. the CAPPFU and the PPCFU jointly fulfill the concurrent decision making procedure while the IDFU acts as a coordinator and participates cooperative decision making. Figure 2 shows the relations among these three units.

It has been shown through real application that the architecture makes it possible to dissolve the complexity problem raised in CAPP and PPC integration, to solve the difficulty of how to determine the integration scale and to detail the way of information and functions coupling. What is more, the architecture also leads to a new model combining decision making in layer, in distribution and in concurrent cooperation as well. For a traditional CAPP and PPC integration problem, we can, according to the architecture,

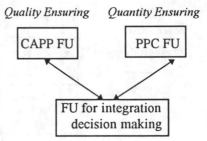

Figure 2 The Principle of an IU

dissolve their original functions and then re-assign and finally put together those functions among CAPPFU, PPCFU and IDFU[2] Therefore the original mutually independent and centralized decision making procedure is now turned to be a concurrently interactive and distributively cooperative one. Further application shows that this kind of solution to CAPP and PPC integration can not only supports the overall optimization of the process plans and the production plans, but also fit the dynamic variation of the environment through coordination and feedback its mechanism.

3 Integration reference model

For a general class of enterprises featured with multiple products and small lot sizes, we here suggest a five-layer reference model for the system integration using the criterion outlined in the previous section. The model narrates the layers to which a CAPP or a PPC should belong (for detail see 2). In addition, for every layer, the functions to be fulfilled are also specified which include the division of an IU on shop floors, cells and equipment levels, respectively in addition to their function range on each layer, as is shown in Table 1 at the next page.

Table 1

Integration layer	IU number	Functions of IUs
products production planning layer	1	products process technique analysis, products production planning,process planning for products orders, construction of products process network, capacity checking ,adjustment of process plan and production plan , *etc.*
parts production planning layer	1	parts process technique analysis,parts production planning, process planning for parts orders, capacity checking,adjusting process and production plan for parts orders, feedback control, assigning and adjusting shopfloor tasks, *etc.*
Job-shop planning layer	depending on the division of the cells	planning process in shopfloors,shopfloor tasks decomposing, job shop scheduling ,capacity checking, process and tasks adjusting in a extend of shopfloors, coordinating shopfloor tasks the shopfloors among IUs, feedback control and adjusting cell tasks.
Job-cell planning layer	depending on the division of the cells	process planning in cells, cells tasks decomposing, scheduling and capacity checking, adaptivity adjusting(including process batching), coordinating cells tasks among IUs, feedback control, and adjusting equipment tasks.
Equipment planning layer	as needed	Job-equipment sorting, process adaptivity adjusting, coordinating tasks for equipment, timing and feedbacks among IUs.

From the model, we can see that the five layers are in a tight connection. For example, except the production plan layer, every other layer can make its own integration decision only based on the results of the decision made on its directly upper layer. And each layer can recursively solve the problem handed upward from its directly lower layer through feedback when the latter must but can not, due to its function limitation, the problem. In this way, the decision logic within each integration layer is similar though each has its own limitation of functions. As sketched in Figure 3, this arrangement of the decision process, we call it a *bottleneck-driven decision-making logic*, can efficiently optimize the process plan and the production plan in the integration of CAPP and PPC.

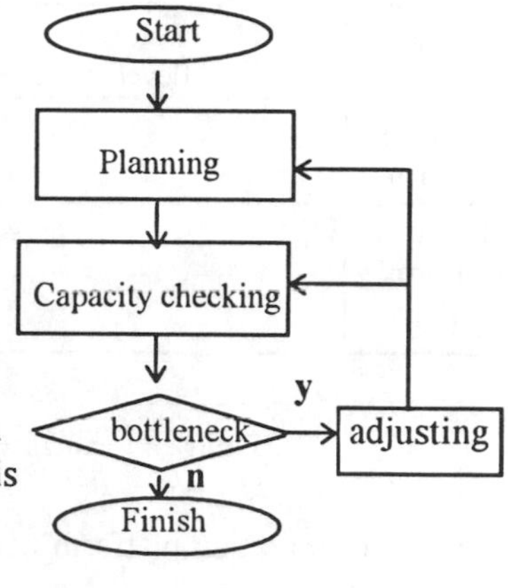

Figure 3 The bottleneck driven logic

4 Integration-platform-based system implementation

The above description defines only the basic framework of an integration system and its corresponding function domain for each of its IUs. For the system implementation, we propose a method based on an integration platform which is shown in Figure 4. The kernel parts of the integration platform is composed of the following four parts:

(1)   a product data model: includes the modeling of parts features, specification of the products structure and products data maintenance as well as data transformation, *etc.*;

(2)   a manufacturing environment model[4]: includes the definition of the control structure of the factory, manufacturing resource modeling, process and manufacturing knowledge expressing as well as the function description and implementation of shop floor data management;

(3)   an integration information model[5]: defines both the plans structure generated by the integration system and the information structure used for intermediate decision making; and

(4)   a general data access and management: controls and administrates and maintains the fundamental databases integrity, completeness, security, *etc.*.

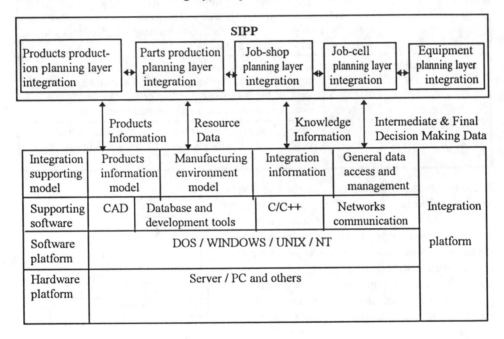

Figure 4   An Integration platform supporting SIPP

Such an integration platform will offer the following conveniences:

(1)   Manage in a centralized way all the process decision-making information raised during CAPP and PPC integration. Since the efficient FI for CAPP and PPC should be assured through each of the CAPP and PPC making full use of other party's abundant procedure decision-making information, how to rapidly and correctly access all these information becomes compulsory. The procedure information structure of the integration information model in the integration platform provides us with a means to visit all the key message during the integration decision making on the current layer and all the results of the decision making on the non-current layers.

(2)   Provide for CAPP and PPC a unified data environment. Traditionally, CAPP and PPC possess their own databases and data structures, which results in at least four drawbacks. One is that data would not be mutually sharable. Secondly the manufacturing environments data which comprises the basis for their own decision makings is usually inconsistent. The third is that each procedure is unable to make quick response to the variation of the data in the other procedure. And the fourth is that the relations between a process planning and an order can not be expressed in the data structure. That is why we establish an unified data model on the integration supporting layer of the integration platform. For example, in a manufacturing environment model, all the static and dynamic resource information is organically collected for CAPP and PPC sharing, while in the integration information model, the process plan and the production plan are tightly combined with each other through an integration plan structure, and thus forms a connection between process plans and orders, which supports the overall optimization.

In addition to the above merits, the platform technology has another distinctive feature that the information model can be efficiently separated from the function model[6], so the information model is highly re-usable, which reduces the hardness to develop or improve a system of integrated process planning and production planning and control (or briefly, SIPP) and also provides a handy integration interface between SIPP and other systems or functions .

5   System realization

As an example, here we give a concrete realization of an IU for a product planning layer (see Figure 5). The functions of all the blocks in the figure can be addressed below.

*   Process technique analysis (P.T.A.): It is the first step in process planning. Since the accurate time when the process planning is being put into use is still not predictable at this stage, the purpose of process technique analysis is mainly to produce, in consideration of their techniques, the process scheme or scheme set for all involved products, components or even parts, and then assign each scheme a priority according to some criteria (e.g., the minimal machining cost, the shortest machining time, the least process number, *etc.*).

• Products production plan (P.P.P.): This is obtained by means of taking into consideration both the predicted and practical products orders. It is only a tentative plan reflecting the market or user demands because no process planning has been formed at the moment.

• Process planning for products production plan layer ($P^5L.$): This is based on the analysis results for process techniques and the preliminary products production plan. And then according to certain criterion, the first process planning for each order is selected, i.e., the process planning for products orders.

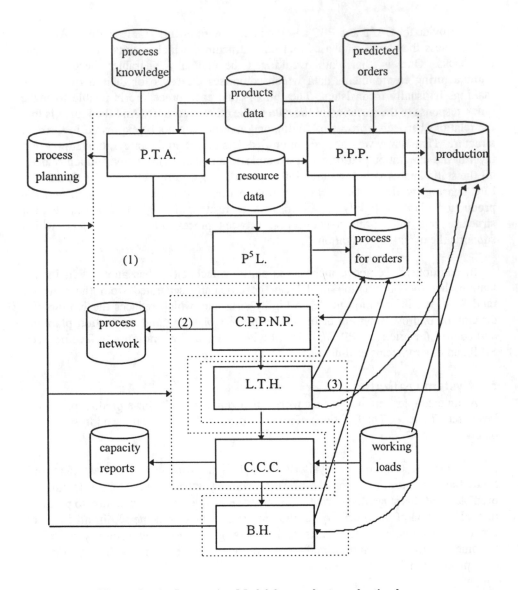

Figure 5   An Integration Model for product production layer

• Constructing progress plan network for products (C.P.P.N.P.): Since each the working load of each equipment might be different, it would be necessary to compute for each equipment its planned working load at different periods of time. To this end, the products production plan and the process plan for an order are decomposed and transformed into a progress plan network.

• Lag time handling (L.T.H.):   According to the progress plan network , it would be easy to find out those processes in progress plan networks that will fall into the lagging time segments. The purpose for the lag time handling is to avoid this situation by means of adjusting the process planning for orders or modifying the due time in the production plan.

- Cycle capacity checking (C.C.C.): This is to get the status regarding the capacity of the equipment as well as the bottlenecks of resource. For this purpose, the progress plan network, the production plan and the structure tree of the products are utilized to calculate, in a specific period of time, the planned working loads of the equipment within various time segments.

- Bottleneck handling (B.H.): This is realized by scheduling processes (moving up some processes, for example), adjusting the process for the orders, and/or modifying the production plan.

It is noted that the integration decision making procedure of the production plan layer can further be divided into three levels: (1)the planning level cares the process technique analysis, products production planning and construction of the process plan for the products production planning layer; (2) the capacity checking, the second level, involves the establishment of the progress plan network for the products as well as the checking of the cycle capacity for the plan; and finally (3)the adjustment level handles lag time and bottleneck. It is also noted that there are feedbacks among the three levels. Through such feedbacks, the control of the execution frequency of the planning level and adjustment level becomes possible.

6 Conclusion

In this paper, we put forward a new solution to a the CAPP and PPC function integration, which takes into consideration not only the factors of dynamic variation of the manufacturing environment but also making reasonable use of the enterprises resource . Therefore, the overall optimization in the CAPP and PPC integration becomes realistic. To this end, a framework of a multiple-layer, concurrent and distribute integration architecture is constructed, in which the key is how to decompose and re-group the functions of CAPP and PPC by their multiple-layer distribution hierarchies. An integration reference model is henceforth proposed and it is enhanced with a small but illustrative system realization.

References
1   Larsen N.E., "Methods for Integration of Process Planning and Production Planning", *Int. J.* **Computer Integratein d Manufacturing,** Vol. 6,(1993) 152 ~ 162
2   Wu Dezhong, Yan Junqi and Wang Liya, "Research of CAPP and PPC Integration concurrency and distribution", *J.* **Shanghai Jiao Tong University,** Vol. 30, No. 12, (1996)1 ~ 6 (in Chinese)
3   Li Bohu, The Guide of the Convention, Standards and implementation of Computer Integrated Manufacturing Systems, **Armament Industry Press**, Beijing, (1994) (in Chinese)
4   Wang Liya, Yan Junqi, "Modeling of Manufacturing Environments in a Object Oriented Method", *J.* **Shanghai Jiao Tong University,** Vol. 30, No. 2, (1996)110 ~ 116 (in

Chinese)
5   Wang Liya and Yan Junqi, "An objet-oriented alternate process data model for CIMS", **Proceedings of International Conference on Intelligent Manufacturing**, Wuhan, China, June , (1995)149 ~ 160
6   Jin Zunhe and Wucheng, The Idea and Development of Integration Software Platform in Factory Automation, **CIMS-China'94**,(1994)6-44 ~ 6-49(in Chinese)

Product Renewal By Using IPD

Robert Bjärnemo

Department of Machine Design, Lund Institute of Technology, Sweden

Abstract. The origination of product ideas for new products is usually the result of a systematically executed product planning process. In smaller companies, most often with no access to such a product planning function, the product ideas are usually generated through less formal procedures. However, regardless of how these product ideas are generated, there is a strong demand for improving these activities in a way which contributes to an increased effectiveness and efficiency of the subsequent product development process and thus of the overall industrial innovation process. Here, the *product renewal procedure* is proposed as effective means for fullfilling these objectives, when the *Integrated Product Development procedure* is utilized for the subsequent product development process. A description of the background to and an example of the practical use of the product renewal process are also included in the paper, as well as a brief outline of the future development of both the product renewal procedure and of the IPD procedure.

1 Introduction

In order to successfully meet the demands originating from the rapid globalization of the economy, most manufacturing companies around the world have been forced to introduce major improvements in their approaches to the development of new products in order to maintain or to improve their competitiveness on the international market. The most significant of these improvements are undoubtedly those focusing on a combined increase of the efficiency as well as of the effectiveness of the *product development process*. The objectives set out in order to reach these improvements are often of a conflicting nature. One such conflict is created when the demand for a reduction of the development resources arises and there is a simultaneous demand for an increase of the quality of the product–to–be.

A prerequisite for creating more efficient and effective development procedures is the access to similarly effective and efficient methods and techniques to support these procedures. An often neglected fact is that the development of these methods or techniques is equally dependant on the existence of an overall development procedure model as the development of the product development procedure model is dependant on the access to, or the possibility to develop, suitable methods and techniques. By omitting this dependency it is more than likely that the method or technique selected or developed instead contributes to an increase instead of decrease of resource consumption, if not proven completely unsuitable for its purpose. The increase of resource consumption is most frequently the result of the need for an adaptation of product–related data to fit the demands from a subsequent method or technique in the development process, such as the frequent need for transformation of product geometry data from a CAD program to fit the demands on the geometry representation by a general purpose finite element program.

In other words, the *product development procedure model* provides the information on *which* problems should be solved during the process and *when* (in which order). The *methods* and *techniques,* on the other hand, provide the means for *how* a given problem, derived from the development procedure model, should be solved. As the main objective here is to present the use of an existing *product development procedure model* in *product renewal,* supporting methods and techniques will not be addressed in any detail, if at all.

In industry the product development process is in turn embedded in a more comprehensive process, the *industrial innovation process.* This innovation process encompasses all of the activities preceding the adoption of a new product on the market, such as basic and applied research, design and development, market research, marketing planning, production, distribution, sales and after sales service. In other words, the innovation process comprises more than the product development process – see Roozenburg and Eekels [13].

Preceding the product development process in the overall innovation process is the *product planning process.* In the perspective of developing new products, the main objective set out for this process is to generate and to decide upon which *product idea(s)* to be developed. For a successful realization of the product development process, the result obtained during the product planning process is of the utmost importance; especially with reference to the development and design of the product–to–be.

In order to facilitate an effective "link", or more precisely, an *integration* between the product planning process and the product development process, with the overall objective to contribute to an increased performance of both processes, the *product renewal process* is proposed here. The concept of product renewal is not confined to those, usually larger companies, in which product planning is an established function, but also to smaller companies lacking this function.

In spite of the technically oriented approach underlying the concept of product renewal, none of the additional aspects taken into account during a traditional product planning process are omitted in the concept.

Before elaborating on the concept of product renewal and the use of *Integrated Product Development (IPD)* in this process, it is convenient to start by introducing the concept of *(engineering) design methodology* as this forms the basis for the *IPD procedure* and thus for the product renewal process.

2 Design methodology and product development

In the late 60s, but more commonly during the 70s, many manufacturing companies introduced *design methodology* as the mean to improve the *(engineering) design process* as well as the product–to–be. In design methodology, the focus is set on the *technical realization* of the product, which means that virtually every phase of the technical realization process is covered – from *why* and *how* a design project is initiated until the completion of the documentation necessary for the materialization (manufacturing) of the product–to–be is achieved.

One of the most well–known of these early design methodologies is the one by the German Professors G. Pahl and W. Beitz. This methodology was first introduced in a series of articles in the German journal *"Konstruktion"* between 1972–74 and later, in 1977, published in a book. Since then, the methodology has been updated and enlarged, thus still holding a foremost position in the current choice of design methodologies – see G. Pahl and W. Beitz [14,15].

In order to give some insights into the contents of a design methodology, the design procedure model according to G. Pahl and W. Beitz is presented in Fig. 1.

As long as the technical performance of a product constituted the single most important competitiveness factor, design methodology alone represented the most effective and efficient means of increasing competitiveness. However, as additional factors (adaptation to business, design, market/consumer, production (primarily manufacturing), sales etc.) also became equally important in handling the competition on the global marketplace, new methodologies focusing on the

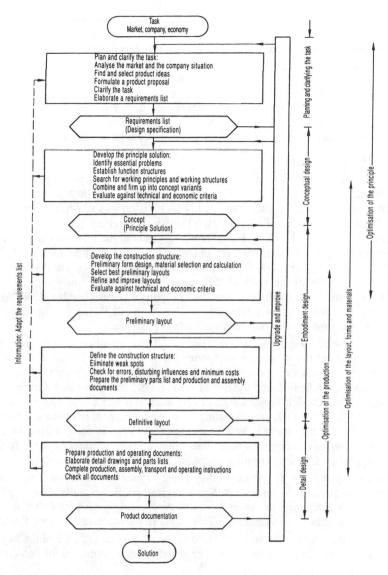

Fig. 1. The design procedure model according to G. Pahl and W. Beitz [15 p 66].

overall development process, the *product development process,* were developed. Note, that the creation of product development procedure models represents an *extension* of the models provided for within design methodology. The design procedure becomes, in other words, embedded into the product development procedure model.

The trends towards the development of *new,* more efficient and effective, development–/design–oriented, *product development approaches,* can be traced back to the beginning of the 80s. The most successful of these new approaches are undoubtedly *Concurrent Engineering (CE)* – see D. Clausing [3, 4], *Simultaneous Engineering (SE)*– see C. W. Allen [1], B. Evans [6] and F. Gordon and R. Isenhour [7] and *Integrated Product Development* – see L. Hein et al. [8], F. Olsson et al. [11], M. M. Andreasen and L. Hein [2] and K. Ehrlenspiel [5].

The common denominator in these new approaches is that the development process is carried out in close interaction between a number of the *functions* or *departments* of the company, such as *design, market, production* and *sales*. Regarding *which* functions to be included in a specific approach is dependent on both *type* (CE, SE or IPD) and *version* of the selected approach. Another characteristic of these procedures is that the work is performed by a *multifunctional product development team (MFPDT)*. Note that these categories/types of product development procedures represent but a fraction of all currently available procedures. One example of an additional approach to and decomposition of the product development process is the managerially–oriented, sequential, *new–product development process* presented in P. Kotler et al. [9].

3 The IPD procedure

In 1969 the late Prof. F. Olsson introduced a procedure model for the product development process or, as he named it at the time, the *Product Origination Process*, [10]. In 1982, this model was revised and renamed. This revision was undertaken in close cooperation between Prof. Olsson and the two Danish researchers Prof. M. M. Andreasen and Technical Director L. Hein. Together they presented the new procedure model under the name *Integrated Product Development* or simply *IPD,* at the *Design Policy Conference* in London in 1982, [8]. Today, IPD is best known as a *general concept* for product development. From the original IPD procedure, two different versions have been derived – a Swedish one by F. Olsson in 1985 [11] and a Danish one by M. M. Andreasen and L. Hein in 1987 (originally presented in Danish in 1985), [2]. In Fig. 2, the overall, functional, structure of both procedure models is presented.

In both IPD procedure models the procedure model is decomposed into *five stages.* In the Swedish model the procedure integrates the *four "parallel" functions* or *subprocesses: market, production, (engineering) design* and *business/finance.* The main difference between the Danish and the Swedish model is that the *business/finance* function is excluded in the Danish version. The difference is due to Olsson's ambition to create a model well–suited for use in the setting up of new companies or subsidiaries, where finance and the establishment of new business are in focus. In the Danish model, on the other hand, primarily companies with an established product development facility and a well–established financial situation are addressed.

The main *integration* in the process takes place between the *four* or *three "parallel" functions/subprocesses.* When *integration* is addressed, it is usually this *"vertical"* integration which is referred to. Apart from this type of integration it is also necessary to mention the *"horizontal"* integration. This refers to the integration *along* the time–axis of the project and *within* each and every one of the subprocesses. One of the key subprocesses is *design and development,* which explains the close connections between design methodology and the IPD procedure models.

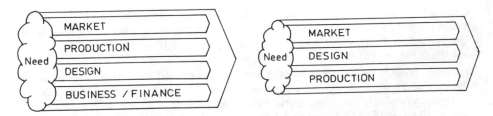

IPD procedure according to Olsson　　　　*IPD procedure according to Andreasen and Hein*

Fig. 2. The overall structure of the IPD procedure models by Olsson [11] and Andreasen & Hein [2].

Finally, in order to give a more complete picture of the structure of the IPD procedure utilized in the *product renewal process,* a detailed procedure model is presented in Fig. 3.

4 Why product renewal?

Since the mid 60s industry–related product development projects have been carried out at the Department of Machine Design, Lund Institute of Technology, Sweden. In the majority of these projects, the focal point has been on the development of *new* products. We have progressed from the initial projects, with a duration of 7–14 weeks, to the one–year projects of today, representing not only an increased cooperation with Swedish industry, but also a continuous development of new procedures and methods/techniques to carry out these projects.

From the experience obtained during these and other industry projects in which the IPD procedure model by Olsson (see Fig. 3) has been used, the following observations represent some of the most important arguments in support of the product renewal procedure:

- The *product ideas* created during the product planning process are often provided in a form very close to a *conceptual solution* (= the working principle of the whole product and its most significant subsystems are determined). The product idea is in other words given in such a concrete form that the possibility to question the selected working principle(s) in the subsequent conceptual design phase is practically omitted. As the product ideas also frequently are created by non–technicians, this may result in major obstacles in the subsequent development in the IPD process.
- The product planning process is mostly dominated by or, in the worst cases, solely performed by non–technicians. This often results not only in situations in which technically insoluble problems are neglected or not identified, but also in situations in which solutions cannot be perceived.
- Product planning is often regarded as a "stand alone" function within the company, focusing on control rather than on participation in the overall innovation process. This is in no way in accordance with the overall objective of creating the best possible conditions for an effective and efficient organization to handle the development of new products.

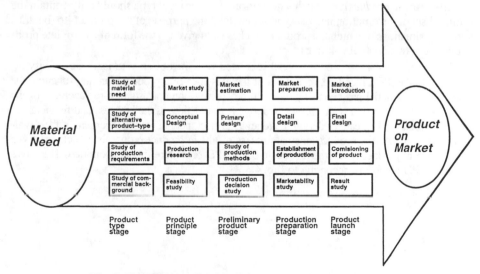

Fig. 3. The IPD procedure model according to F. Olsson [11].

Regardless of the size of the company, the decision to initialize a product development project must always be based upon a solid foundation of facts and figures. This implies that there is an equally important need for handling these issues in smaller companies, without the benefit of a product planning function, as there is in larger companies with access to such a function. As the original IPD procedure model is somewhat unclear as to how the procedure model is to be used in the absence of a product planning function, there is a demand for supplying additional procedural support in order to "substitute" the product planning process in smaller companies.

5 The product renewal procedure model

As the product renewal procedure model encompasses major parts of the activities constituting the product planning process, it is necessary to further elaborate on the contents of the product planning process, before presenting the product renewal procedure model.

The common denominator in the majority of the current product planning procedure models, is the existence of the two major activities *formulation of product policy* and *idea finding*.

The *formulation* of *product policy* consists of two parts, the proclamation of goals and the product market strategy by which the company stipulates which kinds of products are to be developed now and in the future.

The objective set out for the *idea finding* is to breed one or a number of promising ideas which later or immediately will be cultivated into detailed plans for new business activities.

As the product planning process is of a continuous nature, the activity of *controlling* the subsequent development process(s) is often also a part of the planning process. One such implementation is found in the Danish IPD procedure model, since this IPD procedure does not encompass the "built–in" control function provided in Olsson's model by the *business/finance* function.

Like the IPD procedure, the product renewal process should, if possible, be performed by a *team* representing at least the following functions: *market, design, production* and *product planning*, if existing as a company function. Instead of resulting in a *product idea* as in the product planning process, the final results obtained during the renewal process should be a well–defined *product type*. Such a definition of the product type should only contain the minimal set of information necessary to understand the purpose of the product–to–be, thus avoiding situations in which the product idea has been given in the form of a complete product concept (conceptual solution).

As the product renewal procedure model is created with the objective to facilitate an easy integration to the IPD procedure, regardless of the existence of a product planning function, the IPD procedure model has been slightly revised. The revision consists of a replacement of the *business/finance* function with a *project management* function, as the business and financial aspects are already taken care of during the renewal project, but not the management of the development project. As the product renewal process results in product proposals, which are given in the form of product types, the first stage of the IPD procedure model is omitted. This gives a revised IPD procedure according to Fig. 4.

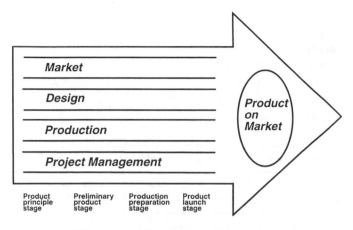

Fig. 4. The revised IPD procedure model.

The four constituent activities of the product renewal procedure model are as follows:

- **Preparation and setting the goals**

  During this activity the following subactivities are to be carried out:
  - □ Clarification of goals, inventory of resources available, critical examination of company "strengths and weaknesses"; the company history might also be utilized.
  - □ Study of existing product history, product assortment as well as the current market and technology status.
  - □ Establishment of product policy and goals and criteria (demands and wishes) for product generation and product evaluation and selection.

- **Initial product generation and primary product selection**

  For the initial product generation the objective is to come up with as many product proposals as possible. One way of achieving this is to utilize intuitive idea generation methods such as brainstorming etc. A search strategy based on *need, resources, situation* and *technology* might also be useful. After the generation of the product proposals, an evaluation of each and every one of the proposals is carried out as a preparation for the subsequent decision on which proposals to develop further.

- **Elaboration of product proposals and final selection of product proposals**

  The most promising of product proposals obtained during the previous activity (= selected proposals) are elaborated upon, thus creating the conditions necessary for the final evaluation and selection of product proposals to be developed (by the IPD procedure).

- **Establishment of product renewal programme**

  For the product proposals derived during the previous activity, a complete renewal programme(s) is to be established. In order to create this programme, time–plans, financial–plans etc. are derived for each and every one of the product proposals, after which an evaluation with reference to six factors such as risk involvement, resources needed, time available etc. is performed. Based on the outcome of this evaluations, a ranking of the proposals is made. This ranking is in turn utilized in creating the renewal programme, in which all of the product development projects are defined.

In Fig. 5, the interaction between the product renewal procedure model and the subsequent IPD procedure model is illustrated.

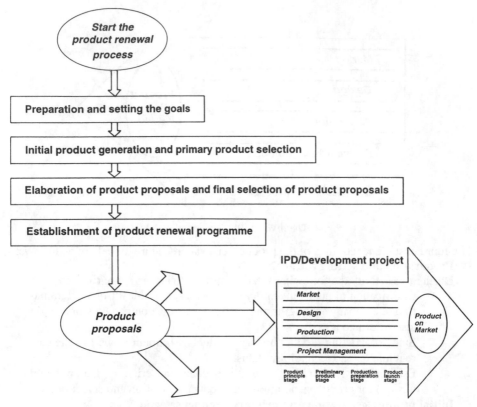

Fig. 5. The interaction between the product renewal and the IPD procedure models.

In a company with a well–established product planning function, the product renewal procedure is most efficiently used in those instances an urgent need for a renewal of the product assortment is at hand. However, this does not exclude the possibility to replace the current product planning function with a modified version of the product renewal procedure. If this concept is chosen, it is necessary to modify the renewal procedure and/or the organization to also address those issues handled by the present product planning procedure on a regular basis. Note that the need for this kind of modifications is only confined to those companies with a well–established product planning function.

6 Product renewal in practice

As previously described, the present state of the product renewal procedure model is the result of the practical experience obtained during a number of renewal projects in Swedish industry. The most important source of observation in this development activity is undoubtedly those projects performed by the students following the *Product Development and Design* curriculum at the Department of Machine Design, Lund Institute of Technology, Sweden.

To give a short insight into the nature of these projects and the feed–back from industry, the 1997 renewal project is briefly outlined. It should be noted that the renewal project forms the basis for a subsequent IPD project, during which only the early phases of the design process are performed. The objective is in other words not to perform a complete product development project, but to supply an insight into the practical use of the product renewal procedure as well as to illustrate the interaction between the renewal project and the subsequent IPD procedure.

This year's product renewal project was performed at *Granuldisk AB* in Malmö, Sweden. *Granuldisk AB* is one of 6 companies belonging to *GS Development AB. Granuldisk* is developing and selling a new type of dishwashers used for potwashing in restaurants and other commercial kitchens.

To remove food remains that have burnt on to trays or pots manually requires a great deal of hard work, patience and strong chemicals. The *Granuldisk* idea or working principle for their machines to remove food remains is copied from nature; the sea laps against stones and rocks and polishes them smooth, but instead of grains of sand *Granuldisk* utilizes specially designed pieces of plastic which scour kitchenware without scratching or damaging.

Granuldisk AB's business idea is based on: *"Improving efficiency and contributing to improving environment, hygiene and economics of all existing potwashing locations with customized systems solutions of the highest quality and reliability".*

An example of a dishwashing machine is given in Fig. 6.

The project to be undertaken by the students was primarily to generate new product ideas based on the working principle utilized in present dishwashers. No further demands or constraints were imposed by the company, which gave the students rather "free hands" to generate new and, hopefully, revolutionary ideas. In practice the students were also free to question and change the current product policy and the company business idea.

The students were divided up into 16 groups with 3 to 5 students per group. In all, 62 students participated in the project. The total time available to each group to perform the project was five weeks – part time.

The results obtained are rather promising. According to an evaluation of the project results performed by Product Manager T. Pehrsson [12], all of the groups have generated at least one product proposal which has the potential to become a future product of the company. This indicates that the students have successfully managed, in a very short time, to acquire the necessary facts and figures to generate and to evaluate and select promising product proposals. However, before a final evaluation of the project results can be performed, the results from the subsequent design activities are also to be considered. As these projects are currently in progress, a more complete evaluation is not currently available. Note that a more specified evaluation needs to involve additional company proprietary information, and is thus, for obvious reasons, not publishable.

Fig. 6. Dishwashing machine *GD 90S*.

7 Conclusion

The results obtained from numerous projects utilizing product renewal in combination with the revised IPD procedure are very promising. However, this does not exclude the need for additional modifications of the present models, including both the renewal procedure and the IPD procedure.

At present the development of a new version of the IPD procedure model, the *A–IPD (Adaptive – IPD)* procedure is in progress and thus also an updated version of the product renewal procedure model. The main reason for these efforts is the need for a more flexible product development procedure model, which can more easily be adapted to the different forms of organizing a company now available. Another influencing factor is the rapid development of different tools to be used during the development process.

Especially the implementation of computer based tools calls for more flexibility to be effective and efficient, to continuously be able to introduce new methods and techniques in the process. An additional factor of interest is the possibility to spread the different functions of a company as well as to facilitate access to expert help by utilizing the possibilities provided by *Internet* and *Intranet*.

8 Acknowledgements

The author gratefully acknowledges the support given by Swedish industry in the development and evaluation of the product renewal process. A very special thanks to Managing Director Per Sandberg and Product Development Manager Lars Lundström at Granuldisk AB for allowing us to perform this year's product renewal projects at your company! I am also indebted to Product Manager Thomas Pehrsson, also at Granuldisk AB, for his help in evaluating the product renewal projects.

References

1. C. W. Allen, Simultaneous Engineering: What? Why? How?, in *Simultaneous Engineering – Integrating Manufacturing and Design,* edited by C. W. Allen, SME (1990), pp 63–68.

2. M. M. Andreasen and L. Hein, *Integrated Product Development,* IFS (Publications) (1987).

3. D. Clausing, World–Class Concurrent Engineering, in *Concurrent Engineering: Tools and Technologies for Mechanical Engineering,* edited by E. J. Haug, NATO ASI Series, Series F: *Computer and Systems Sciences,* Vol. 108, Springer–Verlag (1993), pp 3–40.

4. D. Clausing, *Total Quality Development – A Step–by–Step Guide to World–Class Concurrent Engineering,* ASME Press (1994).

5. K. Erhlenspiel, *Integrierte Produktentwicklung,* (in German), Carl Hanser Verlag (1995).

6. B. Evans, Simultaneous Engineering, in *Simultaneous Engineering – Integrating Manufacturing and Design,* edited by C. W. Allen, SME (1990), pp 3–4.

7. F. Gordon, F. and R. Isenhour, Simultaneous Engineering, in *Simultaneous Engineering – Integrating Manufacturing and Design,* edited by C. W. Allen, SME (1990), pp 14–16.

8. L. Hein, M. M. Andreasen and F. Olsson, Integrated Product Development, in the *Proceedings of the International Conference on Design Policy,* Vol. 2, edited by R. Langdon and S. Gregory, The Design Council (1984), pp 86–90.

9. P. Kotler et al, *Principles of Marketing,* Prentice Hall Europe (1996).

10. F. Olsson, *Produktframtagning,* (in Swedish), Lic. Eng. thesis, Institutionen för maskinkonstruktion, Lunds Tekniska Högskola (1969).

11. F. Olsson et al., *Integrerad Produktutveckling,* (in Swedish), Sveriges Mekanförbund (1985).

12. T. Pehrsson, *Private communication,* Granuldisk AB (1997).

13. N. F. M. Roozenburg and J. Eekels, *Product Design: Fundamentals and Methods,* Wiley, (1995).

14. G. Pahl und W. Beitz, *Konstruktionslehre – Methoden und Anwendung,* (in German), Springer–Verlag (1993).

15. G. Pahl and W. Beitz, *Engineering Design – A Systematic Approach,* Springer–Verlag, (1996).

Determination of Assembly Sequence for Hull Block in Shipbuilding

Hongtae Kim · Byeongse Yoo · Jaemin Lim

Shipbuilding System Department, Korea Research Institute of Ships Ocean Engineering,
KIMM, 171, Jang-Dong, Yusong, Taejon, Korea
E-mail : kht@mailgw.kimm.re.kr

Abstract. When constructing a ship it is necessary to establish a process planning of shipbuilding, which includes fabrication, assembly, and erection of blocks, which are basic units in construction. The assembly process planning determines a production method, a production sequence, equipment and manpower to use, and working condition of block assembly. And the determination of an optimal assembly sequence using positional relationships among the components of a block is one of the most important parts in the assembly process planning. In this paper we developed a simulation program which adopts a neural network to determine a block assembly sequence for shipbuilding. The program derives for a block an assembly sequence based on the Hopfield network, which is created through precedence relationship and the quantification of difficulties in assembling the components of the block. In order to evaluate the performance and the effectiveness of the simulation program developed, the upper deck block of a container ship was selected, and the assembly sequence obtained shows a satisfactory result.

1. Introduction

In spite of rapid growing in the shipbuilding industry the process planning and the production control have been rather lagging behind in progress, resulting in low productivity and high production cost in shipyards. To solve this problem the industry has been concentrating on the research for CIM (Computer Integrated Manufacturing) for the total integration of the design, production and planning.

Currently, however, unnecessary and questionable works are being required in shipyards because of lack of correctly defined job sequence and order. This deficiency is primarily due to the fact that detailed and quality necessary for the production scheduling is not generated in the product design. As the design precedes the production scheduling, the former process naturally cannot produce enough and detailed information for the latter process.

To solve the problem, it is obvious to minimize the difference and it can be achieved through a so-called integrated system, which integrates both the design and production processes. And a considerable attention has been paid to automatic process planning called CAPP (Computer Aided Process Planning), which integrates CAD and CAM. Thus the automatic process planning in concerned with integration of entire production processes such as the production method, production sequence and resources. These are obtained based on the technical knowledge concerning with materials, part, and products through the entire production processes.

To realize automatic process planning, assembly planning is the most important field. Automation of assembly planning begins with analysis of assembly and part models of the product to assemble. Some constraints are frequently involved in automation of the assembly planning. These constraints are acquired from analysis of two models. Automation of assembly planning contains tool/fixture selection, assembly route selection and assembly sequence determination. Among them the determination of assembly sequence is a main part. To find an optimal assembly sequence is closely related to TSP (Travelling Salesman Problem).

This paper investigated an assembly process planning system, in which covers determination of assembly sequence is determined using Hopfield network and precedence-constraint function.

2. Assembly Process Planning

Up to these days, HBCM (Hull Block Construction Method) is widely used in shipyards. HBCM is a method which divides hull into many blocks and builds them up in dock (Storch and Bunch(1995)).

Usually one hull block is composed of several sub-assemblies or part-assemblies that also consist of several pieces or units of pieces.

Block division includes two stages. In the first stage, size of super blocks (pre-erection block) is determined based on the key plan (general arrangement, midship section, lines, construction profile and machinery arrangement). In the second stage, each super block is divided into assembly block considering assembly capacity.

In HBCM, it is practical to plan hull construction in seven levels as shown in Fig. 1. Starting with the block level, work is subdivided down to the part fabrication level to optimize workflow.

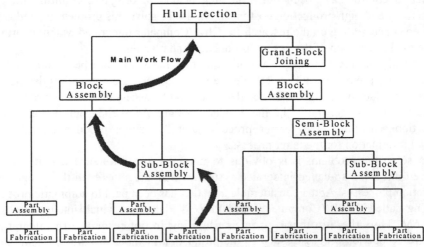

Fig. 1. HBCM Manufacturing Level

962

Information about block assembly to perform assembly process planning contains part relationship, cutting length, fitting length, welding length, welding posture, assembly area, facility and equipment.

Fig. 2 shows analysis of relationships between assemblies; it is a kind of BOM (Bill of Material). As shown in Fig. 2, there are two types of relationship in block analysis. The first, horizontal relationship is determined in design stage. The next, vertical relationship is determined considering economy of assembly in process planning stage.

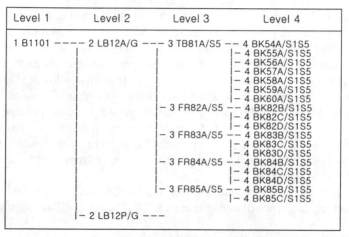

Fig. 2. Block Analysis

3. Determination of Block Assembly Sequence Using Neural Network

3.1 Determination of assembly sequence

Generally, assembly sequence represents transition from the initial state, in which all components are separated, to the goal state, in which all components are in consolidation. A state transition is related to an assembly task, and represents an assemblage of two sub-assemblies.

In assembly process planning, determination of assembly sequence is a very difficult problem, because there are numbers of feasible sequences as the number of component increases. And even an experienced assembly planner does not know all feasible sequences about complex assembly.

Five basic principles are considered to determine assembly sequence in shipbuilding area.

① Avoid a interference between part and sub-assembly by shape and position.
② Balance work load in production line.
③ Minimize transfer and change of direction in production process.
④ Maximize a stability of sub-assembly.
⑤ Maximize a fillet welding task.

To find an assembly sequence considering above principles, the assembly process planning is an essential process.

Various geometric informations have to be generated by computer in performing an automatic assembly process planning. But, up to these days, the assembly process is defined manually using part list and assembly drawings.

In this paper, we made an assembly tree using relationship between assemblies and determined assembly sequences in each level from the highest (level 1) to the lowest (level 4), as shown in Fig. 2.

There are many reviews of generating assembly sequence plans on robot assembly task. Bourjault (1984) considered the assembly of the product and proposed a method to build all the liaison precedence-constraints relative to an assembly. De Fazio and Whitney (1987) also took into account the assembly of the product and proposed a simplified method to elaborate the liaison precedence-constraints. They used Bourjault's concepts but made certain simplifications allowing the practical extensions of the technique to assemblies with more than Bourjault considered. Hommem de Mello and Sanderson (1990) suggested a decomposable production system in which the problem of disassembling one product is decomposed into distinct program. Ko and Lee (1987) presented a system that automatically generates an assembly procedure. They use a virtual link concept to represent relations between components of an assembly.

There are few reviews of application on shipbuilding area. Cho (1995) is reviewed a block assembly sequence planning using a CBR (Case Based Reasoning) method.

3.2 Neural network for assembly sequence

3.2.1 Hopfield network

It is the combinatorial optimization problem to find an optimal sequence among feasible sequences. And this problem is similar to TSP.

TSP is simple to state but very difficult to solve exactly. A set of N cities is given, and the shortest closed path, which passes through each city exactly once, must be found. If an initial arbitrary ordering of the ten cities A, B, C, D, ..., I, J is given, a solution to the TSP is represented as 10×10 permutation matrix.. As shown in Table 1, optimal path would be BAGFCJEDHI, and the total length of this path would be $d_{BA} + d_{AG} + d_{GF} + d_{FC} + d_{CJ} + d_{JE} + d_{ED} + d_{DH} + d_{HI} + d_{IB.}$

Table 1. TSP for 10-Cities

City \ Sequence	1	2	3	4	5	6	7	8	9	10
A	0	1	0	0	0	0	0	0	0	0
B	1	0	0	0	0	0	0	0	0	0
C	0	0	0	0	1	0	0	0	0	0
D	0	0	0	0	0	0	0	1	0	0
E	0	0	0	0	0	0	1	0	0	0
F	0	0	0	1	0	0	0	0	0	0
G	0	0	1	0	0	0	0	0	0	0
H	0	0	0	0	0	0	0	0	1	0
I	0	0	0	0	0	0	0	0	0	1
J	0	0	0	0	0	1	0	0	0	0

In Hopfield network, each entry (X, i) (where X denotes the city and i the position) in this permutation matrix is represented by the voltage output V_{xi} of a single neuronal amplifier. The output of each neuron Xi is connected to the input of each other neuron Yj through a resistance of value $T_{Xi,Yj}$. While neural voltages are allowed to vary continuously, the final result is interpreted by replacing each entry with either a 1 or a 0, depending on whether the neuron's value is high or low (Wilson and Pawley (1988), Hopfield and Tank (1988)).

To calculate Solutions to the TSP, network neurons are represented by an energy function considering the requirements. This requirement can be split into ensuring that stable states corresponding to a permutation matrix are favored and ensuring that matrices representing short paths are favored.

Energy function of TSP is

$$E = \frac{A}{2} \sum_{X=1}^{N} \sum_{i=1}^{N} \sum_{j=1,j\neq i}^{N} V_{Xi}V_{Xj} + \frac{B}{2} \sum_{i=1}^{N} \sum_{X=1}^{N} \sum_{Y=1,Y\neq X}^{N} V_{Xi}V_{Yi} + \frac{C}{2} (\sum_{X=1}^{N}\sum_{i=1}^{N} V_{Xi} - n)^2$$

$$+ \frac{D}{2} \sum_{X=1}^{N} \sum_{Y=1,Y\neq X}^{N} \sum_{i=1}^{N} d_{XY}V_{Xi}(V_{Y.i+1} + V_{Y.i-1}) \tag{1}$$

$$X = successor, \quad Y = predecessor$$

where V_{xi} is the neuron state of the Xth row and ith column, and d_{XY} the cost between two cities. A, B, C, and D are parameters which characterize the performance of the neural network.

The first term will be zero if each row corresponding to a city contains a single 1, with all the other values being zero. The second term is zero if each column corresponding to a position in the tour, contains a single 1. The third term is zero only when the total number of 1's in the network is N, the number of cities. The fourth term describes the length of the tour.

The connectivities $T_{Xi,Yj}$ of the Hopfield network are found to be

$$T_{Xi.Yi} = -Ad_{XY}(1-d_{ij}) - Bd_{ij}(1-d_{XY}) - C - Dd_{XY} \tag{2}$$

$$d_{ij}, d_{XY}: Delta\ Function$$

3.2.2 Consideration of assembly precedence-constraints

There are precedence relationships between parts in assembly tasks. These relationships have to add into the energy function.

Chen (1990) proposed the concept of precedence-constraint assembly sequence. To solve precedence-constraint assembly sequences using Hopfield network, an additional term is added into Equation (1). An additional term is

$$E_r = \frac{F}{2} \sum_{i=1}^{N} \sum_{j=1,i<j}^{N} V_{Xi}V_{Yj} \tag{3}$$

$$X = successor, Y = predecessor, F : positive\ constant$$

The successors and predecessor of precedence condition correspond, respectively, to the right and left sides of the "$\Rightarrow$" operator. That is, the mapping of the precedence-constraints to the Hopfield network can be done by setting the corresponding weights into negative values.

The investigation of parameter and initial conditions affecting the convergence of the Hopfield network has been widely studied.

3.2.3 Cost matrix for assembly task

While the cost matrix in TSP represents the distance between cities, the cost matrix in the assembly planning represents performance of the assembly.

There are two factors determining performance of assembly process. The first is qualitative factor (cost and man-hour), and the second is quantitative factor(stability of assembly, direction of welding). Welding posture is a very important because welding process has influence on both factors.

The types of welding posture are listed as follows:

① Down welding

② Horizontal welding

③ Vertical welding

④ Overhead welding

Assembly planning must try to minimize difficult welding postures like overhead or vertical welding.

In this paper, we developed a method for the cost matrix to quantify the cost for difficulty in assembling (DIA) when performing two tasks continuously. And assembly sequences are determined using assembly man-hour considering welding posture.

3.3 Simulation for determining assembly sequence

Performance of proposed model was tested on real ship. Double bottom block of container ship (D/W 7000 ton-under construction in shipyard A) was used in the simulation (see Figure 3). Construction of grand block "1101" includes part assembly, sub-block assembly and block assembly. In this paper, we concentrate on sub-block assembly and block assembly.

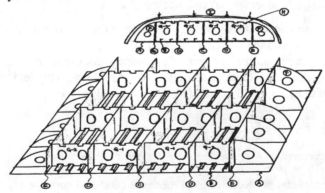

Fig. 3. Grand Block "1101"

As shown in Fig. 3, the drawing of grand block "1101" indicates a symmetry type centering on a center girder. And Table 2 shows a structure of grand block "1101".

Table 2. Structure of Block "1101"

No.	Sub-Assembly		Code
1	Tank Top Plate	101	A
2	Tank Top Longitudinal	102	B
3	Center Girder	103	C
4	No.1 Side Girder	104	D
5	No.2 Side Girder	105	E
6	Open Floor	106	F
7	No. 1 Solid Floor	107	G1
8	No. 2 Solid Floor	108	G2
9	Shell Longitudinal	109	H
10	Shell Plate	110	I

As shown in Table 2, structure of grand block "1101" is composed of ten sub-block assemblies. And Fig. 4 represents fourteen tasks for ten sub-block assemblies.

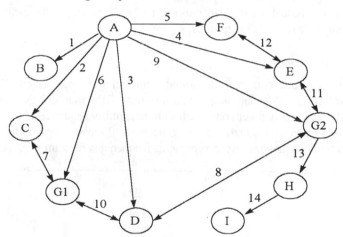

Fig. 4. Assembly Graph of Grand Block "1101"

Assembly sequences are generated by applying Hopfield network that contains precedence-constraints. And cost matrix can be obtained by using a concept of DIA. The actual assembly sequence for block "1101" is shown as follows:
(A) Tank Top Plate ⇒ (B) Tank Top Long. ⇒ (C) Center Girder ⇒ (G1) No.1 Solid Floor ⇒ (D) No.1 Side Girder ⇒ (G2) No.2 Solid Floor ⇒ (E) No.2 Side Girder ⇒ (F) Open Floor ⇒ (H) Shell Long. ⇒ (I) Shell Plate
Table 3 shows precedence-constraints of assembly based on the 14-assembly tasks. And Table 4 shows DIA matrix for block assembly.

Table 3. Precedence-Constraints of Assembly

No.	Preceding Task	Succeeding Task
1	1	5, 8, 9
2	2	5, 8, 9
3	3	5, 8, 9

Table 4. DIA Matrix for Block Assembly

Job	1	2	3	4	5	6	7	8	9	10	11	12	13	14
1	0.0	0.3	1.0	1.0	1.0	0.3	0.3	0.6	0.6	1.0	1.0	1.0	1.0	1.0
2	0.3	0.0	0.3	0.6	0.6	0.3	0.3	0.6	0.6	1.0	1.0	1.0	1.0	1.0
3	1.0	0.3	0.0	1.0	1.0	0.3	0.6	0.3	1.0	0.6	1.0	1.0	1.0	1.0
4	1.0	0.6	1.0	0.0	0.3	0.6	1.0	0.6	0.3	1.0	0.3	0.3	1.0	1.0
5	1.0	0.6	1.0	0.3	0.0	1.0	1.0	1.0	0.6	1.0	1.0	0.3	1.0	1.0
6	0.3	0.3	0.3	0.6	1.0	0.0	0.3	0.3	0.3	0.6	1.0	0.6	0.6	1.0
7	0.3	0.3	0.6	1.0	1.0	0.3	0.0	0.3	1.0	0.6	1.0	1.0	0.6	1.0
8	0.6	0.6	0.3	0.6	1.0	0.3	0.3	0.0	0.6	0.3	0.3	1.0	0.6	1.0
9	0.6	0.6	1.0	0.3	0.6	0.3	1.0	0.6	0.0	1.0	0.3	0.3	1.0	1.0
10	1.0	1.0	0.6	1.0	1.0	0.6	0.6	0.3	1.0	0.0	1.0	1.0	1.0	1.0
11	1.0	1.0	1.0	0.3	1.0	1.0	1.0	0.3	0.3	1.0	0.0	0.6	0.3	1.0
12	1.0	1.0	1.0	0.3	0.3	0.6	1.0	1.0	0.3	1.0	0.6	0.0	1.0	1.0
13	1.0	1.0	1.0	1.0	1.0	0.6	0.6	0.6	1.0	1.0	0.3	1.0	0.0	0.3
14	1.0	1.0	1.0	1.0	1.0	1.0	1.0	1.0	1.0	1.0	1.0	1.0	0.3	0.0

DIA values in Table 4 are divided into easy(0.3), moderate(0.6) and difficult(1.0) cases in performing two tasks continuously.

3.4 Experimental results

Based on the input data just described, simulation is performed. Simulation program is written in MS-C language, running on an IBM Pentium-PC which takes assembly model as input and generate the feasible assembly sequences.

A simulation program is carried out to generate 470 feasible assembly sequences for 10000 assembly sequences. Fig. 5 represents frequencies for sum of DIA.

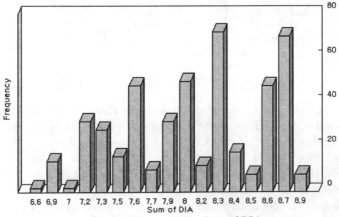

Fig. 5. Frequencies for Sum of DIA

As shown in Fig. 5, the result shows that two assembly sequences have minimum value for sum of DIA among all feasible sequences. Two assembly sequences are shown as follows:

① $1 \Rightarrow 2 \Rightarrow 7 \Rightarrow 6 \Rightarrow 3 \Rightarrow 8 \Rightarrow 9 \Rightarrow 4 \Rightarrow 5 \Rightarrow 12 \Rightarrow 10 \Rightarrow 11 \Rightarrow 13 \Rightarrow 14$

② $2 \Rightarrow 1 \Rightarrow 7 \Rightarrow 6 \Rightarrow 3 \Rightarrow 8 \Rightarrow 9 \Rightarrow 4 \Rightarrow 5 \Rightarrow 12 \Rightarrow 10 \Rightarrow 11 \Rightarrow 13 \Rightarrow 14$

Table 5. Assembly Man-hour and Welding Posture

Task	Assembly	MH	Welding Posture	Tas	Assembly	MH	Welding Posture
1	A ⇒ B	32	Down	8	D ⇒ G2	12	Down
					G2 ⇒ D	24	Vertical
2	A ⇒ C	4	Down	9	A ⇒ G2	24	Down
3	A ⇒ D	8	Down	10	D ⇒ G1	12	Down
					G1 ⇒ D	24	Vertical
4	A ⇒ E	8	Down	11	E ⇒ G2	20	Vertical
					G2 ⇒ E	20	Vertical
5	A ⇒ F	20	Down	12	E ⇒ F	12	Down
					F ⇒ E	20	Vertical
6	A ⇒ G1	24	Down	13	G2 ⇒ H	20	Vertical
7	C ⇒ G1	12	Down	14	H ⇒ I	32	Down
	G1 ⇒ C	20	Vertical				

To determine a final assembly sequence, calculation of assembly man-hour is performed for two sequences using assembly man-hour table in shipyard A (Table 5). As shown in Table 5, tasks(#1, #7, #8, #10, and #12) have a different man-hour depending on assembling orders, which determines welding posture.

Table 6 shows that two sequences (#1, #2), which have the same sum of assembly man-hour, are the final block assembly sequences.

Table 6. Assembly Man-hour for Sequences

No.	Assembly Sequences	Sum of Man-hour
1	A, B, C, G1, D, G2, E, F, G, H, I	252
2	A, C, B, G1, D, G2, E, F, G, H, I	252

In Table 6, #1 sequence is exactly same as the one used by a shipyard shown in Sect. 3.3, while #2 changes the order of B and C sub-blocks from #1 sequence. The best assembly sequence for grand block "1101" is determined that tank top longitudinal, center girder, #1 solid floor, #1 side girder, #2 solid floor, #2 side girder, open floor, shell longitudinal and shell plate, based on the tank top plate, are to be assembled sequentially.

4. Conclusion

In this paper we developed a simulation program which adopts a neural network to determine a block assembly sequence for shipbuilding. The program derives for a block an assembly sequence based on the Hopfield network which is created through the quantification of difficulties in assembling the components of the block.

In order to evaluate the performance and the effectiveness of the proposed model, the upper deck block of a container ship is tested. The result shows that the proposed model can provide a good planning model for the block assembly process. This simulation program is applied as sub-module of integrated hull process planning system that contains assembly planning.

Acknowledgement

The research reported in this paper has been supported by Ministry of Science and Technology of Korea.

References

1. R. L. Storch and H. M. Bunch, Ship Production 2nd edi., Cornell Maritime Press (1988) pp. 67-93.
2. A. Bourjault, Contribution a une approche méthodologique de l' assemblage automatisé: Elaboration automatque des séquences opératiores, PhD thesis, L' Université de Franche-Comté (1984).
3. T.L. De Fazio and D.E. Whitney, Simplified Generation of all Mechanical Assembly Sequences, *IEEE J. Robotics and Automation*, **RA-3**, 6 (1987) 640-657
4. L. S. Homen de Mello and A. C. Sanderson, And/Or Graph Representation of Assembly Plan, *IEEE Trans. Robotics and Automation*, **6**, 2 (1990) 188-199.
5. H. Ko and K. Lee, Automatic Assembling Procedure generation from mating conditions Field Annealing, *CAD*, **19**, 1 (1987) 3-10.
6. K. Cho, Development of Computer Aided Process Planning System for Hull Assembly Shops, *IE Interface*, KIIE, **8**, 2 (1995) 41-53 (In Korean).
7. G. V. Wilson and G. S. Pawley, On the Stability of the TSP Algorithm of Hopfield and Tank, *Biological Cybernetics*, **58** (1988) 63-70.
8. J.J. Hopfield and D. W. Tank, Neural Computation of Decision in Optimization Problems, *Biological Cybernetics*, **52** (1985) 141-152.
9. C. L. Philip Chen, Neural Computation for Planning And/Or Precedence- Constraint Robot Assembly Sequences, *Proceedings of International Joint Conference on Neural Networks* (1990) pp. 127-142.

Computer Aided Systems
for Industrial & Engineering Design
- An Attempt to their Integration

Dieter H. Müller, Claus Aumund-Kopp

Department of Mechanical Engineering, Bremen University
Badgasteiner Str. 1, 28359 Bremen, Germany
e-mail: bik@bik.uni-bremen.de

Abstract. A method is presented which describes an integrative collaboration of both industrial and engineering designers.
A survey on requirements for both industrial and engineering design systems is given and the features and functions of common engineering CAD systems and those of „Industrial Design Systems" are compared.

1. Introduction

The question of industrial design nowadays has become more and more important not only for utility goods but also for industrial products. This is a result of technological developments and market trends which can be summarised as follows:

- due to the trend to miniaturisation (e. g. electronic control systems) and due to necessary coverings (safety) the functions of technical products are no longer visible from outside and therefore have to be made clear to the user by means of industrial design

- the heavy competition in supply industry and capital goods industry and the growing internationalisation of these industries are responsible for the fact that no longer only technical features of a product are responsible for the customers decision to buy the product. These days especially features of industrial design are taken into account more and more.

Although quite obvious - these facts have only been realised and put into action by a few companies yet. This understanding has already been used successfully in some industries, e. g. producers of machine tools. Within other branches industrial design is sometimes used as a subsequent improvement of an existing design only. Thus no Integration of engineering and industrial design takes place and therefore no product improvement is to be expected.

Doing this it has to be kept in mind that the design of a product is strongly effected by the following items /1/:

- technological and physical requirements
- ecological requirements
- manufacturing requirements
- economical requirements

Due to these demands an integration of engineering and industrial design will only be possible, if there is a close co-operation of both the engineer developing the product and the industrial designer determining its appearance. This co-operation should last during the whole phase of development already starting with defining the specifications of the product. Then a product can be expected which fulfils its requirements and which is not only given a functional but also an attractive form and appearance.

Following this procedure does not mean priority of industrial design but rather its integration. Main objective is still the fulfilment of specifications and perfect function of the product.

2. Integration of Product Development and Industrial Design

Qualitative Market Research and Demand Analysis

With the assistance of a method called "Conjoint-Measurement" (a further development by Chaselon & Henning, University of Bremen) an instrument for market research first used in the U. S. /2/ a prospective solution for future product development and product design was found in co-operation with technicians, designers and marketing experts (Fig. 1).

This methodology makes it possible to compare the benefit of use and the image of a product before it is actually brought to market.

A study having pilot character was carried out covering qualitative interviews with specialist customers and users with the help of assessment displays. These displays represent possible configurations of the future product influenced by the factors "brand", "price", "design" and "technology/material".

The evaluation of these assessment displays with the aid of the "Conjoint-Measurement" method and an additional examination of existing offers (products) of competitors then allowed a statement about the relatively most useful and most economical design of the product.

Results of this analysis were market demands, lists of faults of competitive products and a list of wanted requirements supplied by the asked costumers and users. This was used as a basis for the definition of further specifications.

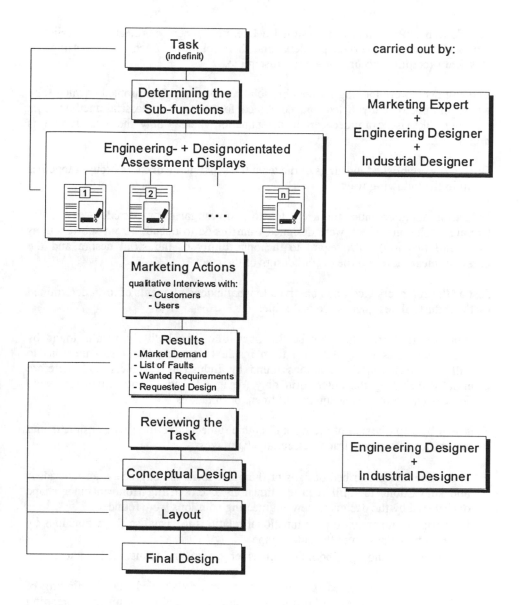

Fig. 1: Integrated Product Development /3/

Integrative Co-operation in Design

Starting from clear technical requirements the conceptual solution for the task was worked out during an integrative co-operation of engineers and designers. According to the VDI-rule 2222 part 1 /4/ especially according to the "Procedure for the Creation of new Products" the design was made.

In Addition a phase had been inserted deliberately where engineers and designers separately dealt with certain problems and their solutions. Each of them using the classical procedures according to their discipline.

The engineers were looking for possible solutions of individual problems which then where morphologically listed and combined following the classical method. The designers likewise examined certain components whereas only the outer form was prescribed.

During the established meetings of the team the separated work was brought together again in the following way:

Industrial designers and engineers together investigated practicable variants for technical solutions which were designed from inside to outside (from sub-functions to overall function) with regard to the opportunity of industrial design and the possible integration into the overall design of the study.

Just as the engineers vice versa observed to what extend elements of form determined by the industrial designers were realisable.

Summarised this procedure can be described like this: both the results found by engineers and thus being designed from inside to outside (from sub-functions to overall function) and the solutions found by industrial designers and therefore essentially including the outer form only (designed from outside to inside) were examined concerning the adequate fulfilment of functions.

This step-by-step division of the work of both engineers and industrial designers and its integration afterwards lead to essential advantages:

- encouraged by prescribed designs of the outer form the engineers got new ideas and suggestions to fulfil required functions i. e. starting from an outer shape determined by the designers new engineering solutions were found
- industrial designers used pure functional relations to visualise these functions by an appropriate design of the outer shape
- mutual understanding of both disciplines (engineers and industrial designers)

This procedure was repeated several times and clearly showed that the finding of new ideas was not influenced but the engineers developed a certain understanding for the overall design of the outer shape and the designers understood technical necessities to provide functionality.

At the end production documents were generated and a prototype was manufactured and tested.

3. Technology and Design

As a rule design, the formal image of new products, nowadays is not yet a result of interdisciplinary teams and true co-operation of engineer and designer. Still to be found are industrial designers working more or less independently from the process of engineering design creating their own concepts and ideas of the shape and appearance of the new product. It is quite obvious that in this way no mutual enrichment among engineer and designer can come about. Thus it is not surprising that the task of industrial design is also not connected to the schedule of the usual engineering design process.

A direct result of the fact described above is that a formal image is prescribed from the start as a solution of the task of industrial design and engineers have to comply with the appearance of the product. Vice versa industrial design is adapted to a finished or almost finished technical solution.

Using the method of integrated product development this problem is taken into account by the fact that the task of industrial design is already situated at the very beginning the procedure (Fig. 2).

Because of that engineers for example benefit from representation techniques of industrial designers:

- new ideas can be translated visually
- the formal impression of new solutions can be judged early
- different 3-dimensional representations economise manufacture of samples and prototypes
- industrial designers bring in unconventional but visually ascertainable and therefore "understandable" ideas

On the other hand designers are kept away from following unrealistic "dreams of styling" too long.

Mutual criticism and help lead to a useful and optimised result with lowest waste of time and money.

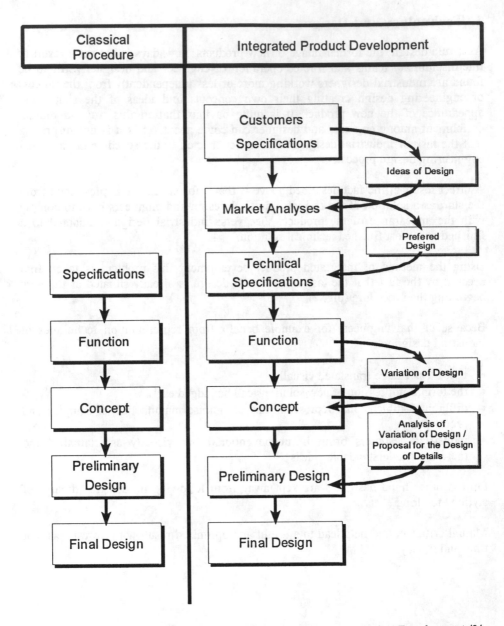

Fig. 2: Comparison of Classical Design Procedure and Integrated Product Development /3/

4. Demands upon a CAD System for Engineers and Industrial Designers

As we have seen engineers and designers can or rather must work together. Nevertheless they still work on different features of a product and follow different ways to work. Thus there are different demands upon the CAD systems they apply.

CAD systems for engineers have to enable their users to work precisely and they have to make clear representations possible.

Looking at the systematic way an engineer develops a product: from the determination of its function (requirement, specification) via finding an effect carrier (part, component) to the complete product, these kinds of systems also have to support their user by finding out further technological data. These additional system functions can be an FEM-module for example determining load and stress/strain behaviour of a part or it can be sub-routines to check the kinematic behaviour of a part in an early stage of design.

A system especially unable for industrial design has to include features concerning the development of the outer shape and appearance of a product because a major part of an industrial designers work deals with this „cover". Doing this the usual way of an industrial designer is the description and representation of an overall function towards the elementary functions and their effect carriers.

Thus the external form and design of a product is divided into separate parts and developed one after another. At first such system for the purpose of industrial design can do without precision. However it principally has to possess extended functions to edit and adapt shape, colours, textures and graphic features of a product.

Software given the attribute „industrial design" primarily means systems that exceed the ability of common CAD systems concerning the features mentioned before. These systems particularly provide:

- 3-dimensional modelling concepts to generate geometry, including freeform geometry
- facilities to generate and to work on graphic features
- modules for photorealistic presentation and visualisation

5. Comparison of Features of Common CAD Systems and a Computer Aided Industrial Design System

As an example we have taken a software bought as an „CAID" system (Computer Aided Industrial Design System) which is capable of 3D-modelling and photorealistic visualisation /5/, /6/.

The procedure to generate the form and design of the product takes place in the following steps - analogous to the separate parts of the external form and design of a product (Fig. 3):

- Generation of geometry
- Shaping
- Colouring
- Texturing

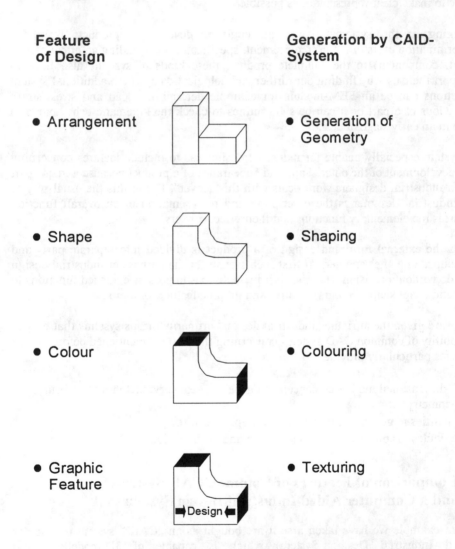

Feature of Design

- Arrangement

- Shape

- Colour

- Graphic Feature

Generation by CAID-System

- Generation of Geometry

- Shaping

- Colouring

- Texturing

Design

Fig. 3: Analogy of Product Design

In contrast with usual systems for engineers which functions are often limited and which only have predefined basic geometries like cylinders or rectangular solids to generate geometries the examined industrial design system enables its user to model freeform shapes intuitively. Doing this the designer can abstain from precision but he does not necessarily have to. The generation of freeform shapes here is done by rotation or extrusion of any cross section for example without being limited to simple circles or lines as basic elements.

On the other hand the abilities of the examined industrial design system concerning common features as measuring and hatching functions are quite restricted. There are also no special functions to determine technological data.

For the further forming of design the shapes of geometries generated as described before can be edited in multiple ways.

Formed geometries can be trimmed (when intersecting each other), they can be blended or they can simply be arranged in new positions. All this is supported by sophisticated preview modes which enable the designer to regard the geometry in different ways e. g. in illuminated scenes and surfaces or with varying colours where other systems often offer only rotated views. Thus the user can have a close look at his model and get a first impression of its overall appearance.

Of particular importance with systems used for industrial design are their functions for defining colours and the appearance of materials as well as functions for texturing and graphic processing. In these systems many more features of that kind are implemented than in „simple" systems for mechanical engineering only. These are especially possibilities to mix colours or to define and assign physical effects like reflection, roughness or refraction and corresponding parameters to surfaces.

Functions to define 3-dimensional textures like „marble" and to assign them to surfaces of geometries in that way that even edges look like having been cut off considerably reduce the expenditure which would usually be necessary to the get same effect.

Furthermore the system being tested here enables its user to import graphic features like drawings, pictures or patterns, to edit them and to attach them to one or more surface elements. For example as a tag in the centre of a surface or mapped around the surface element or elements in total.

Particularly for the creation of photorealistic images the tested software possesses modules to define the complete scene of a picture including bottom, fore- and background, perspective of the camera, lights and shades etc.

Comparing the features of common CAD systems and those of special systems for industrial design it can be seen that most of the different requirements have been taken into account during the system development.

Nevertheless and due to the modular architecture of present CAD systems there is no sharp boundary between systems merely for mechanical and engineering design used to produce drawings and technological data or systems for industrial design mainly visualising the outer form and appearance of a product. Many engineering systems nowadays can be upgraded by adding modules performing computation of kinematic data on the one hand and modules visualising products as photorealistic images on the other hand.

References

1. H. Seeger, Design technischer Produkte, Programme und Systeme: Anforderungen und Lösungen. Berlin, Heidelberg: Springer Verlag (1992)

2. J. Henning, F. Chaselon, Abschlußbericht „qualitative Marktstudie und Bedarfsanalyse für das HATEDE F & E Verbundprojekt", unveröffentlichter Bericht, Universität Bremen (1992)

3. D. H. Müller, K. Oestermann, Integrative Zusammenarbeit von Konstruktion, Marketing und Design, Proceedings of 9th International Conference on Engineering Design, The Hague (1993), 1197-1204

4. VDI-Richtlinie 2222, Blatt 1, Konzipieren technischer Produkte, Düsseldorf VDI-Verlag (1977)

5. C. Aumund-Kopp, Design Systeme - Nutzen und Anwendbarkeit für Ingenieure, - Bremen, Universität, Bremer Institut für Konstruktionstechnik, Studienarbeit (1994).

6. C. Aumund-Kopp, Integration des Designsystems "DeskArtes" in die Prozeßkette eines Yacht- und Schiffskonstruktionsbüros, Universität Bremen, Bremer Institut für Konstruktionstechnik, Diplomarbeit (1995)

Integrated System of CAIP and Quality Forecasting for Mold Manufacturing in CE Environment

Chen Kang Ning, Gao Guo Jun, Chen Shan Hong, Zhang Jun Mei

School of Mechanical Engineering
Xian Jiaotong University, Xian, P.R.China

Abstract. This paper describes a integrated system of the CAIP and quality forecasting for sculptured surface of dies and molds in ball-end milling process. In the system, the machining process planning, inspecting planning and machining error prediction are carried out concurrently. Five interrelated modules are being developed to construct the proposed integrated system. They are: (1) part accuracy feature module, (2) the platform for concurrent development, (3) automatic machining process planning module, (4) automatic inspecting planning module, (5) machining error predication module. The machining error predication module is established based on contributory factors of machining errors. The optimum matching method between the measurement data and CAD-base design data is presented. The machining accuracy evaluation of sculptured surface is described. The predicting results of the dimensional errors to determine the number of the measurement points and their distribution are discussed.

1. Introduction

The traditional approach for the mold manufacturing is in sequence and series fashion. The machining process planning, the NC code generated , the machining execution , the inspection planning and the evaluation of inspecting results are carried out sequentially. The modules are of separation and there is no information exchange between these processes. Once the inspection result is beyond the scope of specification, it is necessary to repeat the processes sequence again. In this paper , a integrated system of CAIP and quality forecasting for mold manufacturing in CE environment is presented. The system is consisted of the accuracy feature module, platform for concurrent development, automatic machining process planning module, automatic inspection planning module and machining error predication module. The client/server pattern is used in the system. On the basis of analysis major error source, the machining error predication model is established. In order to eliminate errors caused with the difference between the machining datum and the inspecting datum, the optimum iteration progress is presented. The variation of inspecting results associated with the change of sampling scheme is closely related to the curvature of surface profile and process capability. The predicting results of the dimensional errors to determine the number of the measurement points and their distribution are discussed.

2. Framework of the System

The Fig.1 shows the structure of the system.

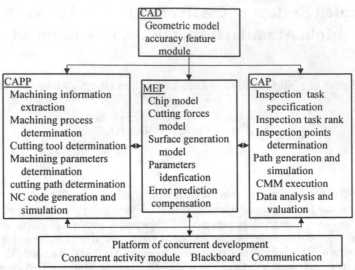

Fig.1 Structure of the system

• CAD module Based on the commercial software the accuracy feature model of the part was developed by use of the neutral file. The CAPP and CAIP accept the accuracy feature model of the part as input and extract concurrently the machining and inspecting information .

•CAPP module The following tasks are implemented in the module. They are: (1) selection machining fashion on the sculptured surface based on the code rules, (2) determination of cutting tools, (3) determination of machining parameters, (4) determination of cutting paths, (5) generation of NC code and the simulation.

• Machining error predication module According to the cutter information, machining parameters and cutting path from CAPP module , the chip geometry model, the cutting force model , the tool defection model and surface generation model are established. The predication result of machining error is fed back to the CAPP module to modify the machining paths. At the same time, the predication results are used to modify the number of inspection point and their distribution in CAIP module. The inspecting results is used to modify the parameters of
the predication model of machining error. So that the machining error predication module is integrated with the CAPP module and the CAIP module.

• CAIP module According to the accuracy feature model of CAD and the predication results of machining error, the six functions are developed in the CAIP module. They are : (1) the definition of inspection information and information acquisition, (2) align part, selection probe and inspection sequence, (3) determination the number of inspection points and their distribution, (4) determination the inspection planning and simulation, (5) CMM execution , and (6) analysis and evaluation.

• Platform for concurrent development In order to support the resource management, control and eliminating the conflict between the tasks, the platform is consisted of concurrent activity model, blackboard and communication. The function of platform is shown in Fig. 2.

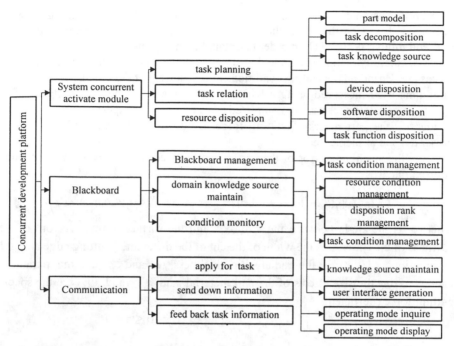

Fig. 2 The function of platform

concurrent activity module supports the functions, including the task planning ,task relation establishing and resource disposing. The concurrent activity model provides the relationship between the tasks and the task decomposition. When the tasks are accurately modeled , the optimum control of the system is realized.

Blackboard provides the system management and control, domain knowledge source maintain and system condition monitoring. To ensure the information insistent and safety, the communication between the tasks is completed through the blackboard. The condition monitoring shows various task condition in real time.

Communication To ensure effectively transfer information between various tasks, the double-action communication are built through the platform for concurrent development. The double-action function is as following: task number, disposition information, feed back information and operating condition.

3. Machining Error Prediction Module

The machining error prediction module accept the parameters of the cutter , machining condition and cutting paths from CAPP module and the number of inspection points and their distribution from CAIP module as inputs to predict machining errors, while CAPP module and CAIP module accepts the predication results of machining error as inputs to modify cutting paths, the number of inspection points and their distribution. It is very important to build up the accuracy prediction model of machining error.

Compared with the flat-end milling process the cutting process for sculptured surface using ball-end milling is more complicated due to the complicated and constantly changing cutting geometry. the predication model of machining error was presented based on the result that the tool deflection is the dominant factor in the ball-end milling process[1]. In order to obtain the instantaneous cutting force model, a series of carefully randomized model-building experiments was performed. The cutting forces are depend on the cutting conditions of the part-cutter system and in the X and Y directions were measured using a Kister model three-component piezoelectric force dynamometer. It is necessary to completed a large amount of experiments. An approach to identify the parameters of cutting force using the result of machining errors is presented and implemented.

the machining error of the surface normal vector at a designed point can be approximated as

$$e = (\sum_{z=z_{min}}^{z_{max}} (\sum_{i=1}^{n} (K_T(z)[t_i(\theta,z)]^{M_T}[-\cos\theta_i(\theta,z)] + K_R(z)[t_i(\theta,z)]^{M_R}[\sin\theta_i(\theta,z)])) \cdot h) \cdot \sin\phi \Big/ k \quad (1)$$

where ϕ is the angle between the tool axis and the surface normal vector, k is the stiffness of the cutter, h is the width of the cut of the differential cutting edge along the z direction, $t_i(\theta,Z)$ is the radial chip thickness, M_t and M_r are the parameters characterizing the size effects of the part material, and $K_t(Z)$ and $K_r(Z)$ are the specific cutting force coefficients.

$$K_T(z) = a_0 + a_1 \cdot (\frac{z}{R}) + a_2 \cdot (\frac{z}{R})^2 + a_3 \cdot (\frac{z}{R})^3 \quad , \quad 0 \le z \le R$$

$$K_R(z) = c_0 + c_1 \cdot (\frac{z}{R}) + c_2 \cdot (\frac{z}{R})^2 + c_3 \cdot (\frac{z}{R})^3$$

$$K_T(z) = a_0 + a_1 + a_2 + a_3 \qquad\qquad Z > R \quad (2)$$

$$K_R(z) = c_0 + c_1 + c_2 + c_3$$

The variables in the function (1) can be divided to two categories. A part of the variable can be accurately calculated through geometric relation such as ϕ, $t_i(\theta,z), \theta_i(\theta,Z)$ and the others have be obtained through the method of parameter identification based on the machining errors which are measured by CMM.

We use the developed software to calculate the variables of first category and the variables of second category are determined to use the method of system identification. For exempla, based on the results of measured errors , a numerical fitting procedure is devised to evaluate the model parameter. The tests were carried out and details of the cutting conditions are shown as following.

Test model: One set slope model and one 2-D extended surface model;

Cutting strategy: Cutting along -X direction;

Machine tool: MAHO MC50 3 coordinate horizontal machining center;

Cutter: high speed steel ball-end mill with 5 mm diameter, two right-handed flutes with 35 helix angle;

Part material: aluminum;

Cutting depth: 2mm; Spindle speed: 1000rpm; Feed: 350mm/min;

Without coolant: The identification coefficients of cutting force $K_T(z)$ and $K_R(z)$ are:

$$K_T(z) = 1360 - 2082 \cdot (\frac{z}{R}) + 622 \cdot (\frac{z}{R})^2 + 278 \cdot (\frac{z}{R})^3$$

$$K_R(z) = 458 + 324 \cdot (\frac{z}{R}) - 1664 \cdot (\frac{z}{R})^2 + 1001 \cdot (\frac{z}{R})^3$$

$$0 \le z \le R$$

$K_T(z) = 178$,

$K_R(z) = 119$

$z > R$

$M_T = 0.86$

$M_R = 0.85$

In order to verify the prediction ability of the developed approach, a machining sculptured surfaces experiment was performed . the inspection results was listed in Table 1 and in Fig,3. It can be seen that the correspondence between the prediction data and measured data is well.

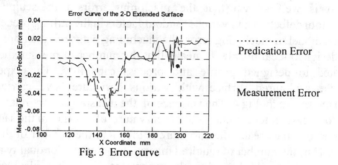

....
Predication Error

_____
Measurement Error

Fig. 3 Error curve

Table 1 Inspection result

	X Coordinate (mm)	Y Coordinate (mm)	Z Coordinate (mm)	Predict Error (mm)	Measuring Error (mm)
Predict Point 1	131.7077	21.2	-31.9336	0.000	-0.0056
Predict Point 2	138.2839	21.2	-30.3808	-0.0132	-0.0225
Predict Point 3	146.4617	21.2	-24.5949	-0.0387	-0.0468
Predict Point 4	151.1904	21.2	-15.9133	-0.0592	-0.0623
Predict Point 5	152.0111	21.2	-9.5795	-0.0326	-0.0402
Predict Point 6	154.7824	21.2	-3.9356	-0.0312	-0.0368
Predict Point 7	169.9032	21.2	-2.0000	0.000	0.0042
Predict Point 8	184.5253	21.2	-2.8131	0.008	0.018
Predict Point 9	188.7286	21.2	-5.3661	0.011	0.0084
Predict Point 10	191.7029	21.2	-9.3394	0.0072	-0.0023
Predict Point 11	193.0710	21.2	-15.9944	0.0116	0.021
Predict Point 12	194.4556	21.2	-19.5208	0.223	0.0224
Predict Point 13	203.2589	21.2	-24.2875	0.01	0.019

4. Inspection Planning Module

The inspection planning module aims at obtaining accuracy valuation for inspected part based on less measurement points. The number of the required inspection points, their distribution and the accuracy valuation of the measurement data are primarily concerned in this module. Owing to the complicated dimensional measurement , the number of required points for obtaining effective quality assurance is unknown (4). The accuracy of sculptured surface is defined as the profile error. It is necessary to use other datum of part when the part is machined and inspected. The errors which is caused the difference between the using the machining datum and inspecting datum are larger and it usually leads to exceed the tolerance limit. In this paper, the number of measurement points and their distribution in close relationship to the curvature of the surface profile and process capability is investigated, and for the accuracy valuation of the measured data an optimal match algorithm is discussed.

- The Number of Measurement Points and their Distribution

A surface profile is within tolerance if the deviation at any point on the surface is within the specified bound. Cases may occur when sampled deviations are within bound while non-sampled deviations are in fact out of tolerance. How to select a set of minimum number of measurement points for achieving the best possible accuracy it is necessary. It is well known that the machining errors of the sculptured surface mainly suffer tool defection and various random factors in the milling process. Through the prediction model of machining error, the machining paths of cutter are modified to unify with designed ideal profile. It means that the number of measurement points is mainly related to designed profile and process capability. The experiments of machining the sculptured surface with various curvature have carried out. The relationship are shown in Fig 4. The variance of the measured errors relates closely to the number of measurement points and the curvature of profile. In the same variance of measured errors, the greater the curvature is the more the number of the inspecting points needs. Thus the number of required measurement points for quality assurance of the surface profile not only depend on the process capability but also depend on the curvature of profile. In the Fig 5, curve(a) represents the result of using equally spaced sample and curve(b) represents the result of the sample interval to relate to curvature. It is shown in order to obtain equal variance of the measurement error , that the equally spaced sample needs more measurement points.

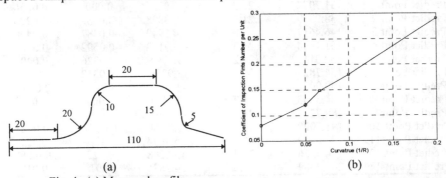

(a)   (b)

Fig. 4   (a) Measured profile
            (b) Relationship between curvature and number of inspection points

- Error Analysis and Valuation

Owing to the difference between the using the machining datum and inspecting datum and various factors in the machining process, two part errors were contained in the measurement data of the surface profile, that is systematic errors and random errors. In order to obtain the systematic error from total errors, the principle of used algorithm is minimization of the sum of squared distance of the measured coordinates from the surface with respect to the six variables of a rigid body transformation. The program flowchart of algorithm is shown in Fig. 6. The measured data and transferred data after using the algorithm are shown in Fig 7. It is obvious that before the algorithm the greatest normal error and the object function F is 0.257mm and $1.225mm^2$ respectively, but after the algorithm the greatest becomes 0.059mm and $0.038mm^2$. This algorithm can be adopted for the use of locating design coordinate system and matching measurement data with design data as well.

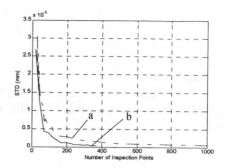

Fig 5 Result of deferent sampling scheme

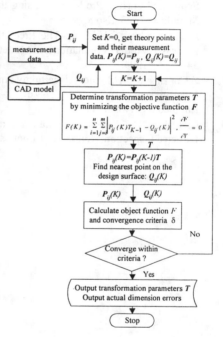

Fig. 6 Flow diagram of optimal matching algorithm

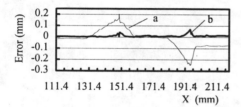

Fig.7 Measured result (a) and transferred result (b)

5. Conclusion

In this paper we construct a integrated system of the CAIP and quality forecasting for mold manufacturing in CE environment. The various information and resource in the system is shared. The concurrent operation and feed back control between modules is realized. In order to obtain conveniently accuracy prediction model of machining error, the method of parameters identification based on the machining errors which are measured by CMM is effect. Based on experiment and analysis, the number of required measurement points for quality assurance of the surface profile not only depend on the process capability but also depend on the curvature of profile. When the process capability is higher and the errors were compensated according to the prediction model of machining error , the number of required measurement points mainly depend on designed part profile.

6. Acknowledgment

The authors would like to thank the National Science Foundation of China and the Doctorate Academic Foundation of Education Committee for support to this work.

Reference

1. E.M.Lim, H.Y.Feng, C.H.Menq and Z.H.lin. The Prediction of Dimensional Error for Sculptured Surface Productions Using the Ball-End Milling Process. Int.J. Mach. Tools Manufact. Vol.35.No.8.1995

2. H.Y. Feng and C.H.Menq. The Prediction of Cutting Forces in the Ball-End Milling Process-1. Model Formulation and Model Building Procedure. Int. J. Mach. Tools e. Vol. 34. No.5. 1994

3. Z.C.Yan and C.H.Menq, Uncertainty Analysis for Coordinate Estimation Using Discrete Measurement Data.   MED-Vol. 2-1/MH-Vol. 3-1, Manufacturing Science and Engineering ASME 1995

4. C.H.Menq, H.T.Yan and G.Y.Lai, Automated Precision Measurement of Surface in CAD Directed Inspection. Journal of Robotics and Automation . Feb. 1990.

Rapid Prototyping

Development of Complex Plastic Parts with Simulation and Rapid Prototyping

Michael Wilmsen

TransCAT GmbH, Karlsruhe, Germany

Abstract. As the development and construction of plastic parts specially for the automotive industrie needs to proceed faster, certain CAE-moduls have been implemented into 3D-CAD-Systems to optimise this proces. One possibility is the simulation of the production from parts for injection moulding. So it is possible to analyse the part during the construction. the best injection points can be chosen, welding lines and airtraps can be prevented. The following step is the production of a „Rapid Prototyping" part. With this model it is possible to produce an easy injection mould for first parts in thermoplastic material within 2 or 3 weeks. The following paper shows samples of products which have been produced in the way described.

1 Introduction

In the last two years, the most CAE software producers spread the slogan of "better - cheaper - faster", which should be a statement for the more efficient use of their programs. Thinking about these points means not only to buy a good CAD system for the development and construction of parts, but also thc possibility of using the models for several following steps.

The production of plastic parts (e.g.: injection moulding) allows the engineer to design free surfaces for the best visual and technical effect. That means as well that complex models, which are produced in a 3D CAD have to be designed only once. After designing it should be possible, to take these data and use other CAE systems to optimise, simulate or produce prototypes.These following steps need a special system and a specialist to handle it. Figure 1 shows the possibilities in developing plastic parts and the required injection moulds. Because of this knowledge, that each module contains, it is not possible, to develop an universal CAE system. Some CAE developer have already recognised this circumstances and stopped their own research on special programs like for example the simulation of injection moulding.

The reason is the big knowledge that expert systems, for example simulation software needs, to copy the real part forming process. So we reach the most important points for an effective CAE system:

◆Using several expert-systems for the different steps of a product development.

◆Combining these special software products in one CAD system.

Only the attention of these two points gives the user a helpful system for developing and constructing parts perfectly. Paying attention to these points, we must recognise, that in the last years, the expert systems got better and better, but they had only few possibilities, to transform part information from a CAD system to an expert system without loosing important information of surfaces or anything else.

So the second step, after developing and during optimising expert systems, is the connection of these systems to the CAD. That means designing the part structure only once and using this geometry for all the following steps as shown in figure 1.

Let us have a look to two samples of these modules. At first we see the different possibilities in simulating a mould injected part and interpreting the results. The second module is the realisation of first parts with rapid prototyping and rapid tooling.

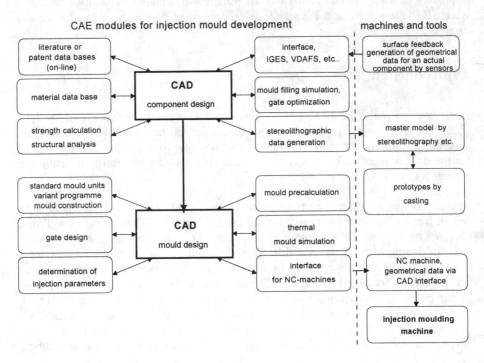

Fig. 1. CAE modules for developing plastic parts and injection moulds

2 Simulation

Complex plastic parts should be designed in 3D CAD systems (figure 2). Then you have a lot of surfaces or a solid part. Either, the first step is the transformation of the model in a part, which can be simulated. As we know that the filling process of an injection mould is a complicate rheological and thermal process, it is necessary to use simplifications for describing the process in a simulation. For example, we cannot consider the thermal conductivity in the direction of flow without raising the simulation time drastically. So it can happen, that the results have to be interpreted critically.

The most important simplification is the use of shell elements for simulating the flow path. That presents the problem of the preprocessing. Either, we take one of the part surfaces (outside or inside) or we need the middle plane. In each case we have to change the 3D part and make new surfaces, mesh it and optimise the mesh. (figure 3).

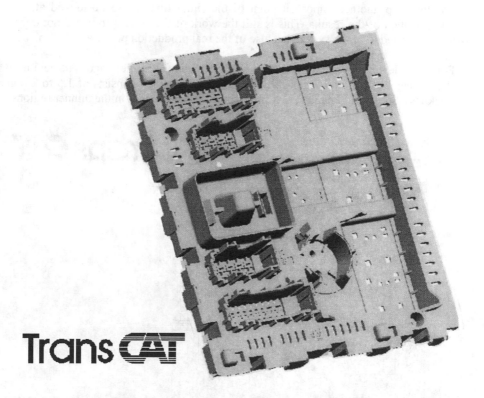

Fig. 2. 3D part (platine) before generating a shell mesh (CATIA; picture: TransCAT GmbH)

Meanwhile, designing this geometry can be realised in the connected CAD system. As we see the steps in figure 2 and 3. As you can imagine, this additional work keep a lot of designers away from using this expert system. What are the possibilities in the future of optimising this procedure:

◆Simplification of the meshing (e.g. meshing the 3D modell and producing automatically the middle plane shell mesh)

◆Meshing the 3D part with shell and tetraeder elements (depending on the geometry).

◆Meshing the 3D part with a tetraeder mesh and simulating the formfilling in 3D.

The three points show the steps of developing the system. Meanwhile it is possible to mesh the part from a 3D mesh to a shell mesh. The other steps will be realised within the next years, depending on the performance of the hardware. These points stand for the commercial use of the system.

At this step another important point of the simulation has to be looked at: The interpretation of the results. This is still the work of the user, who needs not only a good knowledge of the system but also of the real production process.

For example taking the platine that you have seen on the further page. The geometry shows regions of only 0.7 mm thickness. The main region consists of 1.5 to 2 mm thickness. So we can imagine, that filling problems will occur in the thinner regions.

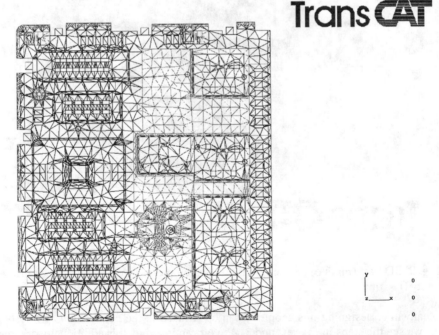

Fig. 3. Shell mesh of the platine (CATIA MESH; picture: TransCAT GmbH)

As the simulation shows (figure 4), it happens that the part cannot be filled completely, because the melted material, a glasfibre reinforced Polybutylentherephtalat (PBT), needs more pressure near the gate to fill the region. Further away of the gate, the flow front has nearly filled the thicker walls, the pressure in the part has been grown high enough and the other thin regions can be filled. But now the melt at the thin plate near the gate has already been frozen, the filling is no more possible. With this part we can see a typical case of the limit between simulation and reality. The result is realistic, as we can see in the part, which has been filled incompletely at first. But at the end of the optimisation at the injection moulding machine, it was possible to fill the part with high pressure.

This effect appears in few cases, because simulation software cannot consider the rheological fluidity in the regions of very slow filling, recognising that a very thin frozen layer at the flow front can be broken through with high pressure. With the right background knowledge it is possible to recognise this software problem in advance and give the right result to the client. Nevertheless the simulation of plastic parts is an important tool for developing and designing complex plastic parts in shorter times.

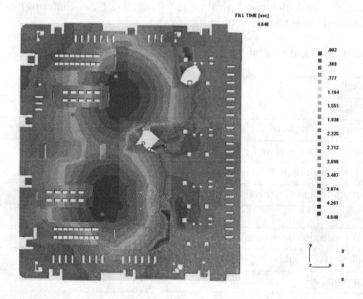

Fig. 4. Simulation result (flow front) of the platine.

3 Rapid Prototyping

The next developing step of a plastic part after the optimisation is the production of first prototypes. Here we have to distinguish between:

◆first prototype

◆design model

◆samples for first testing (similar material, e.g. thermosets)

◆samples for specified testing (original material)

Meanwhile for the first prototype, several "rapid prototyping" systems are existing on the market (e.g. stereolithographie). Herewith it is possible, to build a realistic part out of a 3D CAD system within some hours. If we look at the five best known prototyping systems, we recognise the pros and cons of these systems (table 1). All systems work with producing slice by slice of a part. The system work either with a bath of liquid thermosets (Stereolithographie), a small extruder for thermoplastic material (Fused Deposition Modeling), a thermoset on a solid ground (Solid Ground Curing; Cubital), thermoplastical pellets (Selective Laser Sintering) or paper sheets (Laminated Object Manufacturing).

Table 1. pros and cons of the different rapid prototyping systems.

Property	STL/SLA	Cubital	SLS	LOM	FDM
Thermoset	++	++	0	0	0
Thermoplastic material	0	0	++	0	++
Parts with high elasticity	+	-	-	+	++
Parts with high strength	+	-	+	++	++
Short production thin parts	++	+	-	+	+
Easy finishing	+	+	-	-	+
Short production thick parts	+	++	-	++	+
Production of moulds	+	-	-	++	-
Each geometrie	++	++	+	-	-
Design model	+	++	-	++	+
Test model	+	-	+	++	+
Simple using of the machine	-	-	-	+	++
Tolerance	++	++	+	(+)	-

0   not possible; -   bad results; +   good results; ++ best results

Regarding table 1, the possibilities of the rapid prototyping systems are in the first two, maximum three model variations (see points above). But what happens, when a part of origin thermoplastic material is needed in a very short time (e.g. the production of an injection mould takes mostly 2 months).

Therefore the following step is to produce an injection mould with these systems or with a rapid tooling material.

4 Rapid Tooling

Therefore a mould can be produced with a prototype (figures 5, 6). This happens with a aluminium filled thermoset in the following steps:

◆preparing the prototype as needed (e.g. coating or polishing)

◆preparing the different inserts for the mould (figure 5)

◆positioning the part in the mould to get a good ejection

◆preparing the separation zone

◆filling the different mould parts with thermoset around the prototype, positioned in an aluminium frame.

◆fitting the aluminium frame in a prepared injection mould out of steel (figure 6)

◆installing the ejectors.

Fig. 5. Prototype with prepared cores for the mould

Producing the mould takes about 2 weeks. It is possible to produce between 50 and 500 parts in origin material (e.g. Polyamid 6.6 with 30% glasfibre). Also in this process it is possible, to integrate the CAE modules. The contours of the mould inserts can be prepared with the milling mashine, taken directly from the system.

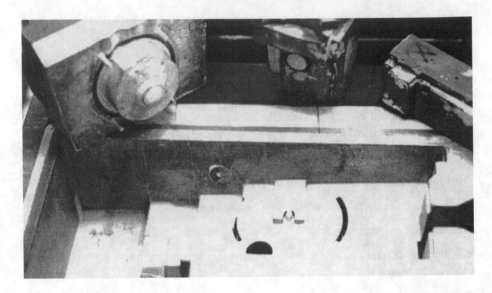

Fig. 6. Complex mould with several cores, with aluminium frame and steel mould

6 References

1.   Mit Kunststoffen zu neuen Produkten; Symposium FhG Karlsruhe 1997

2.   Michael Wilmsen: Seminar: Rapid Prototyping and Rapid Tooling 1996

3.   TransCAT Firmenschriften: Kunststofftechnologie, 1996

4.   Moldflow Firmenschriften

5.   Gebhardt, A.: Rapid Prototyping, Carl Hanser Verlag München 1996

6.   M. Wilmsen, u.a.: Bauteil-Entwicklungszeit mit ..., Kunststoffe 10, 1995

7.   M. Wilmsen:
        Wettbewerbsvorteile durch CAE-Anwend., Kunststoffberater, Nov.1996

Reverse Engineering Physical Models For Generation Of Surface Models

Liang-Chia Chen, Grier C. I. Lin

Centre for Advanced Manufacturing Research
School of Engineering, University of South Australia
The Levels, Pooraka, SA 5095, Australia

Abstract. This paper presents an innovative approach to reverse engineering physical models for generation of subdivision surface models. The approach integrates stereo vision detection, Delaunay Triangulation, and smooth subdivision surface refinement scheme into an automatic surface reconstruction process. In the approach, a coordinate measurement machine equipped with a touch-trigger probe is integrated with a CCD camera, in order to enhance digitising efficiency while high accurate measurement is also obtained. The proposed self-driven digitising process guarantees that the subdivision surface generated reaches user-specified digitising accuracy. By integrating these capabilities, the approach is aimed to reconstruct smooth arbitrary topological surfaces efficiently and accurately. Consequently, the problems encountered in reverse engineering processes can be reduced to a minimum.

1. INTRODUCTION

Reverse engineering has played an important role in obtaining accurate CAD surface models and shortening product development lead time. The product design process often needs to apply physical models for initial conceptual and aesthetic design, product prototyping, performance testing and model modification to obtain an optimal product design. Reverse engineering is an extremely important tool in reconstructing physical models into mathematical CAD models, in terms of the speed of reconstruction and the accuracy of CAD models generated.

However, there are still many problems in effectively implementing reverse engineering. Optical digitisers such as laser range finders, stereo image detectors, moire interferometers and structured lighting devices are very popular due to their advantages in measuring speed and non-contact property. Unfortunately, their scanning accuracy to date cannot be guaranteed to meet strict requirements. For example, laser triangulation probes are poor in terms of surface edge measurement and the measured errors sometimes may be up to 50 μm. These kind of large digitised errors would be very difficult to eliminate in further data processing. Moreover, the huge amounts of data generated from these fast digitisers is irregular, unformatted, error prone, and require extensive processing for generating smooth surfaces[1]. Consequently, the surfaces reconstructed from these digitised data might lose their surface characteristics and become invalid for further product development.

Some engineers apply coordinate measurement machines (CMMs) with touch-triggered probes for their reverse engineering applications. CMMs with contact probes are excellent in measuring accuracy. The problem is that the scanning rate is much slower than 3D optical digitisers due to the inefficiency and limitation of the measuring principles. Although the automatic digitisation of a CMM can be easily achieved by scanning objects along rectangular patches, it is still time consuming and difficult to digitise a surface with an arbitrary topological type.

This paper proposes an integrated reverse engineering approach to reconstructing smooth subdivision surfaces in a coordinate measurement machine (CMM) equipped with a computer vision system and a touch-triggered probe (TTP). A smooth subdivision surface is defined as the limit of a subdivision process applied to a control triangular mesh. Subdivision surfaces can be used to model everywhere smooth surfaces of arbitrary topological types using numbers of unconstrained parameters[2]. This kind of surface representation is also capable of modelling surface discontinuities such as creases and corners which would cause difficulties in fitting Non-Uniform Rational B-Spline Surfaces (NURBS). Furthermore, the proposed strategy aims to integrate the data digitisation and surface modelling into a single self-driven surface reconstruction process so that surface reconstruction can be implemented accurately and automatically[3,4]. Therefore, the major problems encountered in reverse engineering processes can be solved effectively.

2. OVERVIEW OF THE PROPOSED APPROACH

The proposed approach integrates two steps, namely the initial vision-driven surface triangulation process (IVSTP) and the adaptive subdivision-surface refinement processing (ASSRP), into a single automatic process of surface reconstruction. The IVSTP generates an initial triangular patch by using data filtering, stereo image detection and constrained Delaunay triangulation. The ASSRP which integrates surface digitisation refinement and subdivision surface generation, is then used to reconstruct piecewise smooth subdivision surface and control the digitising accuracy within the required digitising tolerance.

The flow chart of the proposed approach is shown in Figure 1. In the IVSTP, a 3D stereo detection method is used to detect 3D surface boundaries[3]. Multiple images are taken by a CCD camera and appropriate image processing methods, such as noise filtering, are applied to reduce image noise to an acceptable level for further processing. A Laplacian edge detection operation is then used to extract image edges effectively. To obtain an accurate boundary measurement, a CCD camera calibration method is also used to establish a camera model which can adequately transform the image coordinates of the surface boundaries to the CMM world coordinates and reduce the transformed errors to a minimum. Furthermore, in order to save the computing time for further 3D stereo image detection, only characteristic points on the surface boundaries are extracted to represent their original contours by using a data filtering strategy developed by the authors[3]. These extracted points can then be constructed into an initial triangular patch by applying a constrained Delaunay Triangulation[3].

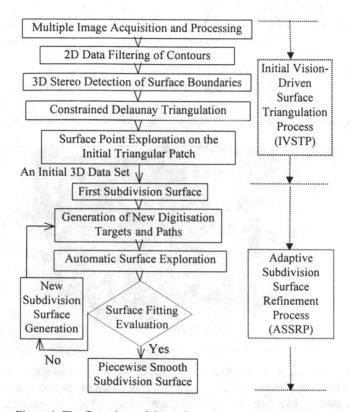

Figure 1 The flow chart of the surface reconstruction algorithm

Following the IVSTP, the ASSRP is then used to refine surface reconstruction by digitising more surface points along their collision-free digitising paths[4]. The ASSRP continues until the digitising accuracy satisfies the user-specified tolerance. As a result, piecewise smooth subdivision surface can be obtained. The detailed strategies will be discussed in the following sections.

3. INITIAL VISION-DRIVEN SURFACE TRIANGULATION PROCESS (IVSTP)

3D stereo vision detection is increasingly being used to enhance the efficiency of various manufacturing tasks due to excellent characteristics such as fast scanning capability and non-contact property. In order to facilitate a CMM with a visual capability for fast digitisation, a CCD camera is integrated with a TTP in a CMM for the purpose of rapidly detecting the object's position and measuring its surface boundary coordinates. The system hardware setup (shown in Figure 2) is designed to achieve the above aim. The system includes a Brown&Shape CMM with a Renishow touch trigger probe and a vision system from the Matrox company. For some applications in which the measuring accuracy is not a major concern but the scanning rate is more important, a laser triangulation probe can then replace a TTP as the digitiser in this system. This indicates that the approach has flexibility to adapt to different digitising devices.

The main aim in the IVSTP is to rapidly detect all surface boundaries, such as edges,

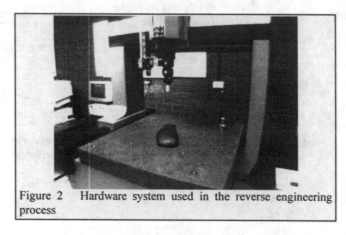

Figure 2   Hardware system used in the reverse engineering process

corners and creases, and triangulate these surface features into initial surface triangular patches. The strategies involved in the IVSTP are camera calibration method, object's edge detection, data reduction of surface edges, three-dimensional stereo image detection, and initial surface Delaunay Triangulation[5].

The flow chart of the IVSTP is shown in Figure 3. In the process, a vision system is used to detect 3D surface boundaries by using a 3D stereo detection method. Prior to the stereo image detection, multiple images are taken and appropriate image processing methods, such as noise filtering, are applied to reduce image noise to an acceptable level for further processing. A Laplacian edge detection operation is then used to extract 2D surface boundaries effectively. To obtain an accurate boundary measurement, a CCD camera calibration method is also used to establish a camera model which can adequately transform the image coordinates of the surface boundaries to the CMM world coordinates and reduce the image system errors to a minimum. Furthermore, in order to save the computing time for further 3D stereo image detection, only characteristic points on the surface boundaries are extracted to represent their original contours by using a data reduction strategy developed by the authors. These extracted points can then be constructed into an initial triangular patch by applying a constrained Delaunay Triangulation method. The detailed principles for the IVSTP have been described in references 3 and 5.

Figure 4 shows an example using the IVSTP to detect the initial surface triangular patches of a toy clay model. The 3D initial triangular patch of a reconstructed toy model was rapidly constructed by the IVSTP. In this case, 75 characteristic points were extracted to represent the shape of the model. The triangular patch was constructed by the Delaunay surface triangulation method. Following this, this initial patch is used in

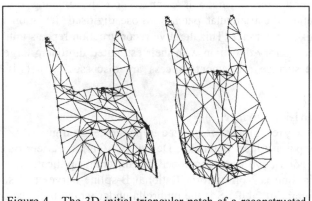

Figure 4   The 3D initial triangular patch of a reconstructed toy model (the left picture is the right ISO view and the right picture is the left ISO view)

the following surface refinement process.

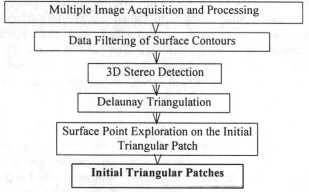

Figure 3   The Initial Vision-Aided Surface Triangulation Process

4 SMOOTH SUBDIVISION SURFACE REFINEMENT PROCESS

The main idea of the digitising strategy is to digitise the surface points following the surface local curvature trend and to automate the digitising process. At first, an initial subdivision surface model can be reconstructed by parameterising the currently available surface points. This model can then be provided to calculate the midpoints between the triangular edges in the initial triangular patch. Moreover, the calculations of the geometric centre of each triangular area and its normal direction provide probe exploration paths to the surface refinement process. The surface digitising process which explores the model-based midpoint estimates on the existing triangular patch can continue to refine the triangular patches whose digitised deviation exceeds the user-specified digitising tolerance. This iterative reconstruction repeats until all new digitised points have satisfactory deviation with their estimated digitising targets. The principles applied in the surface refinement process are discussed in details in the following sections.

4.1 Subdivision Surface

Surface modelling techniques are applied to parameterise digitised points into a surface model. In Computer Aided Geometric Design (CAGD), a variety of curve and surface modelling methods have been developed for different applications. Among these, the most influential one is Non-uniform Rational B-spline curves and surfaces (NURBS). Because the format of NURBS is the umbrella under which both the spline, Bezier, and conic format fit comfortably[6], NURBS has become the standard representation for complex smoothly surfaces.

However, the main disadvantage of using NURBS is that control points should be in a regular rectangular grid. Due to this limitation, NURBS can only be used to represent limited surface topologies rather than arbitrary topologies[2]. Therefore, this approach applied another surface representing form, called subdivision surfaces, to overcome the above problem. Subdivision surfaces are not only easy to implement, but also can parameterise surfaces with arbitrary topological type. Moreover, initial surface triangular patches generated from the IVSTP which detects surface features such as surface boundaries, creases and corners can be directly applied to this type of surface representation.

Subdivision Surfaces were first introduced by Catmull et al. [7,8]. A subdivision surface S(M) is defined as the limit of a subdivision process applied to a control mesh (M) (In our approach, the vertices of the surface triangular patches can be considered as M). There are many subdivision schemes to date. Among these, Loop's scheme using a triangular scheme provides a generalisation of the subdivision surface scheme to generate tangent plane continuous surfaces of arbitrary topological type[9]. Hoppe further modified Loop's subdivision rules to parameterise sharp surface features such as edges and creases[10]. Although subdivision surfaces lack a closed form, some recent research has contributed to the knowledge of how to calculate exact properties such as surface points and surface normals. These breakthroughs have made it possible for subdivision surfaces to be applied in this approach.

4.2 The Adaptive Subdivision-Surface Refinement Process (ASSRP)
The main idea of the ASSRP is to digitise the surface points following the local surface curvature trend and to automate the digitising process. The digitised vertices on the initial surface triangular patches can be used to construct the first subdivision surface by using Hoppe's subdivision rules. The subdivision surface S(M) associated with a control mesh M = (K, V) (shown in Figure 5) is defined as the limit of a refinement process applied to M [9,10]:

$$M, \ M^1 = R(M), \ M^2 = R(R(M)), \ \dots \tag{1}$$

where

K is a *simplicial complex* representing the connectivity of vertices, edges, and faces, thus determining the topological type of the mesh M;

$V = \{v_1, \dots, v_m\}, \ v_i \in R^3$ is a set of vertex positions defining the shape of the mesh M in R^3.

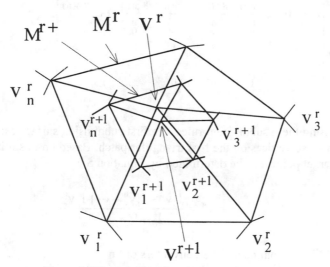

Figure 5   The neighbourhood around a vertex v^r

The refinement procedure R divides each triangle into four subtriangles. The vertices of the new generated mesh are determined by summing weighted averages of the vertices in the existing unrefined mesh as follows[10]:

The new vertex points and the new edge points are computed by the following subdivision rules[10]:

$$v^{r+1} = \frac{\alpha(n)v^r + v_1^r + \cdots + v_n^r}{\alpha(n) + n} \qquad (2)$$

$$v_i^{r+1} = \frac{3\,v^r + 3\,v_i^r + v_{i-1}^r + v_{i+1}^r}{8}, \quad I = 1,\ldots,n \qquad (3)$$

where $\alpha(n) = \dfrac{n(1 - a(n))}{a(n)}$, with $a(n) = \dfrac{5}{8} - \dfrac{(3 + 2\cos(2\pi / n))^2}{64}$

The matrix form of Equations 1 and 2 can be expressed as[9,10]:

$$(v^{r+1}, v_1^{r+1}, \ldots, v_n^{r+1})^T = S_n(v^r, v_1^r, \ldots, v_n^r)^T$$
$$= S_n^{r+1}(v^0, v_1^0, \ldots, v_n^0)^T \qquad (4)$$

where

$$S_n^{loop} = \frac{1}{8}\begin{bmatrix} 8 - 8\,a(n) & \dfrac{8\,a(n)}{n} & \dfrac{8\,a(n)}{n} & \dfrac{8\,a(n)}{n} & \dfrac{8\,a(n)}{n} & \cdots & \dfrac{8\,a(n)}{n} \\ 3 & 3 & 1 & 0 & 0 & \cdots & 1 \\ 3 & 1 & 3 & 1 & 0 & \cdots & 0 \\ \cdots & & & & & \cdots & \cdots \\ 3 & 1 & 0 & 0 & \cdots & 1 & 3 \end{bmatrix}$$

Therefore, by using these subdivision rules, the first subdivision surface estimate can be generated from these vertices on the first triangular patch. Based on this initial estimate, new digitised target points can be determined by Equation 5 [9, 10]:

$$v^{\infty} = \frac{l_0 v^0 + l_1 v_1^0 + \cdots + l_n v_n^0}{l_0 + l_1 + \cdots + l_n} \qquad (5)$$

where $(l_0, \ldots, l_n)$ is the dominant left eigenvectors of Sn

The normal vectors can also be computed by Equation 6:

$$\vec{n} = u_1 \times u_2 \qquad (6)$$

where:
$$u_1 = c_1 v_1^0 + c_2 v_2^0 + \cdots + c_n v_n^0$$
$$u_2 = c_2 v_1^0 + c_3 v_2^0 + \cdots + c_1 v_n^0$$
$$c_i = \cos(2\pi\, i / n)$$

The ASSRP then explores the target points which locate on the midpoints of each edge of each refining triangle. Following this, the adaptive process estimates how well the new digitised points fit with the corresponding target points and determines whether the next exploration procedure will be taken on each refining triangular area. The estimated deviation is computed as Equation 7.

$$E_s(\Delta) = \sum_{i=1}^{3} \| \overrightarrow{E P_i} \| \tag{7}$$

Where $\| \overrightarrow{E P_i} \|$: the length of the deviation vector of each vertex of an exploration triangle (Δ)

If the estimated deviation is larger than the user-specified tolerance, a new subdivision surface model is generated by parameterising all currently available digitised points located on the triangular patches. The new surface estimate can then be used to calculate the new exploration target points for the next loop. The ASSRP which explores the model-based midpoint estimates on the existing triangular patches, continues to refine the triangular patches in which digitised deviation exceeds the specified digitising accuracy. The ASSRP continues the automatic data exploration process until the digitising accuracy reaches the user's specified deviation tolerance. Consequently, the user-specified digitising accuracy can be satisfied in the surface reconstruction process.

5.   SUMMARY AND FUTURE WORK

An innovative reverse engineering approach has been proposed to reconstruct piecewise smooth subdivision surfaces. The approach has many advantages in terms of high digitising accuracy, digitising efficiency enhancement and capability to digitise arbitrary topological surfaces. By integrating an accurate contact probe with a fast digitising CCD camera and the proposed digitising strategies, the deficiency of a touch-trigger probe, namely low digitising rate, can be minimised while high accurate measurement still remains. Moreover, tangent plane continuous surfaces (C^2 quartic triangular B-Splines) of arbitrary topological type can be generated. The drawbacks in using NURBS surface modelling such as limited surface topologies, lack of local refinement procedure and problems in modelling surface discontinuities can be avoided by applying subdivision surfaces. Therefore, the proposed approach provides a good solution to most problems encountered in reverse engineering processes.

The system for the IVSTP has been successfully implemented in our hardware system setup. The ASSRP will be implemented for the integration of the whole approach in the next step. The convergence of the ASSRP will also be investigated to ensure the stability of the whole approach. Some industrial examples will then be used to test the feasibility of the approach.

ACKNOWLEDGMENTS

The authors would like to convey their appreciation towards the Australian Federal Department of Employment, Education and Training for supporting this project under its Targeted Institutional Link Program. Assistance from Study advisers in University of South Australia is also appreciated.

REFERENCES

1. Milroy, M. J., Weir, D. J., Bradley, C., Vickers, G. W. : Reverse engineering employing a 3D laser scanner: case study. Int. J. Adv. Manuf. Technol. **12** (1996) 111-121.
2. DeRose, T., Duchamp, T., Hoppe, H. and McDonald, J.: Subdivision surfaces [Online, accessed on April,1996]. URL:http://www.cs.washington.edu/homes/derose/grail /Projects/ subdivision.html
3. Lin, Grier C. I. and Chen, L. C., "A Vision-Aided Reverse Engineering Approach to Reconstruct Free-Form Surfaces", In Int. Conf. of *CAD/CAM Robotics and Factories of the Future*, London, England, August 14-16, (1996), 854-859.
4. Lin, G. C. I. and Chen, L. C.: An integrated reverse engineering approach to reconstruct free-form surfaces in coordinate measurement machines. In Int. Conf. of *Automation '96*, Mechanical Industry Research Laboratories, Industrial Technology Research Institute, Hsinchu, Taiwan, R.O.C., July 8-11, (1996), 239-246.
5. Lin, G. C. and Chen, L. C.: An intelligent surface reconstruction approach for rapid prototyping manufacturing. Fourth Int. Conf. on Control, Automation, Robotics and Vision, Singapore, Dec. (1996), 43-47.
6. Ma, W: NURBS-based CAD modelling from measuring points of physical models. *Ph.D. thesis*, Faculteit der Toegepaste Wetenschappen, Departement Werktuigkunde, Katholieke Universiteit Leuven, (1995).
7. Catmull, E. and Clark. J. : Recursively generated B-spline surfaces on arbitrary topological meshes. Computer-Aided Design, **10** (1978) 350-355, Sep.
8. Doo, D. and Sabin, M. : Behaviour of recursive division surfaces near extraordinary points. Computer-Aided Design, **10**, 6 (1978) 356-360, Sep.
9. Loop, C. : Smooth subdivision surfaces based on triangles. Master's Thesis, Department of Mathematics, University of Utah, Aug. (1987).
10. Hoppe, H. : Surface Reconstruction from Unorganised points. Ph.D thesis, Department of Computer Science, University of Washington, (1994).

Model-enhanced Rapid Prototyping Using Range Imaging

YY Cai[1], AYC Nee[2], HT Loh[2] And T Miyazawa[3]

[1]Institute of Systems Science
[2]Department of Mechanical and Production Engineering
[3]Sony Precision Engineering Center (Singapore)
National University of Singapore
Kent Ridge Crescent, S119597, Republic of Singapore
Tel: (65)7726724 Fax (65)7744990 Email: yycai@iss.nus.sg

Abstract. Models are essential in computer-aided design and manufacturing (CAD/CAM). They are important for object representation, manipulation, and data communication, etc. With fast and explicit availability of 3D object coordinates, range imaging is increasingly used in various engineering applications such as measurement, inspection, etc. We use a laser digitiser to capture object geometry by rotation scanning. Modelling of the digitised range object is therefore one of the fundamental tasks in this work.

Two modelling approaches are presented here. A NURBS method is first developed to model the sculpture object by employing the NURBS theory. The smooth model generated is then converted to SLA model for stereolithographic rapid prototyping. We also propose a novel approach to modelling range objects using a feature-centred method. That is realized by range image understanding to extract qualitative geometric features of range objects. The feature information can be applied to rectify the errors introduced during the processes of object scanning and numerical calculation. The qualitative feature information is served to enhance various processes in rapid prototyping application including optimized model re-orientation, etc.

1. Introduction

Models are essential in CAD/CAM domain. In a computer integrated manufacturing (CIM) environment, scan-paths for rapid prototyping or probe-paths for coordinate measuring machine (CMM) inspection are traditionally derived from CAD models of the parts. Typically, CAD models are designed using CAD tools and are retrievable from CAD databases. They provide an avenue for object representation, manipulation, data communication, etc. However, there are complex industrial parts that may not be available in CAD databases. In car or shipbuilding industries, for instance, manufacturing of quite number of objects often starts with a crafted master model. With the advent of high quality scanning equipment, a master CAD model can now be generated and manipulated based on the digitisation of a physical master model. This allows engineers to modify or analyse the master model with the aid of CAD tools without the burden of having to recreate the master model. Rapid prototyping of laser scanned object can reap the benefits of shortening the life circle of the product/process development, and improving cost effectiveness, producibility and product quality.

Model reconstruction from digitised object is one of the important tasks in geometric modelling. Approaches to object reconstruction from scanned images for CAD/CAM applications have long been investigated. Triangulating and free-form surface fitting are commonly used in model reconstruction. Delaunay graphs were used by many researchers in order to generate more uniform triangular approximation for range objects [1-2]. Chen and Schmitt [3] discussed range surface modelling by constrained triangulation on discontinuity edges. [4] investigated the free-form surface reconstruction using interpolating techniques for scattered range data. With distinct characteristic of smooth interpolation, NURBS model is widely used in designing and manufacturing applications. It is therefore a basic requirement in this work to have free-surface modelling functionality.

Features are of paramount importance in design and manufacturing contexts, as they are a natural choice for capturing and describing the identity of individual objects, as well as establishing relationships between certain objects [5]. Objects in different levels of abstraction demonstrate different feature information. For rapid prototyping applications, certain feature information may enhance the rapid prototyping processes like error reducing, optimized model re-orienting and re-scaling, etc., to achieve high production quality. One of the common drawbacks of current approaches to object reconstruction from scanned digital data is the lack of the ability of geometric feature understanding. It is also difficult for any CAD-based feature recognisers to detect geometric features from such digital-originated reconstructed models. Form features, like cylindrical bosses, are likely to be represented by a set of polygons or control polygons in the reconstructed models. Besides, due to the nature of imaging, noise-free scanning is usually not possible. For manufacturing applications, the intolerance of down-stream processes to noise in the model makes the use of simply reconstructed models from scanning unsuitable.

In the next sections, we first present the 3D range object capturing using depth mapping. This is followed by the discussions on object reconstruction using NURBS approach and feature-centred approach. The model-based stereolithographic rapid prototyping is then developed.

2 3D Range Object Capturing Using Depth Mapping

To capture the depth information of an engineering part to be measured, a ray of light generated by a laser beam is projected onto the spatial object. With known directions of the source and a displaced detector, the range value of a point spotted by the light on an object can be deduced very quickly. By measuring sequential points and producing a standard list of XYZ coordinates, the laser scanner interfaces to vision and CAD systems with 3D depth maps. Using a unique triangulation technique instead of complicated imaging optics and sophisticated detector-array, the ranging system provides high resolution and accuracy while maintaining a small, fixed triangulation angle across a large work volume [6].

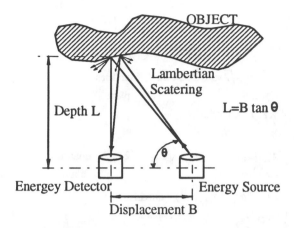

Figure 1. Depth measuring using laser scanner (source: [6])

3 NURBS Modelling For Range Objects

Many engineering applications require generating smooth model to represent objects. Several techniques in computer-aided geometry design (CAGD) have been developed to model the free-form surface objects [7]. In the following, we will apply NURBS theory to to reconstruct complicated free-form surfaces for range objects. If the XYZ coordinates of the data points $\{P_{ij} = (x_{ij}, y_{ij}, z_{ij}), i=0,...,m, j=0,...n\}$ are captured via laser scanning, a B-spline surface using the given points as the control net can be easily generated with the standard B-Spline surface function. However, since such a reconstructed surface usually does not pass through the measured points or the control net, the resultant free form surface is not accurate. To generate a more accurate free-form surface which passes through all the given points, it is therefore required to calculate a real control net $\{V_{ij}, i=0,...,m, j=0,...n\}$ so that a B-spline surface based on the new control polygon is able to interpolate all the given points. A fast algorithm using the bi-cubic B-spline approach is developed here to get the desired surface with the constraint that the resultant surface will interpolate all the measured points. A linear system can be derived from the constraint by considering both boundary and internal conditions of the control nets. Our program is designed to solve the boundary and internal control vertices with the given scan data. Smooth NURBS model can then be obtained by solving the linear systems. Figure 2 shows a reconstructed smooth model from range image using a NURBS approach.

4 Feature-assisted Model Reconstruction From Range Images

Objects in range images are expressed by digital depth values. Homogeneous pixels can be clustered and grouped into object shells, which are features at a global level. We will concentrate on the quadratic features in this work. A fuzzy clustering approach has been developed to partition the shell features by range image understanding [8]. The technique developed can segment and fit shell features in range images simultaneously. In addition, different shells in range images can be partitioned concurrently. Clustering the quadric and planar shells in 3D space and extracting their prototypes from range images are of our particular interest as those information are important to our model-enhanced rapid prototyping application.

Given image points $\{(x_k, y_k, z_k), k = 0,1,..., K-1\}$, the squared distance of the

k-th point to l-th primitive can be defined as

$$d_{lk}^2 = d^2(x_k, y_k, z_k, \pi_l) = (U_l', V_l')M_k\begin{pmatrix} U_l \\ V_l \end{pmatrix} \tag{1}$$

where π_l: $S \cdot U_l + T \cdot V_l = 0$ (called the prototype of the l-th primitive),

$$S = \begin{cases} (x^2, y^2, z^2, \sqrt{2}yz, \sqrt{2}zx, \sqrt{2}xy), & \text{quadrics} \\ (x, y, z), & \text{plane} \end{cases}$$

$$T = \begin{cases} (x, y, z, 1), & \text{quadrics} \\ (1), & \text{plane} \end{cases}$$

$$U_l' = \begin{cases} (a_{00}^{(l)}, a_{11}^{(l)}, a_{22}^{(l)}, \sqrt{2}a_{12}^{(l)}, \sqrt{2}a_{20}^{(l)}, \sqrt{2}a_{01}^{(l)}), & \text{quadrics} \\ (a_0^{(l)}, a_1^{(l)}, a_2^{(l)}) & , \quad \text{plane} \end{cases}$$

$$V_l' = \begin{cases} (2a_{03}^{(l)}, 2a_{13}^{(l)}, 2a_{23}^{(l)}, a_{33}^{(l)}), & \text{quadrics} \\ (a_3^{(l)}) & , \quad \text{plane} \end{cases}$$

$$M_k = \begin{pmatrix} B_k & C_k \\ C_k' & D_k \end{pmatrix}$$

$$B_k = S_k' \cdot S_k, \quad C_k = S_k' \cdot T_k, \quad D_k = T_k' \cdot T_k$$

$$S_k = \begin{cases} (x_k^2, y_k^2, z_k^2, \sqrt{2}y_k z_k, \sqrt{2}z_k x_k, \sqrt{2}x_k y_k) \\ (x_k, y_k, z_k), \end{cases}$$

$$T_k = \begin{cases} (x_k, y_k, z_k, 1), \\ (1), \end{cases}$$

The notion of distance d_{lk} is a similarity measure to indicate how far the k-th point can be classified into the l-th cluster. The shell thickness $J(\pi_l, \mu_l)$ of the l-th primitive is the membership-weighted squared distances of all the points to the l-th particular cluster and the total shell thickness $J(\pi, \mu)$ can therefore be defined as:

$$J(\pi, \mu) \equiv \sum_{l=0}^{C-1} J(\pi_l, \mu_l) = \sum_{l=0}^{C-1}\sum_{k=0}^{K-1} (\mu_{lk})^m d_{lk}^2 \tag{2}$$

where   const $m > 1$,

μ is a (m+1) x (n+1) fuzzy membership matrix, and

$$\sum_{l=0}^{C-1} \mu_{lk} = 1 \text{ (for all } k \text{) and } 0 < \sum_{k=0}^{K-1} \mu_{lk} < K, \ \mu_{lk} \in [0,1] \text{ (for all } l \text{ and k).}$$

Partitioning of a finite set of points in a three-dimensional Euclidean space into fixed C fuzzy clusters is to determine the membership grades μ_{lk}, ($l=0,1,...,C-1$, $k=0,1,2,...,K-1$) of each element. From the viewpoint of optimisation, the partition of K points into C clusters is equivalent to minimise the above total distance. In order to avoid trivial solution of optimal shell clustering, necessary constraints need to be introduced. A unit normal constraint in planar case and an invariant constraint of $\|U_l\|^2=1$, ($l=0,...,C-1$) in quadric case are chosen here [2]. This leads to a constraint

optimisation problem $\displaystyle \operatorname*{MIN}_{|U_l|^2=1, \ 0 \le l < C} \left\{ \sum_{l=0}^{C-1} J(\pi_l, \mu_l) \right\}$. By applying the Lagrange

multiplier method, the above constraint optimisation is transferred into a non-constraint problem

$$MIN\left\{\sum_{l=0}^{C-1}\left\{\left(U_l'\ \ V_l'\right)\begin{pmatrix}E_l & F_l \\ F_l' & G_l\end{pmatrix}\begin{pmatrix}U_l \\ V_l\end{pmatrix}+\lambda_l\left[1-\left(U_l'\ \ V_l'\right)\begin{pmatrix}I & 0 \\ 0 & 0\end{pmatrix}\begin{pmatrix}U_l \\ V_l\end{pmatrix}\right]\right\}\right\} \tag{3}$$

where $\quad E_l = \sum_{k=0}^{K-1}(\mu_{lk})^m B_k$, $F_l = \sum_{k=0}^{K-1}(\mu_{lk})^m C_k$, $G_l = \sum_{k=0}^{K-1}(\mu_{lk})^m D_k$,

$$I = \begin{cases} \text{6x6 unit matrix, quadrics} \\ \text{3x3 unit matrix, plane} \end{cases}$$

Differentiating (3) leads to following equations:

$$\begin{cases} V_l = -G_l^{-1}F_l'U_l \\ A_lU_l = \lambda_lU_l \end{cases} \tag{4}$$

where $\quad A_l = \left(E_l - F_lG_l^{-1}F_l'\right)$, $(l=0,1,...,C\text{-}1)$.

It is not difficult to prove A_l is positive semi-definite, and thus, the optimal shell clustering problem is to find the minimal eigenvalue and its corresponding eigenvector of each shell eigensystem of matrix A_l, $(l=0,1,...,C\text{-}1)$. The eigensystem A_l has three eigenvalues for planar shells and six eigenvalues for quadric shells. A convergent membership matrix μ can be iteratively obtained using Bezdek's updating role [9].

5. Model-enhanced Rapid Prototyping

5.1 Invariant-based Qualitative Feature Identification

The intrinsic relationship between qualitative and quantitative analysis of shape feature can be explored by employing its geometric invariance property. The classical theory of invariance focused on polynomial invariants that are independent of certain admissible transformations. For any quadrics in algebraic form, there are basic independent to three-dimensional general transformations. Based on the basic invariants, the eigenvalue invariants, spherical invariant and revolutional invariant can be defined [8]. We have developed a mechanism to classify the quadric features and represent explicitly the feature parameters using the above invariant quantities of any quadratic shells. As a result, the spherical invariants and the revolutional invariant can be used for inspection or measurement of spherical and revolutional features, respectively. The quadric primitives can be classified according its invariants values [8]. Furthermore, these invariant quantities can be used to explicitly represent the parameters of quadric features such as cylinder radii, cone angles, etc.

The integration of the fuzzy clustering and invariant analysis makes the identification of the qualitative features very efficient. Figure 3 illustrates a range object with three quadratic features and Figure 4 shows three extracted shell from the range image using the partitioning algorithm (the top half part only due to symmetry). The individual shell features can be classified based on its invariant values.

Figure 2. A NURBS model reconstructed for a spiral range object

5.2 Feature-centred Model Rapid Prototyping

Using the NURBS approach we reconstruct the range object. We then utilise the quadratic and planar feature to help achieving better quality of RP manufacturing. In fact, in the presence of noise in 3D data, the geometrical features extracted by range image understanding usually have various shape variations. The designed spherical and revolutional invariants can be used to monitor the variations. By checking the values of the spherical invariant or the revolutional invariant, one can determine whether the variations of the spherical feature or revolutional feature are within allowable ranges. Tolerance information from range images can also be detected utilising invariants. For instance, a sphere is most likely deformed into an ellipsoid-like shape. The maximum and minimum spheres associated with the ellipsoid can be computed using the three calculated radii of the ellipsoid.

Once the model is available, CAD tools can be applied to manipulate the model (rotating, translating and scaling, etc.). Good orientation of the object, however, may lead to good quality of the model rapid prototyping. It is not easy to find a better orientation for the NURBS-based model by purely interactive mainuplation. We propose a novel method to determine its good orientations by referencing the partitioned qualitative features. This is indeed a localisation process for feature identification. It is not difficult to determine its axial or symmetric directions for all quadratic surfaces (Figure 5). These directions, together with the feature areas, determine the final orientation for better RP quality and a model re-orienting operation can then be performed before STL formating. For sculpture model without any quadatic or planar features, however, one has to use the other way to estimate a good orientation which may be highly relied on users' experience. Finally, we convert the NURBS model into a STL format. The layered STL model is then fabricated with a stereolithographic RP system. Figure 6 shows the spiral object manufactured using the SOLID CREATION system at the SONY Precision Engineering Centre (Singapore).

Conclusions. Addressing the needs of manufacturers moving towards variety-based and time-based competition, the newly emerged RP technology provides tools that can significantly streamline the design and product development process, while substantially reducing costs. The basic RP principle allows a conversion of CAD models into layered information used for process control in different layered. To reduce the relatively lengthy process of interactive design and manufacturing, this work discusses the rapid prototyping of range object captured by laser scanning. Range-based modeling techniques are developed to reconstruct physical models. A NURBS modelling method and a feature-centred modelling method are presented. With the reconstructed model, high quality rapid prototyping can be expected. Advanced vision-systems like CT or MRI can be employed to capture the full structure information of 3D objects. Extraction of the boundary surface from CT/MRI volume is one of the challenge tasks. Range-based modelling and rapid prototyping techniques developed may be further applied to this area.

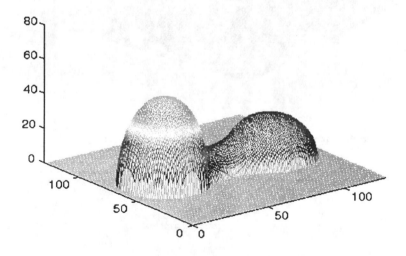

Figure 3. A top view of a range object

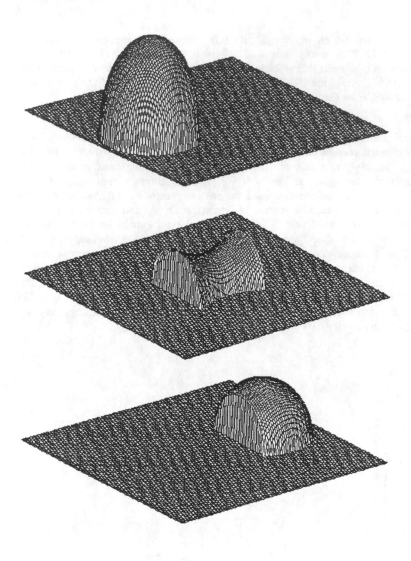

Figure 4. Three quadratic shell features extracted for the range object

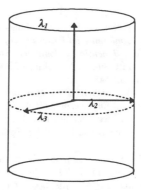

Figure 5. Eigenvectors and eigenvalues with a quadric invariant system

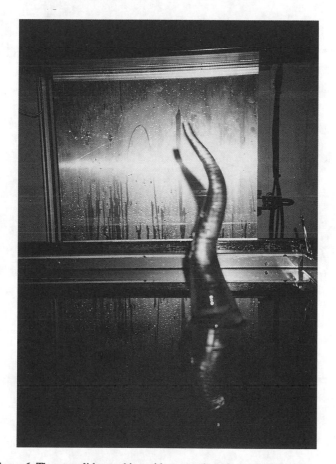

Figure 6. The stereolithographic rapid prototyping of the spiral range object

References

1. J. D. Boissonnat, Representing 2D and 3D Shape with the Delaunay Triangulation, *Proc. of IEEE International Conference on Pattern Recognition*, 1984, pp. 745-748.

2. O. D. Faugeras, *Three-dimensional Computer Vision*, MIT Press, Cambridge, 1994.

3. X. Chen and F. Schmitt, Surface Modelling of Range Data by Constrained Triangulation, *CAD*, 26(8), 1994, pp. 632-645.

4. C. Bradley and G. W. Vickers, Free-form Surface Reconstruction for Machine Vision Rapid Prototyping, *Optical Engineering*, 32(9), 1993, pp. 2191-2200.

5. M. J. Pratt, Aspects of form feature modelling, in *Geometric Modelling Methods and Applications* (Eds: H. Hagens and D. Roller), Springer-Verlag, New York, 1991, pp. 227-250.

6. R. Jarvis, Range Sensing for Computer Vision, in *Three-dimensional Object Recognition Systems* (Jain A. K. and Flynn P. J., Eds.), Elsevier Science Publishers, 1993, pp. 17-56.

7. G. Farin, *Curves and Surfaces for Computer Aided Geometric Design: A Practical Guide*, Academic Press, Boston, 1988.

8. Y. Y. Cai, H. T. Loh, and A. Y. C. Nee, Qualitative Primitive Identification Using Fuzzy Clustering and Invariant Analysis, *Image and Vision Computing*, 14(1996), pp. 451-464.

9. J. C. Bezdek, *Pattern Recognition with Fuzzy Obective Function Algorithms*, Plenum Press, New York, 1981

Numerical Simulation of Cavity Filling of the Powder-Binder Mixture in a Rapid Prototyping Process

Lin Bilon and Gong Haiqing[*]

School of Mechanical and Production Engineering,
Nanyang Technological University
Nanyang Avenue, Singapore 639798

Abstract. The analysis and simulation of the cavity filling process of the powder-binder mixtures with shear-induced powder migration in a powder injection rapid prototyping process was presented in this paper. A fully-implicit time discretization scheme was applied and extended such that the velocity, pressure, temperature, powder and flow front location were iterated and solved simultaneously. The purpose of this work is to develop a simulation technique to help design engineers to understand the physical process, detect non-homogeneous powder distributions and improve the quality of the final parts.

1   Introduction

Rapid prototyping (RP) is a computer-controlled, layer by layer, additive process for fabrication of three-dimensional solid objects without tooling. Application of RP technology is becoming more popular due to its tremendous time saving in the building of complex 3D parts. There are many types of the rapid prototyping techniques, including a laser to print photopolymer layers, mask photopolymer exposure with wax support, laser sintering ceramic powder with a liquid binder, ballistic particle formation, and fused extruded filament formation. For RP technique using powder-binder mixture as building materials, the numerical simulation of cavity filling of the powder-binder mixture is important to the entire rapid prototyping process, in which, the effect of non-homogeneous powder distribution is a critical factor to the quality of the final products.

Flow simulation of the powder-binder mixtures in the cavity filling were carried out by many authors (Lee, et al 1989, Liaquat, et al 1991, Rhee, et al 1992 and Wang, et al, 1992, 1994). The rheological characteristics, thermal properties and mechanical properties of powder-binder mixture were utilized by Fox, et al (1992) to simulate two-dimensional flow through a thin cavity. However, the current flow simulations were all based on the homogeneous powder distribution.

In the present study, a new numerical scheme based on finite element and finite difference method was developed for modeling the cavity filling process with the non-isothermal, non-Newtonian powder-binder mixtures in the rapid prototyping process. Shear induced migration of powder was modeled in terms of powder diffusion equation.

---

[*] To whom all correspondence should be addressed

2   Powder Injection Rapid Prototyping Process

A new rapid prototyping process called Powder Injection Rapid Prototyping (PIRP) process was developed by the authors. The working principles of the powder injection rapid prototyping can be summarized as follows:

- A single component resin is injected uniformly into the working platform by the injection head to form a uniform plate as a outer geometry of the model part.
- The powder-binder mixture is heated and then injected into the cavity formed by the resin at previous step through a computer-controlled nozzle by a cylinder-piston system to create a layer of the model part. The binder in the mixture solidifies when it comes into contact with the platform or the previous layer.
- The next layer of the resin solidifies and adheres to the previous layer, then the outer boundary of the model part is cutted and formed, and the second layer of the powder-binder mixture is injected into the new cavity to form the model part which adheres to the previous layer. The part is built layer by layer until it is complete. The outer support part (resin part) is then removed by heating

3   Governing Equations

A typical filling problem in the thin cavity can be reduced to the two-dimensional. The pressure is a function of x and y, and velocities u and v are the functions of x, y and z. The governing equations can be reduced to the following forms:

3.1   Continuity Equation for the Powder-Binder Mixture

$$\frac{\partial[h\bar{u}(x,y)]}{\partial x} + \frac{\partial[h\bar{v}(x,y)]}{\partial y} = 0 \tag{1}$$

3.2   Momentum Equations for the Powder-Binder Mixture

$$\frac{\partial p(x,y)}{\partial x} - \frac{\partial}{\partial z}\left[\mu_s \frac{\partial u(x,y,z)}{\partial z}\right] = 0 \tag{2}$$

$$\frac{\partial p(x,y)}{\partial y} - \frac{\partial}{\partial z}\left[\mu_s \frac{\partial v(x,y,z)}{\partial z}\right] = 0 \tag{3}$$

where $h = h(x,y)$ is the gapwidth of the cavity, which depends upon the x and y co-ordinate, $u(x,y,z)$ and $v(x,y,z)$ are the velocity components in the x and y direction, and $\bar{u}(x,y)$ and $\bar{v}(x,y)$ are the mean velocities averaged over the cavity gap, which is given as,

$$\bar{u}(x,y) = \frac{1}{h}\int_0^h u(x,y,z)dz, \quad \bar{v}(x,y) = \frac{1}{h}\int_0^h v(x,y,z)dz \tag{4}$$

3.3   Shear-induced Powder Diffusion Equation

The hydrodynamic interaction among neighboring powder particles often creates a non-homogeneous powder distribution in flowing suspension of the powder-binder mixture that were initially well mixed. In concentrated suspensions undergoing shear, a flux of particles is usually reduced from regions of high shear to low and from regions of high particle concentrations to low. The powder diffusion equation is obtained by balancing the fluxes due to bulk convection, gravitational sedimentation and shear-induced diffusion.

$$\frac{\partial \Phi}{\partial t} + u(x,y,z)\frac{\partial \Phi}{\partial x} + v(x,y,z)\frac{\partial \Phi}{\partial y} = \lambda \frac{\partial}{\partial z}\left[\hat{D}_c(\Phi)\dot{\gamma}\frac{\partial \Phi}{\partial z} + \hat{D}_s(\Phi)\frac{\partial \dot{\gamma}}{\partial z} + \frac{2}{9Fr}\Phi f(\Phi)\right] \tag{5}$$

where the scalars $\hat{D}_c(\Phi)$ and $\hat{D}_s(\Phi)$ are shear-induced diffusion coefficients, $\dot{\gamma}$ is the shear rate; the modified Froude number $Fr = \mu_b U/((\rho_p - \rho_b)gL^2)$; $\lambda = (a/L)^2$, the square of the ratio of the particle size a to the characteristic length scale L, ρ_b and ρ_p are the densities of the binder and powder, respectively. Based on dimensional analysis and the mechanisms of shear-induced diffusion, for the unidirectional flow of suspension, these two coefficients are given as:

$$\hat{D}_c(\Phi) = 0.43\Phi + 0.65\Phi^2 \frac{1}{\mu}\frac{d\mu}{d\Phi} \quad , \quad \hat{D}_s(\Phi) = 0.43\Phi^2 \tag{6}$$

The relative viscosity μ is modeled as,

$$\mu = \frac{\mu_s}{\mu_b} = \left(1 - \frac{\Phi}{\Phi_m}\right)^{-1.82} \tag{7}$$

where Φ_m is the solid volume fraction beyond which the suspension can no longer flow, as in the studies by Leighton, et al (1986, 1987), Schaflinger, et al (1990), Phillips, et al (1992) and Raman, et al (1994). The value of Φ_m depends on the properties of the powder and binder. μ_b and μ_s are the viscosity of the pure binder system and the mixture, respectively. $f(\Phi)$ can be expressed as follows:

$$f(\Phi) = \frac{1-\Phi}{\mu} \tag{8}$$

3.4   Energy Equation for Powder-Binder Mixture

Viscosity of the filling process changes drastically with temperature and convection plays an important role in heat transfer during the filling process; therefore, flow and heat transfer are strongly coupled. Moreover, the energy equation remains strongly three-dimensional. Heat conduction is dominant in the gapwise direction, while the convection terms are dominant in x and y direction. The energy equation may be written as follows:

$$\rho C_p\left(\frac{\partial T}{\partial t} + u(x,y,z)\frac{\partial T}{\partial x} + v(x,y,z)\frac{\partial T}{\partial y}\right) = K\nabla^2 T(x,y,z) + Q(x,y,z) \tag{9}$$

where $Q(x,y,z)$ represents the viscous dissipation, which is given as follows:

$$Q(x,y,z) = \mu_s\left[\left(\frac{\partial u(x,y,z)}{\partial z}\right)^2 + \left(\frac{\partial v(x,y,z)}{\partial z}\right)^2\right] \tag{10}$$

4   Formulation of the Solution

4.1   Derivation of the Pressure Equation Using the Hele-Shaw Approximation

If the gapwidth $h(x,y)$ maintains constant along the flow direction, integrating Eq. (2) and Eq. (3) with respect to z and using the boundary conditions, velocities can be obtained as follows:

$$u(x,y,z) = -\frac{\partial p}{\partial x}\int_z^h \frac{\hat{z}d\hat{z}}{\mu_s} \qquad (11)$$

$$v(x,y,z) = -\frac{\partial p}{\partial y}\int_z^h \frac{\hat{z}d\hat{z}}{\mu_s} \qquad (12)$$

Integration of the continuity equation over the gapwidth with the assumption that the velocities vanish at $z = 0$ and $z = h$.

Flow quantity $S(x,y)$ is introduced as follows:

$$S(x,y) = \int_0^h \frac{z^2}{\mu_s}dz \qquad (13)$$

From the continuity equation, the mean velocities can be derived as follows:

$$\bar{u}(x,y) = -\frac{S}{h}\frac{\partial p(x,y)}{\partial x} \qquad (14a)$$

$$\bar{v}(x,y) = -\frac{S}{h}\frac{\partial p(x,y)}{\partial y} \qquad (14b)$$

Combining with continuity equation, we may derive a flow equation in terms of pressure.

$$\frac{\partial}{\partial x}\left(S\frac{\partial p}{\partial x}\right) + \frac{\partial}{\partial y}\left(S\frac{\partial p}{\partial y}\right) = 0 \qquad (15)$$

4.2   Boundary Conditions

The pressure along the advancing flow front can be assumed as constant. For the sake of simplicity, this pressure at the flow front is set to zero. The heat loss from the flow front to the ambient is small in comparison with the other heat transfer process and is neglected. Powder diffusion between the flow front and the ambient is very small, therefore the volume fraction gradient of the powder can be considered as zero.

At the entrance into the cavity, the flow rate is specified as a function of time. The inlet temperature is held at a constant temperature across the gap and taken as either the melt temperature or the calculated bulk temperature at the end of the delivery system. The volume fraction of powder is also treated as a constant.

At any impermeable boundary, the pressure gradient in the normal direction to the boundary is zero. The powder fluxes into any impermeable boundaries, such as wall boundary and the symmetrical boundary are required to be equal to zero. The heat conduction in x and y directions may be neglected in the energy equation, so that the temperatures on the walls of the mold have to be given as a function of time.

5   Modeling of Material Properties

Without considering the elastic effect, the flow behavior of powder-binder mixture can be represented by the generalized Newtonian fluids. Viscosity of the binder is calculated in terms of a power law, i.e.,

$$\mu_b = K_0 \exp(\frac{E}{RT})\dot{\gamma}^{n-1}$$ (16)

where $\dot{\gamma}$ is the shear rate, n is the constant of the viscosity, K_0 is an empirical constant, E is the activation energy flow, R is the universal gas constant, and T is the absolute temperature. Thermal conductivity and heat capacity of the powder-binder mixture are calculated in terms of the mixture law according to the properties of the component.

6   Numerical Implementation

The computations were performed using the successive finite element meshes generated on the filled part of the mid-surface. A finite element and finite difference model was used for the non-Newtonian non-isothermal powder-migration flow analysis based on the generalized Hele-Shaw flow approximation. In the present approach, a single finite element mesh on the mid-surface of the cavity was used to solve for the moving flow front. The two-dimensional pressure field was solved by the finite element method. The temperature and powder distributions at each node along the gapwise direction were solved by the finite difference method.

An efficient front tracking algorithm based on a new FEM/VOF technique and a global/local co-ordinate transformation technique (Couniot, et al 1986, Subbiah, et al 1989, Friedrichs, et al 1993) was extended in this simulation. Based on the previous methods, a local re-meshing technique restricted to the area close to the free surface while keeping most of the elements unchanged was developed to treat the mesh generation and re-mesh of the moving solution domain.

The bilinear quadrilateral elements and local co-ordinate system are used in the solution of the pressure equation. Eq. (15) is multiplied by the shape function and integrated over the solution domain. After integration by part to eliminate the higher order derivatives, this will leads to the following equation:

$$\int_\Omega \nabla N \cdot S \nabla p d\Omega + \int_{\partial\Omega} NS\nabla p \cdot \bar{n} dl = 0$$ (17)

The pressure can be expressed approximately as follows:

$$p(x,y) = \sum_{i=1}^{n} p_i N_i(x,y)$$ (18)

where $N_i(x,y)$ is the shape function

A system of non-linear algebraic equation is obtained:

$$[A] \cdot [p] = [R]$$ (19)

where $[A]$ is the stiffness matrix for the system, while $[p]$ represents the vector of unknown pressures and $[R]$ represents for forcing vector depending upon the given flow rate and the pressure boundary conditions, which are given as follows:

$$[A] = \sum_{i=1}^{n} A_{ij} = \sum_{i=1}^{n} \int_\Omega S\nabla N_i \cdot \nabla N_j d\Omega \quad \text{and} \quad [R] = \int_{\partial\Omega} N_i S\nabla p \cdot \bar{n} dl$$ (20)

Once the nodal pressure is solved, u and v can be computed by differentiating the interpolation function for pressure according to Eq. (11) and Eq. (12).

The diffusion and energy equations were discretized using the finite different method in the gap direction. The time derivative of powder distribution was expressed by forward finite difference. The space derivative at z-direction was expressed as by central difference. The grid points of the powder distribution in the cavity might be equally spaced. The shear-induced diffusion and convection terms were evaluated using the results from the previous time step. This method allows powder distribution as a function of gapwise position from node to node.

The successive finite element meshes were applied in the simulation, which were generated on the filled part of the midsurface. Starting from an initial finite element mesh developing from the initial fluid domain, the pressure equation was solved using an iterative procedure. Then, the velocities of the mixture were obtained using the Eq. (11) and Eq. (12). Then, the powder and temperature distributions can be iterated based on TDMA method. The flow front was moved to a new location with a small, stable time increment. A new mesh was then generated over the new flow domain using the previous mesh as a first guess. The mesh size was enlarged in the flow direction.

7   Results and Discussions

The simulation was performed on a simple cavity as shown in Figure 1. The material employed in this study is an iron/wax-based binder mixture, and the average diameter of the iron powder is 4 micron. The viscosity of which is given as in equation (16), in which $R = 8.3143 J / (mol.K)$, $n = 0.90$ and $E = 14.5 KJ / (mol.K)$. The inlet velocity of the powder-binder mixture is set constant at 0.142 m/s. Powder volume fraction is assumed as constant at the inlet section, i.e., $\Phi_{in} = 0.50$. The computation was performed under the simulation conditions given in Table 1.

Table: Input parameters for the simulation

Range of Melting Temperature	360 –400 K
Initial Mold Temperature	310 –330 K
Injection Temperature	380 K
Density	4500 kg/ m³
Special Heat	1300 J/(kg K)
Thermal Conductivity	2.0 W/(m K)

In order to simplify the simulation and analysis, the mold thickness remains constant.

Top view

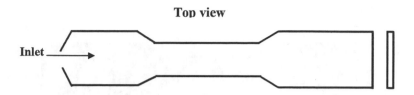

Inlet

Figure 1   Schematic of the cross section of a simple cavity for simulation

Figure 2 shows the transient numerically generated meshes over the flow domain at some chosen time steps during the filling simulation. The number of the meshes increases as the flow domain expands, so that it can reduce the computational cost. Figure 3 shows the typical results for pressure profile near the completion of the filling stage. To represent effects of the temperature and powder distribution, the relative gapwidth Z is introduced. Z stands for the ratio of the z coordinate to the gapwidth. The three-dimensional temperature distributions are displayed in Figure 4 as a set of contour plots in the x-y plane for four different layers along the cavity gapwise direction: (a) the central plane Z=0.50, (b) Z=0.625, (c) Z=0.75, (d) Z=0.875. The temperature distributions along the centerline of the cavity in the gap-wise direction are shown in Figure 5. The three-dimensional powder distributions are displayed in Figure 6 as a set of contour plots in the x-y plane for five different layers in the mold gap-width: (a) the central plane Z=0.50, (b) Z=0.625, (c) Z=0.75, (d) Z=0.875, (e) the cavity surface Z=1.0. Figure 6 indicates that the powder distribution change greatly at the different layers of the cavity during the filling process. This non-homogeneous powder diffusion is mainly caused by the shear-induced powder diffusion and flow convection. The gapwise powder distributions along the centerline of the cavity are also shown in Figure 7, which shows that the powder volume fraction along the gapwise direction changes greatly.

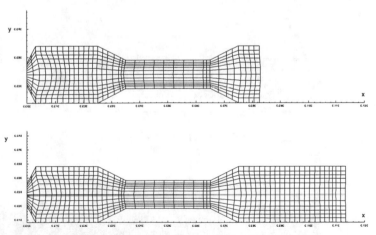

Figure 2   The transient numerically generated meshes over the flow domain at some chosen time steps during the filling simulation

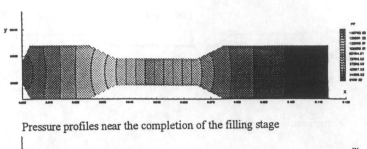

Figure 3   Pressure profiles near the completion of the filling stage

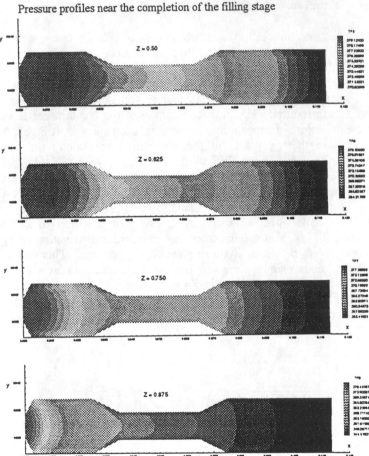

Figure 4   Temperature distributions of the powder-binder mixture at some selected planes: Z=0.50, Z=0.625, Z=0.75 and Z=0.875

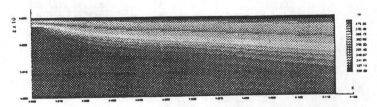

Figure 5   Temperature distribution along the centerline of the cavity in the gap-wise direction

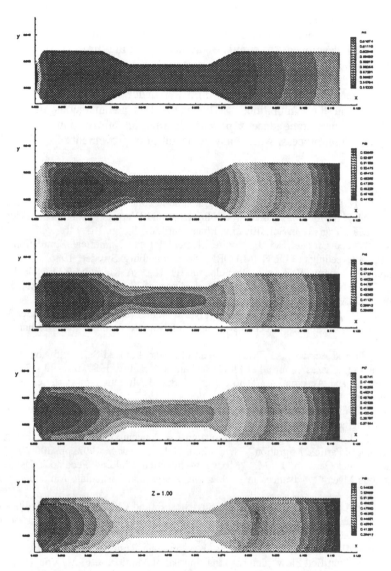

Figure 6   Powder distributions of the powder-binder mixture at some selected planes: Z=0.50, Z=0.625, Z=0.750 Z=0.875 and Z=1.0

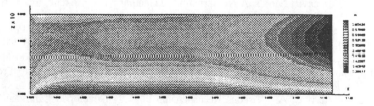

Figure 7   Powder distribution along the centerline of the mold in the gap-wise direction

8   Conclusions

A computational method based on the finite element and finite difference hybrid method was developed to simulate the cavity filling process of non-Newtonian powder-binder mixture in the rapid prototyping process. A new grid generation approach to the analysis of the cavity filling process was also presented to reduce the computational cost.   Shear-induced diffusion under non-isothermal conditions was investigated in the simulation.   This method is advantageous in analyzing and detecting the non-homogeneous powder distribution of the powder-binder mixture during the filling process, which may be useful to improvement of quality of the final product.

References

Chiang, H. H. and Himasekhar, K., "Integrated simulation of Fluid Flow and Heat Transfer in Injection Molding for the Prediction of Shrinkage and Warpage", HTD-Vol. 175/Md-Vol. 25, Heat and Mass Transfer in Solidification Processing, ASME, pp. 133-146.

Couniot, A. and Crochet, M. J., "Finite Element for the Numerical Simulation of Injection Molding", Proceedings of the NUMIFORM' 86, Gothenburg, Sweden, 1986.

Fauchon, D., Tanguy, P. A. and Dannelongue, H. H., " A Rigorous 2-D Mold Filling Model Using The Finite Element Method", Proceedings of the NUMIFORM' 89, Fort Collins, Colorado, USA, 1989, pp. 259-263.

Friedrichs, B. and Guceri, S. I., "A Novel Hybrid Numerical Technique to Model 3-D Fountain Flow in Injection Molding Processes", Journal of Non-Newtonian Fluid Mechanics, 49(1993), 141-173.

Leighton, D. and Acrivos, A., "Measurement of Shear -Induced Self-Diffusion in Concentrated Suspension of Spheres", Journal of Fluid Mechanics, Vol. 177, 1987, pp. 109 -131.

Leighton, D. and Acrivos, A., "The Shear - Induced Migration of Particles in Concentrated Suspensions", Journal of Fluid Mechanics, Vol. 181, 1987, pp. 415 - 439.

Liaquat Alinajmi and Lee, D., "Modeling of Mold Filling for Powder Injection Molding", Polymer Engineering and Science, Mid-August 1991, Vol. 31, No. 15, pp. 1137 -1148.

Phillips, R. J., et al, "A Constitutive Equation for Concentrated Suspension that Accounts for Shear-induced Particle Migration", Phys. Fluids A 4(1), January, 1992, pp. 30 - 40.

Raman, R. and German, R. M., "Mixing of Injection Molding Feedstock Using Batch and Continuous Mixers", Advances in Powder Metallurgy and Particulate Materials, Vol. 4, 1994, pp. 1 - 14.

Rhee, B. O., Cao, M. Y., Zhang, H. R., Streicher, E. and Chung, C. I., "Improved Wax-Based Binder Formulations for Powder Injection Molding", Advances in Powder Metallurgy and Particulate Materials, Vol. 2, 1991.

Schaflinger, U., Acrivos, A. and Zhang, K., "Viscous Resuspension of a Sediment within a Laminar and Stratified Flow", International Journal of Multiphase Flow, Vol. 16, pp. 567 - 578, 1990.

Subbiah, S., Trafford, D. L. and Guceri, S. I., "Non-isothermal Flow of Polymers into Two-dimensional, Thin cavity Molds: A Numerical Grid Generation Approach", International Journal of Heat and Mass Transfer, Vol. 32, No. 3, pp. 415-434, 1989.

Wang, C. M., Carr, K. E. and McCabe, T. J., "Computer Simulation of the Powder Injection Molding", Vol. 4, Advances in Powder Metallurgy and Particulate Materials, pp. 15-26, 1994.

Wang, C. M., Leonard, R. C., Posteraro, R. A. and McCabe, T. J., "A Finite-Difference Computing Model for 3-D Simulation of the Powder Injection Molding Process", Proceedings of the 1992 Powder Injection Molding Symposium, pp. 435-449, 1992.

Zhang, K. and Acrivos, A. "Viscous Resuspension in Fully Developed Laminar Pipe Flows", International Journal of Multiphase Flow, Vol. 20, No. 3, pp. 579-591, 1994.

Rotary Axis Compensation for a 5-axis Micro Milling Rapid Prototyping System

Ng Peow, Fang Wei, Ngooi Chan Siang, Li Hui, Li Bo, Zou Jiguo and Gong Haiqing*

School of Mechanical and Production Engineering, Nanyang Technological University, Singapore

Abstract. In view of the 'staircase or stepped effect' in the present commercial rapid prototyping (RP) systems, a new approach is proposed and is currently under development to produce true stepless parts. The new RP system is a combination of 5-axis motion control technology and state-of-the-art micro milling. As in any rapid prototyping system, accuracy is one of the many key issues. This paper presents the rotary axis compensation of the system to improve the overall machining accuracy.

1 Introduction

In today's highly competitive industries, product quality, time-to-market and cost are the critical factors for success. Faster development cycle requires the generation of highly accurate model directly from the CAD data. Rapid prototyping, or RP technologies offer the precise solution for this purpose.

RP technology was commercially introduced in 1987 with the presentation of StereoLithography (SLA) [1,2,3,4]. Several processes are now commercially available in the United States, Europe, and Asia. To name a few, like the Solid Ground Curing from Cubital [2,5], Fused Deposition Manufacturing (FDM) from Stratasys [2,6] and Laminated Object Manufacturing (LOM) from Helisys [2,7,8]. Although different techniques are used in these RP systems, there is a common drawback. Parts produced by the current commercial systems have *staircase*, or *stepped* slanted surfaces. This inspires the search for a new RP technology, which will produce *truly accurate stepless* parts. The system currently under development is a new 5-axis CNC micro milling system, which consists of three linear and two rotary axes. The two rotary axes enable tangent cutting, which eliminates the staircase effect.

2 5-Axis Micro Milling for Rapid Prototyping

Like the other RP systems, the 5-axis micro milling RP system 'decomposes' a CAD model into boundary contours of thin slices. Each slice is then 'cut out' by micro milling, and built one layer on top of the other, as in LOM. Figure 1 shows the picture of the system currently under development, it consists of a x-y machine table and a z-axis. At the end of the z-axis is a robot wrist (i.e. the roll and pitch axes). It is similar to a 3-axis CNC milling machine, but with the addition of two rotary axes to enable tangent cutting. The cutting method, i.e. micro milling, is similar to conventional slot milling, except the tool diameter is much smaller.

---

* To whom all correspondence should be addressed

The building of each layer in this system requires 4 processes as listed below:
1. Laying of photopolymer on the machine table.
2. Solidify the photopolymer with UV (the solidification process is beyond the scope of this paper).
3. Micro-milled the boundary contour of the model.
4. Final face milling for cosmetic purpose.

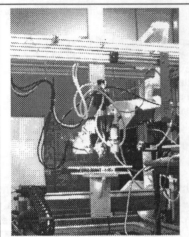

Figure 1: Picture of the 5-axis Micro Milling RP Machine

After the final face milling, it is ready to build the next layer.

Current commercial RP systems use a reasonably thin build layer to minimize the 'staircase effect'. Take LOM for example, each build layer thickness is around 0.0038" to 0.01" (0.0965 to 0.254mm), and a fixed thickness is used throughout the whole building process [7]. Consequently, to build a part of 15mm height will require 60 to 150 layers! The 5-axis micro milling RP system on the other hand, does not have a fixed build layer thickness. The thickness of each build layer depends on the complexity of the boundary contour, and is maximized whenever possible. This approach minimizes the number of build layers required, and subsequently speeds up the building process.

3 Layer Transfer Interface (LTI) Format

Presently, there are three main techniques in slicing the STL file, they are:
1. *Fixed slicing*: Uniform layer thickness is applied throughout the part. The drawback of this technique is that flat surfaces that lie in between two slicing planes may be missed out.
2. *Semi-fixed slicing*: This is a modified fixed slicing technique. Special attention is given to those 'problematic flat surfaces' to ensure that they are not missed out during the slicing operation.
3. *Adaptive slicing*: The layer thickness is not fixed. It depends on the complexity, or difference between the two adjacent slice contours.

Among these techniques, the adaptive slicing technique is the best as far as building speed is concerned. However, its implementation can be difficult. For instance in SLA, variable build layer thickness implies that the laser intensity must always vary accordingly. Furthermore, the benefits of adaptive slicing can only be fully utilized in a system that allows tangent cutting. For system that only allows normal cutting, to minimize the staircase effect, the system has to revert to thin layer thickness for slanted surfaces, as illustrated in figure 2. The STL slicer for the 5-axis micro milling RP machine adopts the adaptive slicing technique. The sliced data is presented in a new file format known as the LTI (Layer Transfer Interface) format[1] Unlike the other 2D sliced

---

[1] LTI format is a new data format specially developed for the 5-axis micro milling RP system by a group led by Dr. Thomas Gong Haiging in the Nanyang Technological University

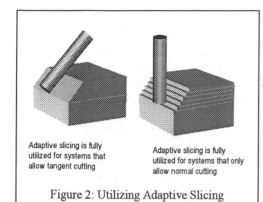

Adaptive slicing is fully utilized for systems that allow tangent cutting

Adaptive slicing is fully utilized for systems that only allow normal cutting

Figure 2: Utilizing Adaptive Slicing

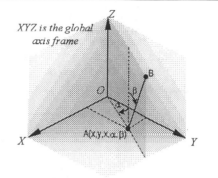

Figure 3: The LTI Format

data formats such as the SLC [9] and CLI [10], the LTI format includes additional data, like the slope information at every point, to orientate the cutter for tangent cutting. In this format, every point has 5 coordinates as shown in figure 3. 'x', 'y' and 'z' are the cartesian coordinates, while 'α' and 'β' are the roll and pitch coordinates respectively.

The LTI format is a 'two ways data format', the data can be transferred into the STL facetted object for modifications. This is a feature which neither the SLC nor CLI format can provide. Similarly, the LTI format can be converted into the SLC and CLI formats, but the reverse is not possible. Although the LTI format contains more information than the other two formats, its file size is considerably smaller as it uses the adaptive slicing technique to minimize the number of slices required to build a part. This reduces the storage size and speeds up the exchange between the computer and RP system.

4 Tool Diameter Offset

In CNC milling, it can be cumbersome and difficult for the NC programmer to program the cutter's path based on the size of the milling cutter being used. Also if the cutter size changes, it would be infeasible to change the program based on the new cutter size. For this reason, tool diameter offset allows the programmer to ignore the cutter size as the program is written. However, in this system, the approach is slightly different, instead of specifying the tool diameter offset at the tool path generation stage, the offset is specified during the STL file slicing stage. That is, the tool diameter is already set before the file is sent for tool path generation. The reasons for such approach are as follows,

1. During the STL to LTI format translation, some of the data are purposely ignored to keep the LTI file size small. Take a plane that is formed by any two consecutive segment points for example, to offset the tool diameter, we need to know which side of the plane should we place the cutter. Consequently, additional information must be added for each and every segment point. This will result in a much larger file size and eventually slow down the processing time.

2. The STL slicer, besides slicing the STL files, is also a off-line process simulation software. The tool diameter is an essential parameter for this application.

3. As mentioned earlier, the STL slicer adopts the adaptive slicing technique to minimize the number of build layers required to build a part. To do this, the STL slicer needs to know what is the maximum slicing thickness allowable. This slicing

thickness depends on the diameter of the tool since the maximum depth of cut allowable is two times its diameter. (i.e. depth of cut $_{max}$ = 2 x Tool Diameter)

Due to the additional information for the two rotary coordinates, the 'point' generated by the STL slicer describes more than just a point. It represents a line segment of undefined length. Therefore, from this point onwards, the five coordinates representation, i.e. $P(x,y,z,\alpha,\beta)$ will be referred to as *segment point*, while $P(x,y,z)$ will be referred to as *point*.

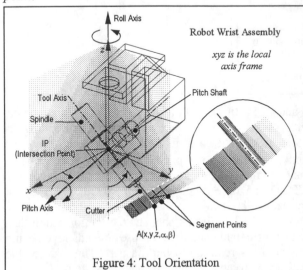

Figure 4: Tool Orientation

Figure 4 shows the schematic of the robot wrist assembly. Given a segment point $A(x,y,z,\alpha,\beta)$, (x,y,z) represents the position of the cutter tip along its centreline. 'α' is the orientation of the pitch axis, it is measured from the X-axis. 'β' is the orientation of tool axis, it is measured from the roll axis. Because the tool diameter offset is already specified in the LTI file, each segment point then represents the position whereby the tool axis must be aligned with.

Therefore, if a tool measured L_{tool} from the tip of the micro mill along its centreline (or tool axis) to the IP position, by replacing |AB| with L_{tool} (refer to figure 3), point 'A' represents the actual cutting (milling) position, while point 'B' represents the IP position. The tool path is formed by joining these IP positions with *straight* line segments.

5 Motion Compensation

To build a RP machine with high accuracy and precision, compensation is always required. There are two different ways to compensate the deviations or errors in machines. For instance, the straightness error of a carriage guide can be compensated, either by purely *hardware techniques*, such as scraping the guide, or by *software techniques*, which is also known as the *motion compensation*. The open loop compensation is used in the current 5-axis micro milling RP system, however, the closed loop compensation technique will be employed in the future system(s).

Motion compensation is a three steps procedure:

Step 1: Describe the machine setup by a simplified mathematical model. Two models are required, one for the ideal setup, and the other for the actual setup, which includes all the possible geometrical errors. Because of the geometrical errors, the two models will have different mathematical representations for the same motion transformation. By comparing the difference between the two representations, the compensation function for the machine is established. (This step will be discussed in detail in the coming sections.)

Step 2: Establish an experimental procedure to measure the parameters (i.e. the geometrical errors) in the model for the actual setup. Due to the constraint of the physical setup of the machine itself, direct measurement can be difficult to carry out at times, in such cases, indirect measurement should be used [11].

Step 3: Conduct experiment to verify to effectiveness of the compensation.

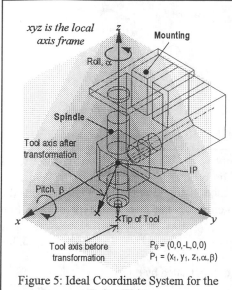

xyz is the local axis frame

Mounting

Roll, α

Spindle

Tool axis after transformation

IP

Pitch, β

Tip of Tool

Tool axis before transformation

$P_0 = (0,0,-L,0,0)$
$P_1 = (x_1, y_1, z_1, \alpha, \beta)$

Figure 5: Ideal Coordinate System for the Robot Wrist

Figure 5 shows the ideal coordinate system of the robot wrist used in the RP system. It has the following characteristics:

1. The roll axis coincides with the z-axis;
2. The roll axis is perpendicular to the pitch axis; and
3. The tool, roll and pitch axes intersect at *one* point.

Consider an arbitrary transformation **[M]**, such that

$$[P1] = [M].[P0]$$

where **[P0]** is tip the of the tool when it is in the vertical position, and **[P1]** is the tip of the tool after the transformation.

[M] consists of a pitch rotation of angle β, followed by a roll rotation of angle α z-axis.

i.e. $[\mathbf{M}] = [\mathbf{R}]_{\text{roll},\alpha} \cdot [\mathbf{R}]_{\text{pitch},\beta}$ …eq.(5.1)

For the ideal system, the transformation $[\mathbf{R}]_{\text{roll},\alpha}$ can be represented by a rotation about the z-axis with an angle α, $[\mathbf{R}]_{z,\alpha}$. Similarly, $[\mathbf{R}]_{\text{pitch},\beta}$ can be represented by a rotation about the x-axis with an angle -β, $[\mathbf{R}]_{x,-\beta}$. Therefore, eq.(5.1) becomes

$[\mathbf{M}]_{\text{ideal}} = [\mathbf{R}]_{z,\alpha} \cdot [\mathbf{R}]_{x,-\beta}$ …eq.(5.1a)

Consider only the cartesian coordinates,

$$\begin{bmatrix} x_1 \\ y_1 \\ z_1 \\ 1 \end{bmatrix} = \begin{bmatrix} \cos\alpha & -\sin\alpha & 0 & 0 \\ \sin\alpha & \cos\alpha & 0 & 0 \\ 0 & 0 & 1 & 0 \\ 0 & 0 & 0 & 1 \end{bmatrix} \cdot \begin{bmatrix} 1 & 0 & 0 & 0 \\ 0 & \cos(-\beta) & -\sin(-\beta) & 0 \\ 0 & \sin(-\beta) & \cos(-\beta) & 0 \\ 0 & 0 & 0 & 1 \end{bmatrix} \cdot \begin{bmatrix} 0 \\ 0 \\ -L \\ 1 \end{bmatrix} = \begin{bmatrix} L\sin\alpha\sin\beta \\ -L\cos\alpha\sin\beta \\ -L\cos\beta \\ 1 \end{bmatrix}$$

where (x_1, y_1, z_1) is the cartesian coordinates of the tool tip after the transformation, L is the tool length (measured from the tool tip to the IP position) and the fourth parameter is the *scaling factor* which is usually included in robotics.

Hence $P_1 = \begin{bmatrix} L\sin\alpha\sin\beta \\ -L\cos\alpha\sin\beta \\ -L\cos\beta \end{bmatrix}$ …eq.(5.2)

5.1 Angular Offsets of the Roll and Pitch Axes

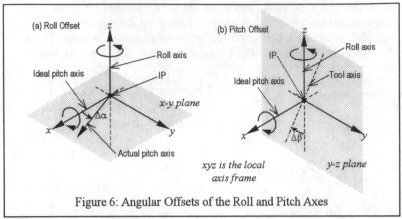

Figure 6: Angular Offsets of the Roll and Pitch Axes

Figure 6a shows the roll axis offset of the robot wrist. When the roll axis is reset to its origin position, ideally, the pitch axis should lie on the xz-plane, or $\Delta\alpha = 0$. However, due to the geometrical and dimensional tolerance of the mounting, deviation from the ideal pitch axis usually occurs. Hence $\Delta\alpha$, in general, is not zero. This type of deviation can be easily compensated as follow,

$$\alpha_{compensated} = \alpha_{nominal} + \Delta\alpha \qquad \qquad ...eq.(5.3a)$$

Similarly, deviation $\Delta\beta$ (see figure 6b) can be compensated by eq.(5.3b)

$$\beta_{compensated} = \beta_{nominal} + \Delta\beta \qquad \qquad ...eq.(5.3b)$$

5.2 Linear Offsets of the Roll and Pitch Axis

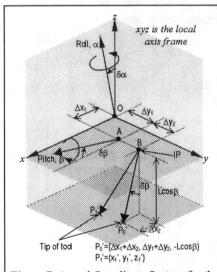

Figure 7: Actual Coordinate System for the Robot Wrist

Compensation of the roll and pitch axes only ensures that both the roll and pitch axes are parallel to the x-z plane. The pitch axis, for instance, may be some distance (Δy_1) away from the x-z plane and at an angle ($\delta\beta$) to the x-y plane as shown in figure 7. The actual IP (intersection point) position, instead of coinciding with O, which is the ideal IP position, may be Δx_1 from the y-axis and Δy_2 from the pitch axis. These are all the possible geometrical errors in the actual setup. Since there isn't any rotary axis along the y-axis, $\delta\alpha$ and $\delta\beta$ cannot be compensated by software as in the case of $\Delta\alpha$ and $\Delta\beta$. However, $\delta\alpha$ and $\delta\beta$ can be minimized through hardware compensation. Therefore, we may assume that $\delta\alpha \approx 0$ and $\delta\beta \approx 0$. If $\delta\beta \approx 0$, $L\cos(\delta\beta) \approx L$. However, the displacement of the tool tip along the x-direction (Δx_2) due to $\delta\beta$ can be significant depending on L, hence, it must also be taken into consideration during motion compensation. With all these geometrical errors, the cartesian coordinates of the tool tip before transformation can be expressed as $P_0'(\Delta x_1 + \Delta x_2, \Delta y_1 + \Delta y_2, -L)$, where " ' " represents the actual coordinate system.

Consider the transformation [M] in eq.(5.1), we may either use the nominal or compensated values for 'α' and 'β' (refer to eq.(5.3a) and eq.(5.3b)). However, if the nominal values are used, then the positional error, or deviation becomes a combined effect of both the linear and angular offsets.

The latter approach is preferred and we shall see its advantages in section 5.3.1. After the angular compensation,

$$\alpha' = \alpha + \Delta\alpha \qquad \qquad \ldots\text{eq.(5.3c)}$$
and $\quad\beta' = \beta + \Delta\beta \qquad \qquad \ldots\text{eq.(5.3d)}$
where $\quad\alpha$ and β are the nominal values.
$\quad\alpha'$ and β' are the compensated values.

With reference to eq.(5.1), for the actual system, the transformation $[R]_{roll,\alpha}$ is equivalent to a rotation about the z-axis with an angle α', $[R]_{z,\alpha'}$. Transformation $[R]_{pitch,\beta}$ is more complicated in the actual system since the pitch axis does not coincide with the x-axis (refer to figure 7). The transformation $[M]_{actual}$ can be expressed as,

$$[M]_{actual} = [R]_{z,\alpha'} \cdot \{[T]_{O \to A} \cdot [R]_{x,-\beta'} \cdot [T]_{A \to O}\} \qquad \ldots(5.1b)$$

Let $P_1'(x_1', y_1', z_1')$ be the coordinates of the tool tip after the transformation, we have

$$
\begin{bmatrix} x_1' \\ y_1' \\ z_1' \\ 1 \end{bmatrix} =
\begin{bmatrix} \cos\alpha' & \sin\alpha' & 0 & 0 \\ \sin\alpha' & \cos\alpha' & 0 & 0 \\ 0 & 0 & 1 & 0 \\ 0 & 0 & 0 & 1 \end{bmatrix} \cdot
\begin{bmatrix} 1 & 0 & 0 & \Delta x_1 \\ 0 & 1 & 0 & \Delta y_1 \\ 0 & 0 & 1 & 0 \\ 0 & 0 & 0 & 1 \end{bmatrix} \cdot
\begin{bmatrix} 1 & 0 & 0 & 0 \\ 0 & \cos(-\beta') & -\sin(-\beta') & 0 \\ 0 & \sin(-\beta') & \cos(-\beta') & 0 \\ 0 & 0 & 0 & 1 \end{bmatrix} \cdot
\begin{bmatrix} 1 & 0 & 0 & -\Delta x_1 \\ 0 & 1 & 0 & -\Delta y_1 \\ 0 & 0 & 1 & 0 \\ 0 & 0 & 0 & 1 \end{bmatrix} \cdot
\begin{bmatrix} \Delta x_1 + \Delta x_2 \\ \Delta y_1 + \Delta y_2 \\ -L \\ 1 \end{bmatrix}
$$

$$
= \begin{bmatrix} (\Delta x_1 + \Delta x_2)\cos\alpha' - \Delta y_2 \sin\alpha' \cos\beta' + L\sin\alpha' \sin\beta' - \Delta y_1 \sin\alpha' \\ (\Delta x_1 + \Delta x_2)\sin\alpha' + \Delta y_2 \cos\alpha' \cos\beta' - L\cos\alpha' \sin\beta' + \Delta y_1 \cos\alpha' \\ -\Delta y_2 \sin\beta' - L\cos\beta' \\ 1 \end{bmatrix}
$$

Hence $P_1' = \begin{bmatrix} (\Delta x_1 + \Delta x_2)\cos\alpha' - \Delta y_2 \sin\alpha' \cos\beta' + L\sin\alpha' \sin\beta' - \Delta y_1 \sin\alpha' \\ (\Delta x_1 + \Delta x_2)\sin\alpha' + \Delta y_2 \cos\alpha' \cos\beta' - L\cos\alpha' \sin\beta' + \Delta y_1 \cos\alpha' \\ -\Delta y_2 \sin\beta' - L\cos\beta' \end{bmatrix} \qquad \ldots\text{eq.(5.4)}$

Comparing eq.(5.2) with eq.(5.4), the deviation in P_1 becomes

$$\Delta P_1 = -\begin{bmatrix} (\Delta x_1 + \Delta x_2)\cos\alpha' - \Delta y_2 \sin\alpha' \cos\beta' - \Delta y_1 \sin\alpha' + L(\sin\alpha' \sin\beta' - \sin\alpha \sin\beta) \\ (\Delta x_1 + \Delta x_2)\sin\alpha' + \Delta y_2 \cos\alpha' \cos\beta' + \Delta y_1 \cos\alpha' - L(\cos\alpha' \sin\beta' - \cos\alpha \sin\beta) \\ -\Delta y_2 \sin\beta' - L(\cos\beta' - \cos\beta) \end{bmatrix} \qquad \ldots\text{eq.(5.5)}$$

such that $P_{1,compensated} = P_{1,nominal} + \Delta P_1$.

Therefore for every nominal segment point $P(x,y,z,\alpha,\beta)$, the corresponding compensated segment point is given by $P'(x',y',z',\alpha',\beta')$, such that,

$$\begin{bmatrix} x' \\ y' \\ z' \\ \alpha' \\ \beta' \end{bmatrix}_{compensated} = \begin{bmatrix} x \\ y \\ z \\ \alpha \\ \beta \end{bmatrix}_{nominal} + \begin{bmatrix} -(\Delta x_1 + \Delta x_2)\cos\alpha' + \Delta y_2 \sin\alpha'\cos\beta' + \Delta y_1 \sin\alpha' - L(\sin\alpha'\sin\beta' - \sin\alpha\sin\beta) \\ -(\Delta x_1 + \Delta x_2)\sin\alpha' - \Delta y_2 \cos\alpha'\cos\beta' - \Delta y_1 \cos\alpha' + L(\cos\alpha'\sin\beta' - \cos\alpha\sin\beta) \\ \Delta y_2 \sin\beta' + L(\cos\beta' - \cos\beta) \\ \Delta\alpha \\ \Delta\beta \end{bmatrix}$$

$$\dots\text{eq.}(5.6)$$

5.2.1 Experiment to determine the linear offsets of the roll and pitch axis

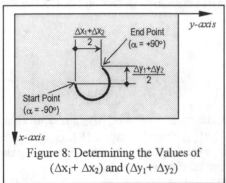

Figure 8: Determining the Values of $(\Delta x_1 + \Delta x_2)$ and $(\Delta y_1 + \Delta y_2)$

From eq.(5.6), the geometrical errors required are $(\Delta x_1 + \Delta x_2)$, Δy_1 and Δy_2. Indirect measurements are used to determine these values. There are only two steps,

Step 1 : Do the necessary angular offsets in section 5.1.

Step 2 : Make a cut on a test surface by rotating the roll axis from $-90°$ to $+90°$. Note that these are the compensated values.

For the ideal system, only a dot will be observed. However, if a semi-circle is observed (see figure 8), it implies that there are some deviations. Since, the angular offsets are already compensated, the deviation must be caused by the linear offsets of the roll and pitch axis.

The horizontal and vertical deviations are due to $(\Delta x_1 + \Delta x_2)$ and $(\Delta y_1 + \Delta y_2)$ respectively. Depending on the position of the start and end points, these deviations may have negative values. Unfortunately, from this experiment, the individual of values of Δy_1 and Δy_2 cannot be determined, but we shall see how they can be deduced in section 5.3.1.

5.3 Implementation of Motion Compensation

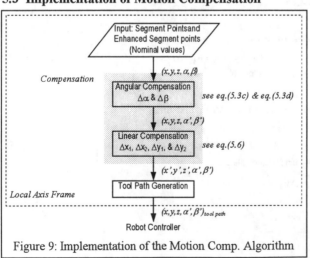

Figure 9: Implementation of the Motion Comp. Algorithm

The implementation of the motion compensation is shown in the flow chart (figure 9).

5.3.1 Verification of Concentricity Compensation

A test file was written to instruct the machine to cut a slot by rotating its roll axis from -90° to +90°. To ease investigation, the smallest micro-mill (∅0.2mm) was used. As milling (cutting) causes deflection in both the z-axis and the cutter itself, it is desirable to reduce these deflections by minimizing the cutting force during the experiment. This is achieved by:

1. Ensuring a shallow depth of cut;
2. maintaining a low feed rate; and
3. using a high spindle speed (at 100,000rpm) for cutting.

The pictures of the slot is captured and analyzed by an image processing software

Figure 10 shows the concentricity error before and after the compensation, different values of Δy_2 (refer to section 5.2.1) are tested until the minimum values of DX and DY are obtained.

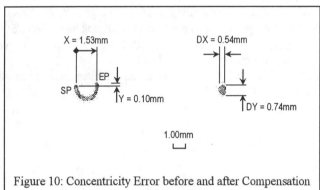

Figure 10: Concentricity Error before and after Compensation

6 Conclusions

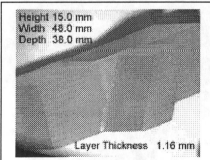

Figure 11: Picture of a 14 Layer Prototype

The concept of 5-axis micro milling for RP application was proposed to eliminate the staircase effect in current rapid prototyping. A prototype system has been built and has proven its feasibility and potential.

A motion compensation algorithm was also developed to improve the machining accuracy. Experiment were conducted to verify the effectiveness of motion compensation, and the outcome was very positive.

For the verifications of the tangent cutting and other building processes, a simplified 14 layers (15mm) prototype was built. The result was acceptable, the concept of tangent cutting to produce stepless part was proven feasible. However, there is still room for improvement in terms of building speed and accuracy.

References

1. Crowell, George. StereoLithography. 1989. Dearborn, Michigan: Society of Manufacturing Engineers.

2. Jacobs, P.F. 1996. StereoLithography and other RP&M technologies : from rapid prototyping to rapid tooling. Dearborn, Mich.: Society of Manufacturing Engineers in cooperation with the Rapid Prototyping Association of SME; New York: ASME Press.

3. Schweikhart, R. 1990. StereoLithography : applications for rapid prototyping. Dearborn, Mich. : Society of Manufacturing Engineers.

4. Welcome to 3D Systems. 1996. [web page]. http://www.3dsystem.com/.

5. Cubital: The Faster, the Better. 1996. [web page]. http://www.iquest.net/cubital/.

6. Welcome to Stratasys, Inc. 1997. [web page]. http://www.stratasys.com/.

7. Michael J.B. 1995. Tele-manufacturing rapid prototyping on the Internet. IEEE computer graphics and applications. IEEE Computer Society: National Computer Graphics Association.

8. Helisys: Laminated Object Manufacturing. 1996. [web page]. http://helisys.com/.

9. .SLC File Format. 1996. [web page]. http://cadserv.cadlab.vt.edu/bohn/rp/SLC.html.

10. Common Layer Interface (CLI). 1994. [web page]. http://www.cranfield.ac.uk/aero/rapid/CLI/cli_v20.html.

11. Sartori, S & Zhang G.X. 1995. Geometric error measurement and compensation of machines. Annals of the CIRP, 44, (2). 599-609.

Generation of Tooth Models for Ceramic Dental Restorations[*]

Sven Gürke

Department of Cognitive Computing & Medical Imaging, Fraunhofer Institute for
Computer Graphics, Darmstadt, Germany

Abstract. Complete CAD/CIM-systems for dental restorations require
user interaction for the construction of tooth inlays in ceramics. An automation is only possible if the typical geometry of the teeth is known
by the system. One realizable way for the solution of this problem is the
restoration of the occlusal surface by adaptation of an appropriate tooth
model.

In this paper, we present a method for developing and generating of tooth
models including dental medical knowledge about the occlusal surface.
Furthermore, our approach includes an adaptation process which can be
extended to an automatic tooth restoration system.

1 Introduction

1.1 Conventional Method

In the last few years, on the field of tooth restorations we can ascertain an
increased acceptance of ceramic inlays besides plastic and gold inlays. Ceramic
restorations possess an inconspicuous appearance like natural teeth and exhibit
similar hardness. Furthermore, in contrast to amalgam ceramic inlays have no
toxical pertinence and a long durability. Unfortunately, most ceramic inlays are
produced in dental labs and therefore the patient must accept at least two visits
at the dentist.

1.2 Alternative

One popular CAD/CIM system is the CEREC-device [8] (see figure 1) of the
Siemens AG. In this system, the dentist has to take an optical impression of the
prepared tooth with an intra-oral 3D-sensor. Then he designs the inlay interactively with an integrated CAD-system. Finally, an integrated milling machine
produces the inlay from a block of ceramics. In the processing pipeline, the interactive construction process represents the difficult part of this method. Often
the surface of the occlusal part of the inlay has to be modeled manually after
the inlay is inserted into the cavity.

---

[*] Sponsored by the Bayerische Forschungsstiftung

1.3 New Technique

The aim of a research project is to minimize the complicated user interaction for the construction of an inlay. An automation of a CAD/CIM system is only possible if the typical geometry of teeth is known by the system. One realizable approach for the solution of this problem is the restoration of the occlusal surface by adaptation of an appropriate tooth model.

Fig. 1. CEREC-device for dental restorations

In this paper, we present a method for the generation of suitable tooth models. Input data are two-dimensional range images. They can be captured by the CEREC scanner or an another intra-oral 3D scanner. After generation of a range image a corresponding tooth model is chosen out of the tooth library. The development of such a tooth model is described in section 3. By a series of deformations and refinements the model adapts automatically to the scanned tooth. Result is a triangular representation of the occlusal surface. The processing pipeline is shown in figure 2.

This step is necessary for quality estimations and improvements of the developed tooth model, but also as a basis process for an automatic dental restoration system.

Due to the deformable 3D tooth model and the included adaptation process this approach differs from other methods as it is proposed in [9, 10] for instance.

2 Structure of the Tooth Library

There are a lot of dental medical analyses [1, 3, 12–14] dealing with the division of different teeth into only a few numbers of types. Siebert [12] shows that for each tooth group like the first upper or first lower molar only few types are needed to cover all forms occurring in the nature. Our tooth models make a great use of

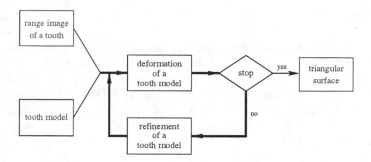

Fig. 2. Occlusal surface restoration using a tooth model (heavy arrows indicate design loop)

this analysis. For each tooth group at least one tooth model is designed. Then, it is guaranteed that our tooth library represents all typical forms occurring in the nature and the task of the adaption process is reduced to fashion the individual tooth anatomy of the patient.

3   The Tooth Model

All the tooth models within the library are built up on the same principle. In this section it is explained how we can derive these main principles and how this information is used for the generation of a tooth model.

The tooth model must fulfill the following demands:

1. It must contain the dental medical knowledge about this tooth type,
2. it must be deformable in an easy way, and
3. it must represent the tooth surface.

3.1   The Geometrically Deformable Model

A model which suffices in particular the above listed demands is the geometrically deformable model (GDM).

Geometrically deformable models were introduced by Miller et al. [2, 7] for the segmentation and visualization of two- and three-dimensional objects. D. Rückert [11] uses geometrically deformable models for the segmentation of two-dimensional medical images.

GDMs present a very intuitive approach to generate models of objects whose shape cannot be described in a simple mathematical form. They can operate on a sparse sampling of the data while still allowing the level of detail extracted to be set by the application. By altering the level of detail, quality models can be generated while still providing a compact representation. This fact is used in the provided adaptation process by reiterating a deformation and a refinement step.

A geometrically deformable model is created by placing an initial simple model in the data set which is then deformed by minimizing a set of constraints. The process starts with a simple non-self-intersecting triangulation of the occlusal surface. Then, the model is deformed based on the set of constraints, so that the model grows and shrinks to fit the scanned tooth. There is an energy function associated with every control point of the GDM determining the local deformation, the properties of the model, and the relationship between the construction features of the tooth. By minimizing these functions the model adapts the scanned tooth automatically.

The dental medical knowledge can leave its remark on the model twice. On the one hand the outer characteristic features of the tooth surface are included in the model by the configuration of the triangulation. On the other hand the relationships between the features are introduced by assigning special energy functions to the control points of the triangulation.

Hence, our proposed tooth model consists of a triangular surface and a set of specially chosen energy functions.

3.2   The Surface of the Model

The crowns of the teeth and especially the occlusal surface consist of hills and valleys. They are the basic elements of the tooth surface [6]. Also the classic technique for producing crowns bases on these elements.

These features are used for the generation of a triangulation of the occlusal surface by assigning each basic element of the tooth some points or lines. These geometrical elements build the frame of the triangulation. Figure 3 shows wireframe representations of the models of the first lower molar and of the first upper molar. Each control point and most edges of the triangulation represent at least one basic element of teeth.

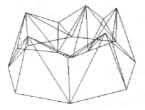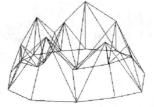

Fig. 3. Model of the first lower and the first upper molar representing main information of the occlusal surface

The extraction of the typical feature points is done by a semi-automatic process. For an undamaged typical tooth the maxima, minima and, saddle points are calculated and entered into the image. After an interactive check of the detected points we get the coordinates of the most important points on the tooth surface. The list is completed by searching a fixed number of points on the equatorial line of the tooth. This can be done by following some automatically detected lines. An example is shown in figure 4.

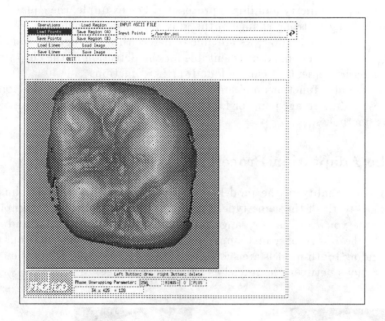

Fig. 4. Semi-automatic extraction of the most important surface points

3.3 The Energy Functions

The energy functions occupy a very important part within the model. Their quality determine the quality of the final restorations.

In contrast to Miller et al. [2, 7] and D. Rückert [11] in our approach every control point cp of the GDM gets its own energy function

$$E_{cp} : \mathbb{R}^3 \to \mathbb{R}.$$

They are defined as a weighted sum of given functions $E_1, \ldots, E_n$

$$E_{cp}(x) = \sum_{k \in I_{cp}} w_k E_k(x) \qquad \text{with} \qquad I_{cp} \subset \underline{n}.$$

Each of these functions E_k focuses on a specific point, curve, or surface characteristic. In combination, they represent a measurement for the quality of a point with regard to a particular tooth feature. As each energy function is minimized, the model will deform while searching for the surface of the tooth, and maintaining its topology.

In order to develop a new tooth model we search for each basic element of the teeth, as the cusp tips or the central fossa, a precise characterization in terms of the differential geometry and image processing. For example, the equatorial line is determined by a strong change of the gray values in the neighborhood of the line. Using a derivative filter, we are able to detect these changes. A function for the judgement of local maxima is suitable for the detection of the cusp tips. For the other basic elements it can be found further functions.

There are also functions measuring the distance between a model point and the corresponding image point or for the preservation of the surface topology or a fixed smoothness.

4 The Adaptation Process

To judge the quality of the model we apply it to a series of range images of undamaged teeth of the same type. Figures 5a) and c) show two examples. On the left side you can see the range images of a first lower molar and on the right side the image of a first upper molar. Wireframe representations of the corresponding tooth models are shown in b) and d). Now, the adaptation occurs by interactions between the model and the range image.

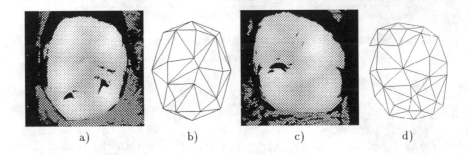

a)   b)   c)   d)

Fig. 5. a) and c) range images of the 1. lower molar and the 1. upper molar, b) and d) the triangular surfaces of the corresponding models

The optimal geometrical approach is realized by two partial processes (see also figure 2):

- In a first step the GDM adapts to the tooth surface by deformations using simulated annealing optimization.
- In a second step the GDM is refined by a subtriangulation process.

These two steps are repeated since a sufficient degree of refinement is reached. The quality of the deformation process is controlled by the above described energy functions. The real deformation takes place by minimizing the energy functions. It has been shown that the segmentation process of Miller [2] is very susceptible to noise and artifacts. The reason lies in the usage of the Hillclimbing process. Rückert provided as an improvement the simulated annealing optimization. It has been shown that this method in theory always finds the global minimum of a function. Due to the enormous complexity of our energy functions we decided the usage of simulated annlealing [4, 5] as our optimization method.

4.1 The Deformation Process

During the deformation process at each temperature T all control points are passed through. If a control point cp is moved to a randomly generated position p is decided by the condition

$$\exp\left(-\frac{E_{cp}(cp) - E_{cp}(p)}{T}\right) < r,$$

where $r \in [0, 1]$ is a random number. A violation of the topology is reduced by the usage of suitable energy functions. The process terminates, if a negligible impovement of the total error is reached.

After the deformation step the whole process stops if the triangulation has reached a sufficient degree of refinement.

4.2 The Refinement Process

Otherwise we subtriangulate the model. Therefore, we are able to reconstruct finer structures of the scanned tooth. We distinguish between four cases. They are shown in figure 6.

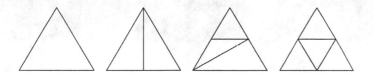

Fig. 6. Possible cases of subtriangulation

As described, the adaptation process is controlled by minimizing of energy functions. After subtriangulation we have to assign suitable energy functions to the new control points. Since the new control points are always in the middle of the old edges they have the common features of the two vertices. Hence, the new control points get the weighted sum of the common functions E_k of the two vertices.

5 Results

In this section we show the results of the adaptation process which judge our generated tooth model. Therefore we introduce two criteria:

- a numerical criterion which rests on the mean distance of a control point to the image and describes the precision,
- a visual criterion which shows the behavior of the construction features within the adaptation process.

The first criterion is a necessary condition of the reconstruction process. A very good adaptation of the occlusal construction features is requirement for a successful implementation of a tooth restoration. This can be judged with the second criterion.

The dental clinic in Munich has put a lot of tooth images at our disposal. We use these images for the development of the tooth models and for the tests of our algorithms. The used 3D-scanner provides range images with a resolution of 30x30x1 μm in x-, y- and z-direction.

Figure 7 shows the results of the deformations in different resolutions starting with the initial model and terminating after four refinement steps with a mean error of 1–2 μm calculated in the vertices of the final triangulation. The resulting triangulation is modeled by 4261 triangles. The whole process has taken 62 seconds on a SPARC 10 workstation. Typical computing times were 60–80 seconds for the calculation of four respectively five refinement steps.

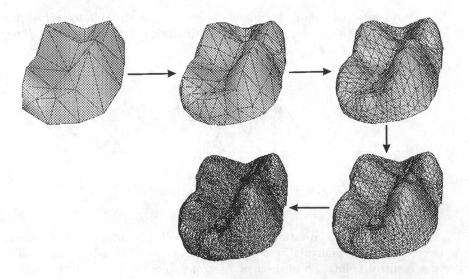

Fig. 7. Result of an adaptation process in form of triangular meshs

Figures 8 and 9 show the the corresponding changes of the range image in the different resolution steps. At the end of the chain we can see the original scanned tooth image for making a comparism. We recognize an effect of smoothing as it is visible particularly in the fissures.

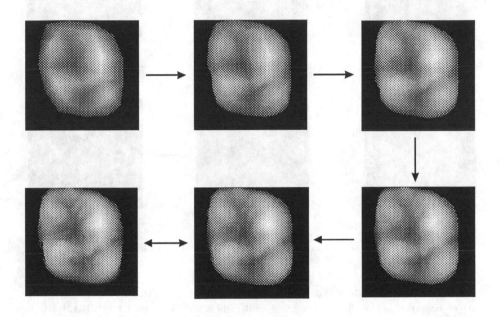

Fig. 8. Result of an adaptation process in form of range images

Figure 9 shows the difference images. Following the images counterclockwise we ascertain an decrease of the distance between the original range image (at the end of the series) and the range images in the different resolutions of the adaptation process. The higher distance values at the border of the tooth are caused by the given range images (see figure 5).

Figure 10 shows the use of the second criterion. The control points of the resulting triangulation which belong to a chosen construction feature as the equatorial line or the cusp tips are entered the rendered range image. Now we are able to judge the quality of the adaptation in the construction features of the tooth. An abnormal behavior as the slipping off the cusp line to the equatorial line is ascertainable in an easy way. However, at the same time the result of the reconstruction shows no losses of accuracy.

6 Conclusion

In this paper, we presented a method for the generation of suitable tooth models for dental restorations. The approach includes an adaptation process which can

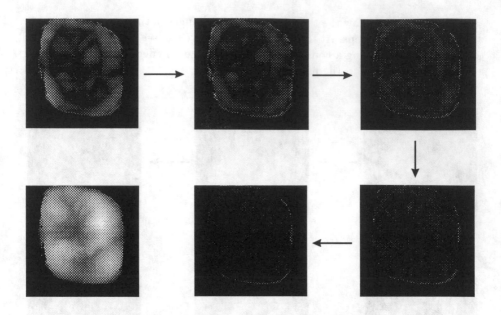

Fig. 9. Result of an adaption process in form of the difference images

be extended to an automatic tooth restoration system. An important topic for future research is the development of methods which lead to a restoration of the destroyed occlusal surface. Therefore, during the adaptation the points above the cavity are detected. A change of their behavior of adaptation will lead to a closure of the cavity according to the dental medical knowledge of the model.

Acknowledgement

I wish to express my thanks to G. Saliger from the Siemens AG in Bensheim, and K.-H. Kunzelmann from the dental clinics of the University Munich for their assistance and preparation of images. Special acknowledgement is due to P. Neugebauer and D. Barnickel, who have made a lot of useful ideas, comments, and suggestions.

References

1. L. Abrams, R.E. Jordan, and B.S. Kraus. *Dental anatomy and occlusion.* The Williams and Wilkins Company, Baltimore, 1969.
2. D.E. Breen, W.E. Lorensen, J.V. Miller, R.M. O'Bara, and M.J.Wozny. Geometrically deformed models: Method for extracting closed geometric models from volume data. *Computer Graphics*, 25(4):217–226, July 1991.
3. T.E. De Jonge Cohen. *Mühlreiters Anatomie des menschlichen Gebisses.* Verlag Arthur Felix, Leipzig, fourth edition, 1920.

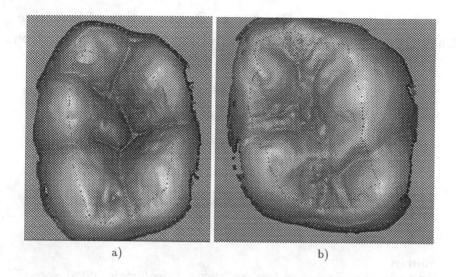

a)   b)

Fig. 10. Construction features of the model after the adaption process: a) 1. lower molar, b) 1. upper molar

4. S. Geman and D.Geman. Stochastic relaxation, gibbs distribution, and the bayesian restoration of images. *IEEE Transaction on Pattern Analysis and Machine Intelligence*, 6(2):721–741, 1984.

5. L. Ingber. Simulated annealing: Practice versus theory. *Mathl. Compu. Modelling*, 18(11):29–57, 1993.

6. William H. McHorris. *Einführung in die Okklusionslehre*. Quintessenz Verlag, Berlin, 1983.

7. J.V. Miller. On gdm's: Geometrically deformed models for the extraction of closed shapes from volume data. Master's thesis, Rensselaer Polytechnic Institute, Troy, New York, December 1990.

8. W.H. Mörrmann and M. Brandestini. *Die CEREC Computer Rekonstruktion*. Quintessenz-Verlag, 1989.

9. K. Myszkowski, V.V. Savchenko, and T.L.Kunii. Computer modeling for the occlusal surface of teeth. In *Computer Graphics International'96*, Pohang, Korea, June 1996. IEEE Computer Society.

10. P.J. Neugebauer. Generating CAD-models of Teeth. In *Proceedings of the Computer Assisted Radiology CAR'96 Conference*, Paris, 26.-29. June 1996. Elsevier Science.

11. D. Rückert. Bildsegmentierung durch stochastisch optimierte Relaxation eines 'geometric deformable model'. Master's thesis, TU Berlin, 1993.

12. G. Siebert. Zur Morphologie der Zähne unter funktionellen Aspekten. *Österreichische Zeitschrift für Stomatologie*, 80:139–147, April 1983.

13. G. Wetzel. *Lehrbuch der Anatomie für Zahnärzte und Studierende der Zahnheilkunde*. Fischer Verlag, Jena, sixth edition, 1951.

14. R.C. Wheeler. *An atlas of tooth form*. W.B. Saunders Company, Philadelphia, fourth edition, 1969.

An Investigation on Parameter Optimization of Selective Laser Sintering Process for Rapid Prototyping Applications

Dr Zhu Chun Bao

CIM Center, German-Singapore Institute, Nanyang Polytechnic
Jurong East, 10 Science Center Rd, Singapore 609079
Email: zhucb@nyp.ac.sg

Leonard Loh, Ng Poey Leong & Jeffrey Chan H C

PE Center, German-Singapore Institute, Nanyang Polytechnic
Jurong East, 10 Science Center Rd, Singapore 609079

Abstract

The quality of solid prototype from the Selective Laser Sintering (SLS) process depends very much upon process parameters selected. The paper describes an investigation on optimal selection of SLS parameters for rapid prototyping application using Taguchi method. The effects of six important parameters, laser power, part temperature, temperature difference between part and piston heaters, laser scan space, layer thickness and powder material composition (e.g., ratio of virgin powder to recycled powder) on part geometrical properties, such as surface finish, dimension accuracy and profile distortion (roundness and curling), have been tested. An L_{16} (2^{15}) orthogonal array was used to accommodate the experiments and to analyze the experimental data.

The paper reports the experiment method used and benchmark designed for part property measurement. Based on statistical regression analysis of the experimental data obtained in the research, the paper indicates the optimal combination of SLS parameters which would result in the desired prototype with respect to several property parameters.

Keywords: Selective laser sintering, Taguchi method, Regression analysis, Rapid prototyping, Parameter optimization.

1 Introduction

Last few years have seen fast development of rapid prototyping (RP) applications. The techniques used in the RP industry are rather complex which involve various aspects of modern scientific and technological developments, such as laser scanning techniques, CAD modeling, CAM, CAD data conversion and transferring, electronic automatic control, etc. Selective Laser Sintering (SLS) is one of the new techniques commonly used today.

The number of factors involved in the SLS process is very large [1] (Zhu CB, Wong D S K, Loh L and Sue H Y, 1996) and some are not thoroughly studied. It is therefore difficult to develop a comprehensive mathematical model for parameter optimization which would result in desired parts. In RP practice, the input conditions for making prototypes are often not optimal. It is particularly true that the best rapid prototyping parts are normally built on a given machine only after multiple attempts, with some tuning or selected process parameters after each iteration. This learning process is very costly and the experience and production expertise gained this way is not systematic, which makes it very difficult to make use of the knowledge for RP industries.

There is a need to investigate the optimal parameter setting for a given rapid prototyping machine and known application requirements. Two major problems should be addressed in order to carry out experimental optimization for the known SLS machine, 1) the number of tests should be small as running the SLS Sinterstation can be very costly and 2) the limited tests should provide enough information for exploring the cause-effect of the SLS process. Taguchi method meets just the above requirements and therefore is used in this research.

2 SLS Part Geometrical Properties

The SLS part geometrical properties can be characteristicsed by a number of parameters, such as deviation of measured dimensions from the desired dimensions determined by a CAD model, dimensional tolerance and trueness of geometrical profile, such as sharpness of corners and edges [2][3], roundness, curling and creep distortions as well as growth. Surface quality of SLS parts can be measured using roughness (R_a), surface texture and aliasing [4]. Considering easy-to-measure nature of these parameters, in this investigation, following are selected for quantification of part geometrical properties.

1)   *Part surface finish R_a* , surface finish of SLS parts may range from 5 to 50 μm[5, 6].

2)   *Dimensions deviation (Δl_i)*, measurement of actual part size of L_i provides data for setting offset/scaling for part dimension compensation due to part shrinkage after or during the sintering process. In the investigation, deviation (Δl_i) of each dimension L_i is determined as follows :

$$\Delta l_i = (L_{io} - L_{im})/ L_{io} \tag{1}$$

Where, L_{io} is design size, L_{im} is the actual dimension measured, normally, it is a mean of multiple measurements on the same dimension.

3)   *Roundness (K),* K typically represents the trueness of a geometrical feature (circularity). It is defined by

$$K = (\Phi_{max} - \Phi_{min})/ \Phi_m \tag{2}$$

Where, Φ_{max} and Φ_{min} are the maximum and minimum diameter measured on the round element, Φ_m is the mean of the multiple measurements of the diameter.

4)   *Curling (β),* Curling is a phenomenon that the SLS part's edges or corners rise up (see Fig.1). During the selective laser sintering process, temperature differences can exit in different areas of the part being built. These temperature differences cause uneven shrinkage, which in turn causes curling.

There are two kinds of curling, i) in-build curling and ii) post-build curling. Severe in-build curling may cause the part to shift in the part bed when the roller passes over the part, while post-build curling generates permanent part profile distortion. In this investigation, curling (β) is quantified as follows:

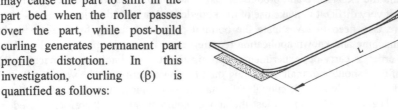

$$\beta = \Delta h/L \qquad (3)$$

Fig.1. Curling and measurement

3 Description of Experiment Design and Implementation

3.1 Benchmark Design

A benchmark (shown in Fig.2) has been designed to provide a test piece with all the necessary geometrical elements that ease measurement of part geometrical properties, e.g., roundness, curling, dimension accuracy, surface finish and growth. The geometrical elements designed are distributed on and supported by a flat base (2 mm in thickness).

This part contains stepped boxes, the dimensional tolerances in X, Y and Z are characterized, as are other factors, such as cone and cylinder circularity, cone angle errors, edge straightness. Flat areas at different locations and in different directions provide the surfaces for roughness measurements. The long stripes in both X and Y direction are used for curling measurement.

During test, each benchmark is numbered which is correspondent to the trial number of the experiment and each geometrical measurement element is also named so that the data collected could be easily stored and analyzed accordingly.

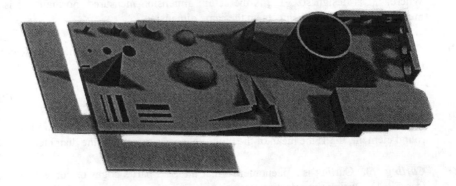

Fig.2. The benchmark designed for testing SLS part properties

3.2 Control Factors

Although there are a number of parameters affecting SLS process [1], only six parameters (see Table 1) are selected as controllable factors in the research due to their effects on the SLS part geometrical properties. For each factor, two levels are considered in the experiment design.

Table 1 Factor & Level Specification

	Factors	Symbol	Units	Level 1	Level 2
A	Powder material composition (Virgin:recycled)	R	-	1:1	1:2
B	Laser power 9% (3W)	P	% (W)	10% (3.5)	9% (3.0)
C	Part Temperature	T	°C	185	186
D	Temperature difference btw part & piston heaters	ΔT	°C	82	80
E	Scan space	d	mm	0.17	0.15
F	Layer thickness	Δa	mm	0.10	0.11

In Table 1, R is the ratio of virgin powder materials to recycled ones mixed in the build. Factor P (laser power) takes two levels, 3.5 (w) and 3.0 (w) which covers most commonly used power setting range for building prototype with Nylon Composite (Laserite Powder, LNC7000). Two temperature factors (T and ΔT) have been selected to investigate their effects on part profile distortion and shrinkage. The scan space (d) and layer thickness (Δa) are used to test their influence on part geometric accuracy and surface finish.

Some other parameters are fixed, for instances, 1) temperatures for powder material pre-heating are set as 80 °C (Left bed), 80 °C (left bed), 80 °C (Part piston); 2) the slice thickness is set at 0.11 mm.

3.3 Experiment Plan

The L_{16} (2^{15}) array is selected to accommodate the experiment. The factor column design is shown in Table 2. As can be seen from Table 2, the factors A, B, C, D, E and F are assigned into Columns 1, 2, 4, 8, 11, 13 respectively in L_{16} (2^{15}) array. This design is of low resolution [7], in which the main factor effects can be estimated but two-factor-interactions are confounded each other. The interaction confounding that exists in this experiment is noted but ignored as the first round of the test is to try to investigate the main effects of the major factors on the output parameters, and the interactions will be investigated in further research.

Table 2 L_{16} (2^{15}) Array Factor Column Design

Col No	1	2	3	4	5	6	7	8	9	10	11	12	13	14	15
Factors	A	B	A×B	C	B×F	B×C	e_1	D	B×E	B×D	E	C×D	F	e_2	B×F

Table 3 is the layout of the experiment plan. As can be seen from Table 3, the powder material composition (the ratio of virgin powder to recycled powder) is assigned to Column 1 as this parameter is the most difficult one to alter. The rest factors are randomly assigned into main effect columns 2, 4, 8, 11, 13 accordingly. Column 7 and 14 are used for error evaluation [8, 9].

Table 3 Experiment Plan and Sample Data (Part Curling Test)

Trial No	1 (A) R (V:R)	2 (B) P (W)	4 (C) T (°C)	8 (D) ΔT (°C)	11 (E) d (W)	13 (F) Δa (mm)	β_{Xi}	β_{Yi}
1	1: 1:1	1: 10%	1: 185	1: 82	1: 0.17	1: 0.10	0.395	1.113
2	1: 1:1	1: 10%	1: 185	2: 80	2: 0.15	2: 0.11	0.017	0.042
3	1: 1:1	1: 10%	2: 186	1: 82	1: 0.17	2: 0.11	0.034	0.170
4	1: 1:1	1: 10%	2: 186	2: 80	2: 0.15	1: 0.10	0.387	0.622
5	1: 1:1	2: 9%	1: 185	1: 82	2: 0.15	1: 0.10	2.794	5.083
6	1: 1:1	2: 9%	1: 185	2: 80	1: 0.17	2: 0.11	0.034	0.085
7	1: 1:1	2: 9%	2: 186	1: 82	2: 0.15	2: 0.11	0.017	0.042
8	1: 1:1	2: 9%	2: 186	2: 80	1: 0.17	1: 0.10	1.535	3.525
9	2: 1:2	1: 10%	1: 185	1: 82	2: 0.15	2: 0.11	0.068	0.643
10	2: 1:2	1: 10%	1: 185	2: 80	1: 0.17	1: 0.10	0.051	0.556
11	2: 1:2	1: 10%	2: 186	1: 82	2: 0.15	1: 0.10	0.017	0.684
12	2: 1:2	1: 10%	2: 186	2: 80	1: 0.17	2: 0.11	0.102	0.512
13	2: 1:2	2: 9%	1: 185	1: 82	1: 0.17	2: 0.11	0.378	0.171
14	2: 1:2	2: 9%	1: 185	2: 80	2: 0.15	1: 0.10	0.017	0.001
15	2: 1:2	2: 9%	2: 186	1: 82	1: 0.17	1: 0.10	0.239	0.042
16	2: 1:2	2: 9%	2: 186	2: 80	2: 0.15	2: 0.11	0.342	0.042

3.4 Experiment Implementation

The experiments were carried out in Rapid Prototyping Laboratory of Precision Engineering Center at German-Singapore Institute, Nanyang Polytechnic. All experiments are implemented using the SinterStation^{TM} 2000 System to explore the effect of the control variables on part geometrical properties. The major characteristics of the system are listed as follows:

Laser	- 25- or 50 watt CO_2CW laser
Max build envelope	- 305 mm in diameter and 381 in height (for Wax Polycarbonate)
	- 203 ~ 254 mm in diameter and 381 in height (for Nylon Composite/Nylon)
Nominal slice thickness	- 0.13 ~ 0.18 mm
Tolerances	- 0.38 mm (for initial part build accuracy)
	- 0.13 (for optimized parameter settings)

4 Analysis of Experiment Data and Discussion of the Results

4.1 Experiment Data and Discussion

A large amount of data are collected from the investigation and systematically analyzed using orthogonal regression method [8, 9]. Here, as an example, two sets of measured data on the part curling distortion is listed in Table 3, where, β_{Xi} is the curling measured on the part in X direction, while β_{Yi} is measured in Y direction. β_{Xi} and β_{Yi} are mean values of multi-readings taken from the parts. Same methods have been used to take readings for other geometrical measurements, e.g., part surface finish $R_a(\mu m)$, dimensions L_i (mm) and roundness (K). Typical graphical representation of level average analysis of the test results is given in Fig.3, where the effects of various SLS parameters on part geometrical properties can be clearly viewed. Within each effect analysis diagram (e.g., curling, roundness, dimensions and roughness), two sets of data are presented, and designations of the data are also given in the diagram.

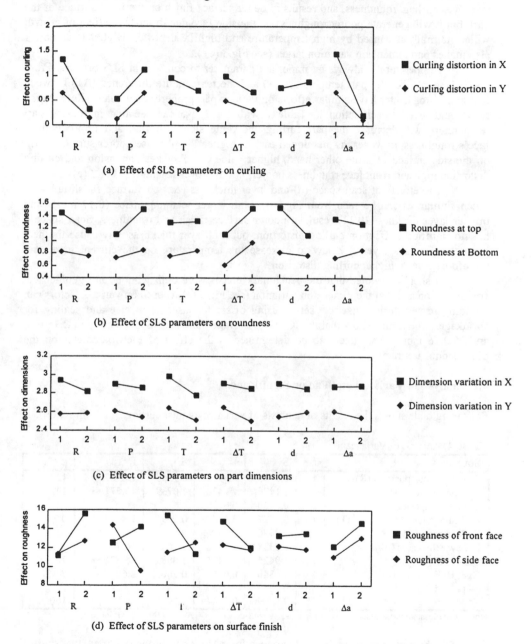

(a) Effect of SLS parameters on curling

(b) Effect of SLS parameters on roundness

(c) Effect of SLS parameters on part dimensions

(d) Effect of SLS parameters on surface finish

Fig.3. Response of part geometrical properties via the major process parameters

As can be seen from Fig.3, R at level 2 (e.g. 1:2, more recycled powder used) generates low curling distortion, better roundness and less dimension variation. But it results in poorer surface finish (Fig.3 (d)). Using higher laser power (Level I: 3.5 W) comparatively improves curling, roundness, and results in better surface finish on the front surface of the part, but it will generate poorer roughness on the side face due to the side-effect of growth which normally is caused by high temperature and high laser power. Higher laser power also makes part dimension variation larger (see Fig.3 (c)).

Temperature is always an important parameter to consider in SLS process. From Fig.3, it seems that the part temperature (T) and the temperature difference (ΔT) between part and piston heater have similar effects on various part property parameters. Refer to the factor and level specification in Table 1 with Fig.3, we can see that increasing part temperature will decrease the part curling and dimension distortion and improve front face's roughness, however, as mentioned earlier, higher T will cause poorer surface finish on the side surface. On the other hand, higher value of ΔT makes dimension and curling distortion high and front face roughness poorer, but improves roundness (Fig.3 (c)).

The effects of scan space (d) and layer thickness (Δa) on surface finish are very much similar, as can be seen from Fig.3(d), both larger scanning space (0.17 mm) and thicker layer thickness (0.11) result in poorer surface finish and roundness. Scan space d has no significant effect on curling distortion, but thin layer thickness may cause the heat from the previous layer(s) below to increase the temperature on the current layer and therefore generate higher curling distortion.

It should be noted that the benchmark pieces were built without any geometrical compensation and that the dimension variation (see Fig.3 (c)) caused by various factors can be compensated during machine set-up using offset for laser beam size and scaling for shrinkage. The results show that the X has an average dimensional inaccuracy of 2.75 mm and Y 0.76 mm. What needs to be determined is the effect of each parameter on the dimensional accuracy.

4.2 Analysis of Variance and Factor Significance

Following table shows the results of analysis of variance against curling (in X direction).

Table 4 Analysis of Variance

Source	f	S	V	S'	F	ρ (%)
A: Powder composition (R)	1	0.9995	0.9995	0.9258	13.567 **	12.2
B: Laser power (P)	1	1.1476	1.1476	1.0739	15.577 **	14.1
C: Part Temperature (T)	1	0.0730	0.0730			
D: Temperature difference (ΔT)	1	0.1327	0.1327	0.059	1.8009	0.7
E: Scan space (d)	1	0.0496	0.0496			
F: Layer thickness (Δa)	1	1.2338	1.2338	1.1601	16.747 **	15.3
CxD, AxF	1	1.9425	1.9425	1.8689	26.376 **	24.6
CxE, BxF	1	0.6340	0.6340	0.5603	8.6058 *	7.38
e - pooled	3	0.221	0.0737			
Total	15	7.5909	0.4744			100

Note: * At least 90 % confidence, ** At least 95 % confidence, *** At least 99 % confidence.

Table 5 is the summary of analysis of variance for all four part property parameters studied in the investigation, which shows how significant each factor affects the output.

Table 5 Summary of Analysis of Variance for Various Factors

Source	Curling		Roundness		Dimensions		Roughness	
	X	Y	Top	Bottom	X	Y	Front	Side
A: Powder composition (R)	**	**	**		**		***	*
B: Laser power (P)	**	**	***					***
C: Part Temperature (T)					***	**	***	
D: Temperature difference (ΔT)			***	**		**	**	
E: Scan space (d)			***					
F: Layer thickness (Δa)	**	**					**	**

Based on the investigation, for a given geometrical property criterion, optimal parameter settings can be obtained (see Table 6). Following are two examples:

(1)   The optimal setting for good prototype profile and dimension accuracy is: A_2 (R =1:2) , B_1 (P=3.5W), C_2 (T = 186 ᵒC), D_2 (ΔT =82 ᵒC) , E_2 (d = 0.15mm) , F_2 (Δa = 0.11).

(2)   The optimal conditions for good part surface finish : A_1 (R = 1:1) , B_2 (P = 3.0W), C_2 (T = 186 ᵒC), D_2 (ΔT =82 ᵒC) , E_2 (d = 0.15mm) , F_1 (Δa = 0.10).

Table 6 Optimal Parameter Settings

Level ╲ Goal / Source	Curling		Roundness		Dimensions		Roughness	
	X	Y	Top	Bottom	X	Y	Front	Side
A: Powder composition (R)	A_2 ⊗	A_2	A_2	A_2 ⊗	A_2	A_2 ⊗	A_1	A_1
B: Laser power (P)	B_1	B_1	B_1	B_1 ⊗	B_2 ⊗	B_2 ⊗	B_1 ⊗	B_2
C: Part Temperature (T)	C_2 ⊗	C_2 ⊗	C_1 ⊗	C_1 ⊗	C_2	C_2	C_2	C_1 ⊗
D: Temperature difference(ΔT)	D_2 ⊗	D_2 ⊗	D_1	D_1	D_2 ⊗	D_2	D_2	D_2 ⊗
E: Scan space (d)	E_1 ⊗	E_1 ⊗	E_2	E_2 ⊗	E_2 ⊗	E_1 ⊗	E_1 ⊗	E_2 ⊗
F: Layer thickness (Δa)	F_2	F_2	F_2 ⊗	F_1 ⊗	F_1 ⊗	F_2 ⊗	F_1	F_1

⊗ Effect of the factor is not significant.

4.3 Regression Analysis of Experiment Data

The main effects of the SLS output parameters normally can be expressed as functions of input operation parameters in the form of empirical/regression formulae [8], namely, in following format:

$$Y = C_o [R]^{n1} [P]^{n2} [T]^{n3} [\Delta T]^{n4} [d]^{n5} [\Delta a]^{n6} \qquad (4)$$

Equation (4) provides a basis for regression analysis of the experiment data. The data collected from the experiment investigation can also be expressed using following matrix:

$$Y = X\beta + \varepsilon \qquad (5)$$

Where, Matrix X is constructed by the factors A, B, C, D, E and F, each varying with two levels. it can be coded using following equations:

$$x_{i,j} = \frac{2\left(Log\,\Psi_{ij} - Log\,\Psi_{imax}\right)}{Log\,\Psi_{imax} - Log\,\Psi_{imin}} \qquad (6)$$

In Equation (6), $\Psi_{i,j}$ (i =1,2,3,4,5,6) represents the Factor A, B, C, D, E, F respectively, $x_{i,j}$ (i=1,2,3,4,5,6) represents the level code for each level. The subscript, j, can be either 1or 2. When j=1, $\Psi_{i,1}$ takes upper level (e.g., Ψ_{max}), when j=2, $\Psi_{i,2}$ takes lower level (e.g., Ψ_{min}).

Computing with the above Equation (6) and the factor's level specifications in Table 1 results in following level coding table (Table 7).

Table 7 Factor's Level Coding

	A: (R)	B: (P)	C: (T)	D: ΔT	E: (d)	F: (Δa)	Code
Samples	x_1	x_2	x_3	x_4	x_5	x_6	
Upper level	1:1	3.5	186	82	0.17	0.11	+1
Lower level	1:2	3.0	185	80	0.15	0.10	-1
Gap	0.5	0.5	1.0	2.0	0.02	0.01	

The coefficients matrix β in Equation (5) can be calculated using following regression analysis model:

$$\beta = (X^T X)^{-1} (X^T Y) \tag{7}$$

As an example, the calculation results using Equation (7) for prediction of roundness (K) are listed in Table 8.

Table 8 Regression Analysis of Experiment Data (Roundness at top)

Code S/N	x_0	x_1 A: R	x_2 B: P	x_3 C: T	x_4 D:ΔT	x_5 E: d	x_6 F: Δa	Roundness k_i	y_i =log (k_i)
1	+1	+1	+1	-1	+1	+1	-1	1.391	0.1433
2	+1	+1	+1	-1	-1	-1	+1	0.604	-0.2190
3	+1	+1	+1	+1	+1	+1	+1	0.932	-0.0306
4	+1	+1	+1	+1	-1	-1	-1	1.764	0.2465
5	+1	+1	-1	-1	+1	-1	-1	1.232	0.0906
6	+1	+1	-1	-1	-1	+1	+1	1.875	0.2730
7	+1	+1	-1	+1	+1	-1	+1	1.103	0.0425
8	+1	+1	-1	+1	-1	+1	-1	2.760	0.4409
9	+1	-1	+1	-1	+1	-1	+1	1.217	0.0853
10	+1	-1	+1	-1	-1	+1	-1	1.556	0.1920
11	+1	-1	+1	+1	+1	-1	-1	0.288	-0.5406
12	+1	-1	+1	+1	-1	+1	+1	1.149	0.0603
13	+1	-1	-1	-1	+1	+1	+1	1.645	0.2162
14	+1	-1	-1	-1	-1	-1	-1	0.867	-0.0619
15	+1	-1	-1	+1	+1	+1	-1	1.003	0.0013
16	+1	1-	-1	+1	-1	-1	-1	1.656	0.2190
$X^T Y$	1.1589	0.8158	-1.2843	-0.2799	-1.1428	1.4349	0.1348		
$X^T X$	16	16	16	16	16	16	16		
$(X^T X)^{-1}$ $(X^T Y)$	0.0724	0.0510	-0.0803	-0.0175	-0.0714	0.0896	0.0084		

Equation (4) can be expressed as its Log format,

$$\log (Y) = \log (C) + n_1 \log (R) + n_2 \log (P) + n_3 \log (T) + n_4 \log (\Delta T) + n_5 \log (d) + n_6 \log (\Delta a) \tag{8}$$

Equation (8) can be simplified as :

$$y = b_0 + b_1 x_1 + b_2 x_2 + b_3 x_3 + b_4 x_4 + b_5 x_5 + b_6 x_6 \tag{9}$$

The coefficients $(b_0, b_1, b_2, b_3, \ldots b_6)$ in the above equation (9) can be calculated using Equation (6) and experiment data (y_i) listed in Table 8, the result is shown below:

$$y = 0.0724 + 0.051x_1 - 0.08x_2 - 0.0175x_3 - 0.0714x_4 + 0.896x_5 + 0.0084x_6 \qquad (10)$$

Using level coding equations (6), following regression equation can be generated finally

$$K = 2.556 \times 10^4 \, R^{0.33876} \, P^{-0.14967} \, T^{-0.01543} \, \Delta T^{-0.075} \, d^{3.2928} \, \Delta a^{0.4071} \qquad (11)$$

The same procedure has been followed to compute regression formulae for prediction of other geometrical properties, e.g., curling, dimension variation and surface finish. Following table (Table 9) lists all the constants C_0 and exponential coefficients n_i ($i = 1 \sim 6$, see Equation 4).

Table 9 Summary of Constants and Exponential Coefficients

Prediction Coefficient	Curling		Roundness		Dimensions		Roughness	
	X	Y	Top	Bottom	X	Y	Front	Side
C_0	1.33E-8	1.61E-7	2556.51	17.041	4.1537	1.0995	388.24	166.72
n_1	0.76446	1.6289	0.3387	-0.05435	0.06068	-0.0029	-0.4174	-0.1193
n_2	-0.3950	0.5251	-0.14967	0.09185	0.00767	0.01034	-0.0501	0.15728
n_3	0.02947	0.04928	-0.01544	0.01059	-0.0137	-0.00896	-0.04831	0.01611
n_4	0.05301	0.20589	-0.07506	-0.09549	0.00417	0.01254	0.04038	0.00867
n_5	5.04136	7.0063	3.29278	0.19604	0.14434	-0.12360	-0.1322	0.2315
n_6	-11.738	-10.620	0.40708	0.98760	0.03937	-0.26563	1.5469	1.3344

5 Conclusions

An experimental investigation has been carried out to explore the relationship between the input parameter and the properties of the SLS parts with an attempt to optimize the input parameters for a given rapid prototyping application. As the behavior of the SLS process on output varies significantly based on what input is selected for building the SLS parts, this results in a huge number of variables that need to be considered in the experimental investigation. To minimize the amount of tests, the Taguchi Method was used to design the experimental plan and to analyze the experimental data. This provides more efficient and less time-consuming method of searching for the ideal parameter setting for rapid prototyping applications.

By doing an analysis of variance for each input property parameter, the significance of the parameter on the output parameters was estimated. From these estimations, an optimal combination of input parameter settings was obtained which resulted in desirable SLS parts. The optimal parameter combinations were found to conform to known trends in selecting parameters. It should be noted that the results are by no means the best setting possible, but rather the optimal setting for the input parameters chosen in the research.

The regression analysis presents the experiment data in the form of empirical formulae which would help to predict possible output with a satisfactory degree of accuracy although the formulae obtained may need to be improved in further research.

6 List of References

1. Zhu CB, Wong D S K, Loh L and Sue H Y, *Parameter interactive architectures and Fuzzy sets analysis of SLS process for optimising rapid prototyping applications,* the proceeding of the seminar of CAD/CAM for Rapid Product Development, 23-24, May, 1996, Singapore, pp102-114.

2. Christian Nelson, Kevin McAlea, Damien Gray, *Improvements in SLS part accuracy,* proceedings of Solid Freeform fabrication Symposium, 9-11 August, 1993, The University of Texas at Austin, Austin, Texas, USA.

3. Uday Lakshminarayan, Christian Nelson and Kevin McAlea, *Rapid prototyping accuracy: a view of recent results,* proceedings of Solid Freeform fabrication Symposium, 9-11 August, 1993, The University of Texas at Austin, Austin, Texas, USA.

4. David Thompson and Richard Crawford, *Optimizing part quality with orientation,* the proceedings of Solid Freeform fabrication Symposium, September, 1995, The University of Texas at Austin, Austin, Texas, USA, pp362-368.

5. Lakshminarayan U, McAlea K, and Booth R, *Manufacture of iron-copper composite parts using Selective Laser Sintering (SLS^{TM}),* the proceedings of Solid Freeform fabrication Symposium, 9-11 August, 1993, The University of Texas at Austin, Austin, Texas, USA.

6. Irem Tumer, David Thompson, Richard Crawford and Kristin Wood, *Surface characterization of polycaobonate parts from Selective Laser Sintering,* the proceedings of Solid Freeform fabrication Symposium, September, 1995, The University of Texas at Austin, Austin, Texas, USA, pp181-188.

7. Phillip J. Ross, *Taguchi Techniques for Quality Engineering, Loss Function, Orthogonal Experiments, Parameter and Tolerance Design,* McGraw-Hill Book Company, New York, USA.

8. Zhu CB, An investigation on grinding power in creep feed grinding using the method combining dimension analysis with orthogonal regression, *Journal of Northeast Univ. of Technology,* December, 1983, No.37, p67-76.

9. Genichi Taguchi, *System of Experimental Design, Engineering Methods to Optimise Quality and Minimise Costs* (Volume One, Volume Two), Unipub by Kraus International Publications, White Plains, New York, and American Supplier Institute, Inc., Dearborn, Michigan, USA, 1989.

CE Environments

Management and Scheduling of Design Activities in Concurrent Engineering[*]

Qian Zhong   Yingping   Zheng

The Institute of Automation, Chinese Academy of Sciences, Beijing, P.R.China

Abstract: The objective of management and scheduling of design activities in concurrent engineering (CE) is to improve the concurrence of design process and minimize the total design time. In this paper, firstly, some pecularities of design activities in CE are analyzed, and the limitations of conventional project management techniques are presented. Then a new feasible and systematic approach for management and scheduling of concurrent design activities is developed, and two kinds of scheduling problem are formulated and their resolutions are given.

1. Introduction

Concurrent Engineering has probed to be a valuable tool for maintaining competitiveness in today's ever changing and expanding world market by providing a systematic approach to the integrated, concurrent design of products and their related processes. The main objective of CE is to shorten the product life cycle, improve the product performances and reduce producation costs[1].

The design process involves a number of design activities and their inter-connections, a limited amount of available resources restrain the simultaneous execution of these activities. Scheduling of these design activities is a very important step to improve the concurrence of design process and shorten the total design time[2]. Some conventional techniques, such as critical path methods (CPM) and program evaluation and review technique(PERT)[3] have been successfully adopted for different types of projects. However, these techniques need to be extended to resolve the management and scheduling problem in CE due to the fact that the concurrent design activities have some special characteristics, such as uncertainty, randomness, concurrence and repetitiveness.

There have been some work of scheduling of concurrent manufacturing projects[4,5,6,7,8], however, it is still lack of a systematic formulation and techniques to this problem. The main purpose of this paper is to present a feasible and systematic approach for efficient management and scheduling of design activities in CE. According to our understanding of CE, we should use a global view to schedule the design activities. Specially, we emphasize the resource and information factors, because in concurrent design process, whether or not one activity can be executed depends on the available resources provided by the system and the available information delivered by the preceding activities.

In the following sections, on the base of new pecularities of design

---

[*] Supported by National 863 High-tech Project Foundation and National Natural Science Fund

activities in CE, an improved and systematic approach is proposed, including the extended method to establish the network of design activities, the more feasible assessment of time distributions, the formulation of scheduling problem and the global optimal scheduling rules.

2. New Pecularities of Design Activities in CE

CE differs from the traditional sequential product development by considering the whole concerns in product life cycle at early design stage and dispatching the activities that are performed according to precedence constraints to be executed simultaneously as many as possible[9]. There are some new pecularities of design activities in CE, such as:

1) CE allows the serial, overlaping or simultaneous execution of activities according to the technical constraints, procedural constraints and information availability. If represented in precedence relationships among activities, there are three types: lead, lag and overlap. But in CPM and PERT, there is only one type of precedence relationship among activities: lead relationship.

2) A main characteristic of CE is to evaluate the design results at any time and correct the design defects timely to obtain good performances, which will cause some activities to be repeated. And the results of evaluation are stochastic, so the processing time of them are not easy to be predicted in advance. Thus it is necessary to use the dynamic scheduling policies in CE. However, in CPM and PERT, the decision are taken in a static mode.

3) The available resources are limited in design process, the simultaneous execution of activities will cause resource conflicts, thus how to assign resources to activities is very important in CE. But, CPM and PERT assume that the relationship between the resources and duration of activities is known a priori for all activities.

4) In CE, the execution of succeeding design activities is based on the information provided by the preceding activities. The most important logic relationship among activities is information dependence relationship. But CPM and PERT methods do not consider the information dependence relationships.

Therefore, from the four respects mentioned above, the conventional techniques need to be extended to satisfy the specific and unusual project scenarios of CE.

3. Model of Design Activities

The basis for management and scheduling of activities is to build the model of design activities to express relationships among activities clearly and evaluate the time distribution of activities accurately.

3.1 Network of design activities

To represent the complex relationships among activities more clearly, such as time order and information dependence, we propose a simple method---extended activity-on-node(E-A-O-N) to establish the network of design activities.

E-A-O-N can be represented as a directed gragh $G = (N, A, C)$, which is developed from A-O-N method. N is a set of nodes, each node indicates a design activity in design process. A is a set of arcs, the arrow of arc drawn from activity 'i' to activity 'j' indicates that the activity 'j' depends on the information provided by the activity 'i'. C is a set of weights of arcs, indicating the time order between the activities. $C_{ij} \in C$, $(i,j) \in N$, $C_{ij} \geq 0$ means that activity 'j' must be executed after activity 'i' finishes with the lag of C_{ij}, $C_{ij} < 0$ means that activity 'i' and activity 'j' has an overlap of $|C_{ij}|$ (see in figure 1).

Adding the weights of arcs in A-O-N can represent the time order and information dependence relationships among the design activities more clearly and accurately.

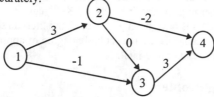

Fig. 1   A network of design activities in extended A-O-N

3.2   Assessment of Activity Time Distribution

It is a key factor in management and scheduling problem. Some design activities are new and unique in nature, it is not possible to predict the duration of them in advance, and the processing time of some design activities are uncertain due to the stochastic evaluation activities. It is a difficult problem to find a good approximation to assess the time distribution of design activities.

Keffer and Bodily[10] have explored the use of a simple k-point discrete approximation to estimate the processing time which appears robust to mis-specification yet make only modest demands upon the assessor. According to our experience, the processing time of one activity must be in an interval (A_1, A_2), which could be divided into several time slots, and the processing time could be in one slot with estimated possibility. For example, the time distribution of activity 'i' could be approximated by the discrete random variables T_i. It is assumed that there are K possible time slots, according to the previous knowledge, the processing duration of 'i' could be in the time slot (A_{ir1}, A_{ir2}) with probability p_{ir} (r=1,2,...K). So the time distribution of activity 'i' can be expressed as:

$$P\left(A_{ir1} \leq T_i \leq A_{ir2} \right) = p_{ir} , r = 1,2, \cdots , K .$$

Each activity may have its own value of K. If K is chosen sufficiently large, it is evident that the approximation will approach the true distribution to any desired degree of accuracy. Normally, K=6 is ok. The estimated processing duration of activity 'i' is: $\overline{t_i} = \sum_{r=1}^{K} (\frac{A_{ir1} + A_{ir2}}{2}) p_{ir} ,$

the estimation deviation σ_i is: $\sigma_i = \sum_{r=1}^{K}(\frac{A_{ir1}+A_{ir2}}{2} - \bar{t}_i)^2 p_{ir}$, and

the standard deviation S_i is: $S_i = \sqrt{\sigma_i}$.

Then the earliest/latest start time and earliest/latest finish time of each activity in the network of design activities can be obtained by forward and backward calculations. Firstly, we must derive distributions corresponding to convolutions and maximization. For two activities 1 and 2, activity 1 has K_1 possible duration and activity 2 has K_2 possible duration. This process gives rise to $L=K_1 \times K_2$. If this procedure is repeated at successive nodes throughout the network, the value of L rapidly becomes large. Ord[11] has proposed a simple and easy-to-implement procedure to reduce the computation by merging successive sets of K_2 , named "collapsing the distribution". Full details have been provided in the literature.

From this way, it can obtain an acceptable degree of approximation and require only minimal computing facilities.

4. Scheduling Problem

The scheduling problem in CE is a resources-constrained scheduling[12] . According to the different requirements in CE, we think there are two types of scheduling problem in design process:

1) Scheduling simultaneous execution of the combination of activities to minimize the sum of tardiness when design activities have different due dates.

2) Sequencing and scheduling design activities to be performed simultaneously such that the completion time of design process is minimized, given the processing time distribution and the resource requirements of each design activity.

Firstly, we give the formulation of such two types of scheduling problem.

4.1 Formulation

To formulate the scheduling problem, it is assumed that every design activity in the network of design activities should obey:

1) Every activity cannot be interrupted once begun.

2) Activities are permitted to wait for resources when the resource conflicts exist.

3) Resources are assumed to be available per period in constant amounts and are also demanded by an activity in constant amounts throughout the duration of activity.

4.1.1   Formulation of the first scheduling problem

The formulation is as below:

$$\min \sum_{i \in A} w_i T_i \tag{1}$$

subject to:
$$f_j - f_i \ge C_{ij} + d_j, (i,j) \in H \tag{2}$$

$$\sum_{S^t} r_{ik}(t) \le R_k(t), t = 1,2,\cdots,f_n \qquad k = 1,2,\cdots,m \tag{3}$$

$$T_i = \max(0, f_i - D_i) \quad \text{where} \tag{4}$$

A: set of activities in design process, $i \in A$, $i=1,2, \cdots, n$

w_i: : weight of activity 'i'.

T_i: : delaying time of activity 'i'

d_i : processing time of activity 'i'

f_i : finishing time of activity 'i'

D_i : due date of activity 'i'

H : set of pairs of activities indicating the precedence constraints

$r_{ik}(t)$: amount of resource type k required by activity i at time t

S^t : set of activities executed simultaneously at time t

$R_k(t)$: total availability resource type k at time t.

The objective of scheduling problem is given in (1). Constraint (2) indicates the time relation of activity i and j. The resource constraints given in constraint (3), indicating that at each time and for each resource type k, the resource amounts required by the activities executed simultaneously cannot exceed the resource availability. Constraint (4) gives the computation of delay time T_i.

4.1.2 Formulation of second problem

To formulate the second scheduling problem, at first, let us consider two dummy nodes, namely 0 and n+1, such that their processing time and amount of resource requirement are 0. The problem can be formulated as:

$$\min f_{n+1} \tag{1}$$

subject to :
$$f_{n+1} \geq f_i, \forall i \in A \tag{2}$$

$$f_j - f_i \geq C_{ij} + d_j, (i,j) \in H \tag{3}$$

$$\sum_{S^t} r_{ik}(t) \leq R_k(t), t = 1,2,\cdots,f_n \quad k = 1,2,\cdots,m \tag{4}$$

The meanings of parameters are as above.

4.2 Scheduling Rules

The two types of scheduling problem above mentioned are usually NP-hard, so it is hardly to find any efficient algorithm to solve it. Normally, there are three kinds of methods to resolve them in the literature[13]. The most effectively method is to develop an heuristic solution procedure for satisfactory solutions[14]. In this paper, we only propose some scheduling rules and show a simple example of heuristic scheduling.

In concurrent design process, a simultaneous execution of a number of activities would cause resource conflicts. The scheduling rules should decide which activity will be delayed. We know, working on the wrong activity could delay the delivery of some important outcome (design information). Thus the critical design activities that influence the total design time should not be allowed to be idle. Proper dispatching at the upstream design activities is a feasible way to ensure an uninterrupted flow of information through whole design stages[5].

In concurrent design process, there are three key factors for successful and timely completion of the design process:

1) Using the limited resources effectively;

2) Scheduling simultaneously as many activities as possible without violating the constraints;

3) Optimizing the flow of information, that means to provide the information firstly to those activities whose information is needed by downstream activities mostly.

Therefore, the scheduling rules in CE should consist of the rules of allocation of the limited resources and optimization the flow of information.

4.2.1 Rules of allocation of resources

Firstly, we give the definition of *set of schedulable activities---S_t.* A set S_t is the set of activities which can be implemented at time t according to the information they need and their preceding design activities have been completed.

In order to deal with the resource conflicts, Demeulemeester and Herroelen[15] proposed the concept of *minimal delaying alternatives*, and proved that *it is sufficient to consider only the minimal delaying alternatives to resolve the resource conflicts.* To our problem, the supplementary sets of minimal delaying alternatives --- S^k, k=1,2,...,L are proposed, named *maximal scheduling alternatives*. A set S^k is a subset of S_t, which is defined as the such combination of design activities from the set S_t that can be performed simultaneously without violating the resource constraints, but the addition of any one more activity in S^k from set S_t would result in resource conflicts. All subsets S^k consist of *a scheduling set--- S(t).* It can be proved that in order to resolve the resource conflicts, it is sufficient only to schedule the maximal scheduling alternatives.

The sets of S^k can be obtained by enumerative procedures.

4.2.2 Rules of information optimization

We propose the concept of *information weighted factor(IWF)* to deal with information priority problem. *IWF,* denoted by λ($0 \leq \lambda \leq 1$), is a measure of the information provided by one activity how to influence its succeeding activities. IWF of activity 'i' could be decided by the numbers of activities that depend on the information provided by activity 'i' . There are unified rules to obtain the value of IWF of each activity by analyzing the information dependence relationships among the activities in *network of activities*. The more activities that depend on the information delivered by activity 'i' ,the larger value of λ_i will be.   So the critical activities in design process are those activities that have large value of IWF .

When a number of activities compete for the resources, the activities with larger values of IWF will be given higher priority and executed earlier, thus to avoid the resource conflicts and assure the smooth flow of information to the the critical activities.

Different scheduling problem will have different scheduling rules, but all feasible scheduling rules in CE should consist of the two rules mentioned above.

4. 3   A Simple Example

We use the example presented in Kusiak's paper[5] to illustrate how to use the rules of allocation of resources and information above mentioned.

Suppose that, the ongoing design phase is D and the set of schedulable activities consists of 5 activities. The design process under consideration consumes 4 types of resources. The status of these four resources is indicated by a status vector. Suppose that the resource status vector at time t is R=(5   5   4   0.9). The known condition are listed in table 1.

Table 1.   Activities with their resource requirement, priorities, and estimated duration

Activities	Resource Requirement	weights	Estimated Duration	Latest StartTime	Latest Finish Time
5	(2 1 1 0.15)	0.10	2	9	11
7	(2 3 1 0.2)	0.20	7	6	13
8	(3 3 3 0.5)	0.10	3	11	14
10	(4 5 3 0.25)	0.30	6	13	19
11	(1 1 2 0.4)	0.10	3	13	16

The main intention is to schedule as many activities concurrently as possible. The scheduling rule we used in this paper is to minimize the sum of maximal waiting time.

The maximal waiting time---D_j is computed as:

$$t^* = t + \max_{i \in S^k} d_i$$

$$D_j = \max\ (\ t^* - LS(j)\ ,\ 0\ )$$

The heuristic scheduling rule is as:

$$Z = \min_{D} (\ \sum_{j \in D^k} (w_j \times D_j)\ )$$

The scheduling procedures are:

Step 1: The schedulable set S_t ={5,7,8,10,11}.   The resource vector at time t is R=(5   5   4   0.9).

Step 2: The maximal scheduling alternatives are S^1 = {5,7,11}, S^2= {10}, S^3= {5,8}.

Step   3: t^*=5+$\max\ d_i$ =5+7=12.   In S^1, Z_1=0.10×max(12-11,0)+0.30×max(12-13) = 0.10.

Perform next iteration, Z_2=1.20 , Z_3=0.40.

All possible subsets S^k and corresponding value of objective funtion are tabulated in Table 2.

Table 2.   The subsets of schedulable activities and their weighted tadiness

S^k	Total Resource Utilized	Z_k
S^1={5,7,11}	(5 5 4 0.75)	10
S^2={10}	(4 5 3 0.25)	120
S^3={5, 8}	(5 4 4 0.65)	40

Step 4: The subset of schedulable activities with minimum weighted tadinesses is chosen, which is found to be S^1 = {5,7,11}, min Z=0.10. Hence, the activities 5, 7, 11 are scheduled simultaneously. The activities 8, 10 are delayed.

The details of scheduling algorithms to the two kinds of scheduling problem in CE will be discussed in other paper. We have developed a software prototype to support the decision making in concurrent engineering environment with Lotus Notes and Visual C++ language under Windows NT LAN, management and scheduling of design activities is a subsystem of it.

5. Conclusions

In this paper, to resolve the problem of management and scheduling of design activities in CE, a feasible and systematic approach is proposed in global view. On the base of new characteristics of design activities in CE, there are limitation of conventional project management techniques to be used in CE scenarios. We give a systematic approach to deal with this problem, including building a model of design activities, formulation of scheduling problem and providing some scheduling rules of allocation resources and information. A simple numerical example is given to illustrate the rules, and the details of concrete scheduling procedures will be discussed in other papers.

REFERENCES

1. Sohlenius, G.,"Concurrent Engineering", Annal of the CIRP, 41, 2 (1992), 645-655
2. Gu, P. & Kusiak, A. (editor), Concurrent Engineering: Methodology and Applications, Elsiver Science Publishers,1993.
3. Moder, J.J., Phillips, C.R. and Davis, E.W.,"Project Management with CPM, PERT and Precedence Diagramming (3rd edition) Van Nostrand Reinhold, New York,1993.
4. Adedeji, B.B.,"Scheduling of Concurrent Manufacturing Projects", Concurrent Engineering Contemporary Issues and Modern Design Tools, (1992) 93-109.
5. Kusiak, A.,"Scheduling Design Activities", Information and Collaboration Models of Integration, S.Y.Nof (ed.), Kluwer Academic Publishers, (1993) 43-60
6. Badiru,A.B.,"Project management in manufacturing and high technology operations", Wiley, New York,(1988,).
7. Badiru, A.B.,"Project Management Tools for Engineering and Management Professionals", Industrial Engineering & Management Press, Norcross, GA, (1991)
8. Badiru,A.B.,"Scheduling of Concurrent Manufacturing Projects", Concurrent Engineering : Contemporary Issues and Modern Design Tools,(1993) 446-464
9. Palmer, B.& Korbley, L. et.al., "Modeling the Concurrent Engineering Process", CERC Technical Report, West Virginia University,(1992).
10. Keefer, D.L., & Bodily, S.E.,, "Three-point Approximation for Continuous Radom Variables", Management Sci.,29(1983) 595-609
11. Ord, J.,"A Simple Approximation to the Completion Time Distribution for a PERT Network", Res. Soc. Vol. 42, 11(1992) 1011-1017
12. Christofides, N., Alvarez-Valdes,R. & Tamarit,J.M.,"Project scheduling to resource Constraints: A Brach and Bound Approach,European J. Open. Res.,29 (1987) 262-273
13. Davis, E.W.,"Project Scheduling Under Resource Constraints: Historical Review ad Categorization of procedures", AIIE Trans., 5, 4 (1973), 297-313
14. Kurtulus, L. & Davis, E.W.,"Multi-Project Scheduling: Categorization of Heuristic Rule Performance", Management Sci,,28, 2(1982), 161-172
15. Demeulemeester,E. & Herroelen W.," A Branch-And-Bound Procedure for the Multiple Resource-Constrained Project Scheduling Problem", Management Sci.,38,12(1992),1803-1818

PIKS: Product Information And Knowledge Servers For Concurrent Engineering Environments

D. Domazet., QZ Yang, YZ Zhao

Gintic Institute of Manufacturing Technology, Singapore

Abstract. In this paper we present the STEP information database server, known as PIKS, developed at Gintic as the result of the PIKS project. PIKS stores and manages STEP AP203 objects by using ObjectStore ODBMS, and exchanges product information with its users (e.g. CAD systems) with STEP Part-21 data files. PIKS provides STEP objects consistency and integrity, as well as concepts that are convenient for concurrent engineering.

1. Introduction

In concurrent engineering (CE) environments, product information needs to be shared among different computer systems/applications. In general, different data models are used for representing product information and problems occur when exchanging product data. STEP standard [1] specifies a standard product data model that can be used as a neutral product representation in heterogeneous computer environments. Standard STEP Part-21 data files [2] may be exchanged between different systems (such as CAD/CAM/CAE systems) if they are provided with adequate STEP translators that converts their proprietary data models into the STEP data model and vice versa.

Product data exchange is not very convenient in collaborative and concurrent engineering environments as it is difficult to control and manage the data consistency, integrity and propagation of design changes. Product Data Management (PDM) systems can distribute and manage STEP files, but do not have a control of their content as they usually do not interpret data files. Applications integrated by a PDM system have exclusive control and responsibility of their content. Therefore, PDM systems have limited control over application (or product) data and its integrity and consistency.

In this paper we present a solution for the management of STEP information that is based on the object-oriented database technology. Instead of data exchange, PIKS (Product Information and Knowledge Server) provides sharing of STEP AP203 [3] information. Main PIKS features are listed in Section 2, and its architecture is presented in Section 3. Database organization of STEP AP203 objects is presented in Section 4 and use cases supported by PIKS version 1.0 are described in Section 5. Section 6 summarizes the presentation of PIKS and implemented concepts.

2. Main Features of PIKS

PIKS version 1.0 provides the following functions:

- *Object persistent storage:* PIKS stores STEP AP203 objects in an ObjectStore object-oriented database [5] whose schema is compliant with STEP standard [3].

- *STEP file check-in/check-out:* STEP AP203 objects are transferred by STEP Part 21 data files [2] when they are checked-in and checked-out. STEP Part 21 files are used only for the transfer of data between PIKS and applications, and are not persistently stored. PIKS re-establishes external references of the checked-out objects with objects that were not checked-out, once they are checked-in back into the database.

- *UoF Check-out data granularity:* STEP data sets may be checked out according to specified Units of Functionality [3].

- *Data consistency and integrity:* Concurrent access to the data, and data consistency and integrity is managed by the ODBMS concurrency control based on the two-phase locking protocol.

- *Support of collaboration:* Support of collaborative group work by providing project-, group-, and personal workspaces. For instance, only users (individual or groups) of a group workspace may access (with write or read privileges) some design objects.

- *Event-driven notification:* PIKS notifies users of a shared (project- or group-) workspace on some user-specified events, such as check-in and check-out of specified objects, or modification of some data objects.

3. PIKS Architecture

PIKS uses client /server software architecture. Figure 1 presents the architecture of the server site of PIKS.

Schema Manager translates STEP AP203 EXPRESS constructs into C++ definitions, installs AP203 and PIKS class libraries to form a STEP-compliant product data model, and creates database sche-mas from this model. *STEP Object Manager* (SOM) analyses objects found in an imported STEP file, puts them into the database, and organizes the database according to the UoF (Units of Functionality) specifications of AP203 [3]. It also gets STEP objects from the database using requested product ID and UoF, and creates an export STEP file that is sent to the user who requested STEP objects. *PIKS Object Manager* (POM) puts or gets meta (control) objects needed for normal operation of PIKS, such as users, groups, projects, assignments, group- and project workspaces, available products etc. (see the Administration module later).It also maintains the data consistency among PIKS meta objects. *GUI* (Graphical User Interface) module provides an end-user interface to PIKS (forms and menus). Current version is based on X11/Motif windows tool-kit.

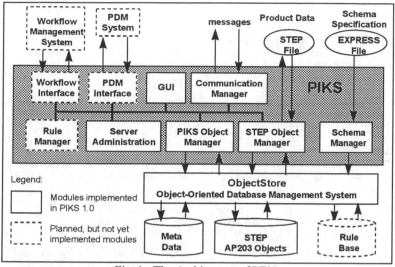

Fig. 1. The Architecture of PIKS

Communication Manager distributes automatically notification messages generated by the system and receives messages from the client site of PIKS. In future versions of PIKS, the *Rule Manager* will handle Event-Condition-Action (ECA) rules [4] in order to provide features of active databases (e.g. automatic database updates, design change propagation, notifications). *Workflow Interface* will provide an interface to a workflow system that will coordinate activities of PIKS users and their interaction with PIKS. *PDM Interface* will provide an integration of PIKS with a Product Data Management system.

PIKS is a C++ class library that uses the ObjectStore ODBMS for storing both application (STEP) and meta (PIKS) data [5] and ST Developer [6] to translate AP203 schema into C++ classes, and to convert data formats from STEP Part-21 to ObjectStore, and vice versa. ILOG View 2.1 is used as a GUI builder [7], and ILOG Broker [8] is used for object distribution, communication, and interaction between the server and client of PIKS in heterogeneous computer environments.

4. STEP AP203 Database Organization

PIKS is a specialized database server for the persistent storage and management of STEP AP203 objects [3]. The STEP application protocol 203 specifies 188 data entities, and many functions, local and global rules related to the configuration controlled design of mechanical products. STEP AP203 standard specifies these data sets in the form of 14 Units of Functionality (UoF). Three of them are shown in Figure 2, and Bill-of-Material (BOM) UoF with Application Interpreted Model (AIM) objects is presented in Figure 3. As it can be seen, the data model for STEP AP203 is very complex. In order to achieve higher data retrieval efficiency of such complex data models, we decided to use the object-oriented database technology for the server's back-end [5].

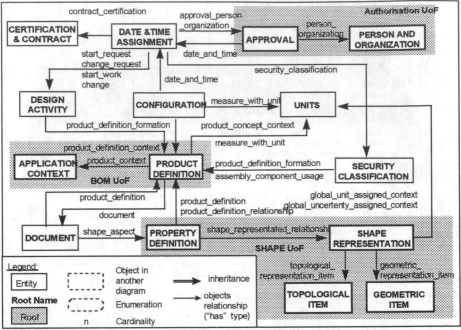

Fig. 2. STEP AP203 entity sets

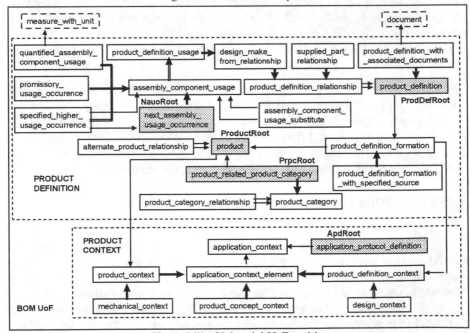

Fig. 3. Bill-of-Material UoF entities

Entities of the Application Interpreted Model (AIM) of STEP Application Protocol 203 are converted to C++ classes that specify the database schema of PIKS. Most of applications (e.g. CAD/CAM systems) that support STEP Part-21 data files, need STEP data sets, and not a single STEP data entity (object). PIKS 1.0 AP203 AIM database is organized according to UoFs, and provides needed UoF-based data sets when PIKS users specify UoFs required for their applications.

Navigational data access provides an efficient mechanism to collect or update needed data objects for each UoF required by a user. As this require identification of the root objects, from which the navigation starts, for each UoF one or more root entities are determined. We will explain this in the case of BOM UoF (Figure 3).

The root is an entry point for retrieval of all objects in the list. For BOM UoF, three database roots are created to ensure every object pertaining to this UoF can be reached by database applications through these root objects (NauoRoot, PrpcRoot, and ApdRoot). For instance, to create a root object pointing to the next_assembly_usage_occurrence OS list, the following OS API will be used:

$$db->create\_root("NauoRoot")->set\_value(next\_assembly\_usage\_occurrence) \qquad (1)$$

We use the following API to do the root updating:

$$db->find\_root("NauoRoot")->set\_value(next\_assembly\_usage\_occurrence) \qquad (2)$$

Through "NauoRoot" root all objects in the next_assembly_usage_occurrence list can be accessed/extracted by another OS API:

$$db->find\_root("NauoRoot")->get\_value() \qquad (3)$$

If a given product need to be extracted , its ID (PID) number is adjusted as a top level PID. Each of three OS lists holding BOM objects will be traversed by three OS cursors. An OS cursor keeps track of the iteration over an OS list to ensure every element in the list to be visited. All objects extracted from these three lists are then moved and saved as a STEP file by calling Rose API [6]: ROSE.saveDesign(). Actually, This API translated objects from the Rose internal data format into STEP Part 21 file format.

If a user requests only BOM objects related to an sub-assembly, an OS cursor will still be used to iterate over an OS list, such as the next_assembly_usage_occurrence list. For each list, a search path must be created to reach a part ID object from any one of objects in the list. Sometimes, this path may be very long and complex. Through this path, a part ID object will be retrieved. Then an evaluation will be performed to check whether an object in the list matches that one given by the user. If it is true, return this object. If it is not matched, next object in the list will be examined. Only requested objects will be moved and written into a STEP file and sent to the user.

5. PIKS Classes

as PIKS is a middleware system component between a STEP object-oriented database and an application or user. It is developed as a client/server system. Figure 4.a shows developed class categories of the server site of PIKS. We use Booch's notation [9] here, as we used Rational Rose OO development tool for design of the system [10]. In the design stage we tried to satisfy criteria for robust design, by separating all classes in three types of objects: interface objects, control objects, and entity objects [11]. We also tried to group classes that share similar properties, or are closer inter-related, into the same class categories with clear interfaces. For instance, if we want to change the underlying database system, we only need to replace DB_Access class category with the new one relevant to the new database system. Similar is valid for the GUI system (Server GUI class category), or for system administration (Administration class category).

Figure 4.b shows currently available classes in the Server Manager class category and their relationships with server's classes in other class categories. PiksServer class is the main control object of the server site of PIKS. For each specified use-case (explained later), a separate sub-class was created. New such classes will be added in future, for new use-cases. PiksAPI is a class with API methods that can be used for the integration of PIKS with other systems, and for communication with the client part of PIKS. Notifier class distribute notification messages to PIKS users.

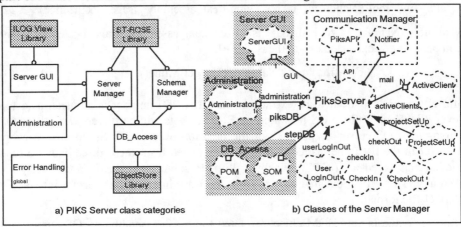

Fig. 4. PIKS server class categories and basic classes

ServerGUI is an abstract class. Its subclasses create different windows, dialog forms and menus. We used IlogViews in development of these classes [7]. The Administration class provides methods for specifying PIKS users, groups, data access privileges, roles, workspaces etc. Figure 5.a shows classes linked with the Administration class that are responsible for storing and managing these meta data. Figure 5.b shows classes of the DB_Access class category. SOM (STEP Object Manager) class manages all STEP persistent objects. For each UoF, a separate sub-class was created, responsible for putting, getting and updating STEP objects of the corresponding UoF. InBox and OutBox classes manage import and export STEP data files, respectively. POM (PIKS Object Manager) class stores and retrieves all PIKS meta objects (such as objects in the Administration class category), that are stored in the ObjectStore database.

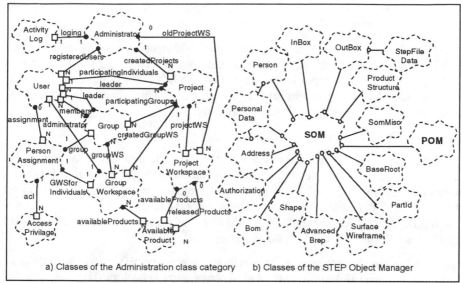

a) Classes of the Administration class category b) Classes of the STEP Object Manager

Fig. 5 Classes of the Administration and SOM class categories

Figure 6 shows PIKS classes of the client site. PiksClt is the main control object. CltDlg is the abstract class. Its sub-classes create different windows, forms and menus that are used on the client site for the interaction with a user. Other classes are entity classes. ServerInfo class provides information about existing PIKS servers, and CurrentSession class - about current session data. PersonalWS class manages the personal workspace for a user. It manipulates with existing STEP files described by U-AvalableProduct

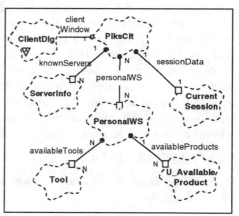

Fig.6. PIKS client classes

objects, and software tools (applications, such as a CAD system) are described by the Tool objects.

6. Use Cases Supported by PIKS 1.0

According to Jacobson's use case analysis [11], we implemented the following use cases in PIKS 1.0:

- *Project setup*: All project related objects (see Fig. 5.a) are specified, such as PIKS users, groups, projects, project and group workspaces, users assignments and privileges, etc.

- *User log-in & log-out:* Provides users registration (log in), log out and open/close/switch a session with a chosen project or group workspace of PIKS.

- *STEP file check-in & STEP file check-out:* STEP objects found in an import STEP file are put in the object-oriented database (check in use case) or are read from the database and put into a created export STEP file sent to the user who requested STEP information (check out use case). These two use cases will be explained in more details in next two sections.

6.1 STEP File Check In

If an imported STEP file contains new STEP objects, the new objects are created in the database. Otherwise, old objects are updated with restored relationships with other objects of the STEP AP203 data model. Using appropriate GUI forms, the user chooses one STEP file from his personal workspace and sends the message fileChekIn to the PIKS server site, i.e. invokes fileCheckIn method of the PiksAPI class. This API method has the following signature:

fileCheckIn (char* userName, char* userRole, char* projectID, chargroupWSName, char* dbName, char* stepFileName, char* checkInPID)

Before checkIn method of the SOM class is invoked, CheckIn class execute some usual servers activities, such as: checks and verifies user's authorization (if rejected, notifies the user) and copies the named STEP files from the user's personal workspace into the directory specified for the InBox server's object. SOM class analyzes the imported STEP file (stepFileName) and updates the ObjectStore database (dbName) by using appropriate UoF-related classes. Previously created database roots are updated and objects found in the STEP file replace old objects. If some of them are new, they are stored as new objects. After completing the check in task, CheckIn class logs the checkIn activity (using the ActivityLog class) and informs all users of the group workspace (groupWSName) through Notifier class.

6.2 STEP File Check Out

When a user needs STEP objects, he/she needs to open appropriate check out GUI form and specify:

- product (by ID number) or a component of a product,
- a desired STEP file name (that will be created at the server's site),
- desired UoF data sets related to the specified product, and
- read or write privilege

Based on the specified data, PIKS client send a fileCheckOut message to PiksAPI class on the server machine (Figure 7). This check out API method has the following signature:

fileCheckOut (char* userName, char* userRole, char* projectID, char* groupWSName, char* dbName, char* productID, char privilage, UOF uof, char* stepFileName)

UOF is a composite object that specifies the list of UoF numbers/names needed for check out. The PiksAPI class invokes checkOut method of the CheckOut class (Figure 10). It verifies the user's, registers him as an active client and invokes checkOut method of the SOM class. This method gets all stored STEP objects related to the requested UoF and product ID, and creates an export STEP file with collected STEP objects. OutBox class sends this file to the user who sent the check-out request and notifies all users of the used group workspace..

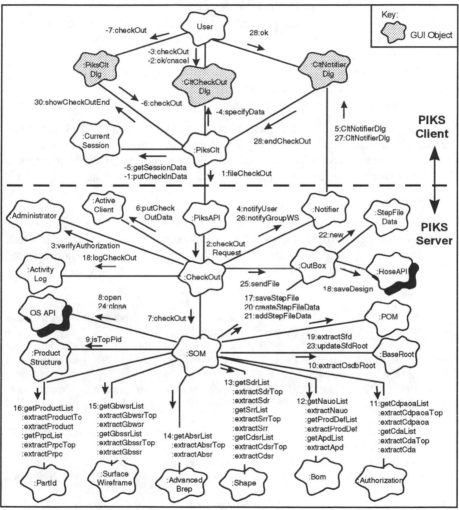

Fig.10 Object interaction diagram for STEP file check-out use case

7. Conclusion

PIKS 1.0 provides the basic functionality for persistence storage and management of STEP AP203 objects, such as project set-up, users log in and log out, and STEP file check in and check out. Its back-end object oriented database (ObjectStore) stores

STEP objects, and STEP files are used only for data transfer between the clients and server and vice versa. The developed query system uses a set of predefined (and programmed) UoF-based queries that the user can choose by using appropriate GUI forms and menus. PIKS can be integrated with other systems by using its API class (PiksAPI class) with the developed API methods.

For next versions of PIKS we plan to add also the following new features:

- Java interface to PIKS (PIKS clients as Java applets),
- use of more advanced and appropriate concurrency control mechanism (e.g. optimistic concurrency control, and new lock and transaction types),
- flexible data access granularity levels (from an attribute, to an UoF),
- STEP SDAI (Standard Data Access Interface) interface to the PIKS database,
- an interface to PDM and workflow systems
- implementation of other STEP application protocols(e.g. shipbuilding industry)

PIKS may be very suitable for inter-enterprise collaboration (virtual enterprise concept). Organizations involved in a joint product development and manufacture can share product related information stored and managed by PIKS, as a common STEP data server that supports heterogeneous computer environments.

Acknowledgment

The paper presents some of results of the PIKS project, the three-year in-house project of Gintic that started in October 1994. Authors wish to express their gratitude to the Gintic Institute of Manufacturing Technology for providing very good research conditions for this project.

References

1. ISO 10303 - 1: Industrial automation systems and integration - Product data representation and exchange - Part 1: Overview and Fundamental Principles, 1994
2. ISO 10303 - 21: Industrial automation systems and integration - Product data representation and exchange - Part 21: Implementation methods: Clear text encoding of the exchange structure, 1994
3. ISO 10303 - 203: Industrial automation systems and integration - Product data representation and exchange - Part 203: Application protocol: Configuration controlled design, 1994
4. Domazet D., Active data driven design using dynamic product models, *Annals of the CIRP* Vol. 44/1/1995, pp 109-112
5. ObjectStore Release 4, *Reference Manual*, Object Design Inc., USA, 1995
6. ST-Developer Version 1.4, *ROSE Library Reference Manual*, STEP Tools Inc, USA, 1995
7. Ilog Views 2.3, *Reference Manual*, Ilog, France 1996
8. Ilog Broker 2.1, *Reference Manual*, Ilog, France, 1996
9. Grady Booch, *Object-Oriented Analysis and Design*, Second Edition, The Benjamin/Cummings Publishing Company Inc., 1994
10. Rational Rose/C++ 2.0.14, Rational, USA, 1994
11. Objectory Version 3.0, Rational, USA, 1996

CONCURRENT ENGINEERING THROUGH DATA SUPPORT AND COMMUNICATION IN COMPUTER INTEGRATED MANUFACTURING ENVIRONMENT

P.K. Jain, S.C. Jain and Tafesse G-S.B

Dept. of Mechanical & Industrial Engineering. University of Roorkee
Roorkee - 247 667 (India)

Abstract. The current status of CE in CIM environment, its impact in manufacturing, its possible application in the future and the need of integrated databases for CE in CIM environment is discussed. A hybrid interfaced/distributed CIM database integration system with global and local data model for integration based on modular and functional approach is proposed. In parallel a KBS and a CE Blackboard module is included in the system architecture to enhance the coordination, cooperation, decision support, data support and communication of CE in CIM environment.

Key Words: Computer Integrated Manufacturing (CIM), Concurrent Engineering (CE), Data Base Management System (DBMS), Design For Manufacurability (DFM)

1. Introduction

Traditional design and Manufacturing of products is a sequential process and it depends on iterative design and manufacturing process. However, the iterative process is time consuming and since the design determines significant amount of the product cost (about 70 to 80% of the production cost), a new approach Concurrent Engineering (sometimes called simultaneous engineering or life-cycle engineering) is practiced [1]. Concurrent Engineering is the simultaneous consideration of production, function, design, materials, manufacturing process and cost, taking into account later-stage considerations such as testability, serviceability, quality and redesign [2]. The implementation of CE by considering the life-cycle factors avoids the possible resources spent for a redesign. CE is characterized by a focus on customers' requirement, moreover it embodies the belief that quality is in built in the product and quality is the result of continuous improvement of a process which, is supposed to be an integral part of CIM considerations from the order receipt till the product life cycle ends. The objectives of CE in CIM include:

- elimination of waste
- reduction in lead time for product development & product delivery
- improvement of quality
- reduction of cost
- continuous improvement

An ideal CIM system applies computer technology to all of the operational functions and information processing functions in manufacturing from order receipt, through design and production to product shipment [Groover M.P., 1987]. CE targets to produce a product which is economically sound, environment friendly, attractive in its design, produced with in shortest lead time etc.[4]

The marriage of CIM into CE is the manufacturing philosophy for futuristic industries, where life cycle considerations including the product quality, lead time, cost effectiveness and recyclability are the basic issues. The CIM environment which, is gradually getting reality through the advancement of CAD/CAM is on the verge of totally implementing the concepts laid by CE through the advancement of information technology in CIM framework. Where, integrated business and manufacturing activities link the need of the customers with the manufacturer.

2. Integrated Database for CE in CIM

The heart of CIM system is a management of information system which processes, handles and controls the shared data for administration, design, planning, scheduling and control. The efficiency of operating CIM depends on the quality and integration of a well-designed information management system[5].

Databases for life-cycle product design considerations in the CIM framework must be integrated to communicate with various interrelated functions of manufacturing activities to secure this target Different approaches for database models has been implemented including: a loose collection of independent databases, a central database, an interfaced database and a distributed database[6].

In CE environment manufacturing industry handles a lot of information needed for administration, design, planning, scheduling & control. There is a close relationship between data which is being processed and used in various activities for example design information is needed for planning, scheduling, machine programming, quality control & so on. For this reason an information system must be designed as an entity which comprises all manufacturing activities in CE frame work. The essential features which are needed to be incorporated for CE in CIM environment, to assure the life cycle issue considerations includes distributed computer systems, a communication system for real time information exchange, a common data & knowledge base which can be reused for different manufacturing activities and decision support module to coordinate and communicate at each levels for strategic planning decision.

CE requires diverse heterogeneous data that addresses various needs of design, planning and manufacturing. Similarly CIM database is composed of CAD databases, accounting and administrative control databases and manufacturing requirement planning databases. However, handling efficiently these separate and heterogeneous databases are very difficult. To challenge this reality CIM database should be capable of sharing data, handling heterogeneous data and avoiding data inconsistency. Meanwhile, different database integration methods have been applied to handle the design and manufacturing activities including[7,8,9]:

A. Developing a single, new, unified homogeneous database system to store and manage the various types of design planning, manufacturing and accounting data, based on a uniform data model by discarding the individual functional database in CIM.

B. Developing an integrated data administration system to manage heterogeneous distributed databases.

C. Developing a frame-work for knowledge-based systems to integrate the heterogeneous multi-database of CIM.

3. Reinforcement of data support and communication for CE in CIM environment

CIM is highly dependent on data communication where, design considerations are weighted against manufacturing considerations, including material availability, manufacturing tolerances, cost considerations etc., within the context of the prescribed quality levels. Subsequently, the design data is transferred to manufacturing requirement planning system to verify the availability of manufacturing resources and to the accounting and administrative control system, to estimate the planned cost of manufacturing to test the feasibility of manufacturing the product[7]. Lastly, the system determines the possible cost trade-offs between such attributes as features, tolerances and quality and feeds this information back to the CAD system. Where the majority of these databases serve multiple applications to designer, system operators and other related personnel.

The CIM system gradually grows with the advancement of the technological and scientific knowledge, while the integration of the manufacturing & business activities need a revolutionary change to merge the concepts of CE in to CIM frame-work. CIM comprises many separate modules or subsystems, there is no generally agreed sub-system structure, not even a generally accepted list of subsystem, that was the main reason why the ESPRIT CIM project has addressed three separate but related goals[13]:

1. To modularize the total CIM into functionally discrete sub-systems.
2. To describe the minimum specifications of each sub-system.
3. To identify the interrelationships that exists between any one CIM sub-system and all other sub-systems.

The proposed CIM system which is targeted to further advancement of integration of CIM modules based on functionality of each module oriented to generalized activities. Communication of CIM needs to be built in such a way that all data communications to take place in real time and it should take into consideration factors like, different hardware configuration, software, mode of processing, vendor's attitude etc.

To solve these integration issues for effective information exchange a communication network protocol must be chosen in addition to the proposed architecture of CIM frame work in CE environment [fig.1]. For all modules to communicate conceptually over the same net, the proposed hybrid distributed/interfaced CIM model in concurrent engineering environment facilitates communication and data exchange at the local and global model.

4. Global and local data model in CIM

A grand scale integration and developing a series of data is the main concern of CIM. Coordinating the whole manufacturing activity for competitive production is the basic target but, because CIM is a complex in the sense that it involves accessing large databases for data storage and retrieval purpose, conventional data processing (i.e. numerical, alphanumeric computations and symbolic processing) and computer decision making. An alternative communication module which enhances seamless access and integration of manufacturing activities in the CIM framework is essential. The emergence of different knowledge-based systems, database management systems and data models can be incorporated for a CIM data model where, subdividing the whole manufacturing activity into specialized groups to form a global and local data model can be a basis for communication. These data models consists of interconnected complex computer controlled manufacturing activities including:

1. Computer aided planning and logistics (CAPL) data model.
2. Computer aided design (CAD) data model.
3. Computer aided manufacturing (CAM) data model.

The engineering information system in such a frame work assists efficient and fast handling of data often in real time. Since function based similar manufacturing data models can be easily handled by the same type of DBMS (for example a CAD data can be transferred using graphical & alphanumeric data model) for different functions in the sub-module.

5. Hybrid Interfaced/Distributed data communication in CIM environment

To integrate the basic concepts of CE, to manage complex data in CIM environment, to handle design, planning, scheduling, administration, controlling and other related data, CIM functions are classified in to CAD database, CAM database and CAPL databases. To simplify the organization of data-structure, handle interrelated heterogeneous data, and to enhance communication between them, a hybrid interfaced/distributed data communication system architecture is proposed for CE applications in CIM environment (Fig.1)The proposed system architecture considers the following facts:

1. To modularize the CIM system into discrete functional CIM subsystems where enhanced communication between each sub modules can be categorized.
2. To integrate different local data management system for communication and data exchange in a neutral data format for goal oriented specifically localized applications.
3. To provide coordination of the distributed data models for cooperative, hierarchical tasks through these modules and to strengthen the link and communication for collaborative works.

The distributed databases are integrated by a post and pre-processor for data exchange and communication. The KBS handles data extraction, manipulation and decision support for CE applications. Functional level communication is performed at the data level with out any additional data modification and the data structure can be custom designed to meet the specific requirement of the manufacturing needs. The CE blackboard system coordinates the collaboration and communication of each database group, where decision support system with a user interface can be embedded into it.

The control blackboard uses data interpreted by a KBS, an additional function of this module is evaluation & controlling of the system by formulating and retrieving hypothesis concerning the system. The decision support system embedded in the global control unit coordinates on line and off line planning, monitoring systems, diagnosis of abnormal statuses and intervention to modify system parameters.

Complex applications like design and manufacturing needs considerations of large collection of graphical and alpha numeric data concurrently. That is the reason why an object-oriented data base management systems, where object-orientation concept such as classes, inheritance, encapsulation and complex objects are applied to improve integrity, flexibility, reliability and scalability of these large and complex data. Additionally object models are semantically rich, and they provide a variety of type, abstraction mechanisms and relations for expressing the complex inter schema relationships and potential conflicts among data at each component system [11]. Based on the facts mentioned above, the object-oriented data modeling is selected for the proposed hybrid interfaced/distributed CIM architecture since, they are the best alternative to model data and to facilitate automation of design and manufacturing functions.

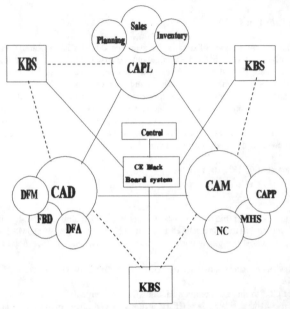

Fig. 1 A hybrid distributed/ interfaced CIM Database Integration system for CE application

CAPL - Computer Aided Planning & Logistics   NC- Numerical Control
MHS- Material Handling and Servicing   DFA- Design for assembly,
CAPP- computer aided process planning   FBD - Feature-Based Design
CAM -Computer Aided Manufacturing   CAD- Computer Aided Design
DFM -Design for Manufacturability   KBS - Knowledge-Based System

The decision support system of the proposed model is typically controlled by the Blackboard system where, information is passed through the KBS. The KBS eases the communication between the local data modules and the distributed DBMS's. It also adds reasoning capabilities to the system which, in turn improves the efficiency of the CIM system. Additionally the KBS, combining knowledge and data driven systems and applying semantic integrity constraints it provides the system with intelligent answers at all levels of the manufacturing activities. The Blackboard system, based upon the dynamic data and knowledge available improves the communication, coordination, cooperation, conflict resolution and decision support for CE in CIM environment.

The proposed Hybrid interfaced/distributed data integration module can be used to solve problems of concurrent engineering. A subset of concurrent engineering approach, i.e. Design For Manufacturability (DFM) is taken into consideration to demonstrate the information flow and the communication net work by the methodology used in our research work in DFM.

6. Application of some concepts of CE in CIM environment

The result of applying the DFM (a subset of concurrent engineering) introduces the adoption of concurrent rather than serial approach to the various product design and process planning functions in the existing manufacturing facility. To disclose the soundness of the proposed CIM architecture, a DFM module for machined part manufacturability analysis is considered, where Knowledge-Based approach is implemented, for integrating multiple sources of knowledge (which are required for DFM evaluation) .

In order to fulfill the requirements set by the manufacturability analysis module of the proposed system the following points are considered as a basis in order to facilitate the DFM analysis of a machined-part in CIM environment

A modular problem decomposition approach, to capture different activities, to organize these modules hierarchialy, to reduce the complexity of the modules, and to enhance easy management and abstraction is used.

During product development, the product passes through different stages of design and manufacturing activities. These activities are by nature modular and a modular approach to capture the intents of each activities will give the best problem solving approach. Modularizing into discrete functional CIM subsystems simplifies the integration, and complexity of the entire system. Because functional level communication is performed at the data level (locally) and additional data modification can be easily custom designed to meet the specific requirement of the manufacturing needs. Modular approach organises design, manufacturing and planning activities as modules which are at the same level and represented by an object-oriented data model to pass message between each modules.

Integration by part representation scheme to enhance the design and manufacturing data exchange in real time and to ensure concurrency.

Designing for manufacturability requires the fusion of the designers intention for a part with a manufacturing requirements for the product . The designers intentions are a set of functions which the product will provide or require. These functions are related to the features of part. Feature technology, provides a better approach to integrate design and applications following design such as: engineering analysis, process planning , machining and inspection. Multiple representation of features depending upon their application, are used to model these different design and manufacturing activities. Combining these different views is difficult, but different techniques has been proposed including: feature refinement, feature conversion, feature mapping, feature translation, feature transformation, and feature relaxation (Solomon et.al.). To achieve the greatest possible flexibility with respect to application-related specification, geometrical and technological attributes (face relations, edge classification, dimensions, location/orientation/size parameters and other non-geometric attributes) are used to model and evaluate manufacturability (Appendix- 2).

Integration by database.

Various heterogenous dynamic data available in the data base of CIM information system can integrate DFM and KBS in CIM environment through a real time active database support. Additionally the data available in the database enhances integration of the design and manufacturing activities, enhances communication between different modules and facilitates integration. Database systems are active integration tools, by which extraction of data, storing of data, and generating rules from the data base is important for manufacturability analysis in CIM environment . As described above part representation using "features " is convenient for implementation of object-orientation concept, and different attributes can be modelled in the CAD database including: design rules for building rules , active values/demons and methods, to represent feature prototypes with objects , classes, methods and dynamic bindings, feature derivation for representing class hierarchy and inheritance. These CAD database can be used also to model manufacturing and planning of production, where it acts as an integration agent for DFM analysis in the design activities. Object-oriented concept that uniformly models any real world entity as an object is applied in manufacturing to model a part, operation process, machine tool fixture and operating parameters as entities, so the relationships between each entity are mostly many-to many and form a very complex graph. DFM analysis is enhanced in CIM environment using both local and global data models and integrating them by an object-oriented-database that views the world as entities defined by their functional characteristics, which supports collaborative works of the team of developers and the customers.

Integration by a communication software to realize the feed back from the customer at the design stage or after operation of the product for a possible redesign of the product or a part to enhance two-way communicability.

Co-ordination of design, manufacturing and planning parallel with a customer oriented production are the core demand for the CIM activities, designed to adapt in the changing product specification. In this particular case, at one end the customer sets his preferences and priorities while the manufacturers at the other end are prepared to change their product in respect to the demands of the customer, using the available material resources or by adapting the resources with out spending major capital expenditure which is the objective of DFM. Real time co-ordination of these factors is difficult, but their consideration is vital, in today's competitive market. That is the basic reason for searching a real time integration software on the top of the existing CAD\CAM software to ensure two way communication between the user and the manufacturer, to practically avoid redesign, give better service, maintenance, and to design custom tailored product for a specific user. Compatibility issues for establishing data format standards, communication network standards, computer architectural standards are among the difficulties hampering integration and real time communication, but the research in these areas are encouraging.

Implementation of the above mentioned integration strategies using features requires the modelling of the part using the object oriented concepts. The part model in our DFM analysis uses the object oriented paradigms objects, classes, instances of classes and etc. to model the part features in the feature hierarchy modeled in Table.1. The KBS are organised for different functions (e.g fixturability analysis, etc.). Other related activities are organised in a modular manner in a predefined hierarchical relationships, to make effective the whole manufacturing activity and to encourage real time communication of the design, manufacturing and planning, as to produce a product which is custom tailored to the needs of the customer.

1. To combine the two activities (Design and Manufacturing) a feature based modelling is implemented with a new class of features (Appendix -2) to capture the basic tasks of the two activities "function" of the part and "manufacturability" of the part.
2. The manufacturability analysis module uses the same feature-based part modelling to represent material charcteristics, tolerances, and surface finish indexes with additional attributes added to represent these product attributes.
3. The module considers a number of process plans which are of a realizable type among which to choose after the manufacturability analysis based on the manufacturability rating. Initially a good plan is considered as the current best plan, and then search of state space of possible choices are accomplished.
4. The fixtureability of the design is checked by the corresponding knowledge base module. Considering the part to be machined and the existing resource model.

The manufacturability of the part is analysed step by step by considering the features in each part, their shape , dimension, and location, the available resource, etc. The final evaluation of the part is based on the manufacturability rating index. The steps-wise description of the DFM task followed in the present work considers the procedures listed in the (Appendix - 1)

Conclusion

The communication and data support issue is vital for CE in CIM environment, that is why the proposed hybrid interfaced /distributed CIM communication module with local and global data model, the KBS and the CE Blackboard control module can be implemented to solve the problems of CIM which includes sharing of data among CIM subsystems and the existence of inconsistencies among these data. Local and global data models and CE Blackboard control module are included in the proposed hybrid interfaced/distributed CIM communication architecture to enhance communication, through data support and to make the decision support method more effective, by extending and improving data application in CIM. To disclose the soundness of the proposed CIM architecture, a DFM module for machined part manufacturability analysis is considered, where Knowledge-Based approach, object-oriented paradigm and distributed databases concepts are implemented, for integrating multiple sources of knowledge which are the important parts for the DFM evaluation methodology. There are still problems to be addressed in the future to solve the communication of information in the CE application in CIM environments and these problems includes modelling events, handling anomalous information, incomplete knowledge and exception handling.

Appendix-1

Step-wise description of the task

No	Description of the task

1. product design from a Feature-Based CAD solid modeller.
2. Feature based stock modelling for stock material representation (user editable) and selection of stock material or near-net shape workpieces by comparing the features of the part with the stock material.
3. Feature based geometric and tolerance modelling and feature-feature relationship modelling
4. Feature by Feature comparison of stock material versus part feature model for cut dimensions depth of cut, number of passes and total machining allowances (t, i, h) and preliminary cutting sequence selection (recommended cutting variables and preliminary cutting sequence selection from the best possible plan stored in the database).
5. Input of the required resources or constraints (user editable) and representing them in the form of feature-cutting operation-resource model.
6. Limitation imposed by the machine horse power, maximum spindle speed and maximum depth of cut will be imposed (user editable).
7. Checking about accessability, rigidity, fixiturability and non-destruction of parts using the knowledge-based module for DFM analysis.
8. Identifying the feature of the part with serious problems and assigning a penalty factor for manufacturability; the penalty factors considers: tolerance, shape(morphological), resource, cost, etc. analysis.
9. Constraint checking for the limitation imposed using feature-based modelling applying feature-cutting operation -machinetool-tool model and feature-feature relationship model, and selection of operational sequences.
10. Corrections for tool approach and over travel made, then calculation of machining time and result display.
11. Amendment f or the values for the hardness of the work material, tool material and size of the cuts,operational sequences, etc...(cost, resource, quality or user defined selection criteria will be available for customization) and assigning penalty factors.
12. Part design is accepted or rejected for manufacturing.

Appendix-2

Part modelling using a hierarchical feature-modelling of the part.

No	Feature type	Description	Purpose
1.	General features	cylindrical, plane, etc.	To represent stock shape, material, overall dimensions
2.	Basic features	External and internal cylindrical surfaces, conical, etc.	To represent the manufacturing attributes including shape, dimension, etc.
3.	Auxiliary features	holes, chamfers, threads, etc.	To represent the functional properties of the feature.
4.	Tolerance modelling features	edges, vertices, faces, chamfers, centre lines, datum features (among the auxiliary features)	To represent the geometry, topology, and tolerance of the part.

References

1. Abdella H.S. & J. Knight (1994), An Expert System for Concurrent product and process design of Mechanical parts, *Instn. Mech. Engrs*, part B, Vol. 208, pp. 167-172.
2. Young R.E.(1992), A Grief and P.O'Grady, An Artificial intelligence based network system for concurrent engineering, *Int. Journal of Prod. Res.*, Vol. 30(7), pp. 1715-1735.
3. Groover M.P. (1987), *Automation, production system, and Computer Integrated Manufacturing*, Prince-hall,India.
4. Bayliss D.C, R. A. Kuesum, R. Parkin and J.A.G.Knight(1995), Concurrent Engineering philosophy implemented using computer optimized design, *Proc. of Instn. Mech. Engrs.*, Part. B, Vol. 209, pp. 193-199.
5. Andrew Kusiak(1988), *Artificial Intelligence for CIM*, IFS/Springer Verlag.
6. Remobold U., Nnaji B, Stor A.(1993), *Computer integrated manufacturing and engineering*, Addison Wesley.
7. Kaiser G.E. and et al.summer 1988, Database support for knowledge based Engineering Environments, *IEEE Expert*.
8. Dills D.H. and W. Wu, Using knowledge-based technology to integrate CIM databases, *IEEE Trans. on Knowledge and data Engg.* Vol. 3(2), 1991, pp. 237-245.
9. Beynon-Davis (1981), *Expert data base system*, McGraw Hill, N.Y.
10. Levent Orman (1988), Functional development of database application", *IEEE Trans. on Software Engg.*, Vol. 14(9).
11. Ahmed Elmagarmid and Evaqqelia Pitoura(Oct. 1995), " Using Object technology for databse system integration, *IEEE Computer*, pp. 67-68.
12. Subramanyam S. and S.C.-Y, Lu(1991), Computer aided Simultaneous engineering for components manufacturing in small and medium lot-sizes, Trans. of ASME, *Journal of Engineering for Industry*, pp. 450-464, .
13. Yemons R., Choudry A., Hagen P.J.W. Ten(1985), *Design rules for a CIM*, NHPC.
14. Don Ralstan & Tony Munton (1987), Computer integrated manufacturing, *Computer Aided Engg J.,Vol.4(4)*, pp 167-174.
15. Knox Charles S.(1987), *Organizing data for CIM application*, Marcel Dekker Inc.
16. Beach et al.(1988), Integration soft ware the key to CIM ,*IFS Springer Verlag*, pp 445-453.
17. Allen D.K(1986), Architecture for CIM,*Annals of CIRP Vol.35(1)*, pp351-355.
18. Martin R. & Dans J.E.(1985), The role of data in CAD/CAM integration, *Computer Aided Engg J. Vol.2 (3)*, pp97-101.
19. Gerzco J.M.& Buchman A.P. (1985), *An object-oriented language for CAD and requied data base capabilities , language for automation Chang S.(ed)*, Plenum press N.Y.
20. Nagel R.A. et al. (1986) Developing and implementing integration ststegies for manufacturing companies, *Annals of CIRP Vol.35(1)*, pp 313-316.
21. Wirt et al. (1989), Requirments on interfaces & data for NC data transfer in view of Computer integrated manufacturing, *Annals of CIRP Vol.38(1)*, pp 443-448.
22. Saleman Roy M. (1985), The Evolution from CAD/CAM to CIM: possibilities, problems and strategiesfrom the future, *Computer and Graphics, Vol.9(4)*, pp 435-439.
23. Salomons O.W., Van Houten F.J.A.M., Kals H.J.J.(1993) , Review of research in feature-based design, *Journal of manufacturing systems Vol. 12(2)*, pp 113-132.
24. Gupta S.K.(March 1997), Using manufacturing planning to generate manufacturability feedback, *ASME Journal of Mechanical Design Vol.119*, pp 73-80.
25. Arimoto S., Ohashi T., Ikeda M., and Miyakawa S.(1993), Development of machining-producability evaluation method (MEM), *Annals of the CIRP Vol . 42/1*, pp 119-122.
26. Juri A.H, Saia A., and De Pennington A.(1990), reasoning about machining operations using feature -based models, *Int J. Prod. Res. Vol. 28(1)*, PP 153-171.
27. Korde U.P., Bora B.C., Stelson K.A. and Riley D.R. 1992), Computer-aided process planning for turned parts using fundamental and heuristic principles, *ASME Journal of Engineering for Industry Vol. 114*, pp 31- 40

The Management of Parts for Complex Assembly System in PC Environment

Yin Wensheng, Wu Yongming, Wang Tongyang and Zhou Ji

CAD Center, HuaZhong University of Science and Technology, P.R.China

Abstract. It is important to manage the parts in complex assembly design system. It will be useful if an effective environment is provided to designers to manage the parts and subassemblies in this assembly or in other assemblies. In these environment designers can save, find and modify the parts or subassemblies hierarchically, orderly and conveniently. It is required for assembly design system to support the top-down design process and to let designers have a clear and complete realization about their own products. The management of assembly design system for parts now generally uses the method of managing files directly or by other database management systems, but it is hard or not convenient for complex assembly systems. A data structure for managing parts and subassemblies in complex assembly system is presented and with this structure one can establish a code rule easily in PC environment and can simply manage complex assembly system that has many parts. It can also be used for the management of other complex systems.

1 Introduction

One of the important purposes of computer design system for assembly is to provide an effective environment to designers. In these environments designers can specify the relative positions of all of the parts, tell the interactions between these parts, and analyse the kinematic properties of the assembly and assembly tolerances. It can also save, find and modify the parts or subassemblies hierarchically, orderly and conveniently. It is under the management of parts and assemblies that we can design a product according to the habit of real process of product design and realize the top-down design in the computer design system. We can make the designers understand their products from the start to the end completely and clearly and help them to realize their design easily and quickly.

Generally a part or a subassembly is saved and opened as a file. In order to help designers to understand the hierarchical relation of a assembly, a hierarchical tree such as featuremanager in SolidWorks is provided in some CAD systems. There are some disadvantages of management of parts in these system and it is hard to manage the large scale assembly system. We use a code management of parts in our assembly system of CDS(Concurrent Design System).

2 Characteristics of complex assembly and management of parts

A real assembly may be a complex assembly and there are a large amount of parts and subassemblies in it. We must manage all the parts and subassemblies and their attributes such as the features of parts, the mates of parts, and etc. With the development of the computer technology, the management of production becomes more and more convenient. We can simply manage all parts and assemblies of a factory easily.

2.1 Assembly design representations

In the assembly system, the use of data structure must adopt a particular application. The first type of assembly structure is the virtual link structure that is a well-known hierarchical assembly representation (Lee and Gossard, 1985) and has influenced the data structure used in other parts. It is a tree that tells us the structure of the assembly and assembly sequence in Fig. 1.

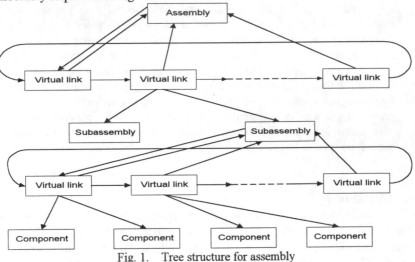

Fig. 1.   Tree structure for assembly

The second type of assembly structure is the general graph structure that more explicitly represents part-to-part relationships than the tree and is particularly well-suited to tolerance and kinematic chain generation (Wang and Ozsoy, 1993) (Kim and Lee, 1989). The disadvantage of the general graph is that it lacks the hierarchical model. The best structure should be the combination of these two representations and will process not only the hierarchical relations but also the mate relations. Fig. 2 shows the relations of parts and subassemblies.

The feature-based part representation is in common use of assembly model and it has more advantages than the old ones. Generally, a feature is a geometric primitive consisting of a predefined pattern of low-level geometric entities, typically a volume or a set of surfaces, that have more information than part geometry and have significance to an application. The features of assembly are provided in some design systems such as in SolidWorks, which represents some assembly operations. Feature-based part representation is used in CDS.

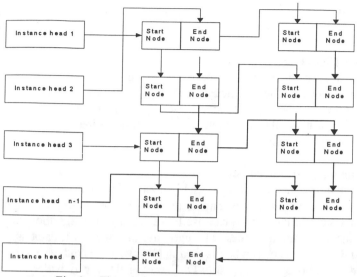

Fig. 2.   The relations of parts and subassemblies

2.2 Simple assembly system

The simple assembly system is the system that has a few parts and the management of this system is very simple. In this system, we insert a file by a name to load the data of a part or a subassembly into the assembly system, and we save the file also by the name to write the data of part or an assembly to the disk. The file name we see is the name of the part or assembly. It is the way that manages parts and assemblies directly by   file management system. It is useful and effective for simple assembly systems but it is required to remember all names that represent the parts and subassemblies.

2.3 Complex assembly system

Generally, a product consists of thousands of parts and subassemblies. If we use simple assembly system the disadvantages are:

It is difficult to inquire the parts or assemblies; There are too many files on the disk to find the one we need; Although we can make a lot of catalogues for the assembly, it is required that we know not only the structure of assembly but also the rule of saving file.

The file name will be different from the name of the part or assembly. If we save the file of part by the same name of the part, it will confuse the management of parts. For example, a part or a component can have only one name, but when it is used in an assembly system, it will have many instances which have the same data structure and different names. The name we see will be the name of the instance but not the name for saving. If we use Chinese name for representing the part in China, we will find it very hard to establish and inquire the Chinese file name for saving and loading the part.

In the real design process, a part has a name and a code. The name consists of English or Chinese word which has a very clear meaning and easy to remember, and the code consists of a series of alphabets which present a certain meaning. Each code of part must obey code rule and present a part exclusively. Designers and workers remember the parts by the name and look for the parts by the code.

2.4 Managing files of part directly

The simple management of parts is the method of using the files directly. Data of each part consists of data files. Although each part has different format of files, it can be read or written as a unit. The function of every computer operation system is the access of files, so it naturally uses the management of files directly. It is very effective when the assembly system is small.

2.5 Managing parts by other systems

In order to manage the complex assembly system, we can use other database management systems to manage the parts. For example, we can use Sybase to create a management system of part or we can use PowerBuild to manage the technological documents, including the part files. The parts are managed as a database unit like float, integer data type. They can be inserted in OLE way in Windows environment if the part and assembly system supports OLE way. This way is widely used in the management of parts and takes an very important role. But there are some disadvantages in it:

Higher cost

If we only have the assembly system, we can not manage the parts. We   have to buy the database management system which is costly.

Lower speed

Use of other database management systems will lower the speed of assembly system. The database management systems will take up a lot of memory and take a lot of time. If the data of parts is processed as a block of OLE, many programs will be processed simultaneously and this will make the speed more slowly.

2.6 Management of featuremanager design tree

A method of management of featuremanager design tree has been proposed in the SolidWorks. It provides a featuremanager design tree to manage assembly, parts, features of parts and mates among parts and subassemblies. The featuremanager design tree allows designers to quickly make changes or evaluate the steps used to create the model and let designers understand the model construction. It is useful for designers to inquire the parts and subassemblies in the assembly but it is hard to know where the part or subassembly is.

2.7 The process of top-down

At the stage of conceptual design of a product development, the designers first generate a functional representation of the design and often take some methods to simplify the design. We often sketch a simple part to present a complex product, and we create a chart of processes to realize the functions of the product. We can get a real product by converting the chart of processes to the parts and subassemblies and detailing the simple parts. It is important for designers to construct a draft of product at the stage of conceptual design and change the parameters to get new designs.

Construction of parts and subassemblies will be determined at the stage of conceptual design and can be described as a code rule of product. Generally, this code rule will be active not only in this product development but also in the other products of the same series of products. It is useful to establish a code rule quickly for the development of products.

3 Managing parts by code

If we use the access of files of operating system, it is necessary for us to remember all the names of parts and subassemblies. If we use the other database systems, it will be costly. If we use the featuremanager design tree, we can find a part in this assembly easily, but we cannot find out the part which is not in this assembly. It is necessary to use a new method to solve this problem, which is not directly associated with the files or position of files. Because designers do not want to know the physical position of parts, their only concern is the logical position of files and this will be good for top-down design process and concurrent assembly. It is proposed that using codes to manage parts is a useful method.

3.1 Detaching of part and its file name

Detaching of part and its file name is very important. If we use part name as the file name for saving the part data, it will be hard to manage the versions of part. It is required that the part name is clear for us to understand the meaning of the part. The name will be long or express in English or other languages. For example, Chinese will be used in China to name the parts. It is not convenient to use these names to save the parts in file management of system. We must detach the name of part and the name of file for saving the data of part, so we can interact with assembly system by its Chinese name or English name which has a meaning for designers to remember. The physical position for saving the part will be determined by assembly system itself, and versions of part will be maintained by the system itself too.

3.2 Management of top-down and bottom-up

A product has been developed according to the following flowchart:
According to this chart the design is a repeating process and will be modified at any stage. It is required that the assembly design system can provide an ability to modify the design at any time. The top-down design is a design that designers make functional representations of product at first and then break down the complex functional representations into simpler functional representations and satisfy the functional constraints gradually. It is widely used in all industrial design. The bottom-up design is a design that designers design the parts or subassemblies at first and then assembly these parts or subassemblies into a assembly. It is hardly to find in traditional design, but it is used in computer aided design (CAD) because the top-down process is hard to realize while the bottom-up process is easily implemented.Computer aided design system is powerful to design a part for its extreme ability of 3D construction and can provide a realistic part which has no error for designers, so it is widely used in the development of product. In order to support

the real design of product, the CAD system must provide the ability of top-down design. It is useful to use the code of product to manage the parts and subassemblies in top-down design and bottom-up design. Using code of product we can easily find the part we want and understand the progress of design. We can easily make functional representations of product and dispatch the tasks to other designers. The functional decomposition of product will be recorded automatically and design can be carried out by different designers simultaneously. The general engineer will know the progress from the messages among the designers. Each modification of part or subassembly can be known by the related designers. Because a CAD system has an ability of bottom-up design and by code of parts it can manage the parts for top-down and bottom-up design.

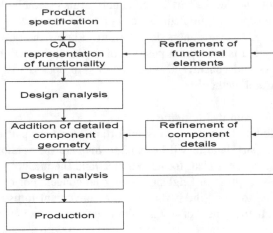

Fig. 3.　Top-down process

3.3 Creating a code rule at run-time

In order to manage the development of a product, designers usually make a code rule for the product. The code rule is not only used in the development of this product but also used in the developments of other products. The code rule is usually designed to manage a series of products or even the whole products of a factory. We can easily make an interface of code rule for a factory, but it is hard to make interface for every factory and when the code rule is changed, it is required to modify the program code. It is proposed that we can make a code rule at run-time and when the code rule is changed we can update the code rule at run-time too. Designers are required to write a text file which describes the code rule by a certain regulation and lets the system to read in. The code rule written by text format is very easy for designers to read and modify.

When the code rule is read by system, the legality of code rule will be checked by the system. If there are some errors, the system will prompt the type of error and indicate the place where there is an error. The designers will not remember the code rule again because the system will give the definition of each code. This is a sample of text file describing the code rule:

```
// Code rule
{       // Code start
// The first code segment
      1, Machine tool, Describe the type of machine tool;
      {      @L,Lathe;
             P,Planer;
             … … }
// The second code segment
      1,Group, Describe the main constructs;
      {
             {              # Machine tool,0,1;
                     {     L; }
                     @6,Boring lathe;
                     … …

                     # Machine tool, 0,2;
                     {     P; }
                     @5,Hydraulic planer;
                     … …

             }

      }
… …

// The fifth code segment
      1, Version, Describe the version of design , , T;
      {              @A,Z, Version;  }
… …

// The twelfth code segment
      2, serial number of part, serial number of part, ,T;
      {              @00,99, serial number of part;      }
}                     // End of code
```

When the designers update the code rule it is necessary to ensure that the new code rule does not confuse with the old one.

The system supports code rules used in most of the factories and companies. It is required that the code rule has hierarchical and left to right structure. The code rule the system can solve is shown as Fig. 4

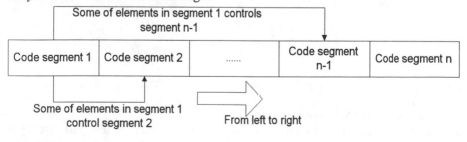

Fig. 4.   Code rule of product that the system can solve

3.4 Various and intelligent requisition

Various and intelligent requisition is supported in this system. We can inquire a part in these ways:

Key in a code completely

Designers input the whole code and the system will match the code with code rule and find the demanding part. If there is not a part which has the same code as the designers input, the system will prompt an error message and provide some parts which have code similar to the input code and let designers select correct one.

Select the code under the navigation of the system

The system displays each code and its meaning from left to right, layer by layer. The designers select a code and come into the next code layer. When all codes of   layers have been selected, the complete code is made. That is the code of part you want.

Key in the code incompletely

Sometimes we know the code partially for we know there is not other part which has the same code fraction, or the code has the default values, or we will search some parts which have the same regulation, we can input the code incompletely and the system will provide all of parts which have the same code we input.

Because some code rules are designed for people to read so it is hard for computers to recognize. This system is of some intelligence and can match most codes like designers.

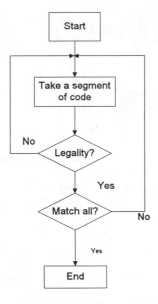

Fig. 5.   the process of code matching

Key in a code which has no regulation

Sometimes we create a part only for temporary use and we save this part directly by file management system. We can input the part name and indicate the path of file without selecting the code rule.

3.5 Need no database

It will be costly for CAD system running in PC environment to have a database management system. Security and other properties of design system will be raised if we use database management, but it will decrease the speed of running of CAD system. Generally, if we use the database management system, the system will consume a lot of memory to run the database management system and the CAD system will run in the OLE way. It is required to increase the cost in PC environment. By the management of code of parts, the database management system is no longer required and this can be written into the CAD system. As part of assembly system, it takes little memory and has high speed to inquire the parts.

3.6 Management of assembly

An assembly has a hierarchical tree which is described as code in product to record the relation of parts and subassemblies. We not only deliver a hierarchical tree for designers to operate on the parts in this assembly but also deliver a way of code of parts to insert a part or subassembly which is not in this assembly before.

BOM is one of the contents of assembly design. This system takes little memory and can run in higher speed. It is easy to produce BOM tables for an assembly and also for all products of a factory.

3.7 Mini PDM system

Product Data Management(PDM) is an important technology. It stores and controls the data that is critical to product design, manufacturing and technical support, so it manages the process of product development. It is required to base on the database management system and becomes very complex. Although our system loses some properties, it is helpful to support conceptual design and top-down design because the code rule implies information of design process. This system is the mini system of PDM and the data structure is already used in the InteMAN system(the PDM system of CAD Center of HUST).

4 Data structure

The assembly system consists of various software components written in C++, and C++-based ACIS solid modeller is used. The structure of code of product has four classes shown in Fig. 6.

The classes of code of product are derived from class of ENTITY of ACIS which is the base class of model. They inherit the features of ENTITY of ACIS and are easy to access data of code. CODE_TREE is the class which describes the whole code rule; class CODE describes the value and meaning of each code segment; class CODE_GROUP describes the data of groups in each code segment; and class CODE_ELEMENT describes the value and meaning of each code element or records the value of code of current part.

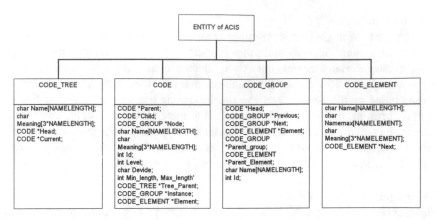

Fig. 6. the components of code of product

5 Conclusion

The data structure for managing parts and subassemblies in complex assembly system is used in CDS. In this system the designers need not load and save the parts or subassemblies directly by their file names. They can use Chinese words to represent the parts or subassemblies, and it is useful for Chinese engineers to design products.

In this system designers can create a code rule at run-time and inquire a part in various ways. One can   key in a code completely, or select the code under the navigation of the system or key in the code incompletely. This system has some intelligence and can search and match the code. If one likes to use file management system, he can key in a code which has no regulation. The code data of product has been built into the system so it makes the modelling system effective.

This system will be helpful to top-down design process and concurrent assembly because the code of product can represent the process of design.

Reference

1. Kunwoo Lee and David C Gossard, A hierarchical data structure for representing assemblies: part 1 , Computer-Aided Design, Vol. 17, No. 1, pp. 15-19, 1985.
2. D.A Beach & D.C. Anderson, A Computer Environment for Realistic Assembly Design , Proceedings of The 1996 ASME design Engineering Technical Conferences and Computers in Engineering Conference August 8-12, 196, Irvine, California. 96-DETC/CIE-1336.
3. Gritt Ahrens, Oliver Tegel, Conceptual product development ,   Proceedings of The 1996 ASME design Engineering Technical Conferences and Computers in Engineering Conference August 8-12, 196, Irvine, California. 96-DETC/EIM-1402.
4. Rajneet sodhi and Joshua U Turner, Towards modelling of assemblies for product design, Computer-Aided Design Volume 26 Number 2 February 1994, pp.85-96
5. Kim, S. H. and Lee, K. An assembly modeling system for dynamic and kinematic analysis, Computer-Aided Design, Vol. 21, No. 1, pp.2-12, 1989.
6. Wang, N. and Qzsoy, T. M., Automatic Generation of Tolerance Chains From Mating Relations Represented in Assembly Models, J. of Mechanical Design, Vol. 115, No. 4, pp. 757-761, 1993.

An Intelligent Mechanism to Support DFX -abilities in Automated Designer's Environment

Wulanmuqi and Jiati Deng

The State 863/CIMS Design Automation Laboratory (DeAu Lab.)
Institute of Manufacturing Systems (720 Institute)
Department of Manufacturing Engineering
Beijing University of Aeronautics and Astronautics, Beijing, 100083, P.R.China

Abstract. The paper presents an intelligent DFX mechanism for product design. The authors solved this problem by taking the product life-cycle issues into consideration during design process from the designer-oriented perspective of view. At first the designer-oriented computer environment *DesignerSpace* is introduced for understanding the designer how to implement design activities and DFX method better. In order to integrate design knowledge from downstream aspects for the optimized decision-making, an intelligent DFX mechanism is developed to incorporate knowledge-base, algorithm-base and monitoring/debugging tools. *DesignerSpace* with DFX abilities is implemented and the further development is strongly intended.

1 Introduction

As we know, time, quality and cost have been three key criterion of product competitiveness in the market, and now the focus is placed on product development time reduction from which other metrics probably can be obtained. The whole product design and development process is carried out along the timeline of product life-cycle. At the stage of design decision making, 70% to 80% of the product life cycle value is formed at the expense of about 20% of product life cycle[1]. So it will be most effective means to get time advantage by continuously improving product design, especially at the early stage - the stage of conceptual design. Concurrent Engineering (CE) is a product life-cycle-concerned methodology. In addition to the functionality that product concerns, CE emphasizes the optimization of both product and product development process. Distinct from the traditionally sequential way, the design is largely depentent on the integrated product development process. So considering DFX(Design For X) into the early stage of design that could be one of the fundamental concerns will affect the optimization of product and process. DFX methodology embodies product design philosophy of CE, where X corresponds to manufacturing, assembly, serviceability, disassembly and recyclability , or other post-manufacturing issues. DFX takes the requirements into account from manufacturing, assembly, marketing, maintenance and recycling, etc. The final result is more efficient product development cycle: less development time, improved quality and less cost. The challenge drives research on design methods of product and product development process, and their implementations. According to S. Sivaloganathan et al.[2], an integrated design system "Design Function Deployment (DFD)" was proposed to incorporate Quality Functions and customer requirements throughout design. Feng and Kusiak[3], Hashemian and Gu[4], Schmitz and Desa[5], Hsu, Lee and Su[6], Boothroyd and Dewhurst[7], Makino, Barkan and Pfaff[8], and Ishii, Eubanks and Marks[9] have given their contributions. Topics on design for manufacturing (DFM) and design for assembly (DFA) are the most common areas.

The authors propose to establish a designer-oriented environment named *DesignerSpace* as an intelligent framework assisting in fulfilling DFX, and also it functions as an organizer to implement each unit activity in product design and development processes. All product design and development activities organized, managed and implemented by *DesignerSpace* are controllable and reconfigurable. *DesignerSpace* establishes the infrastructure for the design and development processes, and can be automated well. So *DesignerSpace* behaves as a knowledge-based and distributed intelligent problem-solving mechanism adopting software agent technology. DFX is treated as a kind of mechanism integrated in *DesignerSpace* for optimizing the product design at the designated activity stage under *DesignerSapce* management. The paper is organized as follows: after a brief introduction in the first section, *DesignerSpace* is described in Section two. Then the intelligent DFX mechanism and its components in *DesignerSpace* are explained for product design. Finally the implementations and conclusions are presented, and the future research is expected.

2 *DesignerSpace*: Designer-Oriented Environment

Product design and development processes[1] are performed by a series of concurrent or sequential engineering activities which have certain inputs, outputs, control and leverage mechanism[2] . So every design activity is organized and carried out within/by *DesignerSpace*, that is, a working space which can help the designer to accomplish the design activities[3] in the way of automation. Such a *DesignerSpace* is developed to incorporate utilities to decompose and coordinate design processes, provide design assistance tools, resources and optimized design methods, and carry out each design activity partitioned from the process under the constraint/control which is specified or configured from the higher level for management of the activities. The framework of *DesignerSpace* illustrated in Fig.1, has one *Management* module and three

---

[1] : Generally there are the following major processes in product design and development: customer requirement collection, planning, design, process planning, production, service and maintenance, etc.

[2] : Their definitions conform to those of IDEF0.

[3] : Design activity is a distinct stage of a design's development whose boundaries are usually characterized by changes in design fidelity[11].

fundamental modules: *Process* module, *Resource* module and *Mechanism* module.

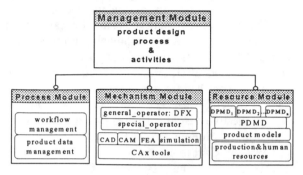

Fig.1. The organization and management of *DesignerSpace*

- *Management* module, functions as a platform integrating *Process* module, *Resource* module and *Mechanism* module, and managing the operation procedure of deign activities. ① For certain activity a_i (i=1,...,n) starting at the time t_i, any creation, modification or improvement on product can be precisely described by the changes of Product Definition Model Data(PDMD) in *Resource* module. So the activity a_i is said only an operation on PDMD. ② It accesses to the existing PDMD created by the previous activity a_{i-1}, and outputs new instance of PDMD generated by current activity a_i within the timeframe from t_i to t_{i+1}, which is the beginning time of next activity a_{i+1}. ③ Using DFX and CAx tools from *Mechanism* module can provide the means to solve problems adaptively. These operations described above is for solving product design problems levered by a knowledge-based and distributed intelligent multi-agent system. ④ Friend user interface is easier for different disciplinary engineers to solve their specific engineering design works.
- *Process* module, like a leader organizing design process in higher level, is responsible for design process decomposition, workflow management and information management. It provides *DesignerSpace* facilities that transmit the designer concrete design task to be completed in the form of task order, and permit the designer to partition a design problem into manageable subproblems. The sample of a task order is shown in Table 1. The leader of a product design team is able to control and monitor the workflow of product development.

Table 1. Task order from *Process* Module

Description	Input	Output	Resource	Coordination
Task_#:	Item:	Item:	Hardware:	Perspectives:
Goal:	From_whom:	From_whom:	Tools:	
Designer(s):	Time:	Time:	Document:	
Constraints:				
Duration:				

- *Resource* module, offers the meta-data of product information models and data concerning product design processes and activities, and specifies the real production system components covering the technological and human resources referenced by product design and development processes. The physical and logical relations between them are defined. According to the workflow management in *Process* module, the resources assigned for each designer to finish his design task are reached from *Resource* module.
- *Mechanism* module, provides CAx design assistance tools required for the designer's easy and effective design activities throughout design processes. As for different design tasks, different product model data are required as inputs and outputs, and different CAx tools are used; for example, CAD/CAM system with 3D modeling, 2D draughting, assembly, NC codes generating and other functions. When a designer receives the definite design task and possesses CAx tools helping finish the task, *Mechanism* module also offers design knowledge about how to solve the problem with optimization. In the following it will be discussed in detail.

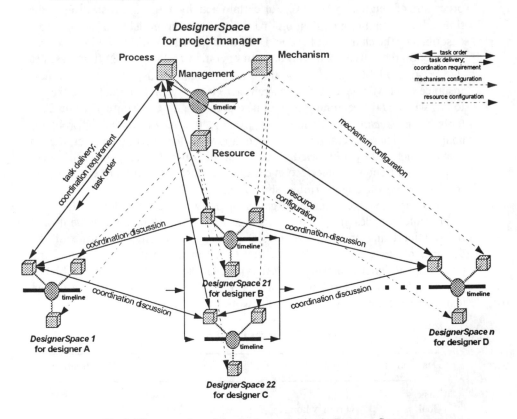

Fig.2. The operation of the design activities in *DesignerSpace*

When a design process is partitioned into various design activities and designated for different team members to finish, *DesignerSpace* is actually worked as a designer working space for fulfilling their activities. One *DesignerSpace* corresponds to a design activity. During the product design process, many *DesignerSpaces* are required to work cooperatively although the contents of different *DesignerSpace* from *Process* module, *Resource* module and *Mechanism* module are not the same at all. The activities of design process carried out within *DesignerSpace* are illustrated in Fig.2.

As seen in Fig.2, *DesignerSpace* is originally constructed for the team leader (such as the project manager) at top level with all configurable resources, mechanism and process timeline, and then is copied to low level *DesignerSpace* (numbered 1, 21, 22, or n) as a template for each team member (A,B,C or D) in specified design activity with configured resource and mechanism. Within each activity that a *DesignerSpace* manages is the operation procedure which requests the access to resources and mechanism. The multiple agents within *DesignerSpaces* cope with the fulfillment of design activities. Project leader at top level has a project-agent to help organizing team, configuring resources and mechanism, and coordinating team activities. At middle level, each team member has engineering-agent help to cooperate with other members and accomplish his own engineering activity. At low level (which is not shown in Fig.2.) the R-agent help each middle level *DesignerSpace* to manage and access its resoures adaptively.

3 Intelligent DFX Mechanism

From the information integration point of view, design activities are the processes of the information handling. *Mechanism* module in *DesignerSpace*, like an operator, operates on the information (which can be thought as the input or output) being processed during certain design activity, as shown in Fig.3. Therefore *Mechanism* is a comprehensive framework which intends to integrate more computational technologies along the whole product design process.

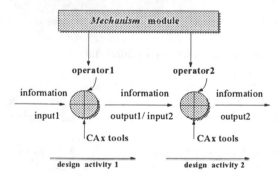

Fig.3. The role of *Mechanism* module in the design information processing

Because of different engineering application domains involved in solving design problem, *Mechanism* offers *general_operator* and *special_operator*. The former is proposed for general mechanical product design, such as DFX; and the latter has definite application scope and needs special design technologies, for example, blankdisc design in aircraft. Any mechanism operator is giving the constraints on design, but CAD system is needed to help construct PDMD under the constraints. *General_operator* or *special_operator* are made up of three basic components: *knowledge_base, algorithm_base* and *monitoring/debugging*. The components of DFX mechanism with *Process* and *Resource* modules in *DesignerSpace* are described in EXPRESS-G in Fig.4.

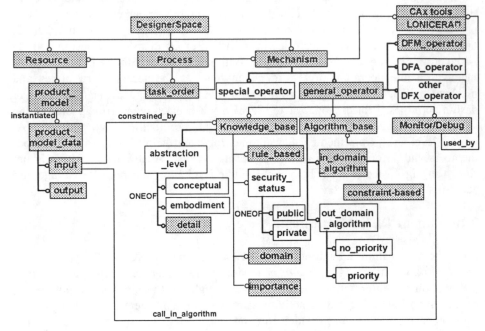

(*): LONICERA is a integrated CAD/CAPP/NCP system based on STEP, developed by DeAu Lab, BUAA.

Fig.4. The components of DFX mechanism in *DesignerSpace*

3.1 Product models and operated product model data

The DFX intelligent mechanism operates on the product model data for obtaining the optimized product model data which are of DFX abilities. The product models used in *DesignerSpace* are structured into two levels: geometric_model and feature_model[12].

- Geometric_model, is the representation of product geometry and topology using geometrical and topological elements based on AP203/STEP[4].
- Seeing that the geometric_model lacks engineering semantics, feature_model describes the geometrical shape with product design and engineering knowledge

---

[4]: AP203/STEP: Part 203, Application Protocol: Configuration controlled design;

1104

in features. Referencing AP214/STEP[5] , the feature-based product model called product_definition_model_data (PDMD) is established, and its transformed model data called domain_product_model_data (DPMD) with a particular domain application are derived from the feature_model.

The product model data operated by DFX mechanism are the instantiated data of the feature_model: PDMD. They are the basis of information integration in the product life cycle. The support of the geometric_model are indispensable.

3.2 Knowledge Base (KB)

In such a DFX intelligent mechanism, design knowledge is involved in multi-disciplines and multi-domains. So *Knowledge_base* is a collection of design requirements representing expertise from different domains covering the product life cycle, such as manufacturing, assembly, marketing, service and recycling, etc. These design requirements need to be classified, abstracted and represented as knowledge written and stored in *Knowledge_base*. For example, from manufacturing point of view, fabrication means (such as machining, casting, forging or moulding), equipment (such as machine tool, cutting tool, or die), and shopfloor production resources and conditions, constrain the product design. *Knowledge_base* is intended to support the design process from conceptual, embodiment to detail design. Therefore each piece of design knowledge is labeled which domain it belongs to, and is abstracted or refined which design level it is in. There are the conceptual level, the embodiment level and the detail level according to different design phases. Such classification of design knowledge is convenient and efficient for information management during design processes. But more difficulties exist in the knowledge representation during the conceptual design and the embodiment design than during the detail design.

Whenever some knowledge is added into or deleted from *Knowledge_base,* the consistency with other knowledge in either the same or different domain will be checked. If it is violated, changes in knowledge are required until the consistency is achieved and remained. The representation of knowledge in *Knowledge_base* is rule-based and object-oriented. The inheritance and poly-morphism support the design information management of *Knowledge_base*.

The general representation of knowledge is described in the following form by reference to [13].

```
{Knowledge:
    {form                    ;the equation or logical relation
     abstraction_level       ;name of the abstraction level
```

---

[5] : AP214/STEP: Part 214, Application Protocol: Core data for automative mechanical design processes.

```
        domain              ;which domain offers this piece of knowledge
        constrains          ;product model's entities knowledge concerns
        importance          ;on a scale of 1 to 10
        parent              ;supertype of this knowledge
        child               ;subtype of the knowledge
        security_status     ;public or private?
        }
};
```

3.3 Algorithm Base (AB)

After *Knowledge_base* about certain kind of product design is established, decision-making methods are needed. *Algorithm_base* in the DFX intelligent mechanism is responsible for it. Since one DFX concerned is not optimized at expense of the others, DFM, DFA, DFS(Design For Service), or DFR(Design For Recycling) should be considered concurrently. *Algorithm_base* provides two algorithms which are *in_domain_algorithm* and *out_domain_algorithm* respectively.

Considering only one DFX, such as design for manufacturing, usually there is an uniform representation in product model and knowledge. We can find a common algorithm. Constraint-based design is one of the effective methods to cope with this kind of DFX problems. So *in_domain_algorithm* offers inference mechanism and search algorithm inside certain domain for one DFX. While taking two or more DFXs into account at the same time, for example, DFM and DFA, the characteristics of DFM and DFA determine that different methods are used for different domain applications. Therefore *out_domain_algorithm* is required and gives compromising the conflicts among more than one DFX seeking for the optimized design result. If necessary, human interactions are in demand. There are *forward_processing* and *backward_processing* methods in *in_domain_algorithm* for domain knowledge inference. For *out_domain_algorithm*, the importance or priority of each rule in domain knowledge and that of each domain knowledge in *Knowledge_base* has to be defined.

3.4 Monitoring/Debugging

Monitoring/Debugging in fact is a visual tool for monitoring the design activity and debugging the design problems. The elements of *Knowledge_base* can be checked; for example, the relationship between rules is able to be visualized in network diagram. Any correction or change in rule network will be reflected in *Knowledge_base* written in text form. During inference processing, the operation of which rule is active or not are displayed. So the designer can set or clear breakpoints dynamically. It is very easy for the designer to operate the design activity. Then the satisfactory and optimized result about part or component can be displayed in CAD/CAM system from *Mechanism* module.

4 Implementation and Conclusions

The intelligent DFX mechanism is developed within the framework of *DesignerSpace* and partially implemented to achieve optimized mechanical part/component and its process design during the detailed design phase. A prototype system *Intelli-DFX* coded in C++ on HP workstation can be demonstrated now. The core part is a STEP based integrated CAD/CAM system-LONICERA (also the registered trademark), which is being developed in our institute. LONICERA system has integrated a 3D parametric feature based modeler on AP214/STEP and a STEP model data access environment , and has integrated workflow management module. The implementation of DFX mechanism in *DesignerSpace* framework is briefly illustrated in Fig.5.

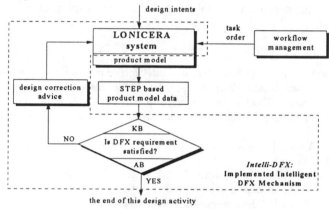

Fig.5. Implementing DFX mechanism within *DesignerSpace*

Designers write design requirements and knowledge concerning their design activities into the knowledge base in their own *DesignerSpace* during the process of requirement analysis. DFX mechanism works while their design tasks being in progress. Once the product model data, which could not be the end required output, are produced, *Intelli-DFX* is provided the designer to check the manufacturability, and informs him of design problems in the form of text or visualization for correction.

The main idea and implementation of the designer's design environment *DesignerSpace* with intelligent DFX abilities have been presented in this paper. *Process, Mechanism* and *Resource* modules integrated in *Management* module constitutes the computer architecture of *DesignerSpace*. *Mechanism* provides the part/ component/product design with intelligent abilities. The implementation helps us find out problems to be improved and promote the further development.

1. A single function of DFX, like only DFM or DFA, has the single objective besides satisfying the functional requirements. Research in the past was mostly involved in one separate DFX. In fact, a product design problem is always constrained by more than one objective at the same time. So we are developing

further to integrate DFM and DFA in simultaneous consideration of manufacturing and assembly.

2. High performance of DFX mechanism is tightly related with those of the other parts in *DesignerSpace*. So the fine integration of *Process, Resource* and *Mechanism* modules and development of product data management will increase the efficiency of design more.

3. Since the designers in a design team concern different design knowledge and may also be geographically distributed, a distributed intelligent design cooperative system is needed for the operation of *DesignerSpace* using multi-agent technology. Upon the developed *DesignerSpace*, the research supported by National 863/CIMS Hi-Technology Research and Development Project (865-511-706) is underway concerning about distributed artificial intelligence for virtual product design environment based on Internet/Infranet.

References

1. P.O'Crady, D.Ramer, J.Bolsen, Artificial intelligence constraint nets applied to design for economic manufacture, *Computer Integrated Manufacturing*, **1**, 4 (1988) 204-209.

2. S.Sivaloganathan, N.F.O.Evbuomwan, A.Jebb, The development of a design system for concurrent engineering, *Concurrent Engineering:Research and Applications*, **3**, 4 (1995) 257-269.

3. C.X.Feng., A.Kusiak, Constraint-based design of parts, *Computer-Aided Design*, **27**, 5 (1995) 343-352.

4. M.Hashemian, P.Gu, A constraint-based system for product design, *Concurrent Engineering:Research and Applications*, **3**, 3 (1995) 177-186.

5. J.Schmitz, S.Desa, The development of virtual concurrent engineering and its application to design for producibility, *Concurrent Engineering:Research and Applications*, **1**, 1 (1993) 159-169.

6. W.Hsu, G.Lee, S.F.Su, Feedback approach to design for assembly by evaluation of assembly plan, *Computer-Aided Design*, **25**,7 (1993) 395-410.

7. G.Boothrody, P.Dewburst, Product design for manufacturing and assembly, *Manufacturing Engineering*, 4 (1988) 42-46.

8. A.Makino, P.Barkan, R.Pfaff., Design for serviceability, *The 1989 ASME Winter Annual Meeting*, 1989, USA.

9. K.Ishii, C.F.Eubanks, M.Marks, Evaluation methodology for post-manufacturing issues in life-cycle design, *Concurrent Engineering:Research and Applications*, **1**, 1 (1993) 61-68.

10. J.T. Deng, Product design and development *technical report*, Institute of Manufacturing Systems, Beijing University of Aeronautics and Astronautics, 1996

11. M.A.Hale, J.I.Craig, F.Mistree, D.P.Schrage, DREAMS and IMAGE: a model and computer implementation for concurrent,life-cycle design for computer systems, *Concurrent Engineering:Research and Applications*, **4**, 2 (1996) 171-186.

12. J.T.Deng, LONICERA- a STEP based CAD/CAM integrated system, *Proceedings of Conference on CALS*, Nov. (1996), Singapore.

13. Navin chandra, Mark S. Fox, Eric S. Gardner, Constraint management in design fusion, *Concurrent Engineering: Methodology and Applications*, edited by P. Gu and A. Kusiak, Elsevier Science Publishers B.V. (1993) 1-30

FACTORY AUTOMATION

Shopfloor Integration

JAVA Implementation of Shopfloor Access and Monitoring of A Manufacturing System

Ong Chee Wee, Osamu Kaneko, Kang Siew Hwa
Japan-Singapore Institute, Nanyang Polytechnic

Abstract

The software developing environment has been changing very rapidly with the explosive growth of the Internet as well as the Intranet over the last two years. The development of the JAVA language by Sun Microsystems is one of the most significant event in the Internet world. In this paper, we would like to introduce the feasibility and possibility of using the JAVA language for implementing shopfloor access and monitoring of a manufacturing system.

1 Introduction

An automated manufacturing system as shown in figure 1 is currently being developed in Nanyang Polytechnic. Its major system building blocks are four robot assembly cells complete with conveyor systems, an automated guided vehicle (AGV) and an automated warehouse (AWH). A shopfloor control and monitoring network system is used to coordinate and schedule the activities among the various equipment in the automated assembly system. This 10BaseT Ethernet network system consists of a host computer, a network hub and a number of node computers. The various equipment in the automated system are linked to the node computers through serial communication interfaces. There is also a gateway node computer which is connected to the Internet to link the system to the external world. This setup will provide an Intranet for borderless access and monitoring of the automated manufacturing system.

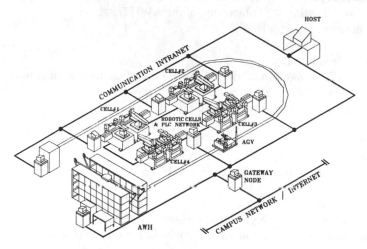

Figure 1. The Automated Manufacturing System in Nanyang Polytechnic

This borderless access and monitoring of the automated manufacturing system can be achieved in the following main domains as shown in figure 2.

i. **Automated Manufacturing System**
   - Flexible Assembly using 4 Robot Assembly Cells With Conveyor Systems
   - Automated Material Storage/ Retrieval Using An Automatic Warehouse
   - Intelligent Material Transportation Using An Automated Guided Vehicle

ii. **Host Computer**
   - Scheduling & Production Planning
   - Data logging & Statistical Data Processing
   - Database for Automatic Warehouse, Products and Production Output
   - Remote Control & Monitoring of Assembly Cells, Automatic Warehouse and Automated Guided Vehicle
   - Communication Error Detection
   - Contingency Procedures

iii. **4 Cell Node Computers With Vision Systems**
   - Uploading/Downloading of Robot Application Programs
   - Control & Monitoring of Robots (2x) during operation
   - Communication with Host Computer

iv. **PLC Network Node Computer**
   - Uploading/Downloading of PLC Application Programs
   - Control & Monitoring of PLCs (4x) during operation
   - Communication with Host Computer

v. **AWH Node Computer**
   - Control & Monitoring of the AWH through its controller unit
   - Editing, Updating & Maintenance of the AWH Database
   - Communication with Host Computer

vi. **AGV Node Computer**
   - Remote Control & Monitoring of the AGV through its controller unit
   - Scheduling of the AGV operations
   - Communication with Host Computer

vii. **Gateway Node Computer**
   - Remote access and monitoring terminal
   - Communication window to the external world via the Internet

viii. **Fast Ethernet**
   - Data communication highway in real-time (100Mbps)
   - Transmission of audio and video signals for remote monitoring

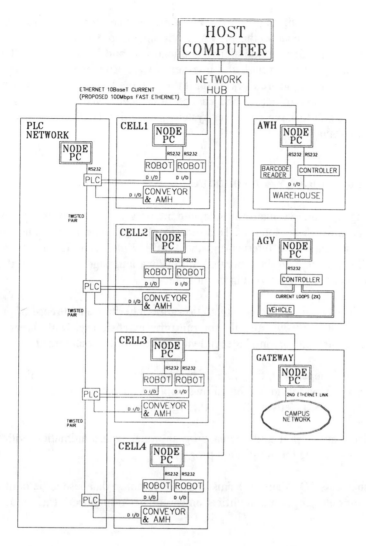

Figure 2. Block Diagram of The Automated Manufacturing System Incorporating A Borderless Access and Monitoring Through The Internet

The system would demonstrate the use of current networking technology in a factory automation environment providing shopfloor supervisory control and monitoring, including SCADA and MRP functions through an Intranet which is connected to the Internet. This method of access, control and monitoring of an automated system through a borderless network is the current trend in factory automation. It would also provide a platform for the development of solutions for industrial automation systems as well as research in the area of remote control, access and monitoring through the Internet.

JAVA language is growing in popularity at lightning speed for applications through the Internet. This is because the JAVA environment provides the development platform for independent stand-alone applications and protocol handlers for distributing dynamic data through the Internet. JAVA language is currently being improved, and many people are working on it through the Internet. In addition to this, JAVA language is easy to access and can be developed seamlessly across various platforms from small to large computer systems, including UNIX, Window NT and Window95 environment.

2 Features of The JAVA Language

The Java language allows platform-independent program code to be dynamically loaded and run through the Internet. This functionality provides revolutional changes to the passive nature of the Internet and the World Wide Web. The followings are some of the features that are incorporated in the Java language that make it the most promising protocol for the Internet in the near future.

Java is an Object-Oriented language that allows the inheritance and reuse of program code both statically and dynamically for run-time extensibility. Multithreading are provided at every level, allowing independent task control by each thread.

Java is platform-independent, therefore its program can run on any machine that has the Java interpreter ported to it. This allows application programs written in Java to be executed in an Open Network System through Internet.

Java language also support multimedia programming, where animation with sound can be incorporated into the application programs.

Java can also access SQL database. This allows a common database to be built within the network for data access and modification can be performed from each node.

3 The Experimental System

The experimental system, as shown in figure 3, has been setup to demonstrate the capabilities of borderless remote access and monitoring. It consists of a Sun workstation, which acts as a host computer, and is configured as a DNS server, as well as a SQL Database server using mSQL. There are two node computers which are used to link the Automatic Warehouse and the PLC network to the host computer. One PC has been assigned as the gateway to the external world through the Internet. Its Web's page shown in figure 4. The Host computer controls the production scheduling and coordinates the AGV's and AWH's schedules in parts/products transportation between the AWH and the Robot Assembly Cells. The hardware of the network is 10baseT Ethernet.

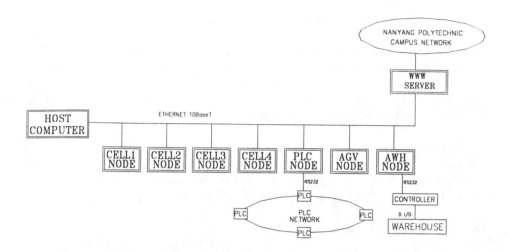

Figure 3. Block Diagram of The Experimental Borderless Access and Monitoring - Manufacturing System

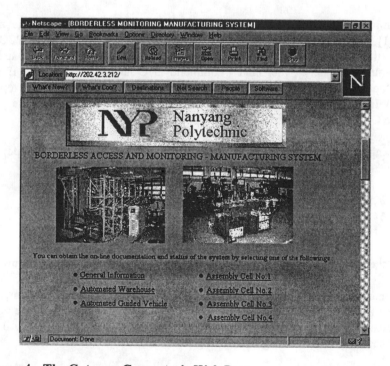

Figure 4. The Gateway Computer's Web Page

3.1 Device Level Control through JAVA Language

JAVA language is well known as a development tool for WWW home page, therefore it is strong in graphic user interface, animation and networking. However, for low level control of devices, such as PLC or Robot, the RS232C serial communication is commonly used. Here, we discuss how to incorporate the RS232C serial communication in JAVA application software.

We used Runtime and Process classes in "java.lang" package. First, prepare an executable file compiled in C language, for the basic RS232C control and communication. This executable file can then be called using Runtime and Process classes in the JAVA application program as shown below.

```
String command[]=new String[2];
try{
    Process process = Runtime.getRuntime().exec(command);
    inFromCommand = new DataInputStream(new
    BufferedInputStream(process.getInputStream()));
    String line = inFromCommand.readLine();
}catch(IOException e){}
```

Here, "command" is String array and command[0] is executable file name and command[1] is outgoing data through RS232C. The response from the RS232C is stored in the "line" string. JAVA can pick up the response by using "process.getInputStream()", the related command in C language is printf(). If the device is "dead" or not connected, the program will check device status and report "NO POWER". Without this process, the JAVA application will hang the PC, because the C executable program will be out of control.

The following program segment can be used to detect the status of the local device that is linked to the node PC through its RS232C port after powered up.

```
if((0x10 & inportb(0x2fe)) == 0x00)
 {  printf("%s",err);
    exit(1);  }
```

Generally, the procedure to initialize the system is to turn on the Host then follow by the node PCs. The Host has to recognize the status of the PCs continuously. The "ping" command can be used to detect the most low level node PC conditions, for example, no power or no connection. We realize continuous monitoring of the node PCs by executing the getStatus() method of SyncClient class program every 5-10 seconds as shown in the following JAVA program listing.

```
class SyncClient{
DataInputStream in;
PrintStream out;
Socket client;
boolean flag=true;
Status status;
String command[]=new String[3];
String response;
int port;
String node;
public SyncClient(int p,String n,Status s){
    status=s;
    node=n;
    port=p;
    command[0]="/usr/sbin/ping/";
    command[1]=node;
    command[2]="1"; }
public boolean getStatus(){
try{
    Process process = Runtime.getRuntime().exec(command);
    DataInputStream dis = new DataInputStream(process.getInputStream());
    response = dis.readLine();
    }catch(IOException ioe){ System.out.println(ioe.getMessage());
                            return false; }
if(response.substring(0,2).compareTo("se") == 0)
    { try
        { client = new Socket(node,port);
          out = newPrintStream(client.getOutputStream());
          in = new DataInputStream(client.getInputStream());
          }catch(IOException e){ flag = false;
                                status.toNotReady(); }
      if(flag == true) { status.toReady();
                         status.onPower(); }
      else{ status.noPower(); }
      return true; }

public void close(){
try{
    out.close();
    in.close();
    client.close();
      }catch(IOException e){ System.out.println(ioe.getMessage());}
  }
}
```

3.2 SQL database

Database management is important for networking in an automated manufacturing system. We used mSQL, which is available over the Internet. The mSQL language offers a subset of the features provided by ANSI SQL. mSQL is a shareware product authored by David J. Hughes, and is free to educational and research organizations. Sun Microsystems provides JDBC Database Access API for JAVA. The JDBC drivers for mSQL are also available through the Internet.

The Main Databases for our system are as follows.

Autowarehouse : Products/Parts in the warehouse
Product_part : Products/Parts ID codes and Barcodes
Production : Production Data, ie. Lot number; Yield; Time; etc.
Error Log : Production Error, ie. Error code; Node number; Time; etc.

The Host controls the overall production schedules, and therefore, these databases are updated or modified by Host only, except the Error log. The node PCs can get information from these databases. They enter data into the Error log when error occurs. The two examples below, show the use of the SQL. The first example is to create table Product_PartDB, the second is the use of the SELECT command.

msql.Query("CREATE TABLE Product_PartDB
(ID int primary key,name char(20),barcode char(5))");

*result = msql.Query("select * from AWH ORDER BY position");*

3.3 Typical JAVA Network Program Between Node PC and Host Computer

The network is based on socket connection of TCP/IP. In the case of this type of connection, the Server or Client status does not depend on whether it is a Host or Node PC, but depend on the network program. A server application program listens to a specific port waiting for connection requests from a client. When a connection request arrives, the client and server establish a dedicated connection. We prepare three connections between Host and each Node PC. The Node PC has two server programs and the Host has one server program. The Thread program is used for server application. The related instances are transferred to these communication programs to do some jobs in accordance with the information from remote side.

The followings are examples of the JAVA communication classes that have been developed in the Host Computer and the Node PC for Client/Server applications.

Cell Node PC Server Programs

SyncClientHandler : Receive the access from Host and send back the Node's status.

The status data for testing are random numbers between 1-50 generated to simulate production data. It is periodically accessed, for example every 5-10 sec. Using "Thread" to enable it to accept a call at any time.

AsycClientHandler : Receive the access from Host and execute some jobs.

In this case, get an instance of Content class(extends Applet), and do some jobs. Using "Thread" to enable it to accept a call at any time.

Host Server Program

ClientHandler : Receive the access from Node and execute some job.

In this case, issue a musical chime to call for attention. This has an instance, "audio"(instance of class Audio). Using "Thread" to enable it to accept a call at any time.

Host Client Programs

SyncClient : This has an instance, "status"(instance of class Status)

This is used to get the status information from Node and set its current status on the Host's graphic user interface.

AsyncClient : This has an instance, String "sendout"

This is used to send out data to Node. It has a function for sending out only.

Node Client Program

Client: This is used to call the attention of the Host.

In this case, send the help command, when the Help button is pushed.

Using these JAVA communication classes, the Client/Server applications that have been established between the Host and each of the Node PCs are illustrated below.

Host Computer		Node PC
SyncClient(status)	⟶	*SyncClientHandler( )*
ClientHandler(audio)	⟵	*Client( )*
AsynClient("updateDB")	⟶	*AsyncClientHandler( )*

3.4 Open Network

We prepare one Node PC (Windows NT) which serve as the gateway to the external world through the Internet. This Node PC has two network cards, one is connected to the local Intranet and another is connected to our campus network. The WWW server program that has been installed is the "Front Page" of Microsoft Internet Explorer. On this Gateway PC, The WWW server and one JAVA program are running concurrently. The JAVA program accesses the Host program every 150 sec. and update the file in "Front Page" of the WWW server directory. Therefore, it is possible to access the production data and status of the automated manufacturing system through Internet from anywhere by using an Internet browser.

In the JAVA applet program in the WWW server directory, the file can be read and stored in "is" instance as follows, and the content can be shown as graphical way. The HTML(Hyper Text Markup Language) refer the applet class in it.

```
is=DataInputStream;
try{ is=new DataInputStream(new BufferedInputStream(
    new URL(getDocumentBase(), "cell4.txt").openStream()));
    }catch(MalformedURLException e) {}
catch(IOException e){}
```

Conclusion

The experimental system for the shopfloor access and monitoring of the manufacturing system has demonstrated that it is possible to extend the applications of the Internet to the manufacturing industry through an Intranet for remote control and data acquisition. This will enable the production facility and the various departments within a manufacturing plant to be easily linked together to create an information network for Computer Integrated Manufacturing through the Internet using the JAVA language for developing the required Client/Server applications.

References

1. JAVA JDK and tutorial, http://www.sunsoft.com
2. mSQL code, ftp://bond.edu.au/pub/Minerva/msql
3. JDBC API, http://splash.javasoft.com/jdbc

Real-Time Supervisory Control of a Flexible Manufacturing System Using AI-based Method

Zhong Jiangsheng

Wuyi University, Mechanical Engineering Department, Jiangmen, Guangdong 529020
China

Wong Chun Chong

Ngee Ann Polytechnic, Mechanical Engineering Department, 535 Clementi Road
Singapore 599489

Abstract. To be able to achieve cost-performance benefits from a flexible manufacturing system (FMS), the main objective considered in supervisory control of FMS is to increase the utilization of all processing cells, e.g. machining cell, inspecting cell, and assembly cell etc. which a FMS consists of. This paper addresses the key issues involved in design and operation of real-time supervisory control for a FMS that consists of following major components, i.e. a machining cell including one CNC lathe and one CNC machining center, a conveyor, an inspecting and assembly cell, an automated storage and retrieval system (ASRS), and an automated guided vehicle (AGV), using artificial intelligent (AI)-based real-time software tool known as G2. The concept of virtual equipment is introduced. The proposed material tracking method is elucidated with an example, and the event-driven AGV dispatching strategy is addressed.

1 Introduction

Flexible manufacturing systems (FMSs) are collections of machines and related processing equipment linked together by an automated material handling system, and all under computer control. These systems are capable of producing a wide variety of items in small batches and in random order (Williams et al 1994).

A typical FMS consists of the following components:
- computer numerical control (CNC) machine tools, usually equipped with tool magazines and tool changers;
- a part load/unload station, and other stations such as inspection or part wash;
- a set of pallets and fixtures where parts are loaded into fixtures, which in turn are mounted on pallets;
- an automated material handling system (automated guided vehicles (AGVs), conveyors, tow carts, etc.);
- a tool storage area and tool transport system;
- part buffers at machines and a storage area for pallets.

Supervisory control is the most important function in a FMS. It includes the monitoring and control of machine tools, material handling equipment, work transportation devices, and auxiliary equipment, i.e. all components a FMS consists of (Hwang et al 1984). In this investigation an artificial intelligence (AI)-based software tool known as G2 is adopted to code the supervisory control of a FMS, which consists of a machining cell including one CNC lathe and one machine center, a conveyor, an inspecting and assembly cell, an automated storage and retrieval system (ASRS), and an AGV. The concept of virtual equipment is introduced. The proposed material tracking method is elucidated with an example, and the event-driven AGV dispatching strategy is addressed.

2 Description of the FMS

The layout of the FMS for the present investigation is shown in Fig. 1. The salient features of the system are described.

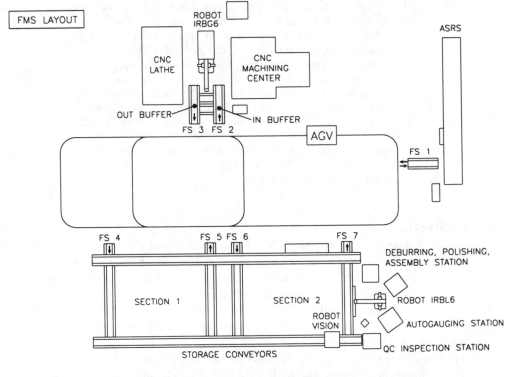

Fig. 1. Layout of a flexible manufacturing system

It consists of two cells and an automated storage and retrieval system (ASRS) connected by an AGV. In the cell 1 (machining cell) there are two machine tools, i.e. one CNC lathe and one CNC machining center, an input buffer and an output buffer, and a robot handling with tools/parts load/unload jobs. The cell 2 (inspecting and assembly cell) consists of an inspection station, an autogauging station, a deburring, polishing and assembly station, and a robot. The local storage, i.e. the section 1 of the storage conveyors is provided for storage parts which need another machining operation after turning/milling operation. At the cell 2 there is also an assembly local storage for temporally store parts to be assembled further. There are seven feeding stations, feeding station 1 to 7 (FS 1 to FS 7), as shown in Fig. 1, for AGV load/unload. Six types of part are processed simultaneously. The six part types are in three product groups, i.e.

> group one: part type 1, part type 2, and part type 3;
> group two: part type 4 and part type 5;
> group three: part type 6.

In the FMS operation all the above parts are machined from the raw materials. The part types 1 and 2 are assembled onto the part type 3. In the group two the part type 4 is assembled onto the part type 5. The part type 6 is an independent part. The system is able to work on different part types as long as the features can be handled by the grippers, fixtures, pallets and other mechanical accessories provided. All workpieces (raw material and semi-finished/finished parts) are held by the pallet in operation. Two pieces of the same type are held in one pallet for the turned parts and one piece for the milled parts. Pallets are being used randomly for the six part types at different stages in the FMS operation. The processing sequences for all six part types are shown in Fig. 2.

The AGV track layout consisting of both unidirectional segments is shown also in Fig. 1. When the AGV completes a delivery task, it waits at home-station provided near the section 1 of the storage conveyors if it is unassigned. The AGV travels only by the shortest route between any two locations.

3 Virtual Equipment

The design of the supervisory control system (SCS) of the FMS begins with the development of classes. The physical equipment (PE) of the FMS in the shop floor are matched into virtual equipment (VE) as classes in SCS. Each VE class is defined with its attributes related to specification and characteristics of its PE and assigned responsibilities for certain tasks required for the supervisory control. As the classes are designed, relationship between classes are defined. The VE classes form tree hierarchy by which a class can inherit attribute(s) from its superior class. For example, for class *Feeding-station* the data structure is defined as:

Class name	*feeding-station*
Superior class	*plant-item*

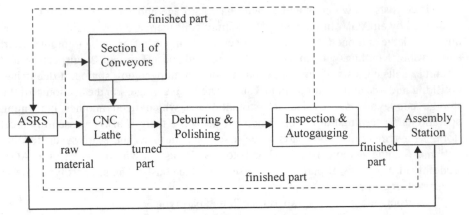

Part types 1, 2 and 4

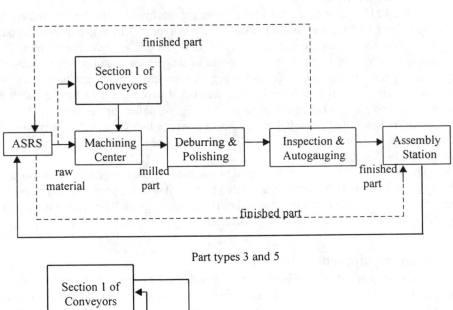

Part types 3 and 5

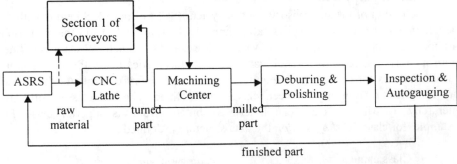

Part types 6

Fig. 2. Processing sequences for the six part types.

Attributes specific to class   *agv-arrival is given by a fms-value*
Inherited attributes   *identity;*
                                       current-action;
                                       current-event;
                                       current-pallet is 0;
                                       last-pallet is 0;
                                       start-time;
                                       expected-duration;
                                       pallet-sensor is given by a fms-value

    In addition to the attribute *agv-arrival is given by a fms-value* specific to this class, the class *Feeding-station* inherits the attributes as above mentioned from the superior class *plant-item,* which is the defined top class for VE. The *fms-value* is another class that is responsible for catching sensor values from shop floor.

    In relation to PE configuration and the number of plant items of the same type the instances of VE classes are connected together logically, as shown in Fig. 3 and Table 1, so that during SCS process data can be correctly and easily transformed bi-directionally, i.e. from SCS to shop floor and vice versa.

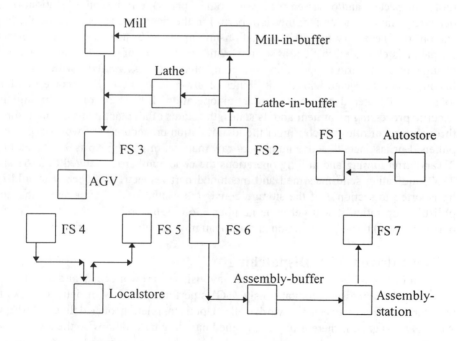

Fig. 3. The connection of virtual equipment

Table 1. Virtual equipment corresponding to physical equipment

Virtual Equipment (VE)	Physical Equipment (PE)
Mill	CNC machining center
Mill-in-buffer	Input buffer for milling operation
Lathe	CNC lathe
Lathe-in-buffer	Input buffer for turning operation
Autostore	Automated storage and retrieval system (ASRS)
AGV	Automated guided vehicle (AGV)
Localstore	Section 1 of storage conveyors
Assembly-buffer	Section 2 of storage conveyors
Assembly-station	Cell 2 including inspection, autogauging, deburring, polishing, and assembly stations
FS 1 to FS 7	feeding station 1 (FS 1) to feeding station 7 (FS 7)

4 On-line Material Tracking

As shown in Fig. 2, in the FMS part(s) can be machined on different machine tools, inspected and/or assembled with other part(s) on the other station(s), if necessary. In order to deliver the right part(s) in the right time at the specific feeding station, it is necessary to track material on-line in the SCS. *Material* (part(s)) placed on pallet is expressed in SCS as a class with an attribute of *pallet code* which will change upon a certain operation has completed. A class *activity* with its many instances is constructed with the attributes of *pre-code and post-code* corresponding to status of material before and after an operation. By means of identifying the specific processing plant item and its status, the status of all plant items are monitored through a set of rules which trigger the identification procedure(s) of work-in-process pallets. For instance, the code number of raw material of part type 6 is 3360 (see Fig. 2), and after turning and milling operations the code number is changed to 3361 and 3362, indicating semi-machined and machined part respectively. Then it should be transported to section 2 of the storage conveyors by the AGV. After deburring and polishing operation it will get code number 3363, 3364 or 3369, if two, one or no part(s) pass(es) through inspection and autogauging operations respectively.

5 Event-driven AGV Dispatching

In a FMS environment a typical part may visit several work cells, and a variety of parts may be processed simultaneously. AGVs perform a variety of delivery tasks to transfer the parts between the work cells. Upon completion of a delivery task, an AGV is reassigned if there is any unassigned handling task; otherwise the AGV is set idle and waits for a task to emerge. The AGV may have to be moved to an alternative position, however, before it is permitted to become idle. If the number of unassigned tasks exceeds that of free AGVs, prioritization of tasks is essential. Mahadevan and Narendran (1990) analyzed the problems arising from multi-AGV systems and strategies for resolving them using analytical and simulation models. Raju and Chetty

(1993) proposed a Petri net-based methodology for modeling and simulating AGVs for FMSs.

In this investigation, AGV dispatching job can be divided into two kind of circumstances during the FMS operation:

1. more than one feeding station request AGV to transport pallets at the same time or during the last AGV trip;
2. No requests for AGV when AGV returns its home after completing the last trip.

When the input buffer of cell 1 is idle or the output buffer is full, or the feeding station of the cell 2 request, AGV immediately takes its trip based on first-come first-served (FCFS) rule. If the requests occur at the same time, then the cell 1 will get higher priority because of its smaller buffer capacity. This event-driven AGV dispatching strategy ensures the machining or assembly process can be continued without interruption.

To maximize utilization of AGV (Dirne 1990), when AGV returns its home with no requests, SCS will assign AGV to retrieve raw material from ASRS, and it will be transported into the section 1 of the storage conveyors according to material priority.

6 Conclusions

The design and operational control strategies of real-time supervisory control for a FMS are presented, that consists of following major components, i.e. a machining cell including one CNC lathe and one CNC machining center, a conveyor, an inspecting and assembly cell, an ASRS, and an AGV. The concept of virtual equipment is introduced, and the proposed material tracking method is elucidated with an example. To this end, the event-driven AGV dispatching strategy is addressed. SCS of the FMS is being tested now with shop floor physical equipment together and the testing results show that SCS works successfully.

Acknowledgment

The authors are grateful to the colleagues in Mechanical Engineering Department and Advanced Manufacturing Workshop, Ngee Ann Polytechnic, who involved in this investigation for their assistance.

References

1. C. W. G. M. Dirne, The quasi-simultaneous finishing of work orders on a flexible automated manufacturing cell in a job shop, *International Journal of Production Research* **28** (1990) 1635-1655.
2. S. -L. Hwang, W. Barfield, T. -C. Chang, and G. Salvendy, Integration of humans and computers in the operation and control of flexible manufacturing systems, *International Journal of Production Research* **22** (1984) 841-856.
3. B. Mahadevan, and T. T. Narendran, Design of an automated guided vehicle-based material handling system for a flexible manufacturing system. *International Journal of Production Research* **28** (1990) 1611-1622.

4. K. R. Raju, and O. V. K. Chetty, Design and evaluation of automated guided vehicle systems for flexible manufacturing systems: an extended timed Petri net-based approach. *International Journal of Production Research* **31** (1993) 1069-1096.
5. T. J. Williams, J. P. Shewchuk, and C. L. Moodie, The role of CIM architectures in flexible manufacturing systems. In *Computer control of flexible manufacturing system,* S. B. Joshi and J. S. Smith (eds) Chapman & Hall (1994) London.

The Description、Verification and Implementation of MMS Interconnection between Heterogeneous Networks

Xiang Fei   Guanqun Gu

Department of Computer Science and Engineering, Southeast University, Nanjing
210096,People's Republic of China,E-mail:xfei@seu.edu.cn

Jieyi Wu   Zheng Li

CIMS Research Center, Southeast University, Nanjing, People's Republic of China

Abstract. In this paper, the protocol specification and service definition of MMS (Manufacturing Message Specification) are briefly introduced. According to the state transition diagram of MMS, a gateway machine used for the interconnection of heterogeneous networks is defined and described. Then ,based on the Petri nets, the MMS interaction between heterogeneous networks is described and analyzed. At last, a MMS communication system in CIMS environment is implemented.

1 Introduction

The core of CIM(Computer Integrated manufacturing) is integration, the basis of which is computer network. The Manufacturing Message Specification(MMS) is the ISO standard communication protocol specific to manufacturing, the purpose of which is to meet communication requirements among various programmable devices in CIM environment. MMS is stemmed from MAP, but it is independent of specific network platform. In a real CIM environment, there exist heterogeneous networks, such as Ethernet and Fieldbus, because of various communication requirements. Therefore it is necessary to study the MMS interconnection in such a case. Fig 1 shows the framework of heterogeneous networks in a typical CIM environment.

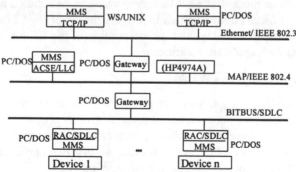

Fig 1. The Framework of MMS Communication System

In order to correctly understand and implement MMS services and MMS interconnection between heterogeneous networks, it is essential to develop a formal methodology to describe the various service units in MMS, based on which the MMS protocol machine can be implemented conformed to international standard. Such a formal methodology is further used to design the structure and functions of MMS gateway , describe the behavior of MMS interactions and test the correctness of MMS interconnections, so that the MMS integrated communication system is implemented successfully with reliability and consistence.

2 System Framework and MMS Overview

According to the communication requirements in today's CIM environment, a framework of *MMS Integrated Communication System* is shown in Fig 1. In this framework there exist three kinds of network products: Ethernet、 MAP and BITBUS. Two gateways are used to interconnect Ethernet/MAP and MAP/BITBUS respectively. HP7974A is a MMS instrument for testing protocol.

MMS mainly consists of two parts: MMS service definition and MMS protocol specification. In MMS service definition, client/server model is used to describe its service interactions with client issuing requests for services to be performed by the server. Object model is used to abstractly describe the services provided by a manufacturing device to a controlling device. A class of object consists of attributes of an object, the operations that can be performed on the object and the behavior of the object when receiving operation requests. Generally there are 10 classes of services in MMS. MMS protocol specification specifies the procedure for the transfer of data and control information in the MMS context, the means of selecting the services to be used by the application entities, and the structure of the MMS PDU used for the transfer of data and control information.

MMPM(Manufacturing Message Protocol Machine) is an abstract machine that carries out the procedures specified in MMS protocol specification.The state transition diagrams are applied in MMPM to describe the protocol. CS(Confirmed Service) is the basis of MMPM and almost all the MMS services have to be handled by CS unit. So in this paper , we will focus on the analysis of CS, the design of the STD (state transition diagram) of CS in MMS gateway, and the description of the CS interactions between heterogeneous network application entities.

3 Description of MMS-GWPM

The main functions of gateway are interconnecting different types of network, providing data associations between application entities and converting heterogeneous protocols, i.e. converting the information constructed by one network protocol to what could be identified by another network protocol, so that two application processes in heterogeneous networks can communicate with each other.

Gateway is the mean to interconnect heterogeneous networks at high level according to the 7-layer reference model. Because there exist apparent differences among the framework and protocols of MAP 、 Ethernet and BITBUS, and MMS lies in application layer, we decide to design the gateway at application layer.

According to the general requirements of gateway, the main functions of MMS gateway should be:

1． Providing associations between different networks at application layer;

2． Converting MMS service primitives and protocols between heterogeneous networks.

3． Having no change on hardware、 protocols and application software of each network;

4． Providing necessary control and management functions such as flow control and route selection.

Referring to the states transition diagrams of CS in MMS protocol specification , the MMS-GWPM(Gateway Protocol Machine) is designed and shown in figure 2.

Where function 2 and 4 are implemented in the state "Converting in Gateway". In order not to change the STD of CS, this MMS-GWPM includes the responses to all the state transitions of MMS requester and MMS responder. For special MMS-PDU such as cancel-PDU, GWPM will response it directly so that the process delay is reduced. Because MMS is connection-oriented, function 1 is accomplished in Environment and General Management Service.

4 Description and Analysis of the Interaction of MMS CS through Gateway

Having designed the STD of MMS-GWPM, we should further consider the description and analysis of the dynamic process when interactions of MMS CS take place through MMS gateway. Compared with the formal technologies such as Estelle

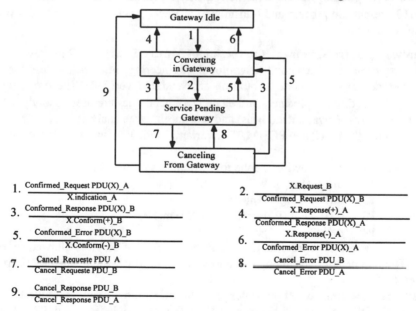

1. $\dfrac{\text{Confirmed\_Request PDU(X)\_A}}{\text{X.indication\_A}}$

2. $\dfrac{\text{X.Request\_B}}{\text{Confirmed\_Request PDU(X)\_B}}$

3. $\dfrac{\text{Conformed\_Response PDU(X)\_B}}{\text{X.Conform(+)\_B}}$

4. $\dfrac{\text{X.Response(+)\_A}}{\text{Conformed\_Response PDU(X)\_A}}$

5. $\dfrac{\text{Conformed\_Error PDU(X)\_B}}{\text{X.Conform(-)\_B}}$

6. $\dfrac{\text{X.Response(-)\_A}}{\text{Conformed\_Error PDU(X)\_A}}$

7. $\dfrac{\text{Cancel\_Requeste PDU\_A}}{\text{Cancel\_Requeste PDU\_B}}$

8. $\dfrac{\text{Cancel\_Error PDU\_B}}{\text{Cancel\_Error PDU\_A}}$

9. $\dfrac{\text{Cancel\_Response PDU\_B}}{\text{Cancel\_Response PDU\_A}}$

Fig. 2. State Transition Diagram of MMS-GWMP

1133

and Lotos, Petri nets have several advantages as protocol engineering: Petri nets can describe asynchronous and concurrent system using the form of figure; it has rigorous mathematical theory which can be used to analyze and verify the behavior of protocol.

4.1 Description of the Interaction of MMS CS through Gateway Using Petri Nets

According to the STD of CS and GWPM, the *Petri Net for the Interaction of CS through Gateway* could be established following the principles mentioned below:

1 . State is expressed by the state place, the message is expressed by the token, the protocol action is expressed by transition;

2 . The service provider of lower layer is characterized by two appositional FIFO queues with reverse direction. Each queue is represented by one place.

3 . The resources in the network are expressed by the number of tokens, including the arrived request, of the outstanding services, i.e., not yet confirmed services, and etc. .

Figure 3 presents the *Petri Net for MMS CS through Gateway*. For clarity, we assume the length of the arriving requests and the number of outstanding services are all one. The actual length of the arriving requests and the number of outstanding services will be shown in section 4.2. At the same time, for simplifying the complex of the Petri net, the net is shown without place *channel*. Place *channel* has an ingoing arc to each of transitions ni (i=1 ~ 5, 6 ~ 10) and an outgoing arc to the same transition ni . The net consists of five parts: the service requester(place Ps1, P1, P8), the responder(Ps3, P4, P5), the gateway(Ps2, P2, P3, P6, P7), the service request queue(Q1, Q2), and the communication network(all other places). t1 ~ t19 present the transition corresponding to the STD of requster, responder and gateway. n1 ~ n5, n6 ~ n10 present the process and transmision at lower layer, tQ presents the arriving of the service requet.

4.2 Analysis of Petri Net for the Interaction of MMS CS through Gateway

According to the viewpoint of protocol engineering, after being described by formal methodology, the model should be analysed and verified . The properties of Petri nets are boudness , liveness and etc. . Two techniques are used to analyze Petri nets: metric analysis(invariant analysis) and reachability tree analysis.

By using the package PESIM, P-invariants and T-invariants of the net are determined.

P-invariants: The following are the four minimum P-invariants.

$$\begin{aligned}
&\text{P1:} \quad m(Q1)+m(Q2)=K \\
&\text{P2:} \quad m(Ps1)+m(P1)+m(P8)=S \\
&\text{P3:} \quad m(Ps2)+m(P2)+m(P6)+m(P3)+m(P7)=S \\
&\text{P4:} \quad m(Ps3)+m(P4)+m(P5)=S
\end{aligned}$$

Where K represents the maximun number of services allowed to enter the CS. The number of outstanding services allowed in MMS is indicated by S. In this Petri net, K=S=1.

Analysis: The meaning of P1 indicates that the maximun of service requests is no more than K, if the request number is more than K, it can not be accepted. P2, P3 and

1134

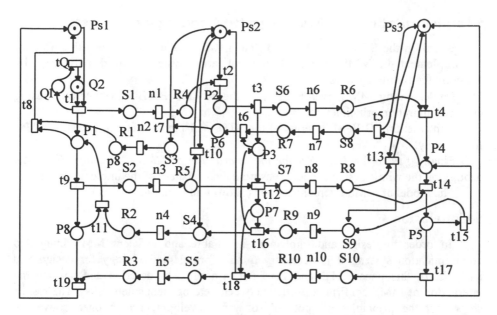

Fig. 3. Petri Net for the Interaction of MMS CS through Gateway

P4 indicate that the subnets of petri net corresponding to the service requester、 gateway and responder are strictly conservative. Furthermore, since all the places are covered by some P-invariants, the petri net is bounded.

T-invariants: T-invariants can be used to determine the normal service operations. A normal service cycle is defined to be any transition sequence which leads the Petri Nets from the initial marking and returns it to the initial marking. There are five T-invariants:

> T1: t1,tQ,n1,t2,t3,n6,t4,t5,n7,t6,t7,,n2,t8;
> T2: t1,tQ,n1,t2,t3,n6,t4,t9,n3,t12,n8,t14,t15,n9,t16,n4,t11,t5,n7,t6,t7,n2,t8;
> T3: t1,tQ,n1,t2,t3,n6,t4,t5,t9,n3,t12,n8,t13,n9,t16,n4,t11,n7,t6,t7,n2,t8;
> T4: t1,tQ1,n1,t2,t3,n6,t4,t5,n7,t6,t7,t9,n3,t10,n4,t11,n2,t8;
> T5: t1,tQ,n1,t2,t3,n6,t4,t9,n3,t12,n8,t14,t17,n10,t18,t19;

Analysis: The T-invariants mentioned above represent five types of service cycle. T1 specifies the type of standard service executions in which no requests of service canceling have been issued by the requester. T2 ~ T4 give the types of unsuccessful canceling service. In t2, canceling requests have been issued by the requester but denied by the responder for some reasons. T3 is cancel error executions in which requests of service canceling from the requester are received by the responder after the services requested have been completed. T4 is also cancel error because the service requested have been fulfilled by both responder and gateway before requests of service canceling are received. Finally, the service executions defined by T5 are successful canceling service execution in which canceling requests have been issued and accepted.

By using the package PESIM, we could get the reachable tree for this petri net, from which we can see that this petri net is live, have no deadlock marking.

5 Implementation of MMS Integrated Communication System

Based on the STD of GWPM and Petri net for the interaction of MMS CS, we can implement the MMS integrated communication system showed in figure 1 by applying the following steps:

Implementing MMS over MAP 、 Ethernet and BITBUS respectively using protocol migration principles.

Implementing GWPM in terms of the STD after the correct configuration of PC for gateway.

Finally , as to the *Petri Net for the Interaction of MMS CS through Gateway*, debugging the whole system, so as to ensure the correctness of the implementation.

The model of MMS interconnection is showen in figure 4.

6 Conclusion

In order to design and implement a reliable and efficient MMS integrated communication system, it is necessary to use formal methodology for design and analysis. In this paper, STD and Petri net is applied to describe and analyze the interaction of MMS confirmed service between heterogeneous networks . The results show that the principles of protocol software development mentioned above are effective and efficient. The work of this paper provides a basis for further analysis, which will improve the system with higher performance.

7 References

1   *ISO/IEC 9506 Manufacturing Message Specification*, Part 1 and Part 2, (1990).
2   Fei_Yue Wang, Performance Analysis of MMS Using GSPN, *Proceedings of the IEEE Conference on Robotics and Automation*, (1991) 1573 — 1578.
3   Guanqun Gu, MMS Communication System in Heterogeneous Network, *Computer Research and Development*, No.3, (1995) 1 — 7.
4   P.Pleinevaux, An Analyis of the MMS Object Model, *IEEE Transaction on Industrial Electronics*, Vol. 41, No. 3, (1994) 265 — 268.

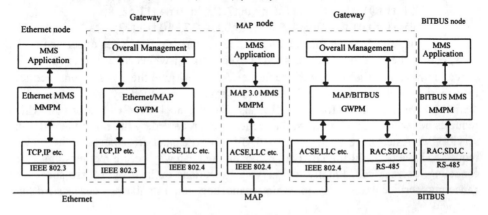

Fig. 4. Model for MMS Interconnection

Assembly Automation

Analysis Method For Automation of A Socket Assembly System - A Case Study

Ng Kok Weng

B.Eng. (Mech & System Eng.), M.Sc (Manufacturing System Eng.)
Production Systems Group, National CAD/CAM Centre
SIRIM Berhad, 1, Persiaran Dato' Menteri, P.O. Box 7035,
Section 2, 40911 Shah Alam, Malaysia

Hamiraj Fahry Abdul Hamid

B.Sc (Industrial Engineering)
Production Systems Group, National CAD/CAM Centre
SIRIM Berhad, 1, Persiaran Dato' Menteri, P.O. Box 7035,
Section 2, 40911 Shah Alam, Malaysia

Abstract. There is a need to automate assembly systems to enable increase in productivity and reduction of manual operation processes. Labour intensive industries are currently facing severe problems of maintaining their skilled labour force as the labour market gets more competitive. Though automation maybe the solution to the current shortages of skilled labour force problem, investments on automated assembly systems require detail and careful study and planning to ensure the well returns of the investments. This means automation of assembly systems must also lead to significant improvement in rate of assembly and a reduction of manpower. Currently there are very few guidelines and assistance to industry in automating their assembly systems. Hence this paper look into the analysis method to be used as a reference platform for Malaysian industry in automating their assembly systems especially on manual intensive assembly processes. Though the technology of this method presented is not something new but the method used is applicable in the process of making decision on automation. Currently, this method was applied to a domestic electrical appliances manufacturer in Malaysia and the results are discussed in this paper.

1.0 Introduction

In the process of automating an assembly system which involves various assembly processes such as inserting of parts, screwing, positioning of parts, etc., deciding which process to be automated is critical. A well analysed and study on the assembly systems is necessary and automating the critical process will lead to significant improvement on productivity and reduction of manpower. In the process of analysing the assembly system, usually weaknesses such as bottleneck stations, high reject processes, manual intensive processes, starvation stations, delicate and stressful processes, hazardous processes, high speed processes, etc. are the main targets of automation. However in most cases, a combination of all these weaknesses occurs. Depending on the severity of the weaknesses, different companies will have different critical weaknesses. Some companies will have significant amount of manual intensive processes and some may have severe bottlenecks' problems but all of them have limited amount of allocation budget. Hence it is vital that the investment in automation will give the biggest impact in improving their productivity.

2.0   Assembly Systems Overview

Basically an assembly system has various activities of assembly. The activities may involve a combination of manual operation, semi-automated operation and fully automated operation [5]. The product involved will be of variants or totally different type. Hence inventories, work in progress, finished goods, etc. will be of variety types too.

Currently in Malaysia, assembly systems that are manually intensive will have operators positioned at various places of the production flow. Each operator will either assemble the complete product or assemble a portion of the product (assemble in modules[6]). In the case of assembling a portion of the product (the assembly is divided into smaller process). Generally the idle assembly process flow would be the cycle time for the work of each operator is equally divided to ensure a balanced assembly line [1]. The line balancing or workstation balancing of any assembly systems will require balancing manual operation processes and machine operation processes cycle time. The balancing between machine and human operators work cycle time requires simulation analysis and is a very difficult process [2]. The machine has the ability to work 24 hours a day (non-stop) with preventive maintenance stoppages occasionally which is planned. The machine will work much faster than a human operator and without varying in their performances.

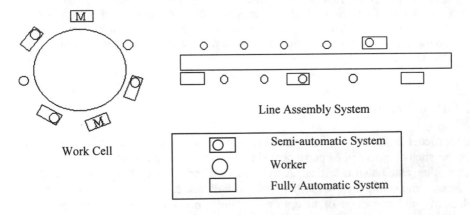

Figure 1 : The Common Layout of Assembly Systems

1140

Based upon this, to balance these two categories of assembly approach, it is preferable to match the human operators' rate to the machine working rate. This is done because, for cost effective systems, machine utilisation must be maximum or 100% (for ideal case) . Figure 1 shows 2 of the common assembly systems layout for a company in Malaysia.

Hence, increase of human operators is carried out in process of matching the work rate of the machine. In local industries, balancing work load is usually neglected hence problems like bottlenecks and starvation occur. Most local (Malaysian) and small industries are interested in automating their assembly processes. However their budget is very tight and assembly systems are usually not well planned and organised. Therefore to automate industries that have these problems and gradual introduction of automation with detail analysis on balancing the system as well as partitioning the system for the process of step-by-step automation. The final aim is to automate the entire system with minimal effect on the output and spread out the cost to avoid sudden financial burden due to heavy investment. The further advantage of this automation approach is as technology and market favour changes in time, the system can adapt to this changes because of the gradual introduction of automation. This lead to a significant reduction of risks involved on the capital investment for the company concerned.

3.0 Analysing Assembly Systems

This automation approach starts with an observation and identification of assembly process which is "generic". "Generic" in this context means applicable to all the product variants that are being produced by the company or can be adapted to other product variants with minimal modification and cost. The "common" process usually involved similar parts or required similar assembly process such as screwing, riveting, etc. Hence the automation process will also be eased or enhanced by application of group technology, design for assembly, etc. onto the products or part of the product itself.

With this in mind, the analysis system will require the follow up of processes like design for assembly, simplification of assembly process, probable redesigning of layout, group technology, etc. Hence after identifying common processes, a work study is required to detail out the movement of an operator's arm and body in the process of assembling the product. Generally each worker has to do a specified movement of hands and body (sometimes even legs) to assemble the product parts. All this micro study on operators in performing their work will be listed down for analysis purposes. This study process is part of a well-known approach known as Method Time Measurement (MTM) where micro motion of worker is used to analyse a particular operation. Hence with MTM, the cycle time for each micro motion of individual worker is obtained.

The time and operation data obtained via observation and MTM must be counterchecked with the real cycle time (obtained by carrying out work study). Appendix 1 illustrates an example of a complete assembly process for a socket (all data are obtained from Industry) and Appendix 2 illustrates the cycle time obtained from MTM analysis in comparison to the actual cycle time and shows how categorisations are made to decide on which process to automate and what to automate. Using process categorisation, the decision on the level of automation can also be made easier and the overall assembly system can be simplified. With this process categorisation, further analysis like cost analysis, automation decision and implementation, stages of automation and various consideration can be made to ensure the company will have a gradual, optimal, efficient and effective application of automation processes over their existing manual intensive system.

The entire approach is very dependent on the accuracy of the data obtained and the observations made. However consideration must also be made on the transfer the product of a non precise transfer system to a highly precise transfer system during planning for automation. The transferring methods are can be various and should be looked into in details.[3, 4]

In short, the entire process is summarised in Figure 2.

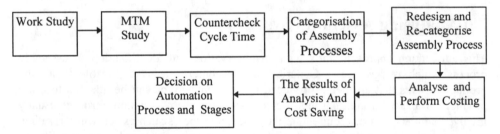

Figure 2 : The flow process of analysis system for automating assembly
systems

In case study of socket assembly, based upon the categorisation process (refer Appendix 3), the categories of assembly process for a socket can be listed as below :

1. Base Positioning
2. Metal Parts Assembly
3. Dolly Switch Assembly
4. Plate Positioning
5. Shutter Assembly
6. Assembling of Plate to Base
7. Screwing Process
8. Final Assembly

Hence by accumulating all the assembly sub-processes under the above mentioned category of processes, the cycle time for each category can be listed as below :

Activities	Cycle Time(s)	% of Total Time
1. Base Positioning	3.10	5.69
2. Metal Parts Assembly	5.84	10.70
3. Dolly Switch Assembly	13.32	24.41
4. Plate Positioning	2.19	4.01
5. Shutter Assembly	5.29	9.70
6. Assembling of Plate to Base	2.55	4.68
7. Screwing Process	16.24	29.77
8. Final Assembly	6.02	11.04

It is important to know that the cycle times of these categories of assembly processe are obtained by adding all the sub-processes under each category. All sub-processe cycle time is obtained via MTM and hence though there maybe some slight error or inaccuracy, the ratio of cycle time among the processes is consistent. Therefore the percentage of total time each category of processes consumed is accurate.

With this data, a chart (refer Figure 3) can be plotted to analyse the assembling process situation more clearly.

Title: **Assembly Activities vs Percentage of Assembly Time Chart**
For Socket E15

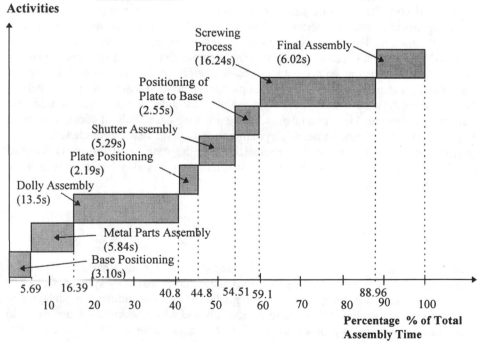

Figure 3 : The Categorising of Assembly Processes For Automation

By referring to the chart above, it is very vivid that screwing process is the most critical and time consuming processes among all the processes involved. The dolly assembling process is the second most time consuming process. It is very important to note that this case study is carried out on an assembly process where one worker assembles a complete socket. So all process above is done by a single worker and the entire line consists of several workers. Hence if there are any processes in this particular system that are to be automated, the most time consuming processes should be the priority. This is because the amount of cycle time for the overall processes will be significant reduced if the slow processes are removed and automated. When these slow processes are automated, the process can be perform in parallel. In this way, with the output of the system will be significantly increase. However it is vital to acknowledge that not all slow processes or bottlenecks can be automated. Automation is influenced by the cost involves and to the complexity of the assembly process concerned.

4.0 Decision and Planning for Automation

All analyse results lead to a clear indication of which assembly processes should to be the priority whenever a company wanted to automate. However the final decision is still dependent on the cost and return (pay back) evaluation. Only cost feasible processes should be automated. The method of automation is to start small and in stages. With reference to the socket case study above, screwing process seems to be the main bottleneck for the socket assembly. If the automation of screwing processes is feasible in terms of cost, then considerations on the cycle time for the prospective automated screwing process must be determined. The key point is to automate with the assembly line well balanced. Hence it is very critical to ensure the objectives of the automation process are met with, because some may want to automate to reduce operator and some may want to increase assembly output , etc. Therefore in the process of planning for automation, a chart to indicate the cycle time of the existing assembly system and the future assembly system must be developed side by side to allow comparison and frequent counter checking as shown in Figure 4 for the case study of socket assembly. Figure 4 shows the entire assembly system and it is vivid that the bottleneck for the overall system is at the testing and label sticking. In this case study, the stress is on the main assembly where 18 workers are working. However it is also important to balance the overall system after improving the main assembly.

5.0 Conclusions

This approach assists significantly in making decision to automate a particular assembly and in this case we tried it on domestic electrical appliance assembly systems. Currently there is no clear assistance, references or guidelines for the industry in Malaysia to evaluate their assembly systems and to decide how and which processes to automate. In fact, during analysis, any processes that can be removed or replaced with a better process that is designed for assembly must be take into account seriously. The product itself should be analysed using Design For Assembly (DFA). The current method mentioned here should not be a main consideration point for automation but should be a complement to other various existing method such as DFA, etc.

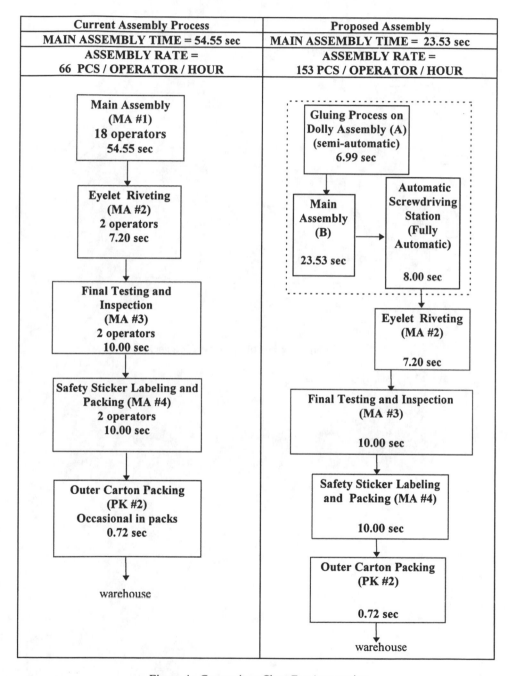

Figure 4 : Comparison Chart For Automation

Appendix 1

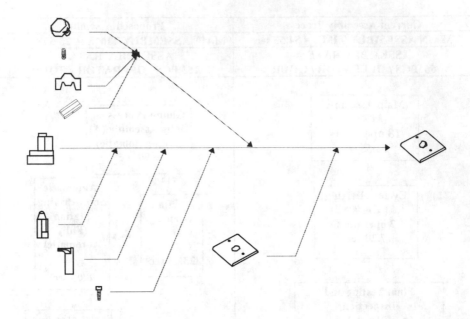

Complete assembly process for a socket

Main Activities	MTM Activities	Total Estimated Time	% of Total	Actual Time
Base Positioning	1. Pick and Place Base (45 T.M.U.) 2. Inspect Base (15 T.M.U.) 3. Turn Base (5 T.M.U.) 4. Place Base Onto Jig (20 T.M.U.)	85 T.M.U. (3.06 sec)	5.69	86.15 T.M.U. (3.10 s)
Metal Parts Assembly	5. Pick and Place SA #1 (55 T.M.U.) 6. Pick and Place SA #2 30 T.M.U.) 7. Pick and Place SA #3 55 T.M.U.) 8. Pick and Place SA #4 (20.T.M.U.)	160 T.M.U (5.76 sec)	10.70	162.16 T.M.U. (5.84 s)
Dolly Switch Assembly	9. Pick and Hold Bridge (45 T.M.U.) 10. Get Stick With Grease (40 T.M.U.) 11. Applying Grease (25 T.M.U.) 12. Place Bridge Onto Base (20 T.M.U.) 13. Pick and Hold Dolly (45 T.M.U.) 14. Pick and Hold Spring (20 T.M.U.) 15. Insert Spring (30 T.M.U.) 16. Pick and Place Yoke (40 T.M.U.) 17. Visual Control (15 T.M.U) 18. Grip Dolly Tightly (10 T.M.U.) 19. Turn Upside Down (5 T.M.U.) 20. Place Onto Base (30 T.M.U.) 21. Test Dolly (10x4 T.M.U.)	365 T.M.U. (13.5 sec)	24.41	369.93 T.M.U. (13.32 s)
Plate Positioning	22. Pick and Place Plate (45 T.M.U.) 23. Visual Inspection (15 T.M.U.)	60 T.M.U. (2.16 sec)	4.01	60.81 T.M.U. (2.19 s)
Shutter Assembly	24. Pick Shutter (45 T.M.U) 25. Pick Shutter Spring (20 T.M.U) 26. Insert Spring Into Shutter (20 T.M.U.) 27. Assemble Onto Plate (30 T.M.U.) 28. Control Motion (30 T.M.U.)	145 T.M.U. (5.22 sec)	9.70	146.96 T.M.U. (5.29 s)
Positioning of Plate to Base	1. Place Plate onto Base (40 T.M.U.) 2. Visual Control (15 T.M.U.) 3. Flip Jig (15 T.M.U.)	70 T.M.U. (2.52 sec)	4.68	70.95 T.M.U. (2.55 s)
Screwing Process	4. Pick and Place Screws (65 T.M.U.) 5. Fasten Screws into Holes (5x4T..M.U.) 6. Go to Next Hole (20x3 T.M.U.) 7. Return Unused Screws (20 T.M.U.) 8. Get Screwdriver (70 T.M.U.) 9. Fasten Screws (30x4 T.M.U.) 10. Transport Screwdriver (30x3 T.M.U.)	445 T.M.U. (16.02 sec)	29.77	451.01 T.M.U. (16.24 s)
Final Assembly	11. Flip Back Jig (15 T.M.U.) 12. Take out Assembled Socket(20 T.M.U.) 13. Test Dolly (10x4 T.M.U) 14. Pick and Place Rivet (55 T.M.U.) 15. Move Socket to Conveyor (25 T.M.U.) 16. Press Counter (10 T.M.U.)	165 T.M.U. (5.94 sec)	11.04	167.23 T.M.U. (6.02 s)

Categorisation of Socket E 15 Assembly Processes

References :

1.   Boothroyd G., and A.H. Redford, (1982), Mechanized Assembly, McGraw Hill Publishing Co. Ltd.

2.   Buzacott, J.A., L.E. Hanifin, (December 1978), "Transfer Line Design and Analysis" - An Overview," Proceedings, 1978 Fall Industrial Engineering Conference of AIIE.

3.   Buzacott, J.A., (1967), "Automatic Transfer Lines with Buffer Stocks", International Journal of Production Research, Vol.5, No.3, pp.183-200.

4.   Groover, M.P., (1987), "Automation, Production Systems and Computer Integrated Manufacturing"

5.   Groover, M.P., (1975), "Analyzing Automatic Transfer Machines", Industrial Engineering, Vol. 7, No.11, pp. 26-31.

6.   Riley, F.J., (1983), "Assembly Automation", Industrial Press, Inc., New York.

Design and Analysis of a Flexible Workcell for Automated Assembly of Electric Motors

Shamsudin H. M. Amin, Mohamad Noh Ahmad, Zaharuddin Mohamed,
Jameel Mukred, Mohammad Ghulam Rahman
Center for Artificial Intelligence and Robotics, Faculty of Electrical Engineering.
University Technology Malaysia, Locked Bag 791,80990 Johor Bahru, Malaysia
Tel: +607-5505119 Fax: +607-5566272
sham@fkeserv.fke.utm.my

Abstract. Current manual assembly process of electric motors has been found to be incapable of keeping pace with the requirements of the manufacturing industry which aims to be globally competitive. The automation of the assembly process is the way to be in the competitive market. In this paper we propose the design of a flexible workcell for the assembly of different electric motors. An analysis of the design has been performed. The Robot Time and Motion (RTM) methodology has been implemented to compute the overall workcell cycle time. This study will be the basis of the implementation of an automated electric motor assembly factory.

1. Introduction

Traditionally the assembly of a composite product (like motors of automobiles) is mostly done through human operation. This type of job is highly repetitive, extremely boring and sometimes tedious. Usually the quality of the job cannot be constantly maintained due to human fatigue after long hours of work. So in order to be in the competitive global market it is necessary to maintain product quality, lower the cost and reduce the lead time ahead of the competitors. This can be achieved by automation of the assembly system. Automated assembly of composite products has been a goal of robotics researchers since the beginning of the field. In this paper a layout design of a flexible workcell for automated assembly of motors is presented.

A great deal of research have been done to develop high-level automatic generation of the assembly sequence plan([1] -[7]), where mating and insertion constraints are considered from only the product component side and assembly sequence has been considered as the reverse of the disassembly sequence of that product ([8], [9]). But in designing a workcell layout for assembling a product task assignment [10], part routing [11], fixturing or gripping, material transferring system, sequencing of the workstations, scheduling of the operations of different machines and other production related criteria are also to be considered.

The following alternative conditions relating to a design process which concern the relations between the product and the system may be identified.
A: Existing product - New system (system design after the product).
B: New product - New system.
C: New product-Existing system(product design adapted to the existing system).

The methods concerning assembly system design may be briefly classified into the following general categories. The relations to the conditions above are shown (A and/or B):

a) Simultaneous design of product and system (B).
b) Methods for total system design (A, B).
c) General problem solving methods relating to assembly system design (A, B).
d) Methods for specific types of assembly systems-mainly flexible automatic assembly system(A,B).
e) Expert system models, often related to the methods in category d (A).
f) Case studies of design and development of assembly systems by companies (A).
g) Human oriented methods often based on technical ideas where the assembly system is regarded as a combination of social and technical systems (A).

A system design can be considered as an integration of two or more criteria. Like System Design Method [12], our design method is a combination of the principle category of b, c, d and e.

An assembly system proposed at Draper Laboratory ([13], [14]) applies only in linear assembly plans (where only one component is added in a single operation to the assembly). An improved design was proposed in [15] where works were being done on different subassemblies in workstations, as shown in Fig.1. Two subassemblies were built and then assembled .

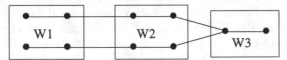

Fig.1. Assembly system with subassemblies

For the transfer of the assembly product, conventional linear assembly line had been well investigated ([20], [21]). A cyclic assembly line where a central conveyor with workstations situated in circular form at outside of the conveyor was proposed in [16] to reduce the work-in-process inventory accumulation as shown in Fig.2. In this paper an assembly workcell design with transfer systems resembling an α-shape with rotary indexing conveyor is chosen.

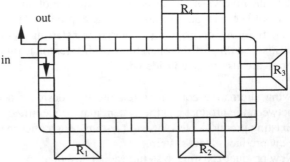

Fig.2 . Cyclic assembly line with robots R1,...,R4

From the exploded view of different motors, process flow for each motor was decided and based on that the required workstations of the workcell were designed. Accordingly a layout design and analysis were performed. The analysis and cycle time computation were performed with alternator. The procedure to achieve our target of study is shown in Fig.3.

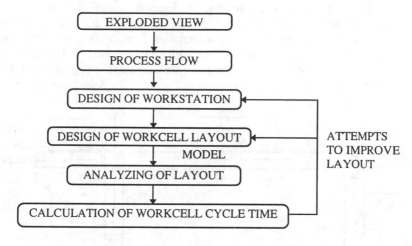

Fig.3. Work Procedure

Robot time and motion (RTM) ([17], [18], [22]) method had been developed for estimating the time that a given robot manipulator model needs to complete a given task cycle. In order to determine the time required to assemble one alternator, RTM methodology had been used. In this paper, design of the workcell is explained in section two, analysis of the workcell design is in section three and results are discussed in section four.

2 . Workcell Design

The main aim is to design a flexible assembly workcell that can be used to assemble different motors used in automobiles. We proposed a workcell that can be used to assemble alternator, starter and wiper motors of automobiles.

The proposed layout of the workcell is based upon the following motors:
1. Ford autolite alternator [23].
2. Typical Ford or American Motors two speed wiper motor [24].
3. Typical positive two engagement movable pole shoe starter.

Following a careful study of each of the above motor's exploded view, preassembled parts were identified and process flow was recognized. According to the motor part's dimension, weight and precedence constraints, a workcell with six workstations was proposed. The number of workstations was estimated from the need of the assembly tasks of the motor that has the largest number of parts. The proposed workcell consists of four robot workstations and two press stations distributed in sequential layout [19] basis.

At the beginning, the designed workcell was a line-shaped assembly process, but it is found that in case of wiper motor, workstation 1 is lightly loaded whereas workstation 6 is overloaded. By following the procedure of Fig.2, ∝-shaped conveyor is designed to perform the assembly process, so that workstation 1 can share the load with workstation 6 through the rotary intersection conveyor part, which indexed during the pallet forwarding from workstation 1 to workstation 2 and during pallet forwarding from workstation 6 to workstation 1. Pallet with fixtures were used to transfer the work-in-process on the conveyor. Complete workcell layout is shown in Fig.4. Design of workstation 1 and workstation 2 are shown in Fig.5 and Fig.6 respectively.

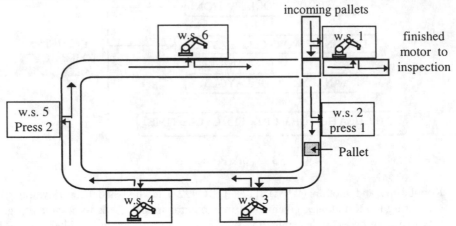

Fig. 4. Workcell Layout

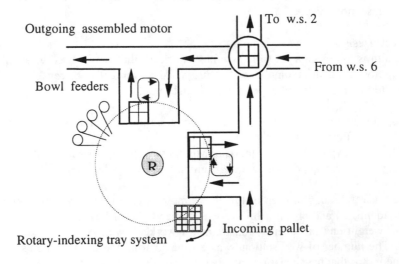

Fig. 5. Workstation 1 layout

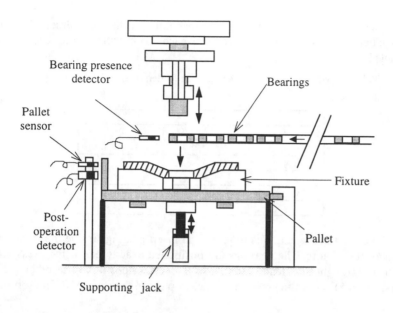

Fig.6. Work station 2 (press station 1)

Trays, bowl feeders(B.F.) and linear vibratory feeder were used in the workstations to supply the different parts to the robots and press units. Large and weighty parts were supplied by trays and smaller parts were supplied by the bowl feeders. Rotary-indexing tray system is used in the workstation to place the trays at certain position within the work envelope of the robot. The linear vibratory feeder placed the bearing at a predetermined position below the press. From the analysis of the exploded views it is found that alternator is made of 14 different types of parts, starter motor is of 26 different parts and wiper motor is made of 36 different parts. The number of trays and bowl feeders required for each motor are shown in Table-1.

Table 1. Trays and bowl feeders(B.F.) for each workstation(w.s.)

	Alternator		Starter motor		Wiper motor	
	Tray	B.F.	Tray	B.F.	Tray	B.F.
w.s. 1	3	0	4	0	3	4
w.s. 2	0	2*	0	0	0	0
w.s. 3	0	4	1	3	3	5
w.s. 4	1	0	3	4	0	6
w.s. 5	0	0	0	0	0	0
w.s. 6	1	3	2	3	2	4

* linear vibratory feeder

In order to assemble all the three types of motors through the same assembly line, maximum number of trays and bowl feeders for each work station were selected as shown in Table 2.

Table 2. Max. number of trays and bowl feeders(B.F.)

	Tray	B.F.
w.s. 1	4	4
w.s. 2	0	2*
w.s. 3	3	5
w.s. 4	3	6
w.s. 5	0	0
w.s. 6	2	4

Workstation 2 i.e. first press station is to press and insert bearings inside the front and rear housings of the motors. The bearings are positioned exactly below the press tool by passing them through an extended tunnel to the correct position. The other press station (workstation 5) is used to press and fit heavy parts to the shaft of the motors.

In each robot workstation Staubli Unimation RX90 industrial robot with multi-gripper end-effector was used to handle the different parts. The Staubli Unimation RX90 industrial robot was selected for its following characteristics:
— Six axes revolute.
— Work-envelope: reach at wrist center 900 mm max. (between joint 2 and joint 5)
  reach at wrist center 290 mm min. (between joint 2 and joint 5)
— Load capacity: 6 kg at nominal speed and 9 kg at low speed.
— Maximum Cartesian speed : 1.5 m/s.

3. Analysis
The proposed layout can be used to assemble any or all of the three types of motors. But analysis was made for only the assembly of alternator. For the assembly of alternator, a pallet with four fixtures was devised to transfer the work-in-process on the conveyor system. All the six work stations were used to assemble alternator. In the robot work station multi-gripper end-effector robot was used to assemble different parts (light parts, heavy parts, nuts, screws etc.). Also turning over of assembly-in-process was done in workstation 4 and workstation 6. In the first press station front bearing was pressed into the front housing and rear bearing was pressed into the rear housing. In workstation 5, fan and pulley assembly was pressed onto the rotor shaft over the assembled parts.

The amount of time required for the work cycle is an important consideration in planning of the workcell. The cycle time determines the production rate of the job which is a significant factor in determining the economic success of replacing the existing system with the new one.

The time taken by the robot to complete an assembly task in a specified workstation was calculated by the RTM methodology. The total time is that consumed in all the workstations and that taken by the pallet to transfer from incoming to the outgoing on the conveyor. Total conveyor length and part transfer time were also determined.

Then the total cycle time for a single alternator was computed. The production rate is calculated with the assumption that the parts transferred in a synchronous way. The RTM calculation for workstation 4 (robot 3) is shown in Table.3.

Table 3. Calculation of cycle time using R.T.M. methodology.

Home position at a height	=	217 mm/s

Robot velocity without load (in arc)	=	800 mm/s
Robot velocity with load (in arc)	=	700 mm/s
Robot velocity without load (in straight line) =		100 mm/s
Robot velocity with load (in straight line)	=	80 mm/s

Seq.	RTM symbol	Weight (lb)	DISTANCE							ELEMENT TIME		Description
			ARC LENGTH			STRAIGHT LENGTH			velocity	for (AL)	for (SL)	
			angle	radius	(mm)	1ST	2ND	(mm)	mm/s	sec.	sec.	
1	R1		45	770	604.5				800	1.2		Move in arc from home position to fixture I
2	R1					770	1110	340	100		3.8	Move straight to fixture I
3	R1					217	22.5	194.5	100		2.3	Move down
4	GR1									0.1		Grasp assembled parts [(3)+(4)+(5)+(2)*3] of fixture I
5	M1d	1.3				22.5	217	194.5			3.0	Unload assembled parts from fixture I >1lb
6	M1	1.3				1110	900	210	80		3.2	Move back straight
7	M1t	1.3									2.0	Turn over the assembled parts
8	M1	1.3	13	900	204.1				700	0.6		Move in arc to fixture III
9	M1d					217	104	113			3.0	Load on assembled parts on fixture III
10	R1					104	217	113	100		1.5	Move up
11	R1					900	770	130	100		1.7	Move back straight
12	R1		108	770	1450.7				800	2.2		Move in arc to the tray system
13	R1					217	30	187	100		2.3	Move down
14	GR1									0.1		Grasp part (1) >1lb
15	M1	2				30	217	187	80		2.9	Move up
16	M1	2	108	770	1450.7				700	2.7		Move in arc to fixture III
17	M1	2				770	900	130	80		2.2	Move straight to fixture III
18	M1d					217	135	82			3.0	Load on assembled parts on fixture III
19	R1					135	217	82	100		1.2	Move up
20	R1					900	770	130	100		1.7	Move back straight
21	R1		32	770	429.8				800	0.9		Move in arc to home position
			Total time for movement in arc				=			18.8		
			Total time for movement in straight line				=				23.0	
			Total cycle time for work station # 4				=			41.7		

4. Results and Discussion

Total time computed for the production of a single alternator is 10.9 minutes. By the use of RTM method, cycle time for each robot work station was computed. It is found that each workstation has different cycle time and workstation 6 (robot 4) took the highest time (187.1 sec) to complete the assigned assembly work as shown in Table.4. So the transfer mechanism was considered synchronous in order to avoid work-in-process accumulation in other workstations.

If 26 working days and 7 active hours per shift per day are considered, and the time span between adjacent two motors is 187.1sec., then:
 production per day =1+(420-10.9)/3.118=132.2 units, and monthly production rate is 3437 units of alternator for a single assembly line. If three shifts per day are considered then the production rate will become 10312 units of alternator per assembly line.

Table 4. Cycle time of the workcell using (R.T.M.) method.

	Cycle Time (second)
1st work station (robot 1)	65.1
2nd work station (press station 1)	4.0
3rd workstation (robot 2)	111.1
4th workstation (robot 3)	41.7
5th work station (press station 2)	3.0
6th workstation (robot 4)	187.1
total transfer time over conveyor	244.7

5. Conclusion

An assembly workcell to assemble different types of motor was proposed in this paper. Selection of robot, workstations were also described in this paper. An analysis of the workcell design was also performed. Even though analysis of the workcell was done for the assembly of alternator, the same line can efficiently be used to assemble starter and wiper motor as well. By the RTM method, the time required to complete the assembly of one alternator was computed. Although this was not done on-line, this investigation can be the basis of the implementation of an automated assembly factory.

References

1.   J. D. Wolter, A Combinatorial Analysis of Enumerative Data Structures for Assembly Planning, *Proceedings of IEEE International Conference on Robotics and Automation*, 1991, pp. 611-618.
2.   S. Chakrabarty and J. Wolter, A Structure Oriented Approach to Assembly Sequence Planning, *IEEE Transactions on Robotics and Automation*, January 1995.

3.	D. F. Baldwin, T. E. Abell, M. C. M. Lui, T. L. De Fazio and D. E. Whitney, An Integrated Computer Aid for Generating and Evaluating Assembly Sequences for Mechanical Products, *IEEE Transactions on Robotics and Automation*, Vol.7, No.1, 1991, pp.79-94.

4.	B. Romney, C. Godard, M. Goldwasser and G. Ramkumar, An Efficient System for Geometric Assembly Sequence Generation and Evaluation, *Proceedings of 1995 ASME International Conference on Computers in Engineering*, pp. 699-712.

5.	R. Hoffman, Automated Assembly in a CSG Domain, *Proceedings of IEEE International Conference on Robotics and Automation*, 1989, pp. 210-215.

6.	M. Goldwasser, J. C. Latombe and R. Motwani, Complexity Measures for Assembly Sequences, *Proceedings of IEEE International Conference on Robotics and Automation*, 1996, pp.1581-1587.

7.	J. Wolter, S. Chakrabarty and J. Tsao, Mating Constraint Languages for Assembly Sequence Planning, *Proceedings of IEEE International Conference on Robotics and Automation*, 1992, pp. 2367-2374.

8.	L S. H. Mello and A. C. Sanderson, A Correct and Complete Algorithm for the Generation of Mechanical Assembly Sequences, *IEEE Transactions on Robotics and Automation*, , Vol. 7, No. 2, 1991, pp. 228-240.

9.	R. H. Wilson and J. C. Latombe, Geometric Reasoning About Mechanical Assembly, *Artificial Intelligence* 71(2), 1994 .

10.	A. Agnetis, F. Nicol'o, C. Arbib and M. Lucertini, Task Assignment in Pipeline Assembly Systems, *Proceedings of IEEE International Conference on Robotics and Automation*, 1992, pp. 1133-1138.

11.	A. Agnetis, C. Arbib, M. Lucertini and F. Nicol'o, Part Routing in Flexible assembly Systems, *IEEE Transactions on Robotics and Automation*, Vol. 6, No. 6, 1990, pp. 697-705.

12.	 M. Lundstrom, M. Bjorkman and C. Johansson, A Method for Assembly System Design Including an Integrated Computerized Design Support, *CompEuro Proceedings on Computers in Design, Manufacturing and Production*, 1993, pp. 52-61.

13.	D. E. Whitney, T. L. De Fazio, R. E. Gustavson, S. C. Graves, C. Holmes and J. C. Klein, *Computer Aided Design of Flexible Assembly Systems*, Report No CSDL-1947, Cambridge, Massachusetts, August 1986.

14.	D. E. Whitney, T. L. De Fazio, R. E. Gustavson, S. C. Graves, K. Cooprider, J. C. Klein, M. Lui and S. Pappu, *Computer Aided Design of Flexible Assembly Systems*, Report No CSDL-R-2033, Cambridge, Massachusetts, January 1988.

15.	V. Minzu, J. M. Henrioud, Systematic Method for the Design of Flexible Assembly Systems, *Proceedings of IEEE International Conference on Robotics and Automation*, 1993, pp. 56-62.

16.	G. Finke and L. Dupont, Combinatorics of Cyclic Assembly Systems, *Proceedings of IEEE International Conference on Systems, Man and Cybernetics*, 1993, pp. 297-301.

17.	S. Y. Nof and R. L. Paul, A Method for Advanced Planning of Assembly by Robots, *Proceedings of SME Autofact*, 1980, pp. 425-435.

18.	S. Y. Nof and H. Lachtmann, The RTM Method of Analyzing Robot Work, *Industrial Engineering*, April 1982, pp. 38-48.

19.	G. Marlin, A Study Case of Critical Analysis, Relating to an Automatic Assembly System, *Proceedings of 6th International Conference on Flexible Manufacturing Systems*, November 1987, pp. 169-176.

20.	A. C. Hax and D. Candea, *Production and Inventory Management*, Prentice-Hall, Englewood Cliffs, New Jersey, 1984.

21. A. Kusiak, *Intelligent Manufacturing Systems*, Prentice-Hall, Englewood Cliffs, New Jersey, 1990.

22. M. P. Groover, M. Weiss, R. N. Nagel and N. G. Odrey, *Industrial Robotics Technology, Programming and Applications*, McGraw-Hill Book Company, 1988.

23. W. K Toboldt and L. Johnson, *Automobile Encyclopedia: Fundamental Principles Operation, Construction, Service and Repair*, Goodheart-Wilcox Company Inc. 1972.

24. W. K Toboldt and L. Johnson, *Automobile Encyclopedia, Fundamental Principles Operation, Construction, Service and Repair*, Goodheart-Wilcox Company Inc. 1975.

Modeling and Estimation of a Pneumatic Positioning System For Lumber Processing

Xiaochun George Wang   and   Dorothy Wong

National Research Council of Canada
Integrated Manufacturing Technologies Institute, Western Lab,
3250 East Mall, Vancouver, B.C. Canada V6T 1W5
Tel.: 01-604-221-3061, Fax: 01-604-221-3001
e-mail: george.wang@nrc.ca

Key words: wood industry, lumber positioning, pneumatic system, position, force compliance, control, modeling, Kalman filtering, adaptive control, self-tuning control, singular pencil models.

Abstract. This paper presents results of a research project on the modeling, estimation and control of a pneumatic lumber positioning system. Such systems are used in machinery in saw mills and other applications. Control of such systems present a major challenge due to the nonlinearity and variation in dynamics. A dynamic model and the simultaneous on-line recursive estimation of the model parameters and internal variables (state variables) are developed, using the singular pencil models. This approach has shown superior performance on the real system than conventional state space model based approaches, and has show potential for improved self-tuning control based on estimated dynamic parameters as well as the state variables.

1 Introduction

This paper is concerned with the position and force control of heavy rollers of a log feeding machine for lumber production. Typically, such a machine employs from 20 to 60 such rollers to position the log for sawing based on laser scanned information about the logs. Both hydraulic and pneumatic actuators are used, often in series connection, for fast positioning and soft handling of lumber. This research focuses on the use of pneumatic actuators to achieve both force compliance and fast and accurate positioning, for reasons of reliability, efficiency and cost reduction. The accurate control of the position, however, is a challenging problem due to the nonlinearity induced by the pneumatic system and the variation of dynamics. A test facility has been built where part of the real machine is integrated with the most difficult roller mounted. The project has undergone through the first phase successfully and is proceeding to the second phase of implementation. The first phase of the project focused on the modeling and system identification of the system. Many set of data was taken under different operating conditions with different input signals such as step, pseudo-random signals etc. Measurement of positions and pressure at different condition was then used for estimating the parameters of a linearized system model. The model constructed from the estimated parameters are then used with the Kalman filter to give the optimal estimation of the state variables. Since the parameters are time varying for such a system, it presents a problem for state estimation and later on the control based on state feedback, since one set of parameters is only good for certain operating condition, not good for other operating condition. All this is complexed with the nonlinearity exist the system.

The problem of parameter and state estimation is fundamentally nonlinear in the state space model and the common input-output model. A new class of model, the Singular Pencil (SP) model is introduced. The SP model contains the input-output and state space model as subsets, but has certain advantages. One of the advantage is shown in the simultaneous state and parameter estimation where, when formulated in the SP model format, the simultaneous state and parameter estimation problem can be formulated into a linear problem. An Ordinary Kalman filter is used for this problem, giving good convergence. Since the parameters and state are estimated at the same time, the estimated states are valid over a wider range of the operating conditions. The optimal adaptive control can be achieved using the estimated parameters and states, in the state feedback optimal design.

2 The Test Facility

A test facility for development of the modeling, estimation and control was setup, as shown in Figure 1. The function of this pneumatic system is to position wood handling rollers at high speed, yet maintaining accuracy and force compliance. It consists of the cylinder, and two valves that control the air flow to and from the top chamber and bottom chamber of the cylinder. The position of the cylinder, read from a position sensor, can vary from 0 inch at the bottom to 12 inches at the top, corresponding to 0.0 V to 10.0 V. This cylinder goes upward when the bottom valve voltage increases and goes downward when the top valve voltage increases, with the two valve voltages ranging from 0.0 to 5.0 V sent by the computer. Due to the non-linearity of the system, there is a delay between the input voltages and the movement of the cylinder, which is present in a form of hysterisis.

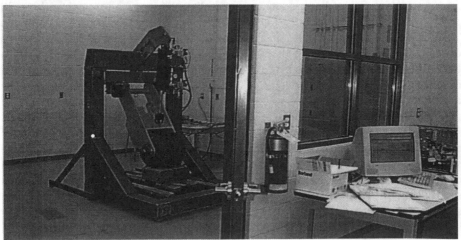

Fig. 1. The test rig and the control system

3. Modeling Of The System

A Singular Pencil Model is used to describe the relationship between the inputs (top and bottom voltages) and output (position of the roller) of the system. It is of the

form
$$[E - FD, G]\begin{bmatrix} x \\ w \end{bmatrix} = P(D)\begin{bmatrix} x \\ w \end{bmatrix} = 0, \tag{1a}$$

where $x \in R^n$ is the *internal (auxiliary) variable* vector, and $w \in R^{p+m}$, with known p and m, is the *external variable* vector, usually consisting of input/output variables of the system. E and F are $(n+p) \times n$ matrices, G is $(n+p) \times (p+m)$, and D is a linear operator. Equation (1a), representing a multi-input, multi-output system, also includes state-space models as a special case. When D is replaced by z, it is written

as
$$\left\{ \begin{bmatrix} A \\ C \end{bmatrix} - \begin{bmatrix} I_n \\ 0 \end{bmatrix} z \right\} x_k + \begin{bmatrix} 0 & B \\ -I & D \end{bmatrix} \begin{bmatrix} y_k \\ u_k \end{bmatrix} = 0, \tag{1b}$$

where $u_k \in R^m$ is the input vector, $y_k \in R^p$ is the output vector, and (A,B,C,D) are the state-space matrices. Another special case is the class of matrix fraction or ARMA (Auto-Regression Moving Average) models of the form

$$\sum_{i=0}^{n} A_i z^{n-i} y_k = \sum_{i=0}^{m} B_i z^{m-i} u_k \tag{2a}$$

the A_i and B_i are matrices and z is the shift operator. This representation can be written in first-order form as

$$\begin{bmatrix} I & & & \\ & \ddots & & \\ & & I & \\ & & & 0 \end{bmatrix} z x_k = \begin{bmatrix} 0 & & & \\ I & & & \\ & \ddots & & \\ & & & I \end{bmatrix} x_k + \begin{bmatrix} -A_n & B_n \\ \vdots & \vdots \\ -A_0 & B_0 \end{bmatrix} \begin{bmatrix} y_k \\ u_k \end{bmatrix} \tag{2b}$$

In general, singular pencil models have the following characteristics which can be some of the reasons to use them:

1. Unification: SP models include as special cases the two types of most commonly used models: state space models and input-output (or MFDs, ARMA) representations.
2. Generality: unlike state space representations, SP models can describe improper systems. PID controllers, tachometers, and inverses of strictly proper systems are some improper systems that are useful both in concept and in fixed stages of design. Futhermore, SP models admit non-uniqueness of the output y and constraints on the input u of the internal variable x of the system.
3. Simplicity: Useful transformations can be simply implemented as sequences of elementary (for sparsity) or orthogonal (for stability) row and column operations on the constituent matrices E,F,G of (1a).

For details of the realization, as well as design and analysis of control systems using such models, the reader is referred to [1,2,3]. The simultaneous estimation of states and system parameters was developed by Salut, Aquilar Martin and Lefebvre[4,5], and Chen, Aplevich and Wilson[6]. The distinction between inputs and outputs in w

is generally unimportant for identification since both types of variables must be measured. For the present development it can even be assumed that the definition of variables in w is unknown or is not unique, since it is not explicitly required. The auxiliary (internal) vector $x \in R^n$ is necessary in order to write the system as a set of simultaneous first-order difference equations. To determine the model structure, the input/output data collected from the open-loop system with sufficient excitation , as shown in Fig 2, is used to fit an ARMAX (Auto-Regression Moving Average with eXogeneous input signal) model of different orders (model structures). It is of the form

$$\sum_{i=0}^{n} A_i z^{n-i} y_k = \sum_{i=0}^{m} B_i z^{m-i} u_k + \sum_{i=0}^{n} C_i z^{n-i} e_k , \qquad (3)$$

where A_i and B_i are matrices and z is the shift operator.

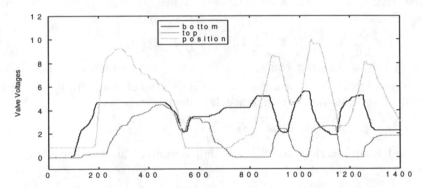

Fig. 2. Input voltages and position

For the pneumatic system discussed in this paper, there are two inputs and one output. If error is accounted for, the ARMAX model is of the form

$$\sum_{i=0}^{n} a_i z^{n-i} y_k = \sum_{i=0}^{m} b_i^1 z^{m-i} u_k^1 + \sum_{i=0}^{m} b_i^2 z^{m-i} u_k^2 + \sum_{i=0}^{n} c_i z^{n-i} e_k , \qquad (4)$$

where y_k is the roller position detected by the position sensor, u_k^1 is the bottom valve voltage, u_k^2 is the top valve voltage, e_k is the equation error, accounting for the noise in the process and measurement, and is assumed to be an i.i.d.(independent identically-distributed) Gaussian process, a_i, b_i^1, b_i^2 and c_i are the parameters of the model, and n and m are the orders of the system. To find the optimal model structure, the performance of the model, judged by many factors such as the variance of the prediction error curve (e_k = *actual position - predicted position*) and the time necessary for the estimation of parameters and states to settle, can be used. Without losing generality, m is made equal to n when comparing models of increasing orders. It was observed that the quantity J_N , the summation of error, didn't improve much beyond $n=4$ and thus the fourth order will be used.

For a fourth order system, the ARMAX model is of the form

$$(z^4 + a_1 z^3 + a_2 z^2 + a_3 z + a_4) y_k = (b_1^1 z^3 + b_2^1 z^2 + b_3^1 z + b_4^1) u_k^1 + (b_1^2 z^3 + b_2^2 z^2 + b_3^2 z + b_4^2) u_k^2$$
$$+ (z^4 + c_1 z^3 + c_2 z^2 + c_3 z + c_4) e_k \qquad (5)$$

This representation can be written in SP form as

$$
\begin{bmatrix}
1 & 0 & 0 & 0 \\
0 & 1 & 0 & 0 \\
0 & 0 & 1 & 0 \\
0 & 0 & 0 & 1 \\
0 & 0 & 0 & 0
\end{bmatrix} z x_k =
\begin{bmatrix}
0 & 0 & 0 & 0 \\
1 & 0 & 0 & 0 \\
0 & 1 & 0 & 0 \\
0 & 0 & 1 & 0 \\
0 & 0 & 0 & 1
\end{bmatrix} x_k +
\begin{bmatrix}
-a_4 & b_4^1 & b_4^2 & c_4 \\
-a_3 & b_3^1 & b_3^2 & c_3 \\
-a_2 & b_2^1 & b_2^2 & c_2 \\
-a_1 & b_1^1 & b_1^2 & c_1 \\
-1 & 0 & 0 & 1
\end{bmatrix}
\begin{bmatrix}
y_k \\
u_k^1 \\
u_k^2 \\
e_k
\end{bmatrix}.
\tag{6}
$$

By simple matrix manipulation, equation (6) can be formulated as

$$
\begin{aligned}
x_{k+1} &= E_* x_k + G_* w_k + C_* e_k \\
0 &= E_o x_k + G_o w_k + e_k
\end{aligned},
\tag{7}
$$

with the matrices being

$$
E_* =
\begin{bmatrix}
0 & 0 & 0 & 0 \\
1 & 0 & 0 & 0 \\
0 & 1 & 0 & 0 \\
0 & 0 & 1 & 0
\end{bmatrix}, E_o = \begin{bmatrix} 0 & 0 & 0 & 1 \end{bmatrix}, G_* =
\begin{bmatrix}
-a_4 & b_4^1 & b_4^2 \\
-a_4 & b_3^1 & b_3^2 \\
-a_4 & b_2^1 & b_2^2 \\
-a_4 & b_1^1 & b_1^2
\end{bmatrix}, G_o = \begin{bmatrix} -1 & 0 & 0 \end{bmatrix},
\tag{8}
$$

and

$$
C_* = \begin{bmatrix} c_4 & c_3 & c_2 & c_1 \end{bmatrix}^T, w = \begin{bmatrix} y_k & u_k^1 & u_k^2 \end{bmatrix}^T.
\tag{9}
$$

Furthermore, the expression $G_* w_k$ can be rewritten as

$$
G_* w_k =
\begin{bmatrix}
-a_4 & b_4^1 & b_4^2 \\
-a_3 & b_3^1 & b_3^2 \\
-a_2 & b_2^1 & b_2^2 \\
-a_1 & b_1^1 & b_1^2
\end{bmatrix}
\begin{bmatrix}
y_k \\
u_k^1 \\
u_k^2
\end{bmatrix} =
\begin{bmatrix}
-y_k & 0 & 0 & 0 & u_k^1 & 0 & 0 & 0 & u_k^2 & 0 & 0 & 0 \\
0 & -y_k & 0 & 0 & 0 & u_k^1 & 0 & 0 & 0 & u_k^2 & 0 & 0 \\
0 & 0 & -y_k & 0 & 0 & 0 & u_k^1 & 0 & 0 & 0 & u_k^2 & 0 \\
0 & 0 & 0 & -y_k & 0 & 0 & 0 & u_k^1 & 0 & 0 & 0 & u_k^2
\end{bmatrix} r_k
$$

$$
= \tilde{G}_* (w_k) r_k.
\tag{10}
$$

where $r_k = \begin{bmatrix} a_4 & a_3 & a_2 & a_1 & b_4^1 & b_3^1 & b_2^1 & b_1^1 & b_4^2 & b_3^2 & b_2^2 & b_1^2 \end{bmatrix}^T$. If noise is ignored, the system can be described by

$$
\begin{aligned}
x_{k+1} &= E_* x_k + \tilde{G}_* (w_k) r_k \\
y_k &= E_o x_k + [G_o + J] w_k
\end{aligned}
\tag{11}
$$

with the matrices being: $G_o = \begin{bmatrix} -1 & 0 & 0 \end{bmatrix}, J = \begin{bmatrix} 1 & 0 & 0 \end{bmatrix}, G_o + J = \begin{bmatrix} 0 & 0 & 0 \end{bmatrix}$ (12)

In the presence of noise, we have
$$
\begin{cases}
x_{k+1} = E_* x_k + \tilde{G}_* (w_k) r_k + C_* e_k \\
y_k = E_o x_k + e_k \\
r_{k+1} = r_k + \pi_k
\end{cases}
\tag{13}
$$

letting s_k hold the states and parameters, i.e. $s_k = \begin{bmatrix} x_k & r_k \end{bmatrix}^T$, the following equations are obtained:
$$
s_{k+1} = F_k s_k + \begin{bmatrix} C_* e_k & \pi_k \end{bmatrix}^T,
\tag{14}
$$
$$
y_k = H_k s_k + e_k,
\tag{15}
$$

where
$$F_k = \begin{bmatrix} E_* & \tilde{G}_* w_k \\ 0_{12\times 4} & I_{12\times 12} \end{bmatrix}, \tag{16}$$

and $\quad H_k = \begin{bmatrix} E_0 & 0_{1\times 12} \end{bmatrix} = \begin{bmatrix} 0 & 0 & 0 & 1 & 0 & 0 & 0 & 0 & 0 & 0 & 0 & 0 & 0 & 0 & 0 & 0 \end{bmatrix}. \tag{17}$

The time-varying matrices F_k and H_k are known because E_* and E_0 are determined from the $\{k_i\}$, and $\tilde{G}_*(w_k)$ and $\tilde{G}_0(w_k)$ are constructed from the measure-ment w_k.

4. Estimation of States and Parameters

Having developed a model to describe the system, the next step is to actually estimate the states x_k and parameters r_k of the model, in other words, the vector s_k. A Kalman filter can be applied directly to give the optimal linear estimate of this vector, by using the recursive equations

$$\hat{s}_{k+1} = F_k \hat{s}_k + K_k (y_k - H_k \hat{s}_k), \tag{18}$$

$$K_k = (F_k P_k H_k^T + S)(H_k P_k H_k^T + R)^{-1}, \tag{19}$$

$$P_{k+1} = F_k P_k F_k^T + Q - K_k (H_k P_k H_k^T + R) K_k^T, \tag{20}$$

with initial conditions

$$\hat{s}_0 = E\{s_0\}, P_0 = E\left\{ \left(s_0 - E\{s_0\}\right)\left(s_0 - E\{s_0\}\right)^T \right\}, \tag{21}$$

where $E\{x\}$ denotes the mathematical expectation of x, and

$$E\left\{ \begin{bmatrix} C_* e_k \\ \pi_k \\ e_k \end{bmatrix} \begin{bmatrix} C_* e_k \\ \pi_k \\ e_k \end{bmatrix}^T \right\} = \begin{bmatrix} Q & S \\ S^T & R \end{bmatrix} \tag{22}$$

where Q, S, and R are noise covariance matrices of the model. Thus, provided the structural integers, the dynamic noise parameters C_*, and Q, S, and R are known, the system auxiliary vector and non-structural parameters can be estimated using an ordinary Kalman filter, which converges under well-defined conditions.

Considerations are taken when selecting the initial values for the above matrices. The state covariance matrix P_k should be symmetrical and large to allow for error at initial conditions. It is chosen as $10000 \times I_{(n+n+2m)\times(n+n+2m)}$ ($I_{a\times b}$ denotes identity matrix with a rows and b columns). Based on measurement of signals in the system and tuning of the matrices for optimal performance, the matrices Q and S are chosen to be identity matrices $I_{(n+n+2m)\times(n+n+2m)}$ and $I_{(n+n+2m)\times 1}$, respectively, and R is selected as 0.1 of the range of deviation in the position, which is around 0.97 when the cylinder is operating in the middle region. Matrix C_*, a measurement of the noise parameters, is chosen to be very small assuming negligible error. It is chosen as

$$C_* = \begin{bmatrix} 0 & 0 & 0 & 0.1 \end{bmatrix}^T. \tag{23}$$

The matrix of variance of the parameters, π_k, is set as 0.5 after a few trials.

Next, performance of the model using the matrices mentioned above is compared with that when matrices Q and S are selected according to equation 22. It can be shown that the estimation error improves slightly with the new Q and S. With $R=0.97$, the values of the matrices used in calculation are summarized as follows:

$$\pi_k = \begin{bmatrix} 0.5 & 0.5 & 0.5 & 0.5 & 0.5 & 0.5 & 0.5 & 0.5 & 0.5 & 0.5 & 0.5 & 0.5 \end{bmatrix}^T \tag{24}$$

$$Q = \begin{bmatrix} 0 & 0 & 0 & 0 & 0.25_{1\times12} \\ 0 & 0 & 0 & 0 & 0.25_{1\times12} \\ 0 & 0 & 0 & 0 & 0.25_{1\times12} \\ 0 & 0 & 0 & 0.097 & 0.25_{1\times12} \\ 0.25_{12\times1} & 0.25_{12\times1} & 0.25_{12\times1} & 0.25_{12\times1} & 0.25_{12\times12} \end{bmatrix} \tag{25}$$

$$S = \begin{bmatrix} 0 & 0 & 0 & 0.09655 & 0 & 0 & 0 & 0 & 0 & 0 & 0 & 0 & 0 & 0 & 0 & 0 \end{bmatrix}^T \tag{26}$$

$$Pk = 10000 * I_{16}; \tag{27}$$

where $0.25_{a\times b}$ is a matrix of a rows and b columns with every entry equal to 0.25. In this section, it has been shown that the two steps of estimating the states and parameters can be done simultaneously by using a linear Kalman filter with the SP model (equation 18-20). For the input/output data shown in Fig 2, the states and parameters $x_4...x_1, a_4...a_1, b_1^1...b_4^1$, and $b_1^2...b_4^2$ estimated using this model are plotted in the following figures, where $b_1^1...b_4^1$ and $b_1^2...b_4^2$ are denoted as b11...b14, and b21...b24, respectively.

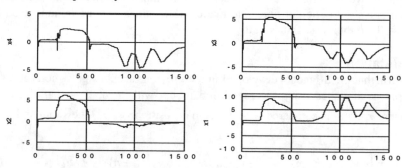

Fig. 3. Estimated states ($x_4...x_1$)

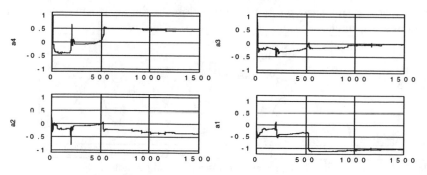

Fig. 4. Estimated parameters ($a_4...a_1$)

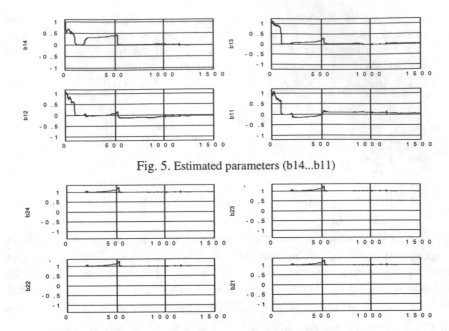

Fig. 5. Estimated parameters (b14...b11)

Fig. 6. Estimated parameters (b24...b21)

From the above figures, it can be observed that the parameters settle to constant values after a short period of time. No matter what values they are initialized at, they should settle to their true values in the end.

5. Validation

To confirm that the SP model developed in the previous sections gives correct results, an output is simulated using the regression model with input u_k and the estimated parameters r_k, and is compared with the actual position y_k. The simulated output y_{sim_k} is

$$y_{sim\_k} = \phi_k \hat{r}_k \tag{28}$$

where $\phi_k = \left[-y_{sim\_k}(k-n) \quad \cdots \quad -y_{sim\_k}(k-1) \quad u_k^1(k-n) \quad \cdots \quad u_k^1(k-1) \right]$. (29)

The following plot shows the simulation compared to the real position. It can be observed that the simulated position is very close to the actual one and thus the estimations are accurate.

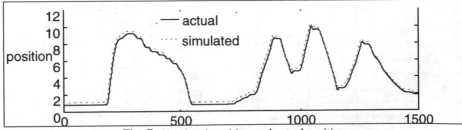

Fig. 7. Simulated position and actual position

Besides being able to provide a way to estimate the states and parameters in one single step as mentioned in the last section, it can be shown that the SP model gives a smaller estimation error ($e_k = y_k - \hat{y}_k$) and takes much less time for the estimation ($\hat{y}_k$) to converge to the real value (y_k) comparing to the State Space Model usually used. Shown in figures 9 and 10 are plots of position estimation and estimation error for the SP model and State Space Model.

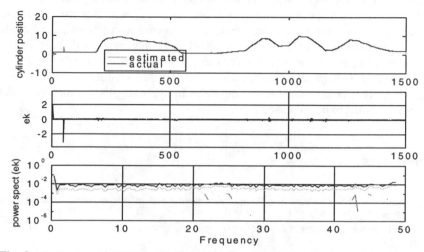

Fig. 8.   Performance of SP Model (Top: actual position and estimation, middle: prediction error, bottom: power spectrum of prediction error)

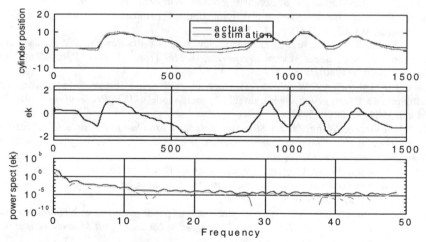

Fig. 9.   Performance of State Space Model (top: actual Position and estimation, middle: Prediction error, bottom: Power spectrum of prediction error)

6. Conclusion and Further Work

This paper presents results on the modeling and simultaneous state and parameter estimation of an industrial pneumatic positioning system for lumber processing machines using the singular pencil model. The dynamics of such systems changes dramatically with load, speed, force, air supply, environment etc. thus a control scheme using the estimated parameters and state variables to maintain optimal performance would be preferable. Such schemes can be the ones discussed in [8,9,10]. Good convergence property is achieved by the use of the linear Kalman filter estimating the state and parameters. A self-tuning control scheme, based on state feedback with the estimated state variables, and using the estimated parameters in the controller design, can promise improved performance over schemes that uses estimated states based on a state space model constructed with pre-estimated model parameters or boot-strap methods. Major challenges, however, still exist in that significant level of nonlinearity is observed in the system, in the form of hysteresis between the input voltage to the valves and the output (the position) of the actuator. This nonlinearity is being compensated and helps in improving the estimation as well as in control (results to be published later).

7. References

1. J.D.Aplevich, Time-Domain input-output representations of linear systems, *Automatica,* vol. 17, no. 3 (1981) 509-521.
2. Aplevich, Minimal representations of implicit linear systems, *Automatica,* vol. 21, no. 3 (1985) 259-269.
3. Aplevich, *Implicit Linear Systems, Lecture Notes in Control and Information Sciences,* vol. 152, Heidelberg: Springer-Verlag (1991).
4. Salut, J. Aquilar-Martin, and S. Lefebvre, Canonical input-output representation of linear multivariable stochastic systems and joint optimal parameter and state estimation, *Stochastica,* vol. III, no. 1 (1979) 17-38.
5. Salut, J. Aquilar-Martin, and S. Lefebvre, New results on optimal joint parameter and state estimation of linear stochastic systems, *ASME Trans. J. Dyn. Syst. Meas. and Control,* vol.102 (1980) 28-34.
6. Chen, J.D. Aplevich and W. Wilson, Simultaneous estimation of state and parameters for multivariable linear systems with singular pencil models, *IKE Proceedings,* vol. 133 (March 1986) Pt. D(2):65-72.
7. Xiaochun George Wang, *Model Structure Selection for On-Line System Identification Using Overlapping Singular pencil Models.* Ph.D. thesis, University of Waterloo, Waterloo, Ontario, Canada (1991).
8. Xiaochun George Wang, Guy Dumont and Michael Davies, Modeling and identification of basis weight variations in paper machines, *IEEE trans. Control Systems Technology,* Vol. 1, No. 4 (1993) 230-237.
9. Xiaochun George Wang, Adaptive pole placement of paper machines with singular pencil models, Proceedings *of the International Conference On Intelligent Manufacturing' 95,* Wuhan, China (June 14-17, 1995) 770-777.
10. Xiaochun George Wang, Self-tuning control with Singular Pencil Models, Proceedings of *1995 IEEE International Conference on Systems, Man and Cybernetics* ,Waterfront Center Hotel, Vancouver, B.C., Canada (Oct 22-25, 1995) 2604-2610.

Inspection and Diagnostic Systems

A Fuzzy Expert System for Monitoring Dimensional Accuracy of Components

[1]Henry Lau, [2]Ralph W.L. Ip [3]Felix T.S. Chan

[1]Advanced Technology Education Centre, Regency Institute, Australia
[2]Dept. of Manuf. Engg. & Engg. Mgt., City University of Hong Kong
[3]Dept. of Ind. & Manuf. Systems Engg., University of Hong Kong

Abstract. Fuzzy expert system uses fuzzy logic control[1] which is based on a "superset" of Boolean logic that has been extended to handle the concept of "partial truth" and it replaces the role of mathematical model with another that is built from a number of rules with fuzzy variables such as output temperature and fuzzy terms such as hot, fairly cold, probably correct. A fuzzy expert system has been implemented in a aluminium die casting company for monitoring dimensional quality of output products. This fuzzy expert system was developed based on the experience of a similar approach for quality control of injection molding components. The die casting parts are supplied to a local car-manufacturing plant and used as assembly components for the motor engines and therefore the dimensional accuracy of them is of utmost importance. This paper presents the implementation of this monitoring system using fuzzy logic theory. Practical examples with descriptions of how the fuzzy rules are fired and the operations of the fuzzy inference engine are also covered.

1. Introduction

The principle of aluminium die casting is to force molten aluminium under certain pressure into a permanent steel mold to produce the required shaped parts. Many intricately-shaped components of various sizes are produced in these molds ranging from small knobs, handles to engine blocks, large appliance castings, automotive crankcase and transmission housings. During operation, high temperature of up to $1150^{\circ}F$ is required to melt the aluminium alloy which is injected into the mold. Studies show that dimensional quality of parts from aluminium die casting operations depends largely on the parameters of the moulding conditions. There are four parameters which may affect the surface finish, mechanical properties and dimensional stability of output components and these include mold temperature, molten alloy temperature during injection, injection pressure, operation time.

To monitor the quality of the products, a monitoring system based on fuzzy logic has been developed and implemented in a car-component supplier in Australia. Fuzzy rules with linguistic terms are set based on alloy die-casting theory, past experience and trial results. New rules can be added anytime whenever some new results are found. Instead of specifying exact numeric value, fuzzy rules support the use of statements with linguistic terms. Just like a classical expert system[2,3] fuzzy expert system consists of knowledge base, rules and inference engine, but instead of dealing with crisp (exact) information, it handles fuzzy (uncertain, imprecise) data as well[4].

2. Traditional Technique

Quality of die cast parts depends largely on the parameters of the moulding conditions as described in the above context. Although these parameters can be set before moulding process starts, the actual conditions may alter due to other factors e.g. blocking of passage for cooling water, unforeseen mechanical or electronic failures. Actual states of moulding conditions can be obtained by using sensing devices and the actual readings can be recorded by instruments or displayed on computer screen if the sensor devices are connected to a computer via some analogue/digital hardware.

Traditionally, if there is suspected dimensional discrepancy occurred in certain components, experts are called in to provide advice for correctional action. In most cases, the parameter settings such as temperature, pressure are to be adjusted in order to rectify the dimensional problem. However, this is not an efficient way to handle the situation and in addition, expert advice is normally based on personal opinion and knowledge.

3. Fuzzy Logic System

A dimensional monitoring system based on fuzzy logic concept has been tested and implemented in a company. This system is designed in such a way that it is able to store knowledge and experience from previous tests and production, with the important feature that expert advice is available whenever it is needed.

Specific components, which require high dimensional and strength quality, are first chosen to be monitored by this system. Take for example the component as shown in Figure 1, it is important to specify which dimensions of the product are to be monitored. As shown in the Figure, the dimensions "a", "b", "c", "d" and "e" of the component are the key dimensions that require high accuracy. Normally, the dimensions of a component depend on two key factors which are the design of the mold and the molding conditions during the process. During the pre-production stage, the mold design (such as draft, runners and gates, side slides) as well as the molding conditions (such as temperature, pressure) are finalised before actual production. However, due to some unexpected reasons, the quality of output components may not be sustained.

A fuzzy expert system is implemented to deal with this problem. The first step of designing the fuzzy expert system is to determine the fuzzy sets[5,6] for the monitored dimensions. Each dimension has to be represented by a fuzzy set which is designed in compliance with the fuzzy logic principle. In this example, the required dimension for "a" is, say, 110.50 ± 0.02 mm. This dimension is then "fuzzified" and represented as fuzzy set shown below :
Undersized - represented by Z function with data "d-0.1" and "d-0.01" (d = 110.50)

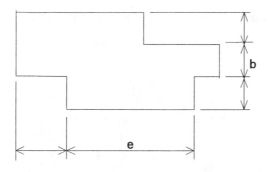

Figure 1 : Sample Component

OK - represented by Π function with data "d" and "0.02"
oversized - represented by S function with data "d+0.01" and "d+0.1"

Figure 2 shows the fuzzy representation of dimension "a". In FuzzyCLIPS[6,7], the above functions can be represented by a template and in this case, it is called Dimension_a:

> *(deftemplate Dimension_a*
> *d-0.01 d+0.01 mm*
> *(undersized (z d-0.1 d-0.01)*
> *(OK (pi 0.02 d)*
> *(oversized (s d+0.01 d+0.1)))))*

For example, the dimension "a" of the component is measured and found to be equal to 110.56 or "d + 0.06" which is obviously oversized. The result obtained is a "crisp" value and needs to be "fuzzified" for fuzzy inference by the fuzzy expert system. In FuzzyCLIPS, the fuzzy set of the actual dimension can be represented as :

(assert (Actual_DimA (- ?Actual_DimA 0.005) 0.0) (?Actual_DimA 1.0)
(+ ?Actual_DimA 0.05) 0.0)

As shown in Fig. 3 , the fuzzy set is :

(Actual_DimA (110.555 0.0) (110.56 1.0) (110.565 0.0))

For the component as shown in Fig. 1, there are five dimensions to be controlled and assuming that there are four parameters to be considered including mold temperature (P1), molten alloy temperature during injection (P2), injection pressure (P3) and operation time (P4), the table for the adjustments of parameters is as shown in Table 1.

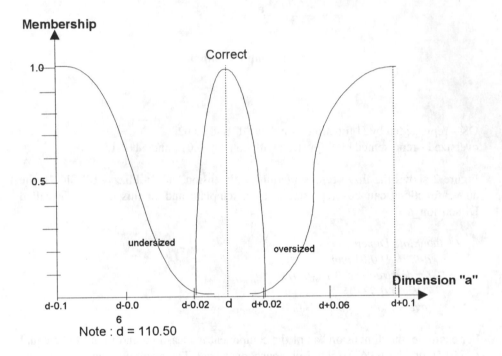

Note : d = 110.50

Figure 2 : Fuzzy Set for Actual Dimension of "a"

4. Modifiers and Linguistic Expressions

A modifier may be used to further enhance the ability to describe the fuzzy sets. Modifiers (very, slightly) used in phrases such as very hot or slightly cold modify (fiʀe tune) the shape of fuzzy sets in a way that suits the meaning of the word used. FuzzyCLIPS has a set of predefined modifiers that can be used at any time to describe fuzzy concepts when fuzzy terms are described in fuzzy deftemplates, fuzzy rule patterns are written, or fuzzy facts or fuzzy slots are asserted.

5. Structure of Rules

A number of rules are established based on the data from Table 1. The antecedents and consequents of the rules contain fuzzy data as shown in the following example.

<If dimension A is oversized and dimension B is undersized>
        Then
<Increase operation time slightly and reduce the molten alloy temperature considerably>

According to Table 1, the recommended action for an oversized dimension "a" is a considerable decrease (CD) of mold temperature (P1). As shown in Fig. 3, the fuzzy pattern of Actual Dimension of "a" cuts the oversized fuzzy pattern at 0.79 (maximum point). According to Compositional Rule for Multiple Antecedents, the result of fuzzy reference is referring to the membership grade 0.79 as shown in Fig. 4.

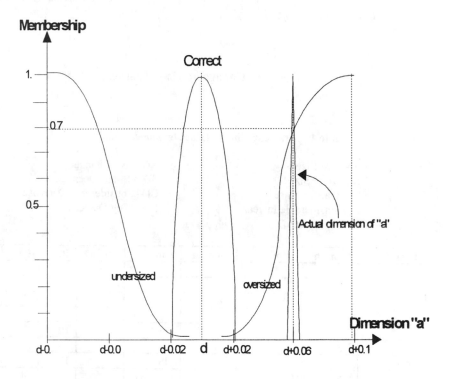

Figure 3 : Fuzzy Set for Actual Dimension of "a"

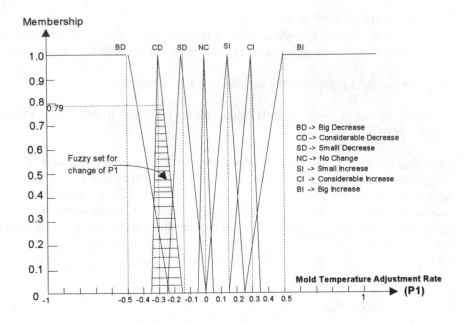

Figure 4 :Change of Mold Temperature Recommended

Table 1. Parameter Change Based on Dimensional Results

OS = OverSized
US = UnderSized
SI = Small Increase
CI = Considerable Increase
BI = Big Increase

NC = No Change
SD = Small Decrease
CD=Considerable Decrease
BD = Big Decrease

Dim. a	Dim. b	Dim. c	Dim. d	Dim. e	P1	P2	P3	P4
OS					CD			
OS	US					CD		SI
		OS					CI	
			US					
				OS		BI	·	
	US		OS				SD	
OS				US		BD		SD
	OS	OS					SI	
OS		US		US		CD		

The fuzzy set for the change of mold temperature is determined and shown in Fig. 4 (shaded area). It should be noted that the outcome of the fuzzy inference process is another fuzzy set[8]. However, in this case, it is essential that only a single discrete action is applied and so a single point that reflects the value of the set needs to be determined. The processing of reducing a fuzzy set to a single point is known as defuzzification[6,9]. FuzzyCLIPS allows users to choose either one by specifying in the codes. In this example, the crisp value is -0.252 i.e. 25.2% reduction of existing value. With this approach, the value recommended for adjustment of one of the important parameter is obtained. Each dimension (totally five in this example) and each parameter (totally four) have its own fuzzy pattern and should be designed based on fuzzy logic principle. The FuzzyCLIPS solution can handle a large number of fuzzy sets and produce the required results to users.

6. Conclusion

This paper presents the application of fuzzy logic in an expert system for the monitoring of dimensional quality of die cast components. Instead of using the traditional technique based on individual's experience and knowledge, the fuzzy expert system provides another approach, which is based on fuzzy control principle, to tackle the problem In this case study, fuzzy sets are used to represent the dimensions of the components. Modifiers can also be used to modify the shape of the fuzzy sets in a way that suits the meaning of the word (very, slightly) used. This system has been used in a company and proved to be quite successful and cost-effective. Full automatic control is under development with the PC interfacing to the automatic sensing equipment and the controlling hardware of the automatic die-casting machines so that whenever there is any dimensional discrepancy, the fuzzy expert system will respond and determine the adjustment of related molding parameters. This is possible due to the fact that FuzzyCLIPS is written in C with source codes provided. Modification of the source codes with some additional interfacing functions allows the direct control of the parameter settings of molding conditions.

References

1. L.A. Zadah, Fuzzy Sets, *Information and Control*, Vol. 8, 1965, pp.338-383.
2. M. Stefik, J. Aikins, R. Balzer, J. Benoit, L. Birnbaum, F. Hayes-Roth, F and E. Sacerdoti, E, *Basic Concepts for Building Expert Systems*. Addison-Wesley, 1983.
3. B.G. Buchanan and E.H. Shortliffe, *Rule-Based Expert Systems*, Addison-Wesley, 1984.
4. J. Giarratano and G. Riley, *Expert Systems: Principles and Programming*, International Thompson Publishing, 1993.
5. K.S. Leung and W. Lam, Fuzzy Concepts in Expert Systems. *IEEE*. September: 43-56, 1988.
6. R.A. Orchard, *FuzzyCLIPS Version 6.02A User's Guide*, National Research Council Canada, 1994.
7. CLIPS Reference Manual, Version 6.0, 1993.
8. J. Buckley and W. Siler, Fuzzy Operators for possibility Interval Sets, *Fuzzy Sets and Systems*, 1987, Vol.22, pp 215-227.
9. T. Chiueh, Optimization of Fuzzy Logic Inference Architecture, *Computer*. May, 1992, pp 67-71.

Fuzzy Classification of Fault Diagnosis Using ANNs

Qian Huang   Dan Pan   GanYing Luo   S. K. Tso

Department of Electronics & Communications Engineering,
South China University of Technology, China

Abstract In recent years, fuzzy logic and neural networks have been gradually introduced in the field of fault diagnosis. However, both fuzzy logic based methods and conversational neural networks have their own disadvantages. One practical way is to integrate fuzzy logic with neural networks. We have presented the feasible method of fuzzy classification of fault diagnosis patterns. In the neural networks using back propagation algorithm, the input vector consists of membership values to fault symptoms while the output vector is defined in terms of fuzzy fault reason class membership values. The effectiveness of the algorithm is demonstrated with the fault diagnosis of a radar equipment.

1 Introduction

Failure is a nearly unavoidable phenomenon with technological products and systems. By fault we mean a dynamic (system) state which deviates from the desired system state. The tasks of fault diagnosis may include diagnosing where the fault occurred and what are the type and kind of the fault, assessing the damage of the fault, and reconfiguring the system to accommodate the fault. In a narrow sense, fault diagnosis partially answers one of the basic issues in system failure engineering: why system failure. Obviously, various symptoms of a system during its operation are essential to implement tasks of fault diagnosis. However, vague symptoms frequently emerge. Fuzzy methodology is a natural tool to incorporate symptoms of this kind and can be also used to deal with vagueness in system models and in human perceptions. Moreover, because fault diagnosis can be thought of as pattern recognition, artificial neural networks (ANNs) are well suited to this task[1].   ANNs using the back propagation algorithm can be used for character recognition. Several studies that examine the use of ANNs for fault diagnosis appear in the literature. Whereas, conventional two-state neural net models generally deal with the ideal condition, where an input feature is either present or absent and each pattern belongs to either one class or another. They do not consider cases where an input feature may belong to more than one class with a finite degree of "belongingness"[2]. While in fault diagnosis, the fault symptoms which we based to implement tasks of fault diagnosis are often vague symptoms[3].However, one practical way is to integrate fuzzy logic with neural networks[4]. The proposed neural network model using the gradient-descent-based back propagation algorithm incorporates concepts from fuzzy sets and is capable of classification of fuzzy patterns.

2 Multilayer Perceptron Using Back Propagation Of Error

The multilayer perceptron (MLP) consists of multiple layers of simple, two-state sigmoid processing elements (nodes) or neurons that interact with weighted connections. The strong functions of the MLP are mainly brought about with the nonlinear property of hidden neurons.[5] At the present, there are many effective studying algorithms for the multilayer percetron, especially back propagation algorithm. The Back Propagation (BP) learning model of MLP converts the problems of input and output in a set of patterns to the problems of the nonlinear optimization, using the most popular least mean square (LMS) algorithm. It is because of the hidden layer that the adjustable parameters of the optimal problem increase so that the accurate results can be gained. If the MLP using BP algorithm is considered as the mapping from the input space to the output space, the mapping is highly nonlinear and can reflect the complex phenomena in real life.

Now consider the network with a hidden layer. Set x_i ($i=1, 2, \dots ,n_0$) and $w_{ij}^{(l)}$ ($l=1, 2$) as the input vectors and the weights of the network respectively. So for respective layer, the inputs are:

$$I_i^{(0)} = x_i \qquad i=1, 2, \dots , n_0 \tag{1}$$

$$I_i^{(1)} = \sum_{j=1}^{n_0} w_{ij}^{(1)} o_j^{(0)} - \theta_i^{(1)} \qquad i=1, 2, \dots , n_1 \tag{2}$$

$$I_i^{(2)} = \sum_{j=1}^{n_1} w_{ij}^{(2)} o_j^{(1)} - \theta_i^{(2)} \qquad i=1, 2, \dots , n_2 \tag{3}$$

Where $w_{ij}^{(1)}$ is the weight of the connection from the jth neuron in the input layer to the ith neuron in the hidden layer, $w_{ij}^{(2)}$ is the weight of the connection from the jth neuron in the hidden layer to the ith neuron in the output layer, $o_j^{(0)}$ is the output of the jth neuron in the input layer, $o_j^{(1)}$ is the output of the jth neuron in the hidden layer, $\theta_i^{(1)}$ is the threshold of the ith neuron in the hidden layer, $\theta_i^{(2)}$ is the threshold of the ith neuron in the output layer, n_0 , n_1 , n_2 is the number of neuron in the input, hidden output layer respectively.

For respective layer, the output are

$$o_i^{(0)} = x_i \qquad i=1, 2, \dots , n_0 \tag{4}$$
$$o_i^{(1)} = g(I_i^{(1)}) \quad i=1, 2, \dots , n_1 \tag{5}$$
$$o_i^{(2)} = g(I_i^{(2)}) \quad i=1, 2, \dots , n_2 \tag{6}$$

where $g(I) = \dfrac{1}{1+e^{-I}}$ is the sigmoid function.

The least mean square error in output vectors, for a given network weight vector W is defined as

$$E(w) = \frac{1}{2} \sum_{\mu=1}^{n_1} \sum_{i=1}^{n_2} (\zeta_{\mu_i} - o_{\mu_i}^{(2)}(w)) \tag{7}$$

where ζ_{μ_i} is the desired output of the ith neuron in the output layer in input-output case μ specified by the teacher and $o_{\mu_i}^{(2)}(w)$ is its actual output. The error $E(w)$ is minimized by the BP algorithm using gradient descent. We start with any set of weights and repeatedly update each weight by an amount

$$\Delta w_{ij}^{(l)}(n+1) = -\alpha \frac{\partial E(n)}{\partial w_{ij}^{(l)}(n)} + \varepsilon \Delta w_{ij}^{(l)}(n), \qquad l=1,2. \qquad (8)$$

where the learning rate $\alpha > 0$ controls the descent, $0 \le \varepsilon \le 1$ is the damping coefficient or momentum and n denotes the number of the iteration currently in progress.

3 Pattern Representation With Fuzzy Concept

In the fault diagnosis, many fault features can not simply be divide by crisp two-value logic but have a certain extent, for example weak, weaker, ordinary, strong and stronger. Therefore, fault feathers should be fuzzed in order to make them closer to the fact in the fault diagnosis .

Now, there is an example about the fault diagnosis of a radar equipment. In the radar equipment, a kind of fault reason usually leads to many of fault features and different fault features contribute differently to a certain kind of fault reason .

Fig. 1. is a typical single stage electrical circuit diagram. Since the parameters of components such as capacitance, resistance value have tolerable errors (usually the tolerable error of ordinary resistance is ±10%) and may give rise to the drifts with the change of environmental temperature and time. Hence, the relations between the features and the reasons are not fully decided. Moreover, the human's diagnosis experiences are usually very fuzzy. So the diagnosis models should be based on the fuzzy system theory instead of the classical mathematics. The voltage of every measurement spot is the continuous variable and all of three fault features are fuzzy values.

There are n fault reasons and m fault features, which are denoted with $X=\{x_1, x_2, \ldots, x_n\}$ and $U=\{u_1, u_2, \ldots, u_m\}$ respectively.

Define $R: X \times U \rightarrow [0, 1]$, $r_{ij}=R(x_i, u_j) \in [0, 1]$, R is the membership matrix. The membership functions $r_{11} - r_{53}$ are shown as below.

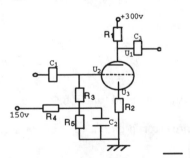

Fig1. the single stage electric

$$\tilde{R}_1 \quad r_{11} \quad r_{12} \quad r_{13}$$
$$\tilde{R}_2 \quad r_{21} \quad r_{22} \quad r_{32}$$
$$\tilde{R}_3 \quad r_{31} \quad r_{32} \quad r_{33}$$
$$\tilde{R}_4 \quad r_{41} \quad r_{42} \quad r_{43}$$
$$\tilde{G} \quad r_{51} \quad r_{52} \quad r_{53}$$

$$r_{12} = r_{22} = r_{52} = \begin{cases} \frac{1}{2} + \frac{1}{2}\sin\frac{\pi}{3}(x - \frac{-39}{2}), & -21 \le x \le -18 \\ 1, & -18 \le x \le -12 \\ \frac{1}{2} - \frac{1}{2}\sin\frac{\pi}{3}(x - \frac{-21}{2}), & -12 < x < -9 \\ 0, & other \end{cases} \tag{9}$$

$$r_{11} = \begin{cases} 1, & 0 \le x \le 2 \\ \frac{1}{2} - \frac{1}{2}\sin\frac{\pi}{2}(x - \frac{6}{2}), & 2 < x \le 4 \\ 0, & other \end{cases} \tag{10}$$

$$r_{13} = \begin{cases} \frac{1}{2} - \frac{1}{2}\sin\frac{\pi}{0.8}(x - \frac{0.8}{2}), & 0 \le x \le 0.8 \\ 0, & other \end{cases} \tag{11}$$

$$r_{21} = r_{41} = r_{51} = \begin{cases} \frac{1}{2} + \frac{1}{2}\sin\frac{\pi}{6}(x - \frac{594}{2}), & 294 \le x \le 300 \\ 1, & x > 300 \\ 0, & x < 294 \end{cases} \tag{12}$$

$$r_{31} = \begin{cases} \frac{1}{2} + \frac{1}{2}\sin\frac{\pi}{15}(x - \frac{235}{2}), & 110 \le x \le 125 \\ 1, & 125 < x < 140 \\ \frac{1}{2} - \frac{1}{2}\sin\frac{\pi}{16}(x - \frac{290}{2}), & 140 \le x \le 150 \\ 0, & other \end{cases} \tag{13}$$

$$r_{32} = \begin{cases} \frac{1}{2} + \frac{1}{2}\sin\frac{\pi}{1}(x - \frac{-1}{2}), & -1 \le x \le 0 \\ 0, & other \end{cases} \tag{14}$$

$$r_{42} = \begin{cases} 1, & x < -150 \\ \frac{1}{2} - \frac{1}{2}\sin\frac{\pi}{5}(x - \frac{-295}{2}), & -150 \le x < -145 \\ 0, & other \end{cases} \tag{15}$$

$$r_{23} = \begin{cases} \frac{1}{2} + \frac{1}{2}\sin\frac{\pi}{2.5}(x - \frac{9.5}{2}), & 3.5 \le x < 6 \\ 1, & 6 \le x \le 10 \\ \frac{1}{2} - \frac{1}{2}\sin\frac{\pi}{2.5}(x - \frac{22.5}{2}), & 10 < x \le 12.5 \end{cases} \tag{16}$$

$$r_{43} = r_{53} = \begin{cases} \dfrac{1}{2} - \dfrac{1}{2}\sin\dfrac{\pi}{0.5}\left(x - \dfrac{-0.5}{2}\right), & -0.5 \le x \le 0 \\ 0, & \text{other} \end{cases} \tag{17}$$

$$r_{33} = \begin{cases} \dfrac{1}{2} + \dfrac{1}{2}\sin\dfrac{\pi}{1}\left(x - \dfrac{9}{2}\right), & 3 \le x < 4 \\ 1, & 4 \le x \le 5 \\ \dfrac{1}{2} - \dfrac{1}{2}\sin\dfrac{\pi}{1}\left(x - \dfrac{11}{2}\right), & 5 < x \le 6 \\ 0, & \text{other} \end{cases} \tag{18}$$

In this example, the fault features U_1, U_2, U_3 have respectively 3, 3, 4 different membership functions because of $r_{21} = r_{41} = r_{51}$, $r_{12} = r_{22} = r_{52}$ and $r_{43} = r_{53}$. So each input feature U_{ji}, $(i=1, 2, 3)$ in the jth input feature pattern can be expressed as the membership values according to different membership functions of the ith feature attaching to different fault reason. Therefore an input pattern $\vec{U}_j = [u_{j1}, u_{j2}, u_{j3}]$ may be represented as :

$$\vec{F}_j = [\mu_{1(u_1)}(u_{j1}), \dots, \mu_{3(u_1)}(u_{j1}), \mu_{1(u_2)}(u_{j2}), \dots, \mu_{3(u_4)}(u_{j2}), \mu_{1(u_3)}(u_{j3}), \dots, \mu_{4(u_3)}(u_{j3})] \tag{19}$$

acting as actual input of ANNs corresponding to the jth input feature pattern $\vec{U}_j$.

According to experts' experiences, decide the continuos fuzzy membership values lying in the interval [0,1] of a input pattern (fault features pattern) belonging to every fault reasons and these values are considered as actual desired output pattern corresponding to the input pattern. Thus a learning pattern set can be gained.

Then, the neural network which includes input layer, a hidden layer and output layer is constructed. The numbers of nodes in the input, hidden and output layer are 10, 8, and 5, $\alpha = 0.3$, $\varepsilon = 0.5$, learning time $=1500$, then $E=0.002$. The result of learning is shown in the Table 1.

Table 1. Input, desired output, and actual output vectors for a set of sample patterns presented to the three-layer neural network

Input Feature				Output Vector				
u_1, u_2 u_3			Input Vector	Desired			Actual	
1	0	1	1 0 0 0 0 0 1 0 0 0	1 0 0 0 0			.97 .02 .01 .01 .00	
294	0	8	0 0 0 0 0 0 0 1 0 0	0 .7 0 0 0			.01 .69 .00 .01 .01	
132	0	3.5	0 0 1 0 1 0 0 0 .5 .0	0 0 .95 0 0			.01 .01 .94 .01 .00	
294	-150	-1	0 0 0 0 0 1 0 0 0 0	0 0 0 .8 0			.01 .02 .01 .79 .01	
300	0	-0.5	0 1 0 0 0 0 0 0 0 0	0 .3 0 .1 1			.00 .03 .01 .09 .97	

In practice, different fault features reflect a certain kind of fault reasons to different extent. In other words, different fault features have different weights for a certain kind of fault reasons. Now, ANN is utilized to gain these weights from the learning patterns and stores them in itself weights. When a new test input pattern (fault features pattern) is inputted, the output value of every neuron in the output layer can be considered as its belongingness attaching to every fault reason respectively.

4. Conclusion

Now, the procedure of the fault diagnosis with Fuzzy ANNs is concluded.

(1) Decide the membership matrix R. According to the experiences, membership function r_{ij} is introduced, namely:

$$r_{ij} = \mu_{x_i}(u_j) \subset [0,1]$$

where $\mu_{x_i}(u_j)$ is the membership function of the fault feature u_j attaching to the fault reason x_i .

(2) Choose the input patterns for learning. In view of every fault reasons, the input patters are typical as soon as possible.

(3) According to experts' opinions, decide the desired output pattern corresponding to every input pattern.

(4) Train the ANN with the learning patterns till success.

(5) Make fault diagnosis utilizing the ANN.

The above-all is an extension to the conventional MLP model that deals with actual input and *crisp* (binary) output values. The proposed model is found to classify fuzzy data with overlapping class boundaries in a more efficient manner. Therefore, it has a promising application of prediction diagnosis.

REFERENCES

1. Won-Yong Lee, Jokn M. House, Cheol Park, et. Fault Diagnosis of on Air-Handling Unit Using Artificial Neural Networks, ASHRAE Technical Data Bulletin, *Fault Detection For HVAC Systems*, (1996): 23-32

2. Sankar K. Pal, Sushmita Mitra, Multilayer Perceptron, Fuzzy Sets, and Classification, *IEEE Trans. on Neural Networks*, Vol3, No.5, Sep.(1992): 683-696

3. Kai-Yuan Cai, System failure engineering and fuzzy methodology an introductory overview, *Fuzzy Sets and Systems*, 83 (1996):135-141

4. Sushmita Mitra, Sankar K. Pal, Fuzzy Multi-Layer Perceptron, Inferencing and Rule Generation, *IEEE Trans. on Neural Networks*, Vol. 6, No. 1, Jan. (1995): 51-63

5. David E. Rumelhart, James L.Mcclelland, And the PDP Research Group, *Parallel Distributed Processing*, The MIT Press, London (1987) pp. 319-328.

Intelligent Visual Evaluation System
for Mirror Finished Surfaces

Kazuhiko Kato*, Toshihiro Ioi*, Masahisa Matsunaga** and Nobuaki Iguchi***

* Department of Project Management, Chiba Institute of Technology, Narashino, Chiba 275
  Japan
** Institution, Chiba Institute of Technology
*** R & D Department, Kuroda Precision Industries LTD, Kawasaki, Kanagawa, 239 Japan

Abstract. This paper describes an automated visual evaluation system for mirror finished surfaces of heat treated steel. The neural network is applied to predict the surface roughness of various kinds of mirror finished surfaces. The surface integrity will be clarified by the learning system if the correlation between the characteristics of each surface image and some optical properties of mirror finished surfaces can be analyzed. It is found that the neural network system is very effective means for evaluating the finished surfaces when image features of mirror finished surfaces are used as the input data and surface roughness and/or kinds of finishing process are derived as the output data.

1 Introduction

The mirror finished surfaces of heat treated steel as block gauge have been made through grinding and lapping processes. These iterative finishing operations and the evaluations of the surfaces after finishing have been performed by expert workers. If the features of mirror finished surfaces are recognized by a vision system and a neural network, the degree of surface roughness and/or kinds of finishing process will be clarified in relation to the image data of surfaces.

The researches[1]-[6] of visual evaluation system for metal surfaces have been reported. These are related to evaluation systems concerning the cross section surfaces[1],[2] after metal cutting, the friction surfaces[3], the surfaces[4] after blast processing, and the surfaces[5] of different metal materials. However, these researches using image processing methods are applied mainly to the evaluation of rough metal surfaces. The visual evaluation systems which can predict the surface roughness for mirror finished surfaces after grinding and lapping processes have not yet been developed, because of the difficulty of measuring methods and the complexity of image processing analysis.

The objectives of this paper are to develop the intelligent visual evaluation system for mirror finished surfaces[6]. It is found that the features of image data of mirror finished surfaces can be decided by texture analyses. The neural network systems are also applied to predict the surface roughness and/or kinds of finishing processes. As the results, it is possible to construct the neural network system which is composed of the features of surface image as input data and the degree of surface roughness and/or kinds of finishing processes as output data.

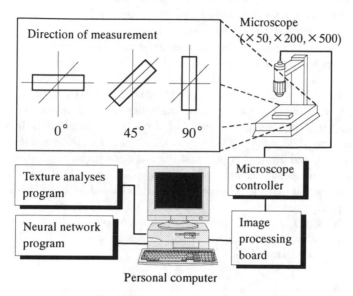

Fig.1. Experimental system

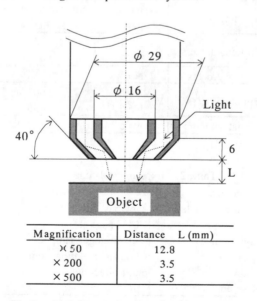

Magnification	Distance L (mm)
×50	12.8
×200	3.5
×500	3.5

Fig.2. Lighting mechanism of a microscope

2 Experimental System and Procedure of Experiments

Fig.1 shows the experimental system which is used in the study. The system is composed of a microscope, a controller, an image processing board and a microcomputer. The mirror finished surfaces which are used in the experiment are the testpiece of standard surface roughness finished by grinding and lapping processes. The light reflection data from a microscope can be analyzed using an image processing board in a microcomputer. The image data are displayed by the multiplication of the magnification of optical measuring system and the digital processing (expansion of pixel data) by microcomputer. The mirror finished surfaces are measured by three measuring directions of 0, 45 and 90 degree for the light axis of microscope as shown in figure 1, and also three magnifications of 50, 200 and 500 are used as the measuring conditions. Fig.2 shows the non-contact lighting head (x 50) of a microscope. A lamp in lighting system is set in a microscope controller, and the mirror finished surfaces can be illuminated by a photo fiber cable of circle type. The uniform illumination for the surfaces is possible using this lighting system. Table 1 shows the surface roughness of each testpiece and Table 2 is the experimental conditions. Fig.3 shows the mirror finished surface images in measuring condition of 200 magnifications of microscope. It is well understood that the different image are obtained by each finishing process; grinding, lapping process 1, lapping process 2 and final lapping. As the result, it is possible to construct the neural network which is compose of the features of surface image as input data and the degree of surface roughness and/or kinds of finishing processes as output data.

Table 1. Finishing process and surface roughness

Finishing process	Number of parts	Surface roughness μ m Ra
Grinding process	No.1	0.462
	No.2	0.432
Lapping process 1	No.3	0.016
Lapping process 2	No.4	0.007
	No.5	0.008
Final lapping	No.6	0.003

Table 2. Experimental conditions

Image size	128pixel × 128pixel
Concentration level	256Gradation
Element for shot	CCD image sensor
Light source	100W halogen lamp
Lighting	Oblique irradiation from multi-direction
Resolution (×500)	1 μ m/pixel

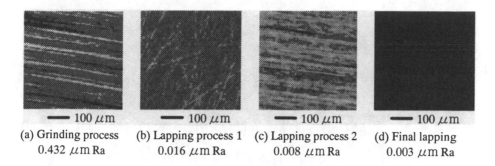

(a) Grinding process	(b) Lapping process 1	(c) Lapping process 2	(d) Final lapping
0.432 μm Ra	0.016 μm Ra	0.008 μm Ra	0.003 μm Ra

Fig. 3. Mirror finished surface images ($\times$200)

3 Texture Analyses for Mirror Finished Surfaces

3.1 Texture Analysis

In general, texture analysis is applied to the statistical treatment of gray level distribution in local area of image processing. The gray level co-occurrence matrix and the gray level run length matrix in the texture analysis[7] are applied to the image processing of mirror finished surfaces. The former analysis is classified into an angular second moment (equation (1)), a contrast (eq. (2)), an energy (eq.(3)) and an entropy (eq.(4)). The others are classified into a short runs emphasis (eq.(5)), a long runs emphasis (eq.(6)), a gray level uniformity (eq.(7)) and a run length uniformity (eq.(8)). Where, $M(f,l)$ is the gray level co-occurrence joint probability, $L(f,g)$ is the distance to each matrix element from the principal diagonal axis of $M(f,l)$, and f and g are the elements of the gray level. $M(f,l)$ is gray level run length probability, and l is the continuous length of the elements. The maximum unit number of the square matrix is 128, and the maximum gray level is set to the value of 256 in the study. It is important in the image processing analysis to decide the scanning direction of texture analyses for mirror surfaces. It is necessary to get the same image features even if the measuring point of microscope, which is faced with mirror surfaces, is rotated for the light axis. If the texture analysis algorithm which is not depended on the scanning directions for surface is developed, it is possible to get the same image features not related to the measuring direction. However, it is difficult to develop the software algorithm. The scanning direction for texture analysis in the research is set to one direction (0 degree) considering the image processing speed, as the results, the measuring condition for CCD camera is set to three directions of 0, 45 and 90 degree for the light axis of microscope as shown in Fig.1. The correlation characteristics between the geometrical properties of mirror surfaces and the image features of texture analyses are investigated in the study. The image features of mirror finished surfaces are indicated by the non-dimensional quantities of gradation. The equation for normalizing is shown in equation (9). Where, X_1 is normalized data, X is gradation value by image feature, X_{max} is maximum value of image feature and X_{min} is minimum value of image feature.

$$I = \sum_f \sum_g \left[\left\{ L(f,g) \right\}^2 \cdot P(f,g) \right] \tag{1}$$

$$C = \sum_f \sum_g \left\{ (f-g)^2 \cdot P(f,g) \right\} \tag{2}$$

$$E = \sum_f \sum_g \left\{ P(f,g) \right\}^2 \tag{3}$$

$$ENT = \sum_f \sum_g \left[P(f,g) \cdot ln \left\{ P(f,g) \right\} \right] \tag{4}$$

$$S_1 = \frac{\sum_f \sum_l \left\{ M(f,l) / l^2 \right\}}{\sum_f \sum_l M(f,l)} \tag{5}$$

$$S_2 = \frac{\sum_f \sum_l \left\{ M(f,l) \cdot l^2 \right\}}{\sum_f \sum_l M(f,l)} \tag{6}$$

$$S_3 = \frac{\sum_f \left\{ \sum_l M(f,l) \right\}^2}{\sum_f \sum_l M(f,l)} \tag{7}$$

$$S_4 = \frac{\sum_l \left\{ \sum_f M(f,l) \right\}^2}{\sum_f \sum_l M(f,l)} \tag{8}$$

$$X_1 = \frac{X - X_{min}}{X_{max} - X_{min}} \cdot 65535 \tag{9}$$

3.2 Results by Texture Analysis for Mirror Finished Surfaces

3.2.1 Characteristics of Image Features by Luminous Intensity

The difference between surface roughness in each finishing process and image features of mirror surfaces must be recognized distinctly as image features to evaluate the surface roughness. The streaks on the surface by grinding processing are recognized as shown in figure 3(a), and the light reflection disperses strongly at the streaks. The other side, the surface after lapping process 2 seems to be smooth as shown figure 3(c), and the image picture is dark. The optimum lighting condition, that the different image features by texture analyses are recognized clearly in each finishing process, is investigated. Table 3 shows the luminous intensity in each lighting level. Fig.4 and 5 show the calculated results of entropy and gray level uniformity by texture analyses. The parameter of the graph is lighting level as shown in Table 3, and 200 magnifications of microscope is applied to the experiment. As the surface roughness increases, these texture features decrease except the surface of lapping process 2. The same feature is obtained in other texture analyses. As the analytical results, it is found that the suitable luminous intensity for the measurement of mirror finished surfaces is from level 2 to level5 at 50 magnifications of microscope, from level 3 to level 6 at 200 magnifications, and from level 5 to level 8 at 500 magnification. The luminous intensity of level 4 at 50 magnification, and level 7 at 200 magnification are set in the experiment.

Table 3. Luminous intensity in each lighting level

Lighting level	Luminous intensity lux		
	× 50	× 200	× 500
Level 1	4580	11700	12600
Level 2	7780	18000	20600
Level 3	12000	27700	32700
Level 4	15800	39100	46900
Level 5	21500	51600	64500
Level 6	28300	69800	82600
Level 7	34700	86000	−
Level 8	47300	−	−

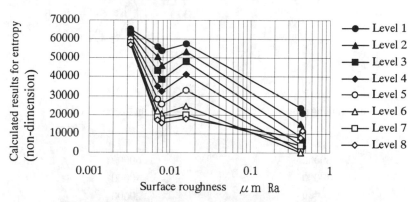

Fig.4. Effect of surface roughness on entropy (× 200)

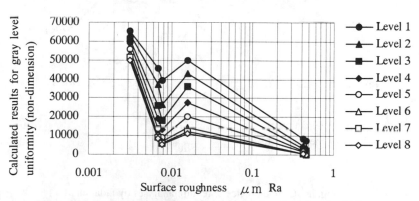

Fig.5. Effect of surface roughness on gray level uniformity (× 200)

3.2.2 Characteristics of Image Features by Magnification of Microscope and Geometrical Property of Mirror Finished Surface

The correlation characteristics between the geometrical property of mirror finished surfaces and the image features of texture analyses are investigated in the chapter. Fig.6 shows the radar charts for the characteristics of image features by texture analyses at 200 magnification of microscope. The parameter is measuring direction of CCD camera shown in Fig.1. Fig.6(a) is the image features by texture analyses for the surface after grinding process, and it is clear that the chart pattern differs slightly by the measuring direction for the regular streaks on the surface. Especially, it is well understood that angular second moment, contrast and run length uniformity depend on the influence of surface geometry. Fig.6 (b) shows the image features for the surface after lapping process 1. The irregular lapping streaks on the surface are recognized,

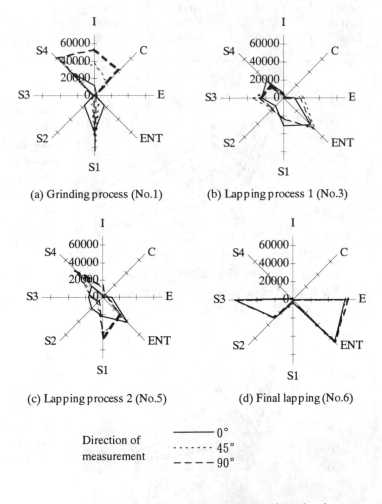

Fig.6. Characteristics of texture analyses by radar chart

and the streaks on the surface are not shown. Therefore, the pattern of radar chart is almost same. Fig.6 (c) shows the image features for the surface after lapping process 2, and it is clear that the chart pattern differs by the measuring direction for regular streaks in longitudinal direction of testpiece. Fig.6 (d) shows the image features for the surface after final lapping process, and it is well understood that the pattern of radar chart is almost same, and the image features do not depend on the measuring directions, because the surface after final lapping is very smooth and there are no thin lapping streaks in condition of 200 magnifications of microscope. The almost same results are obtained for other experimental conditions.

4   Construction of Neural Network and the Verification Results

Fig.7 shows the neural network system developed in the study. The back propagation learning rule of three layers is applied to the system. The input data is the value of eight image features by texture analyses which are shown from equation (1) to (8). The output data is four kinds of surface roughness and/or four kinds of finishing processes. The neural network systems are classified to two types. The one is the system that the measuring direction is set to one direction, 0 or 45 or 90 degree, and as the results, the three kinds of the neural networks are constructed. The another is the system that the measuring direction is set to the total sum of 0, 45 and 90 degree, and the system is learned using these measured data. The magnification of microscope is set respectively to three of 50, 200 and 500. As the results, twelve neural networks are constructed in the study. It is found that the good convergence of square error at the 1,000 to 1,200 iterations in the learning operation was obtained in each network system in case the number of elements of hidden layer is set from 8 to 12. Table 4 shows the verification results for the prediction of surface roughness. Table 4 shows the recognition ratio for the number of testpieces for verification when the microscope

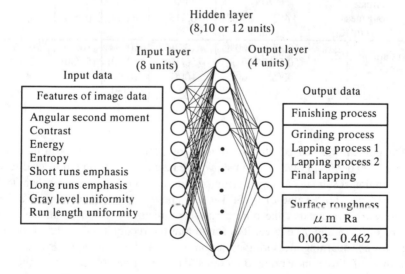

Fig.7.   Neural network system for mirror finished surfaces

Table 4. Verification results of neural network for mirror finished surfaces

Finishing process		Grinding process		Lapping process 1	Lapping process 2		Final lapping
		Recognition ratio					
Number of parts		No.1	No.2	No.3	No.4	No.5	No.6
Surface roughness μm Ra		0.462	0.432	0.016	0.007	0.008	0.003
0°		100% (45/45)	100% (45/45)	100% (40/40)	100% (45/45)	100% (45/45)	100% (40/40)
45°		100% (45/45)	100% (45/45)	100% (40/40)	100% (45/45)	100% (45/45)	100% (40/40)
90°		100% (45/45)	100% (45/45)	100% (40/40)	100% (45/45)	100% (45/45)	100% (40/40)
Learning of three directions	0°	100% (45/45)	100% (45/45)	98% (39/40)	100% (45/45)	100% (45/45)	100% (40/40)
	45°	100% (45/45)	100% (45/45)	100% (40/40)	100% (45/45)	100% (45/45)	100% (40/40)
	90°	100% (45/45)	100% (45/45)	100% (40/40)	98% (44/45)	100% (45/45)	100% (40/40)

Threshold = 0.75

Table 5. Verification results of neural network for mirror finished surfaces

Finishing process		Grinding process		Lapping process 1	Lapping process 2		Final lapping
		Recognition ratio					
Number of parts		No.1	No.2	No.3	No.4	No.5	No.6
Surface roughness μm Ra		0.462	0.432	0.016	0.007	0.008	0.003
Learning of three directions	23°	100% (50/50)	100% (50/50)	100% (50/50)	100% (50/50)	100% (50/50)	100% (50/50)
	68°	100% (50/50)	100% (50/50)	100% (50/50)	100% (50/50)	100% (50/50)	100% (50/50)

Threshold = 0.75

of 200 magnifications is used. Upper column shows the results for measuring direction of 0, 45 and 90 degree. Lower column shows the results for learning of three directions. The recognition results in each measuring condition is very high (greater than 98%) in condition of threshold value of 0.75. Table 5 shows the verification results for the neural network system that the mirror surfaces are measured in condition of three directions of 0, 45 and 90 degree. It is found that the recognition results of the neural network for non-learning measuring directions of 23 and 68 degree is also very high if the features of three measuring directions data are used to construct the neural network.

5 Conclusions

The conclusions drawn from the research are as follows ;

(1) It is found that features of image data of mirror finished surfaces after grinding and lapping processes can be decided by texture analyses.

(2) It is understood that the optimum measuring condition can be decided by the correlation analyses between the measuring conditions and the characteristics of image features of texture analyses. Additionally, the correlation between the influence between geometrical direction of mirror finished surfaces and the image features is cleared.

(3) It is possible to construct the neural network system, which is composed of image features of mirror finished surfaces as input data, and surface roughness and/or kinds of finishing processes as output data.

(4) It is shown that the neural network system has rapid convergence characteristic, and the recognition results of surface roughness of each mirror finished surface are fairly good. The vision system which is developed in the study has great advantages as non-contact, in-line, high speed and free camera positioning system.

References

1. K. Minoshima : Automated distinction cross section surface after cutting using image processing, Journal (A) of JSME, **56**, 525 (1990) 1317. (in Japanese)

2. A. Ishii et al : Automated distinction of surface integrity using a neural network, 6th destruction dynamics symposium, Japan Society of Materials, (1991) 12. (in Japanese)

3. K. Odo et al : Texture analyses for frictional surfaces by image processing, Proceedings of Kyushu branch of JSME, **46** (1993) 136. (in Japanese)

4. Y. Kawaguchi : Application of image processing for evaluation of blasted surfaces, Report of Chugoku industrial technology laboratory, **34** (1990) 21. (in Japanese)

5. Huynh.V.M et al : Texture analysis of rough surface using optical fourier transform, Meas. Sci. Technol., **2**, 9 (1991) 831.

6. K. Kato et al : Development of visual evaluation system for mirror finished surfaces, Proceedings of JSPE, (1995) 253. (in Japanese)

7. For example, H. Ozaki : Image processing - From it's basic to application - , the second edition, (1988) (in Japanese)

A SURVEY OF NON-CONTACT 3D SURFACE IMAGING TECHNIQUES FOR CRANIOFACIAL AND RECONSTRUCTIVE SURGERY

* B T Sim, C K Chua, S M Chou, W S Ng
\# S T Lee, S C Aung

* School of Mechanical and Production Engineering
Nanyang Technological University, Singapore

\# Department of Plastic Surgery
Singapore General Hospital

Abstract. Traditionally, in the planning of cosmetic or reconstructive surgery, the surgeon has to rely heavily on manual measurement of the human anatomy, which is laborious and error prone, and the photographs of the patient. With the advanced development in computer technology, simulation of the surgery is now possible by manipulating the 3D model of the human parts, which is digitised using the non-contact, optical imaging technique. Other benefits include surface measurement, monitoring of pre- and post-intervention results, design and fabrication of prosthesis, etc. A collaborative project between the Nanyang Technological University and Singapore General Hospital is carried out to design a new portable and mobile 3D surface imaging system. This paper presents the results of part of the project: the review of the current state-of-the-art 3D imaging techniques. The laser stripe scanning, the moiré fringe contouring and the holographic method are studied and compared for their feasibility in the digitisation of the human body.

1 Introduction

3D optical surface imaging techniques have been widely used in a variety of applications including apparel design, quality inspection, robot navigation, robotics assembly, gauging, etc. [1]. These non-contact measurement techniques, which offer the ability to measure the shape of the objects accurately in a short period of time, have also successfully found its niche in the biomedical field. Craniofacial surgeons can use 3D surface scanning to obtain a facial model of the patient before surgery, modify the model, and show the patient the anticipated outcome of the surgery. The digitised image will also aid the surgeons in the pre-planning of the reconstructive surgery by making critical anthropometric measurement. In addition, breast augmentation, and the design and fabrication of prosthesis will also benefit form this technology.

The current state-of-the-art optical surface imaging techniques are: laser stripe scanning and moiré fringe contouring. Linney et al have developed a laser-based scanning system for the measurement of the human anatomy [2-4]. Furthermore, active developers like Cyberware Inc. [5], 3D Scanners [6], Laser Design Inc. and Digibotics have commercial laser systems to produce 3D models for clinical application as well as animation industry. Marshall et al [7-9] have developed a 3D clinical facial imager, which utilises the projection moiré fringe contouring method to acquire accurate 3D coordinates from the surface of a patient's face. The Cencit system developed by Vannier et al [10-11] has employed 6 sets of cameras and projectors for whole body surface imaging. For works in apparel design and human engineering, a moiré-based system known as Phase Measuring Profilometry [12] is being developed by the Textile and Clothing Technology Corporation in Cary, North Carolina. Several research groups have also developed their own 3D optical active imaging systems for research and industrial applications [13-18].

This paper describes part of the collaborative project between Nanyang Technological University and Singapore General Hospital, and reviews the current state of the optical based 3D imaging techniques. Three non-contact optical techniques: laser stripe scanning, moiré fringe contouring and holographic method are studied for their feasibility in the digitisation of the human body. These techniques are then compared in detail with reference to parameters such as the acquisition time, accuracy and resolution.

2 System Performance Parameters

Before going into the description of various imaging technique, one must understand the common parameters that are used to specify a 3D imaging system. Based on these parameters, one can define the requirements of a imaging system, and decide the suitability of a particular imaging technique for certain application, such as biomedical imaging. Thus, the application or problem must be properly defined before a solution is determined. The various parameters are depicted in Figure 1.

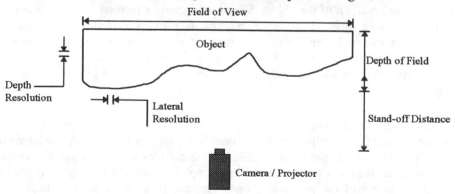

Figure 1. Performance parameters

Another critical parameter, which is especially concerned in the imaging of live subjects, is the cycle time required to capture and display the 3D model. The total time required is usually divided into two categories: image acquisition time and data processing time. The image acquisition time is the time taken to capture the full image of an object. The captured image is then digitised and fed into the computer. The three-dimensional coordinates are then computed by triangulation and the data is processed to display in a different format selected by the user. These processes will contribute to the data processing time. In order to minimise the total imaging time, both the image acquisition process and data processing can work simultaneously by overlapping the data capturing for one measurement with the processing of data for previous measurement. It should be noted that the accuracy of an imaging system can be enhanced by slowing down the acquisition process. Thus, there is a trade off between accuracy and data capturing time. For data processing, the required time will be determined by the data size, processing algorithm and the system hardware (such as the use of a transputer).

In addition to the above parameters, one should also consider the ambient lighting condition. Since most commercial scanners are active systems which project structured illumination onto the object, intense ambient light will increase the noise level and impair the accuracy of the results. Thus, the ambient light should be carefully modulated or the system should be designed to eliminate the unwanted light. Three methods are suggested:

a.) By subtracting the image of the object with a reference image, the projected stripe or other structured pattern will be highlighted. In this case, two images are captured, one with no illumination, and the other with illumination.

b.) By fitting an optical filter in front of the camera, the ambient light can be rejected and allow only the structured light to pass through.

c.) By enclosing the object and the imaging system in a dark chamber.

3 3D Surface Imaging Techniques

In the development of an active optical non-contact imaging system, most of the works have employed the structured light techniques to illuminate and extract the geometric information of the object. Structured light is a method in which a regular geometric pattern (dots, single stripe, multiple lines) is projected onto the object's surface, and viewed by one or more cameras at different positions. In this section, three techniques are discussed: laser stripe scanning, moiré fringe contouring and holography. The first two methods use the triangulation method to compute the 3D data.

3.1 Laser Stripe Scanning

The set-up of the laser stripe scanning system is depicted in Figure 2. A low intensity, eye-safe laser beam is passed through a cylindrical lens so as to project a vertical stripe onto the scanning surface. A charged-coupled device (CCD) camera is used to record the image of the projected stripe. Viewing at an angle to the illuminating source, the laser stripe is distorted according to the object's shape. The laser stripe is then recorded by the CCD camera, and digitised by a frame grabber. The frame grabber is a device that interprets the analogue modulated signals output from the camera as a frame of pixels. The digitised data is stored either in the frame buffer or read directly into the computer memory. To get the full view of the object, the object is rotated on a platform under computer control to obtain a 360° coverage, while the image capturing unit (source + camera) remains stationary. Alternatively, the scanner can rotate around a stationary object. As a result, a complete three-dimensional image of the object is captured. In general, the capturing time varies from 15 to 30 sec, depending on the type of application, working range, data size and the required accuracy.

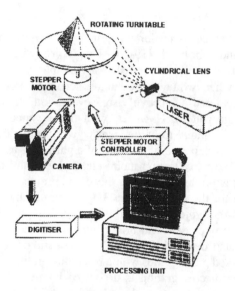

Figure 2 . Laser stripe scanning system. From S Sunthankar, SPIE, vol. 1826, pp 495, (1992).

Next, the image data is processed to obtain the 3D coordinates of the surface points lying along the projected laser stripe. The x and y dimensions can be readily determined by spatial measurement in the image plane, while the third dimension (depth) is computed using the triangulation principle. In addition, a calibration factor is to be included to transform the pixel values to real world coordinates. Thus, a calibration program must be carefully designed and performed to calculate accurate and reliable results.

Finally, the image data file is displayed in surface model, wireframe, etc. depending on the user's selection. Subsequently, for clinical application, further development of more comprehensive software is necessary to manipulate the data. The software will benefit the surgeon in pre- and post-intervention assessment, and the planning of surgical procedures for reconstructive surgery. Besides displaying the 3D model in any desired viewpoint, other graphics facilities should include distance, area and volume measurements, editing of surface shape, and the anatomical registration of images. In addition, the data file can be imported to the CNC machine or rapid prototyping system for the fabrication of 3D models and prostheses. These 3D models will enable the patient to visualise the results of the intervention.

To ensure the accuracy of the scanned data, the projected laser stripe must coincide with the axis of rotation of the turntable. A disadvantage of the laser stripe scanning is the shadowing effect of surface concavities, such as the area under the chin, around the eyelid, the inside of the nostrils, and the wounded surface. Surface infoldings such as the surface behind the ears are still unattainable by all optical scanning system. Furthermore, with an image capturing time of more than 10 sec, this technique is prone to error due to movement artifact, especially with children or infants.

3.2 Moiré Fringe Contouring

The word "moiré" is the French name for a fabric known as watered skill, which exhibits patterns of light and dark bands [19-20]. This moiré effect occurs when two grid patterns of similar period are superimposed. For example, this effect can be seen in two overlapping nets or when two layers of window screen are in contact.

Early work in this field required subjective visual evaluation, or laborious manual analysis of the moiré fringe pattern. However, with the advanced development in computer and peripherals, this moiré technique is revived. In the industry, there are a number of moiré methods developed to fit the need of a particular application, like the shadow moiré, projection moiré, reflection moiré etc. These moiré methods differ in how the deformed grating is produced. Among these, the projection moiré method is found to be more suitable for the application in 3D imaging of live human subjects. The schematic diagram of the projection moiré is shown in Figure 3. In this case, a line grating, which is an array of dark and bright lines, is projected onto the object. The deformed grating is also detected by a CCD camera linked to a computer. By superpositioning the projected grating lines with a reference grating, the moiré fringes are generated. Since the projected grating is deformed by the irregularities on the object's surface, the resulting moiré pattern describes the surface contours, resembling how a topographic map delineates the contours of the land (see Figure 4).

The moiré pattern can only provide a qualitative visualisation, although it contains the depth information of the object represented by the contour interval. In order to obtain the quantitative results (3D coordinates), a technique known as phase-shifting is selected with the aim of automating the analytical process, and improving the system resolution. Phase-shifting involves shifting of the projection grating or the reference grating by predetermined steps (fraction of the grating's pitch), and determining the intensity of the resulting fringes. A minimum of 3 shifted images are sufficient for the evaluation of the moiré fringe pattern. The fringe pattern is then transformed into a phase distribution, which automatically detects the convexity and concavity of the object's surface. The difference in phase together with a calibration factor are used to calculate the depth of the object with respect to a specific plane. The calibration factor is initially obtained by performing a calibration of an object of known depth. Thus, a set of 3D coordinates are derived and displayed in surface models as described in section 3.1.

Using the phase-shifting technique, a precise mechanism is required to shift the grating by a proportion of its pitch. For example, the Auto-MATE system [7] makes use of the piezo-electric translator which is coupled to a precision translation stage onto which the projection grating is fixed. Alternatively, avoiding the use of a mechanical translator, Asundi [21-23] has described a computer-aided method, which shifts a computer generated grating precisely with no moving parts. The computer grating is made up by an array of dark and bright pixels. The dark pixel has a binary value of 0, whereas the bright pixel has a binary value of 1. As a result, the pitch of the grating, which is the sum of the dark and bright line width, can be easily created for a particular application. Moreover, the grating can have any profile like sinusoidal, square or triangular profile. With the help of a liquid crystal display (LCD) projector, the computer grating can be projected, shifted and analysed using the phase-shifting method. The LCD projector, linked to the computer, is able to display the contents on the computer monitor via a suitable graphics adapter. Furthermore, Asundi has also introduced another method to obtain the moiré patterns by superpositioning of the computer gratings using logical operations. For example, the logical AND of two one-bit binary digital gratings will have the same results equivalent to that obtained by physical superposition of two gratings. More detail on this operation is presented in [23].

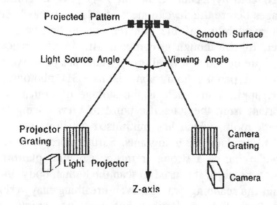

Figure 3. Projection moiré configuration. From P J Besl [1].

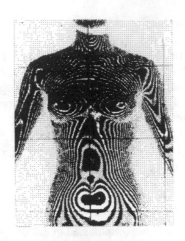

Figure 4. Infinite fringe moiré pattern of a human body.
From H. Takasaki, Appl. Optics 9, 1469 (1970).

3.3 Holography

Holography was discovered by Dennis Gabor in 1947 [24], in connection with 3D viewing of x-ray images before the development of a laser. It has become a practical technique only with availability of the coherent monochromatic light which is obtained from a laser. Its use though growing, is still in its infancy. The lack of maturity of this technique provides several areas which are of interest to researchers for exploration. To most people, holography is like 3D photography. However, besides 3D art work, applications such as the storage of optical information, are probably more important from the scientific point of view. Using the holographic method, a special dark room, which has minimised air flow and sufficient sound proofing, is required to produce the holograms. Furthermore, it is very sensitive to small disturbances and requires a strong frame or isolation platform to eliminate external vibrations. As a result, if it is used to scan the human body, the patient has to remain still throughout the scanning process, even breathing may seriously affect the accuracy of the result. In medical applications, holographic techniques have been used with a certain degree of success in dentistry, urology, otology and orthopaedics [26]. Thus, due to the limitations mentioned above, the holographic method is rejected as the imaging technique for a mobile and portable scanning system.

4 Evaluation of Imaging Techniques

The performance parameters of laser stripe scanning and the moiré topography are summarised in table 1. More detailed comparison of the performance parameters is described the following sections.

Table 1. Laser stripe scanning versus Moiré topography

Performance parameters \ Imaging techniques	Laser stripe scanning	Moiré topography
Accuracy \ resolution	1 μm	1 μm
Depth of field	100 m	10 m
Acquisition time	7 - 15 sec or more	less than 4 sec
Data processing time	30 sec or more	10 sec to few min
Lighting condition	Dim lighting condition	Dim lighting condition
Safety	Eye safe for Class I to Class IIIa laser	Ordinary white light
Field of view	Scanning mechanism required	Vary

4.1 Acquisition time

From table 1, both techniques satisfy the design requirements, and the results are relatively similar. Their suitability for a specific application will depend on the importance of each design parameter. In fact, both the laser stripe scanning and the moiré topography employ the same triangulation principle to obtain the coordinates of the surface points. The laser stripe scanning projects line by line onto the subject during the scanning process while the moiré method projects a series of lines at one time. As a result, the acquisition time for laser stripe scanning is greater than the moiré system due to the horizontal scanning of the laser beam. By adding moving components, the cost of the system will increase significantly for the expensive hardware. In addition, the control of the motion mechanism constitutes another problem, especially in the synchronisation of the moving parts and the rate of data capturing. On the other hand, the acquisition time for moiré system is far superior because there are no moving parts involved. The short capturing time will allow the patient to be scanned during a single breath hold, and motion artifact can be avoided. This will result in a better accuracy for breast scanning, and enable the scanning of young children to be possible. Furthermore, without the mechanical motion required, the errors resulting from vibration, backlash, wear and tear etc. can be eliminated.

4.2 Data processing time

Data processing time for both techniques depends on a few factors such as data size, processing algorithm, types of frame grabber etc. Quick processing time to display the scanned image of the patient will enable the surgeon to visualise the quality of the scan. If the image is distorted due to noise or patient's movement, another scan can be made immediately, instead of asking the patient to come back after a few days, and the scan area can be selected to focus on the critical features. A data processing time of about 1 minute is practical according to the surgeon in SGH. For laser stripe scanning, this requirement is achievable. For example, the facial scanner developed by the University College of London [4] is able to produce the 3D surface model in less than 30 sec with the help of a transputer based processing board, while for moiré topography or patterned light projection, the processing time will depend on the type of fringe analysis algorithm and the method to obtain the interference images or the projected pattern. In general, there is an additional step to identify and locate each stripe of projected light or grid intersection, before applying triangulation to compute the 3D data sets. Marshall et. al. [7] had developed a clinical facial imager, which employs the projection moiré fringe contouring method and has a data capture time of 1.8 sec but a data processing time of about 600 sec. However, the long processing time may be due to the merging of different data sets (from different views) to obtain a full 3D model. On the other hand, the moiré method can produce the instantaneous moiré pattern for qualitative assessment.

4.3 Accuracy and resolution of laser stripe scanning

From table 1, the accuracy and resolution of the laser scanning system can be as fine as 1 μm, which is far superior than the design requirement. However, in 3D surface imaging of the human body, the accuracy and resolution of commercial systems of Cyberware Inc. and 3D Scanners Ltd are claimed to be around 0.5 mm. Keep in mind that high accuracy and resolution can be achieved by increasing the acquisition time and reducing the working range. Thus, a trade-off between these parameters is made to suit the requirements of the application. The resolution of the laser scanning system depends on the size of the CCD array. The standard number of points measured along the laser stripe is either 256 or 512. By reducing the laser spot size or the width of the laser stripe, better resolution can also be obtained. Unfortunately, there are practical limits to the possible reduction of the laser beam's footprint, i.e. the corresponding reduction of the working range. In addition, the laser beam is easy to focus and its intensity is higher than the normal white light employed in moiré topography. Thus, superior quality images can be obtained. Despite the high resolution, a problem arises when there is a sharp change in the surface height (such as at the edge) that occur between two successive laser stripes. The sensor will fail to detect the abrupt change, and the resulting data will be inaccurate. Besides these, the laser stripe scanners tend to be more accurate at the centre of each sweep and less accurate at the fringes.

4.4 Accuracy and resolution of moiré fringe contouring

The moiré system is capable of achieving the same resolution and accuracy of 1μm as in the laser-based system. However, in the 3D digitisation of human body, problems such as the light scattering properties of the human skin, the skin colour and the complexity in the body shape, like the face, will affect the results. The performance of an optical sensor depends greatly on the detection of light scattered from the surface of the object [8]. Human skin is not completely opaque, which allows a proportion of the projected light to be scattered directly from the surface, while a proportion will penetrate through the skin and is scattered back form the flesh immediately beneath [8,9]. This scattered light will reduce the contrast of the projected grating lines, as well as the contrast of the laser stripe for the laser system. To improve the image contrast, one solution will be the use of a more powerful light source. Another solution will require makeup to apply on the surface to enhance the quality of the image captured. Nevertheless, the facial imager developed by Marshall et al [7] has achieved a resolution and accuracy of approximately 0.5 mm. Besides the illuminating source, the sensitivity of the moiré method can be improved by varying the pitch of the grating. With the help of the computer generated grating, the pitch can be changed rapidly and easily, and a minimum pitch can go to as fine as 2 pixels.

4.5 Field-of-view

Another parameter to be considered for a moiré system is the Field-Of-View (FOV) of the light detector. If the sensor has a small FOV, more than one scan will be needed to capture the whole surface area. Consequently, this will cause a registration problem to merge a few sets of data to produce a whole image, and the processing time will increase significantly. Although it is possible to merge the measurement data from two measurements at different views using one set of the projector-camera system, possible movement of the subject between the measurements can introduce serious errors in the merged data. The solution adopted is to make simultaneous measurements using two camera-projector units. The Cencit system [10], for example, has six projector-camera sets. Since the laser scanning system uses motion mechanism to scan the whole area of interest, there will be no problem concerning FOV. However, merging of data sets is still required for whole body scanning as the length of a laser stripe can only cover a certain part of the body. The Cyberware WB4 system [5] requires four scanning units to capture the whole human body.

4.6 Lighting condition

The laser scanning system needs to be operated in a low levels of ambient light because the intensity of the visible laser beam is between 1 to 5 mW. The intensity of the laser beam can be increased to enhance the light detection by the camera, but this may endanger the patient. For eye safe application, the laser intensity must be kept below 5 mW, i.e. Class I to Class IIIa laser. For a moiré system, although the light source is absolutely safe to the human eyes, the intensity of the white light is lower than that of the laser beam, the ambient light must therefore be modulated to increase the signal-to-noise ratio. Other methods commonly used to reduce the stray light problem are described in section 2.

4.7 Multiple view point

In general, for any triangulation system, occlusion is unavoidable. It occurs when the camera's view of the illuminated stripe on the object is blocked by an extrusion part of the object. Another problem of a single view system is that of obscuration, since the camera can only detect the scene that is visible to it. These problem can be overcome by using more than one sensor to capture the image multiple view point. For a moiré system, multiple sensors are required to resolve the FOV problem described above.

5 Conclusion

The current state-of-the-art optical sensing techniques are reviewed and discussed for their application in 3D digitisation of the human body. The three techniques are evaluated with respect to the performance requirements of the new scanner to-be-developed, such as the accuracy and resolution, acquisition time, data processing time, and the scan area. Holography is found to be unsuitable for the designated application. Both the laser and moiré techniques are comparable in performance, and the major differences are image capturing time and the cost of the system. For the commercial systems, the Cyberware WB4 system costs about $400,000, while the PMP moiré-based system is targeted at a price of $50,000[12]. Based on these findings, further development work on the conceptual design of the new scanner, which utilises the selected imaging techniques, can then proceed.

Acknowledgements

The authors wish to thank Miss Fiona Khong Lai Meng for her help in the research work. This project is funded by the National Medical Research Council (NMRC), Singapore.

References

1. P J Besl, "Active Optical Range Imaging Sensors", Machine Vision and Applications, 1: 127-152, 1988.
2. A D Linney, A C Tan, R Richards, J Gardener, S Grindrod and J P Moss, " Three-dimensional visualization of data on human anatomy: diagnosis and surgical planning", Journal of Audiovisual Media in 1993, 16: 4-10, 1993.
3. S Arridge, J P Moss, A D Linney and D R James, "Three-dimensional digitisation of the Face and Skull", J. max. -fac. Surg., vol. 13, pp. 136-143, 1985.
4. J P Moss, A D Linney, S R Grindrod and C A Mosse, "A Laser Scanning System for the Measurement of Facial Surface Morphology", Optics and Lasers in Engineering, vol. 10, pp. 179-190, 1989.
5. Cyberware Inc., "WB4: Whole Body Scanner", http://wwww.cyberware.com
6. 3D Scanners Ltd, "Products", http://www.3dscanners.com
7. S J Marshall, R C Rixon, D N Whiteford and J T Cumming, "Development of a 3-D clinical Facial imager", Proc. SPIE, vol. 1889, pp. 255-262, 1993.
8. S J Marshall, G T Reid, S J Powell, J F Towers and P J Wells, "Data capture techniques for 3-D facial imaging", Computer Vision and Image Processing, ed. A N Barrett, pp 248-275, Chapman and Hall, London, 1991
9. S J Marshall, R C Rixon, D N Whiteford, P J Wells and S J Powell, "The development of a 3-D data acquisition system for human facial imaging", Proc. SPIE Medical Imaging IV, vol. 1231, pp 61-74, 1990
10. Internet document: Surface Imaging Group, "Surface Imaging of the Human Body", Mallinckrodt Institute of Radiology, Washington University School of Medicine.
11. M W Vannier, G Bhatia, P Commean, T K Pilgram and B Brunsden, "Medical facial surface scanner", Proc. SPIE Image Capture, Formatting, and Display, vol. 1653, pp. 177-184, 1992.
12. S Paquette, "3D Scanning in Apparel Design and Human Engineering", http://www.computer.org/pubs/cg&a/apps/g50011.html.
13. K C Ng, B F Alexander, S H Boey, S Daly, J C Kent, D Q Huynh, R A Owens and P E Hartmann, "Biostereometrics - A Noncontact Noninvasive Shape Measurement Technique for Bioengineering Applications", Australasian Physical & Engineering Sciences in Medicine, vol. 17, no. 3, pp. 124-130, 1994.
14. C B Cutting, J C McCarthy and D B Karron, "Three-dimensional Input of Body Surface Data Using a Laser Light Scanner", Annals of Plastic Surgery, vol. 21, no. 1, pp. 38-45,

1988.

15. T T Wohlers, "Reverse Engineering Systems: From Product to CAD and Back again", Cadence, January, pp 45-57, 1993.

16. S M Dunn, R L Keizer and J Yu, "Measuring the Area and Volume of the Human Body with Structured Light", IEEE Transactions on Systems, Man, and Cybernetics, vol. 19, no. 6, pp. 1350-1364, 1989.

17. K Patorski, M Rafalowski and M Kujawinska, "Computer aided postural deforming studies using moiré and grid projection method", Proc. SPIE, vol. 2340, pp. 442-448, 1994.

18. S T Young, S W Yip, H C Cheng and D B Shieh, "Three-Dimensional Surface Digitizer for Facial Contour Capture", IEEE Engineering in Medicine and Biology, pp 125-128, 1994.

19. O Kafri, I Glatt, The Physics of Moiré Metrology, Wiley, New York, 1991.

20. J W Dally, W F Riley, Experimental Stress Analysis, McGraw-Hill, 3rd ed., New York, 1991.

21. A Asundi, "Computer Aided Moiré Methods" Optics and Lasers in Engineering, vol. 18, pp. 213-238, 1993.

22. A Asundi, "Moiré Methods using computer-generated gratings", Optical Engineering, vol. 32, no. 1, pp. 107-116, 1993.

23. A Asundi and K H Yung, "Phase-shifting and logical moiré", Journal of the Optical Society of American A, vol. 8, no. 10, pp 1591-1600, 1991.

24. W Hawkes, OptoElectronics An introduction, 2nd edition, pp. 243-253.

25. Chaimowizz, J.C.A, Holography Technology, pp. 259-270.

26. R Jones and C Wykes, Holographic & Speckle Interferometry, 2nd edition, Cambridge University Press, Cambridge, 1989.

27. P C Mehta and V V Rampal, Lasers and Holography, World Science, Singapore, 1993.

A Condition Monitoring and Fault Diagnosis System for Modern Manufacturing Equipment

Wenbin Hu, Zude Zhou and Youping Chen

School of Mechanical Science and Engineering,
Huazhong University of Science and Technology,
Wuhan,Hubei,430074,P.R.China
Email: wbhu@hotmail.com

Abstract. After analysis of the monitoring and diagnostic tasks, a condition monitoring and fault diagnosis system is designed and implemented, which covers the major aspects in the condition monitoring and fault diagnosis of modern manufacturing equipment. The overall hardware and software design of the system, approach to some functional sub-systems or modules are presented. In particular the condition monitoring, data acquisition, knowledge acquisition and representation, as well as the integrated diagnostic reasoning strategy of the integrated expert system are presented in detail.

1 Introduction

The large-scale automation and integration of modern manufacturing equipment, which has become possible with the development of low-cost digital computers and communication networks, permits more efficiency and flexibility in meeting production schedules and can potentially lead to lower-cost and higher-quality products. This kind of manufacturing equipment such as FMS and CIMS, is being more and more widely used because of its potential to improve the strategic and competitive position of firms.

However, such a manufacturing equipment is very dependent upon the trouble-free operation of all its component parts. When a fault occurs, it is critical to isolate the causes as rapidly as possible and to take appropriate maintenance action. For this reason, corresponding diagnostic techniques and systems are studied extensively with the application and dissemination of modern manufacturing equipment. The condition monitoring and fault diagnosis system for modern manufacturing equipment, presented in the following, is designed to meet this need.

2 Monitoring and Diagnostic Tasks Analyzed

Condition monitoring and fault diagnosis of modern manufacturing equipment means a group of special control functions, namely:
* observation of the machine and process condition by sensors;
* recognition of any incorrect event;

- decision on the necessary control action;
- analyzing the diagnostic information;
- displaying and storing the fault information for maintenance planning.

The fundamental tasks of modern manufacturing equipment is to realize the relative motion of the work-piece and the tool and the auxiliary motions setting out the geometric and technological data of the part program. These motions are to be previously and deterministically planned by the programmer. An unexpected and catastrophic change of the condition may cause the machine or process faults.

In general, the correct operational behavior of manufacturing equipment may be characterized by a series of state transitions of the equipment used during the manufacturing of a product. These state transitions occur because of the proper functioning of causal agents responsible for the transitions. The state is characterized by discrete and continuous variables.

Discrete state variables are not only binary or digital control signals but also the signals of switching sensors observing the motion of auxiliary and positioning mechanisms. The error-free cycle of these mechanisms is the foundation of automatic operation. The tasks of monitoring and diagnosis based on the discrete variables are:
- diagnosis of the CNC functions;
- diagnosis of the integrated PLC's;
- supervision of the right schedule of the cyclical operating mechanisms.

Continuous state variables are the sensor signals measuring the physical state of the machine or process. The tasks of monitoring and diagnosis based on the continuous state variables are:
- monitoring and diagnosis of machining operation;
- indirect monitoring and diagnosis of the most important mechanisms, gearboxes, and functional elements of the manufacturing equipment;
- monitoring and diagnosis of the cutting features of the tools.

3 Overall Hardware and Software Design

Taking into consideration the rapid development of hardware and software tools and the continuously increasing demands of the industry, the design of new, more complex and powerful monitoring and diagnosis systems is necessary.

The vast majority of modern manufacturing equipment have automatic monitoring available for faults characterized by discrete state signals in their controllers. These discrete state signals indicate the machine operating state, by which further diagnosis can be carried out. These signals can be obtained directly by using an on-line linkage between the controllers and the monitoring and diagnostic system computer (This can also be carried out via several information-technical levels using LAN).

However, there is little evidence that the machine and process conditions are continuously monitored inside the manufacturing equipment. Those continuous state variables must be acquired by designing an external data acquisition system. Hence the hardware of the integrated monitoring and diagnosis system with a modular structure is designed as shown in Fig. 1.

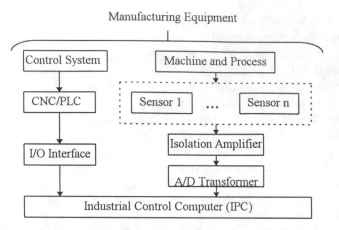

Fig. 1. The hardware block diagram of the system

In order to automate the condition monitoring and fault diagnosis, artificial intelligence in particular expert system is often adopted. However, a highly automatic and integrated manufacturing equipment is a combination of very complex machinery integrated through an equally complex computer system. It is not only limited to just mechanical parts, but will also have electronic, hydraulic, software and human elements, each having different fault distributions. Traditional expert system is far unequal to the task. Therefore, an integrated expert system is designed.

Fig. 2 shows the software block diagram of the integrated monitoring and diagnosis system. This is an integrated expert system with also a modular structure. It is the integration of several functional sub-systems or modules, the integration of numerical computation and symbolic reasoning, etc. All these together form a distributed large expert system. Besides these sub-systems or modules, there is a special sub-system called metal-system, which manages, coordinates and controls the whole system, calls relevant sub-systems or modules to complete corresponding tasks and provides a good environment for man-machine interaction.

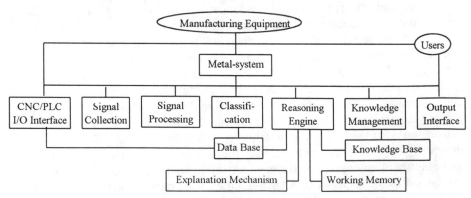

Fig. 2. Integrated monitoring and diagnosis expert system

4 Approach to Some Functional Sub-systems or Modules

In the integrated condition monitoring and fault diagnosis expert system, condition monitoring, data base, knowledge base and reasoning engine are three major functional sub-systems or modules. A more detailed description is given below.

4.1 Condition Monitoring

Early condition monitoring systems relied on the sensing and processing of a single parameter by a single sensor. This kind of monitoring strategy is simple and has poor usability and often brings about false or incomplete diagnosis. This single-sensor and single-parameter strategy is only suited for the monitoring of simple process with a single condition or not frequently changed conditions.

The processes of modern manufacturing equipment are complex and changeable. All parts associated with these processes are closely related to each other. The processes involve a large number factors and the relationship between these factors and processes is very complex and to some extent is fuzzy. To these situations, traditional single-sensor and single-parameter monitoring strategies are powerless. Therefore, a fuzzy hybrid strategy with multiple sensors and multiple parameters must be used to extract multiple integrated parameters from those important parts of the manufacturing equipment, and then hybrid analysis and judgment are carried out so as to reach a significant conclusion from multiple associated parameters.

Parameters Monitored

Considering the sensitivity and easy acquisition, parameters including power, vibration, temperature and pressure are chosen and the monitored objects are spindle, feed axes in three directions (X, Y, Z), hydraulic oil and pneumatic gas (Table 1).

- Power parameters: voltage (U), spindle drive current (I_s), X-axis drive motor current (I_x), Y-axis drive motor current (I_y), Z-axis drive motor current (I_z). Power (P) is calculated by P=UI.
- Vibration parameters: accelerations at the spindle and the three feed axes (X,Y,Z).
- Temperature parameters: temperature of the spindle motor (T_s), oil temperature in the spindle box (T_b), temperature of three axis drive motors (T_x, T_y, T_z).
- Pressure parameters: pressure of the pneumatic gas for claming devices (P_c), pressure of the hydraulic oil for revolving devices (P_r), pressure of the hydraulic oil for feed drives (P_f).

Table 1. Monitored objects and parameters

Objects	Voltage	Current	Power	Vibration			Temperature	Pressure
				X	Y	Z		
Spindle	U	I_s	P_s	a_{xs}	a_{ys}	a_{zs}	T_s, T_b	P_c
X-axis	U	I_x	P_x	a_{xx}	a_{yx}	a_{zx}	T_x	P_r
Y-axis	U	I_y	P_y	a_{xy}	a_{yy}	a_{zy}	T_y	P_f
Z-axis	U	I_z	P_z	a_{xz}	a_{yz}	a_{zz}	T_z	

Feature Extraction

The following two kinds of features are extracted from above parameters:

- Feature that indicates the total average changing trend of the machine or process status at the moment, defined as $\Sigma(k)$. It also represents the deviation of current status relative to its initial status. If the deviation goes beyond a preset limit, the condition associated with the feature is considered to be abnormal. This feature is suited for the situations where faults occur gradually.
- Feature that indicates the instantaneous changing rate of the machine or process status at the moment, defined as $\Delta\phi$, $\Delta\phi=\phi(k)-\phi(k-1)$. If the deviation goes beyond a preset limit, the condition associated with the feature is considered to be abnormal. This feature extraction strategy is suited for the situations where faults occur abruptly.

Thus for each monitored object, extracted features include: U, $\Sigma(P)$, $\Delta\phi(P)$, $\Sigma(I)$, $\Delta\phi(I)$, $\Sigma(a_x)$, $\Delta\phi(a_x)$, $\Sigma(a_y)$, $\Delta\phi(a_y)$, $\Sigma(a_z)$, $\Delta\phi(a_z)$, T. Besides, there is a temperature feature - T_b, and three pressure features - P_c, P_r, P_f. Totally there are 52 features.

4.2 Data acquisition

The data base is a dynamic base which is generated and used by both monitoring and diagnosis. There are three kinds of fault data in the data base, they are: signals in the controller, condition monitoring results and the observed symptom. Fig. 3 shows the acquisition of these data.

4.3 Knowledge acquisition and representation

The knowledge base is another important part of the integrated system. There are two kinds of knowledge: the deep knowledge and shallow knowledge. The former consists of knowledge based on system structure and behavior, based on fault tree analysis and parameter/state estimation; the latter are mainly from engineers' and experts' experience, fault statistics and process history.

In the design of knowledge base, one of the key problem is knowledge representation. In this system, deep knowledge is represented in rule or object-oriented frame, while shallow knowledge is represented as rules or independent facts.

The retailed knowledge acquisition and representation process is described in Fig. 4.

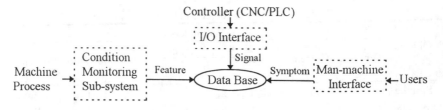

Fig. 3. Data acquisition

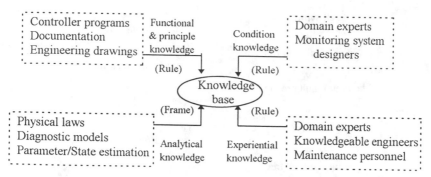

Fig. 4. Diagnostic knowledge acquisition and representation

4.4 Integrated diagnostic reasoning

The reasoning engine is the kernel of the integrated condition monitoring and fault diagnosis expert system. In order to realize integrated diagnosis and imitate the usual fault propagation process of manufacturing equipment, diagnostic knowledge in knowledge bases is also divided into three different levels, that is, functional knowledge (functional decomposition), principle knowledge (decomposition according to the principle of operation) and experiential knowledge (domain experts' or knowledgeable engineers' experience). Meanwhile, all the knowledge bases are in the form of a fault tree, respectively, functional fault tree, principle fault tree and experiential fault tree or rule tree. During the diagnosis of a fault, a human expert or maintenance personnel usually locates the faulty functional modules at first by using of functional fault trees, and then uses the faulty functional modules related principle fault trees to find the rough fault causes, at last obtains the final and more accurate fault causes with the help of corresponding rule trees. Therefore, similar diagnostic reasoning process in the integrated system is carried out as described in Fig. 5.

In Fig. 5, diagnostic reasoning based on the functional fault tree and principle fault tree is performed by the strategy of breadth-first search combined with failure probability of each node in the fault trees. Whether a node is faulty or not is determined by various signals in controllers and the logic relationship among these signals. Reasoning based on experiential knowledge or rule tree is more complicated. For rules associated with machine or process condition, reasoning is performed with the help of condition monitoring results, while for other rules, reasoning is performed as a sequential hypothesize-test cycle. In the cycle, a cost-weighted entropy criterion is used to choose the next part of the rule tree to activate. This entropy criterion helps to select the rule that gives the maximum fault discernment per unit cost in cases where multiple tests might be performed.

Supposing R_{kj} is a node in the kth level of the rule tree and $R_{k+1,1}, R_{k+1,2}, \cdots, R_{k+1,m}$ are nodes in the $(k+1)$th level, cost-weighted entropy of node R_{kj} can be calculated by the following equations:

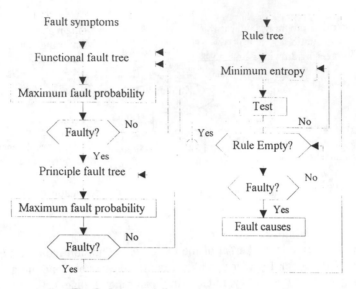

Fig. 5. Integrated diagnostic reasoning procedure

$$H = -\sum_{j=1}^{m} w_j P_j \ln P_j \qquad m \geq 2 \qquad (1)$$

where

$$1 \geq w_j \geq 0, \qquad \sum_{j=1}^{m} P_j = 1$$

The weighting factor, w, is a normalized cost, determined by dividing the actual cost of a measurement operation, by the maximum of the set of measurement costs for all components at the current rule level. P_j is the probability with that $R_{k+1,j}$ is the cause of R_{ki}, under the condition that the test result of node R_{ki} is known.

Cost-weighted entropy is used to select and activate a part of experiential knowledge base or rule tree. It selects the measurement that will give the most discernment at the lowest cost. That is to say, the next rule to be tested, its entropy must be the minimum.

5 Case study

The above integrated system has been implemented on FFS-1500-2 FMS installed at Zhengzhou Textile Machinery Plant in China. FFS-1500-2 FMS consists of a PFZ 1500 FMC, a KBNG85 MC and an automatically guided vehicle (AGV). PFZ1500 FMC is made up of some functional modules such as tool change, adapter change, axis drive and hydraulic drive, etc. Axis drive can also elaborately be divided into spindle drive, X-axis drive, Y-axis drive and Z-axis drive. This kind of decomposition is based on fault tree analysis method. In case a process fault is detected in PFZ1500 FMC, the diagnostic research is shown in Fig. 6 and the faults corresponding to each node are:

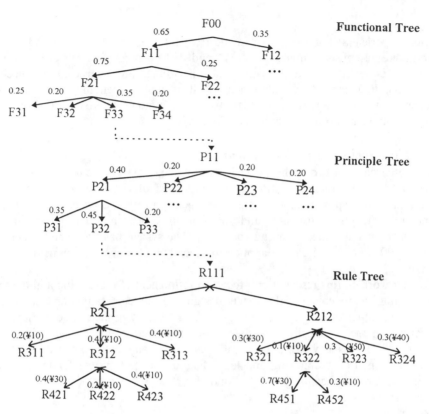

Fig. 6. Sample integrated diagnostic reasoning

F00: FFS-1500-2 FMS	**P22:** X-axis	**R313:** Motor Connection
F11: Machine Tools	**P23:** Y-axis	**R321:** Circuit Connection
F12: AGV	**P24:** Z-axis	**R322:** Power Connection
F21: PFZ1500 FMC	**P31:** Spindle Feed	**R323:** Control Amplifier
F22: KBNG85 MC	**P32:** Spindle Motor	**R324:** Power Amplifier
F31: Tool Change	**P33:** Spindle Transmission	**R421:** Tool Edge
F32: Adapter Change	**R111:** Spindle Motor	**R422:** Cooling System
F33: Machining Process	**R211:** Motor Drive	**R423:** Feed Load
F34: Hydraulic Drive	**R212:** Control Circuit	**R451:** Fuse
P11: Machining Process	**R311:** Mechanical Parts	**R452:** Power Switch
P21: Spindle	**R312:** Motor Temperature	

A first search through functional knowledge base (functional fault tree) leads to F33, which is a terminal functional knowledge node. Principle knowledge base P11 is activated at this point and another search through principle knowledge base (principle fault tree) leads to P32. which is a terminal principle knowledge node. Whether a node is faulty or not depends on signals in the controllers and condition monitoring results.

Then experiential knowledge base (rule tree) R111 is then activated and the cost-weighted entropy is computed for R211 and R212 groups. The rule with minimum entropy is tested first and maintenance personnel are instructed to check the corresponding part of the equipment. If it is faulty, then diagnosis terminates, otherwise the system will check other observed symptoms before backtracking within the functional and principle knowledge bases.

6   Conclusions and future work

The research described in this paper has introduced a systematic methodology for the design of an integrated condition monitoring and fault diagnosis system for modern manufacturing equipment. This is an intelligently integrated system with a modular and reconfigurable structure, and the functions of condition monitoring, fault diagnosis, maintenance planning and so on. The system has been   implemented on FFS-1500-2 FMS and can be applied to other manufacturing equipment with little change.

Further work on this research will focus on refinement of monitoring and diagnostic algorithm, improvement of the system design and implementation, and investigation of the potential of a learning strategy to make the integrated system adaptable to changes in the monitoring and diagnostic environment. Furthermore, in order to improve the efficiency of the system, it is significant to identify a generic strategy for hosting condition monitoring and fault diagnosis expert system for manufacturing equipment on their controllers. This kind of system is called embedded system.

References

1.   E. Ferenc and S. Csongor, Monitoring tasks on boring and milling production cells, *Computers in Industry*, 7(1986) 65-71.

2.   M. Weck and M. B. Hummels, Diagnosis of flexible manufacturing systems (FMS) — aspects of on-line error detection and knowledge processing, *IFAC Fault Detection, Supervision and Safety for Technical Processes*, Baden-Baden,Germany (1991) 361-367.

3.   H. Tang and B. K. Xu. Computer Aided Diagnosis of Mechanical Equipment, Tianjin: Tianjin University Press, 1992.

4.   W. B. Hu. Research on a Quality-Control-Based Fault Diagnosis System in a Flexible Manufacturing Environment. *Ph.D. Dissertation*, Huazhong University of Science and Technology, 1995.

5.   S. M. Alexander, C. M. Vaidya and J. H. Graham, Model for the diagnosis of CIM equipment, *Computer - Elect. Eng.* 19(1993) 175-183.

6.   J. H. Graham, S. M. Alexander and W. Y. Lee. Knowledge-based software for diagnosis of manufacturing systems. *Intelligent Manufacturing — Programming Environment for CIM*, Spinger - Verlarg, 1993.

An Indirect Machine Tool Condition Monitoring System

P. Oksanen, P. H. Andersson

Institute of Production Engineering, Tampere University of Technology, Finland

Abstract. Quality systems according to ISO 9000 standard preconceive that production machinery is to be inspected regularly. Conventional inspection methods are time consuming and the tests need to be performed by skilful personnel. The inspections have also been departed from workpiece measurements. In the *Eureka Maine FMS Maint* project machine tools were measured regularly with the aid of a double ball bar (DBB) device, test pieces and actual workpieces. The Quality level of machining was analyzed by DBB results database and *Polar Check Analysis* software [1]. Correlation was found with temperature conditions, NC compensation parameters, test pieces and actual workpiece measurements. The Introduced maintenance control method is an useful aid for maintenance planning and it gives vital information about state of individual machine tools.

1 Introduction

The key questions in machine tool maintenance are: *when to carry out the needed preventive maintenance actions* and *how to measure and adjust machine's geometry or compensation parameters* [2]. It is evident that maintenance based on calendars only is not the best solution, because machines can be run different shifts or variety of products. The dispersion of the sizes in the final product features is a function of the characteristics of the workpiece, fig 1.

The concept for the integrated machine tool condition monitoring system is shown in figure 2. In order to be able to draw conclusions of the production quality, a large amount of information was gathered and stored in a common database. This database includes information of the basic machine tool structures, characteristics, tools, workpieces and parameters. An important part of the complete system was an automatic data collection (ADC) system. All this information was necessary in order to be able to interpret the measuring data. Information system with databases forms a central and important role of the concept. Measuring methods can be divided to machine tool measurements with conventional devices or quick test methods and production measurements monitoring test pieces or actual workpieces [4]. This work concentrates machine tool measurements with DBB method and result analysis has been made with the aid of the other parts of the concept.

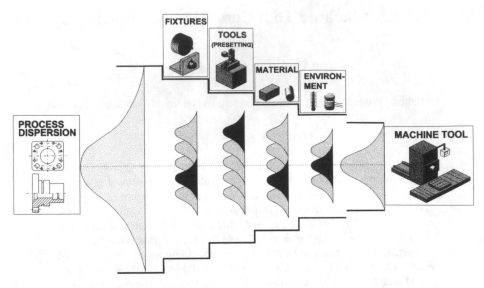

Fig 1. Process dispersion vs. deviation caused by the machine tool.[3]

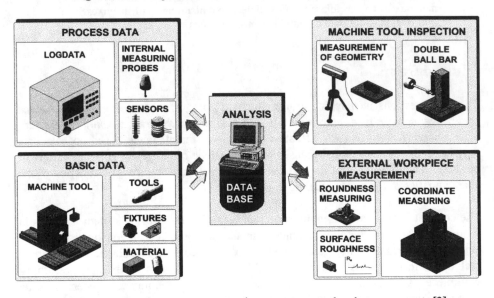

Fig 2. Integrated maintenance system using process control and measurements.[3]

2. Machine Tool Maintenance Control by DBB Method

Before the measurement phase, a comprehensive training session for the pilot factory personnel was made. The initial state of the machinery was examined in a thorough inspection using conventional measuring equipment. Controller parameter changes were made to adjust the machine tools as accurate as possible.

DBB tests were carried out once a month in connection with a test piece run. Workpieces to be measured were selected among those of the highest volume. The features measured were choosed in accordance with indicating the production accuracy. All work and test piece measurements were performed in a coordinate measuring machine.

2.1. Analysis of DBB Measurements

The accuracy of three identical machine tools were evaluated regularly, the time span of the whole project was about three years. All machines were measured with the same NC-program, the position of the measurement was the same. The measuring data and the trends were analyzed with the *Polar Check Analyser* software.

The most important indicator for the quality of machining is the change of the circularity on xy-plane of an individual machine tool. The upper limit of the circularity deviation was set to 15 μm. The trend of measurements were quite similar for all the machines. However, the error sources that caused the circularity deviations varied. Figure 3 shows an example of trend analysis.

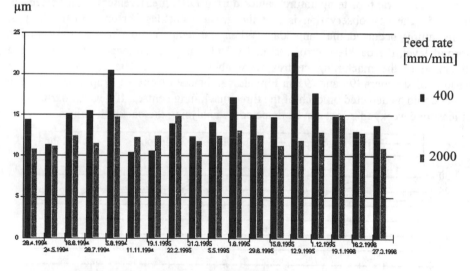

Fig 3. Trend analysis of machine tool errors: xy-plane circularity deviation with DBB method.
[5]

The trend analysis shows that the circularity values of the machine tools were generally between 10 and 15 μm. The upper limit was exceeded occasionally. The measurement results were reanalyzed using individual error parameters. This analysis showed that the greatest circularity deviations were caused by the backlash of the linear axes (Fig. 4) .

1217

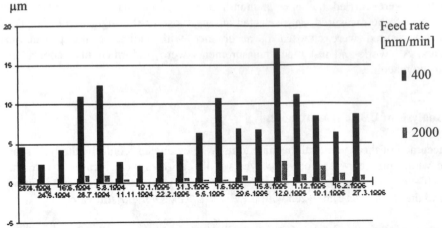

Fig 4. Backlash deviations detected with DBB trend analysis. [5]

As no compensation parameters were changed or major services performed, the reason for circularity error had to be found elsewhere. The temperature analysis of the machining centres showed a relationship between the circularity error and the temperature. Graph of temperature values during DBB measurements can be seen in figure 5. The key observation is that the greatest backlash errors and the highest temperatures occurred the same day. When the temperature changes back to it's normal level, the backlash error reduced without any adjustments. The temperature diversion of the machining centres were about 5... 6°C. The exceptional warm periods of summers '94 and '95 in Finland can be seen from the graphs as well. The effect could be detected on each of the three machining centres. If a warm period lasts at least one week, an effect can be detected by analysing the backlash values.

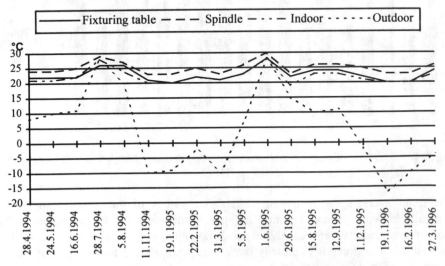

Fig 5. Machine and environment temperatures during the DBB-measurement dates. [5]

Another remarkable observation from the trend analysis graphs was the effect of the NC-controller compensation parameters on the amount of backlash. Also the traverse speed of the machine tool slides has a significant impact on the actual reversal error. The magnitude of the error showed to be greater when using slower feed rates and vice versa.

Conventionally the backlash compensation values are evaluated with laser-interferometer measurements using rapid speed as the positionings during the machining operations are usually performed the same way. The backlash is then properly compensated only in these circumstances. When contour milling using circular interpolation and a slow feed rate, the inertia of the moving masses affects dissimilarly and the effective backlash will be remarkably greater than measured and compensated.

In this project a laser-interferometer and rapid speed was used in parameter compensation in the first phase, later on the method was changed to the DBB and normal machining feed rates. Figure 6 shows the effect of this change.

Only after a major adjustment the deviations could be set to acceptable limits during the slow feed rate operations (overcompensation at higher speeds are however unavoidable). The same pattern could be detected from the test piece measurements.

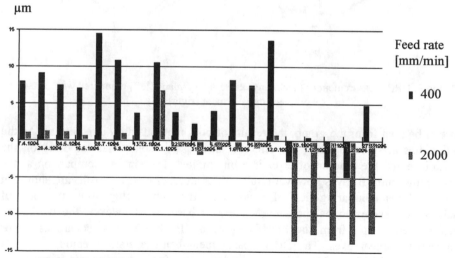

Fig 6. Effect of changes in compensation parameters on DBB trend analysis. [5]

1219

2.2. DBB Analysis Compared with Actual Workpiece Measurements

Actual workpieces produced by the pilot company were measured in the project and the results were analyzed statistically. The results were compared with DBB analysis and the object was to find out how the individual error parameters could be detected with both methods. Especially importance was to know the influence of cutting forces, because DBB measurements are carried out in no load condition.

A comprehensive example of the comparison material is the influence of the backlash error of two different machining centres. Both machines were measured identically. An identical workpiece, with a circular interpolation feature, was machined with both machine tools as well. The result of this comparison is presented in figure 7.

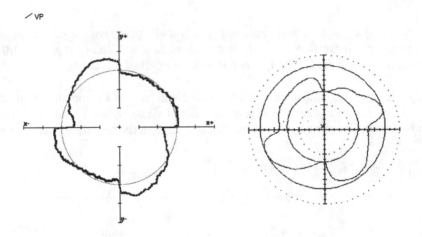

Fig 7. DBB measurement of machining centre A, xy-plane (left) compared with actual workpiece measurement (right). [5]

As can be seen from the graph, the results of both methods are quite similar. As the tool radius acts like a filter, the effect of backlash is somewhat smoother on the workpiece and the amount of error is a bit smaller. The similar comparison at the machining centre B gives however totally different results. Those paths are shown in figure 8. When comparing the DBB results from the both machine tools, the cause of major errors seems to be the same - the axis backlash. However, the workpiece measurement graph from the machining centre B shows great squareness error between the linear axes. The DBB measurement being a static method could not detect the specific deviation. This is due to the cutting forces bending the frame of the machine tool.

By comparison of several test methods could be found that a particular machining centre is individual. Same interpretations from one machine tool can not be generalised to all machine tools, even if they were same type. The DBB method is an useful aid in maintenance control if it is used together with test and workpiece measurements. The DBB could then be a constant control method and other measurements are used as supporting actions when needed.

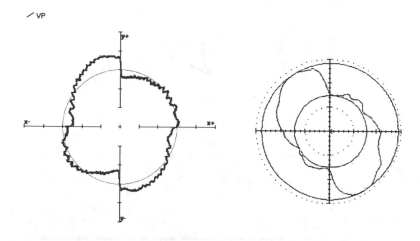

Fig 8. DBB measurement of machining centre B, xy-plane (left) compared with actual workpiece measurement (right). [5]

3. Experiences of the DBB as Part of a Maintenance Control System

As a result of project the DBB can be concluded to be a very effective and quick method indicating trends in machine tool accuracy. This was partly because of very thorough project planning which included static measurement methods. DBB is really a quick test - average measuring time for one machining centre with pallet changes was one hour. Before an effective analysis of measuring results machine tools' dynamic behaviour must be known. Test pieces give a great support for interpretation. The overall concept is visualised in figure 9.

Actual statistical analysis of workpiece measurements should be divided to Statistical Process Control (SPC) of normal production and machine tool condition dependent SPC. If normal SPC gives an error signal, there is no need to make any measurements on the machine tool, but to try to found the error causes from i.q. tooling or materials. Machine tool measurements are not needed until error signals from machine tool condition dependent SPC occurs. Test could then be made by the DBB with the aid of dynamic behaviour information from test piece measurements. Some parameters can be changed by DBB and conventional methods should be used only when major geometrical adjustments or positioning error compensation is needed

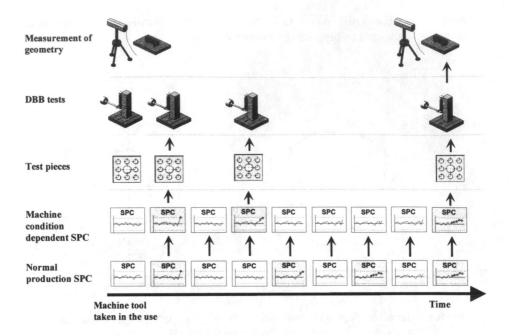

Fig 9. Maintenance control with machine tool, test piece and workpiece measurements. [3, 5]

4. Conclusions

Carefully prepared machine tool maintenance control with the DBB method assisted by the statistical control data from the test and workpieces serve effectively predicting machine tools maintenance and service needs. Almost all the necessary changes in control parameters could made according to the DBB data without any other measurement aid. In some cases the need of a maintenance action was predicted before any parts were broken. This had an positive influence on the machine tool productivity. One of the most important results of the project was establishing the connection between the DBB results and the machine tool temperature behaviour.

References

1. J. Hölsä, Analyse and Simulation of Double Ball Bar Method. Tampere: Tampere University of Technology, (1995). M.Sc. Thesis. 101 p. (Finnish)
2. S. Torvinen, P.H. Andersson, J. Vihinen and R. Milne, Monitoring the accuracy characteristics of the machinery by using a dynamic measurement approach.. In: Winsor, J., Sivakumar, A.I., Gay, R. (eds.) Proceeding of the 3rd international conference computer integrated manufacturing., (1995). Vol 2, 1435-1442.
3. P.H. Andersson, S. Torvinen and J. Vihinen, Indirect maintenance control of machine tools. In: Konepajamies 6 (1994), 54-56. (Finnish)
4. S. Torvinen, P.H. Andersson, J. Vihinen and J. Hölsä, An off-line condition monitoring system for machine tools. In: Hope, A.D., Smith, G.T., Blackshaw, D.M.S. (eds.) Second

International Conference on Laser metrology and Machine Perfomance "LAMDAMAP 95", (1995). 237-250.

5. P. Oksanen, A Machine Tool Inspection and Maintenance Control by Double Ball Bar Method. Tampere: Tampere University of Technology, (1996). Lic.Techn. Thesis. 101 p. (Finnish)

Sensor System for Disassembly of Electrical Motors

Björn Karlsson, Nils Karlsson and Alexander Lauber

Department of Physic and Measurement Technology, Linköping University, Sweden

Abstract The role of reuse and recycling has become more and more important during the last years. There are a lot indications that the producers in the future will have to take care of their own products when they are worn out. During the lifetime of a product it will change to a large extent. To make industrial robots able to perform disassembly in a proper way measurement and sensors will play an important role [1].

In this paper the sensor support of a robotized work station for the disassembly of industrial asynchronous motors is presented. In an industrial motor the copper is situated in the stator windings and in the junction box. There are three steps in the proposed disassembly work. In the first step the functionality of the motor is checked. The second is a manual disassembly part where the terminal box, the shields and the rotor are removed. The third is the robotized automatic part.

In the automatic part the stator winding made of copper will be removed from the stator. There are four main steps in this activity. The first step is to place the motor in the correct position, which is done with a robot guided by a vision system. The second step is to cut off the stator winding at one end of the motor. This is achieved by means of a cutting tool guided by an eddy current sensor and a vision system. The third step is to pull out the stator winding from the stator, which is done by a hydraulic tractive system. After removal of the stator winding a check has to be done to guarantee that all copper has been removed. This test will be performed by an eddy current probe mounted on the robot.

1 Introduction

From a recycling view copper is toxic for steel. The amount of copper in the steel leaving the shredder is typically 0.25-0.3% [2]. If the amount of copper in the steel scrap exceeds 0.02%, the steel scrap can not be used for the production of steel sheets [2]. In electrical motors the copper amount is about 10% of the motor weight. Today small electrical motors when worn out are put in a shredder, due to disassembly problems. If the copper could be removed from the motor before shredding, the usability of the materials out of the shredder would increase significantly. To make the removal of the copper acceptable from economical and ergonomic point of views automatic disassembly supported by sensors should be used.

When a product reaches the scrap dealer a classification has to be done. From the scrap dealer's point of view it is impossible to have knowledge of all products reaching his shop. To guide the scrap dealer an advance sensor and classification system is needed. For an electrical motor, the construction principle is simple but there are a lot of variations in design depending of branch and the field of applications. The classification separates motors in two parts, one for repair and one for disassembly.

The motors classified for repair are sent to repair and the motors classified for disassembly are going to the next step at the disassembly work station. In this step manual disassembly of the shields and the junction box is performed, figure 1. After that it is time for the last step in the chain, the automatic disassembly, which is in focus in this paper. Figure 2 illustrates the test and disassembly system.

2 Test of Electrical Motors

When a motor reaches the disassembly station a condition test of the motor will be done. The different test methods are:

- Vibration measurement for detecting bearing wear, defective rotors and electrical unbalance. The measurements are preferably done with a triaxial accelerometer, threaded or mounted with temporary adhesive, on a horizontal (vertical) machine surface. Vibration measurement should be recorded during start-up and when the motor works without load at nominal r.p.m.[3].
- Acoustic measurement can be used mainly to discover mechanical problems with the electrical motor. Acoustic measurement are to be seen as a complement to vibration measurement.
- Electrical performance [4],[5]
    - Measure of the start-up current characteristics during the first 0.5 s give a lot of information about the condition of the motor, such as stator winding status and unbalance.
    - Measuring stator resistance will give information about broken windings.
    - Perform a high voltage test to get information on the condition of the stator winding.

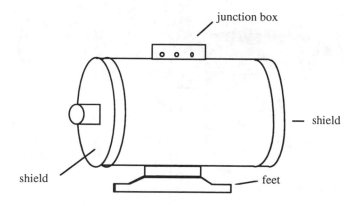

Fig. 1. Schematic drawing of an electrical motor

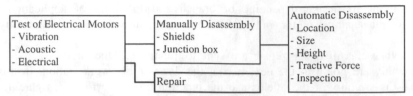

Fig. 2. Block diagram of the disassembly work station

3 Automatic Disassembly

After manual disassembly it is time for the automatic one. The automatic disassembly will be carried out with an industrial robot and for the removal of the stator winding a hydraulic jack will be used. Both the robot and the hydraulic jack will be guided by fused high level information from a sensor system.

3.1 Vision for Location

For location vision is used. One of the bigger problems is how to fuse the information from the vision system with the information from the other sensor systems. To clarify things, the number of parameters detected should be reduced early in the process. The wanted vision parameters are the centre point of the cylindrical part of the motor house, the diameter of the motor house and the location of the feet. These parameters are calculated and the motor is placed in the desired location.

3.2 Eddy Current Measurement for Slot Inspection and Determination of Height

The electric and magnetic properties of the copper winding differ from those of the iron where they are mounted. To measure the height of the motor (figure 3) and to inspect that the winding is successfully removed from the stator after disassembly, eddy current measurements are used, figure 3.

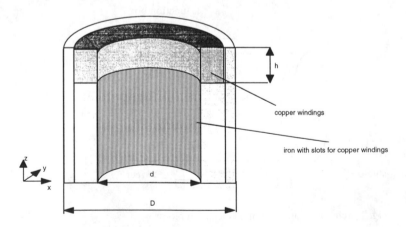

Fig. 3. Schematic picture of the upper part of the electric motor cut in halves as seen from the side.

3.2.1 Principle for Inductive Testing

The inductive testing also called eddy current testing, is based on the law of induction [6]. When a coil is placed in a time-varying magnetic field an electric voltage is induced. The induced voltage can be calculated by Faraday's Law of induction.

$$e_0 = -N \frac{d\Phi}{dt}$$
(1)

N = number of turns of the coil.
Φ = the magnetic flux through the coil.
e_0 = the voltage induced in the coil.

The law of induction is also valid when a solid electrically conducting object is placed in time-varying magnetic field. The solid object can be seen as one short-circuited turn of a coil. The induced voltage now causes currents called eddy currents. Figure 4 shows how a magnetic field is generated by a coil connected to an AC-voltage source. This magnetic field H_p, is the primary field which induces eddy currents in the object. The eddy currents produces in their turn a secondary magnetic field H_s. The magnitude and phase shift of H_s compared to H_p depend on the electric and magnetic properties of the object.

It can be seen that the eddy currents in the test object are concentrated more or less strongly to the surface of the test object, adjacent to the coil causing the primary field. This is called the skin effect phenomenon. This effect increases with increased frequency, test object conductivity and magnetic permeability. The standard depth of penetration of eddy currents, δ, is defined as that depth below the surface of the test object where the current density is 1/e of the current density at the surface. The standard depth of penetration of eddy currents, δ, is determined by:

$$\delta = \frac{1}{\sqrt{\pi f \mu_r \mu_0 \sigma}}$$
(2)

f = the frequency of the magnetic field
μ_r = the relative permeability of the object
μ_0 = permeability of free space ($4\pi*10^{-7}$ H/m)
σ = the conductivity of the test object

With knowledge of the conductivity and permeability of the material of the test object and the thickness of the object a suitable test frequency can be chosen.

3.2.2 Determination of the Height for the Cutting of the Stator Winding

In the disassembly procedure the copper windings have to be cut. The upper part of the windings has to be removed before the copper windings in the slots can be removed. The windings ought to be cut just above the slots figure 3. To be able to find this location in the x/y-plane the vision system is used, and to get the information about the z-position the inductive probe can be used.

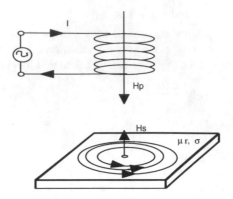

Fig. 4. The production of eddy currents in an object of conductive material placed in the proximity of a coil connected to an AC-voltage [6].

The robot will place the inductive probe, e.g. the coil, just above the motor. The position in the x/y-plane is obtained by the vision system. The edge of the sensor must be near the projected inner edge of the motor. The sensed material at the edge of the sensor is air. The robot lowers the sensor (the negative z-direction in figure 3) until the sensed material is copper windings. The position of the inductive sensor in the x/y-plane might now have to be checked by the vision system (the projection error is smaller at shorter distances) and perhaps adjusted. The coil is lowered until it senses the iron. Then the sensor is moved in the positive z-direction until the sensor indicates copper windings again. This z position is the position to cut the copper windings.

The signal of the sensor that corresponds to iron and the copper windings has to be tested out; it can not be calculated analytically. An approximate result might be simulated by computer programs based on FEM calculations.

3.2.3 Inspection of Stator Slot by Inductive Testing

To check that the removal of the copper windings in the stator slots is successful an inductive sensor will be used. The coil of the inductive sensor will be placed inside the motor with its axis perpendicular to the inner wall of the motor, figure 5. The coil will then be moved (rotated) one full turn around the inner wall.

The sensor output for the coil placed near solid steel is different than the output from the coil when the sensor is placed near an empty slot. As copper has a much higher conductivity than the air in an empty slot the sensor output for copper windings in a slot will be different from the sensor output when all copper has been successfully removed from the slot. If all copper is removed the output of the sensor ought to be the same when passing all slots. The measurement will cause a pattern. If this pattern is periodic the removal of the copper out of the slots has been successful. The expected sensor output can not be analytically calculated. Tests therefore have to be made to get reference signals. Information about the sensor output can however be achieved approximately by computer simulations. The problem of the limited case: which is the maximum possible undetectable amount of copper remaining in a slot must be solved by practical testing.

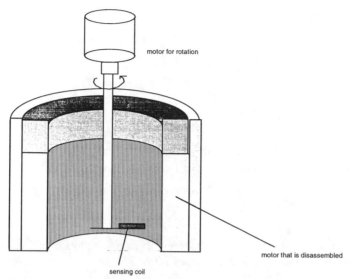

motor for rotation

motor that is disassembled

sensing coil

Fig. 5. Laboratory test set-up for eddy current measurement

To test the measuring principle a test setup has been made in the laboratory, see figure 5.

4 Sensor Data Fusion

To make it possible to fuse information from the different sensors in the system, pre-processing of the data is needed. The great amount of information from the vision system needs to be simplified. The human brain chooses to handle only one part at a time, i.e. in vision, location and shape are handled separately[7]. If we continue to compare with the human a distributed data reduction is performed before the information reach the brain. Every second the human body get more than 40 Mbits of information, but the amount that reaches the brain is only 10-40 bits [8].

The eddy current measurement to detect if all copper is removed or not is made by three vertical placed probes. If all copper is removed the output of the three probes will be the same. Due to disturbances it is important to classify the output. A possible approach is to use a fuzzy set, figure 6 [9].

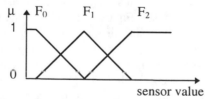

Fig. 6. The fuzzy sets for a sensor

Each fuzzy set is defined as a triangular membership function $F : U \rightarrow [0,1]$ given as

$$F(u;\alpha,\beta,\gamma) = \begin{cases} 0 & u < \alpha \\ \dfrac{u-\alpha}{\beta-\alpha} & \alpha \leq u \leq \beta \\ \dfrac{\gamma-u}{\beta-\alpha} & \beta \leq u \leq \gamma \\ 0 & u > \gamma \end{cases} \tag{3}$$

where u belongs to the overall range of the sensor output and α and γ are these values of u for which the degree of membership of α and γ to F is zero, and β is this value of u for which the membership of β to F is one. The output from the probes is classified with (3) in three different classes:

F_0 = Air with iron on either side of it
F_1 = Copper with iron on either side of it, or iron with air in the slots on either side of it
F_2 = Iron with copper in the slots on either side of it

When measuring both near the slot and near the iron between, for a motor with 24 slots it will be 48 measurements. In figure 7 measurements on three different positions as well as the fused values $f(\mu)$ of the three probes are shown. Equation (4) is used to fuse the measured data, In (4) a value of K=2 is used and N=3 due to that three sensors are used. The K value determine the weighting function, with K=1 we get the usual arithmetic mean value and a bigger K value gives a more unlinear weighting function.

$$f(\mu_1,\ldots,\mu_N) = \frac{1}{K^N-1} \cdot \frac{-1 + K^N \displaystyle\prod_{i=1}^{N} \frac{1+(K-1)\mu_i}{K-(K-1)\mu_i}}{1 + \displaystyle\prod_{i=1}^{N} \frac{1+(K-1)\mu_i}{K-(K-1)\mu_i}} \tag{4}$$

The result out of figure 7 is that in position 1 it is a slot with air, position 2 is between two slots and position 3 is a slot with copper left. The conclusion is that it is still copper left in the motor which has to be removed manually.

5 Results of the inductive measurements

To determine the proper height to cut the copper windings an inductive probe is used. The inductive probe is guided by a vision system in the x/y-plane and positioned by the robot near the inner edge of the sensor. Figure 8 shows the output from the sensor as it is lowered by the robot. Point A is inside the motor, the current through the coil decreases. Point B is the border between the upper winding and the stator iron. The current through the sensor still decreases until point C. The current is stable i. e. the surroundings of the coil do not change with the height. This tells us that the sensor must be in the stator iron region. We turn the direction of the coil and moves it upwards. Point D is the same as B. The increase of the current through the sensor indicates that the region of stator iron is left. This is the height to cut the copper windings. The z-coordinate is got by the use of the robots z-position.

| | Position 1 | | | Position 2 | | | Position 3 | | |
	F_0	F_1	F_2	F_0	F_1	F_2	F_0	F_1	F_2
Probe 1:	0.7	0.3	0	0	0.3	0.7	0	0.8	0.2
Probe 2:	0.6	0.4	0	0	0.3	0.7	0.7	0.3	0
Probe 3:	0.4	0.6	0	0	0.1	0.9	0.5	0.5	0
Fused value:	0.67	0.41	0	0	0.18	0.82	0.37	0.54	0.04

Fig. 7. Classified sensor data from eddy current measurement

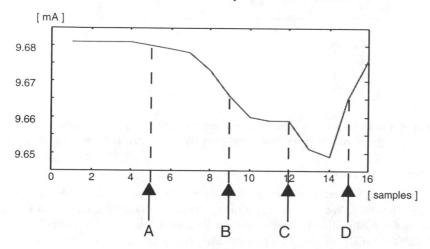

Fig. 8. Output for the inductive sensor for the determination of the height to cut the copper winding. Point A is the height for the upper winding, B is the height for the stator iron, C marks that the sensor has switched directions and D is the height for cutting.

To check that the removal of the copper windings in the stator slots is successfully performed an inductive probe is used. The inductance for the coil used in our experiments has an inductance of 342.4 µH as it is placed near the iron of the stator slot, with copper in the slots on either side of it. As the coil is placed near a slot with copper inside the inductance is 341.6 µH. When copper is removed and the sensor is placed near the stator iron the inductance is 341.7 µH and as it is placed near an empty slot the inductance is 338.6 µH. The output from the sensor for a test of a badly disassembled motor can look as the signal in figure 9. Point A indicates the output for an empty slot, the sensor is moved to the iron point B and so on until point C where the expected notch for an empty slot is exchanged by the output of the copper.

The figures in the result section show that the inductive probe can be used both to determine the correct height to cut the winding and to check that the copper is properly removed from the stator slots. One however has to comment that the sensor signal is rather small and that the sensor has to be carefully positioned to get correct signals. More advanced signal handling can take care of some of these drawbacks but the user should hold in mind that the position and orientation of the inductive probes are important.

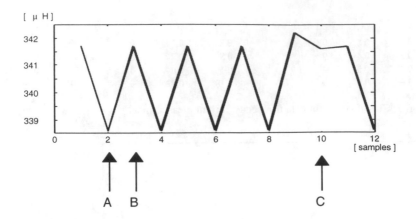

Fig. 9. Output from the inductive probe when it is placed at the iron (point A) and at a slot with air (point B). Point C indicates that it is copper in the slot instead of air.

Acknowledgement

This work has been performed at the Laboratory for Measurement Technology at Linköping University. We take this opportunity to thank technician I. Grahn for fruitful discussions and for building the sensor system. The work is a part of a research project "REMPRODUSE-Cu" in the Programme Environmental and Climate 1994-1998, sponsored by EU Commission, whose support is gratefully acknowledged.

References

1. For a review see B. Karlsson and A. Lauber, The role of measurement in industrial recycling, Proc IEEE-IMTC/96 - IMEKO TC-7, Brussels, Belgium, June 4-6, 1996 pp 1491-1494.
2. P Norrthon, En studie i kretslopp. Hinder och möjligheter för ökad material återvinning (A Study of Recycling), Ecocycle Commission of the Swedish Ministry of Environment, February, 1996.
3. International standard CEI/IEC 34-2: 1972
4. C. Rossi, A. Tonelli, On-line estimation technique for electrical and mechanical parameters of DC and AC motors, *EPE 91/4th European conference on power electronics and applications*, Firenze, September 1991.
5. A. Pouliezos, G. Stavrakakis, C. Lefas, Fault detection using parameter estimation, *Quality and reliability engineering international*, Vol 5, 283-290, 1989.
6. D. H. Libby, Introduction to electromagnetic nondestructive test methods, John Whiley & Sons, Inc. 1971.
7. D H. Ballard et al Deictic Codes for the Embodiment of Cognition, Computer Science Department, University of Rochester, accepted for publication (Copyright 1996: Cambridge University Press)
8. M. Zimmermann, The Nervous System in the Context of Information Theory, *R.F Schmidt & G. Thews (eds.): Human Physiology, 2nd ed.*, Springer-Verlag, Berlin, 1989, pp 166-173.
9. P. Wide and D. Driankov, A fuzzy approach to multi-sensor data fusion for quality profile classification, *Proc IEEE/SICE/RSJ*, Washington D.C., U.S.A., December 8-11, 1996.

AUTOMATED VISUAL INSPECTION APPARATUS FOR CIM SYSTEMS

Dr G Bright
brightg@eng.und.ac.za

Mr R Mayor
mayorj@eng.und.ac.za

Department of Mechanical Engineering
University of Natal, Durban
Private Bag X 10, Dalbridge, 4014, South Africa

Abstract. Automated visual inspection is vital in the production of manufactured products acceptable to world market standards. The volatile nature of the consumer markets is placing increased emphasis on manufacturers to produce cheaper products of higher quality. This paper proposes an unique, cost effective automated visual inspection apparatus suitable for application in the computer integrated manufacturing environment. The inspection apparatus utilises specifically designed camera positioning and part manipulation systems to position a single digital camera at known view points within its inspection envelope. View planning algorithms can then be implemented to perform three dimensional part inspection utilising inspection by multiple views techniques.

1   Introduction

The development of automation technologies has been highlighted as an area of research since the late 1960's [1]. The evolution of group technology (GT), flexible manufacturing systems (FMS), computer aided design and manufacturing (CAD / CAM) and computer integrated manufacturing (CIM) has delivered the technology of manufacturing to the brink of the fully automated factory of the future.

A major retarding factor in the development of the fully automated manufacturing environment has been the markedly slower automation of the quality control and inspection tasks [2]. This paper proposes an innovative, cost effective automated visual inspection apparatus that could be suitable for application in the automated CIM environment. Newman *et. al.* [1] report that the implementation of *in-process verification* techniques, i.e the inspection of the part at multiple steps in the process to assure the operation has been done and done correctly, dramatically reduces the reworking and scrapping costs typical to most manufacturing processes. The inspection apparatus should, therefore, be able to integrate simply and easily with the CIM environment thereby monitoring the development and quality of the part throughout the entire manufacturing process.

This paper proposes the design of unique, dedicated camera positioning and part manipulation subsystems based on established, cost effective mechatronic principles.

Through the coordinated interaction of these two subsystems, the inspection apparatus should be able to position a single digital camera at any arbitrary view point within its *inspection envelope*. This ability to position the camera at known view points should then allow, with the appropriate image processing, the development of reliable three dimensional models of the inspected parts [3].

2   Design Criteria For The Inspection Apparatus

The design of the inspection apparatus was focused by a set of predetermined design criteria. The criteria were adopted to ensure that the inspection apparatus would be able to interact with a flexible manufacturing system that is currently under development at the University of Natal, Mechanical engineering's Mechatronic department. The criteria are listed below:

- The apparatus must be capable of acquiring an image of all 6 major faces of a manufactured part using a single CCD camera. The major faces of a manufactured part (ref Fig 2.1) are defined as top, left side, front side, right side, rear side and bottom, in order of inspection.
- The apparatus must be able to accept a pallet from an entry conveyor at height 0.8 m, manipulate the pallet in any manner necessary for a complete inspection and then return the pallet to an exit conveyor at height 0.8 m. The pallet must be able to accept a pallet of at least 200 x 200 mm in base, but no greater that 325 x 325 mm in base.

3   Conceptual Development Of The Apparatus

The final design concept for the inspection apparatus evolved through an iterative process of conceptual design and critical analysis. This section discusses the evolution of the final design for the inspection apparatus, concentrating on the respective disadvantages and failings of the preliminary design concepts and how these were improved upon in the next concept.

The Gantry System. The initial design concept (ref Fig 3.1 (a)) for the inspection apparatus station was based on a gantry type system similar to those used in coordinate measuring

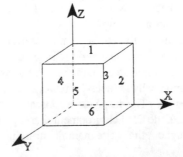

Figure 2.1. A figure defining the six major faces of a 3D part. The coordinate system, XYZ, is also defined where 1 is the top face and 5 is the front face (in the direction of part flow)

machines (CMM's). A large gantry construction was suggested to achieve the required three dimensional positioning of the camera. The disadvantages of the gantry based system are listed below.

- The gantry system could not position the camera below the table rendering the system incapable of acquiring an image of the bottom face of the part.
- The camera positioning system required the movement of a significant structural element, the gantry. The overall mass of the components of the gantry system would cause significant inertial loading on the drive motor and therefore the degree of precision required to correctly position and orientated the camera could not be achieved.
- The risk of damaging the part after the inspection had occurred could not be removed.

The Bridge System. The revised concept (ref Fig 3.1 (b)) replaced the gantry system with a bridge construction. The bridge construction was rigidly attached to the frame of the system and was vertically aligned with a glass / perspex viewing window introduced to the inspection table. The viewing window allowed for the acquisition of an image of the bottom face of the inspected part. A system of flat belt conveyors was introduced in order to control the flow of the inspected parts through the apparatus. The conveyor system was capable of accurately positioning the inspected part in the plane of the camera and therefore the camera positioning system was rationalised to vertical (Z) and horizontal (X) placement in the XZ plane. The improved bridge based system did however have two major failings.

- The bridge system was not capable of capturing all six views. Faces 3 and 5 were not accessible (ref Fig 2.1)
- The movement of the camera along the inner face of the bridge would have required a complex drive mechanism.

The Rotating Cylinder System. This concept (ref Fig 3.1 (c)) considered the replacing of the bridge structure with a cylinder structure to which the camera could be rigidly mounted. With the camera rigidly attached to the cylinder, the camera could be positioned at the four stations indicated in figure 3.1 (c) simply by rotating the cylinder. This rotation of the cylinder could be easily achieved using a motor and a gearing system. A pinion gear could be mounted on the shaft of the motor which would engage the appropriate gear mounted on the outer surface of the cylinder. The costs involved in the fabrication of the rotating cylinder and the machining of the outer gear would have been prohibitively expensive.

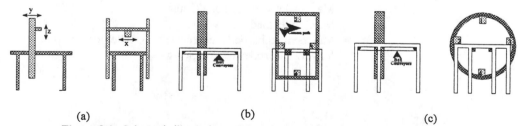

(a)   (b)   (c)

Figure 3.1. Schematic illustrations of the Inspection apparatus design concepts.
(a) the Gantry concept, (b) the Bridge concept, (c) the Rotating Cylinder concept

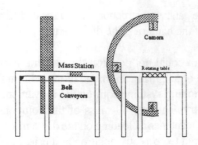

Figure 3.2. The final concept for the inspection apparatus unit, showing the incorporation of the rotary table and the mass station.

The Final Design. The final design concept (ref Fig 3.2) introduced a rotary actuator to the inspection table that enabled the inspected part to be rotated through 360° in the plane of the table. The ability to rotate the part through any angular displacement in the XY plane enabled the acquisition of images of the previously obscured faces of the inspected part; faces 2, 3 and 4. Through the interaction of the rotary inspection table and the camera positioning system (explained in more detail in section 5), the inspection apparatus was now able to fully satisfy the design criteria as presented in section 2.

4   Component Systems Of The Inspection Apparatus

The structure of the inspection apparatus comprised of three interdependent systems; the part positioning system, the part manipulation system, and the camera positioning system.

4.1   The Base Frame

The base frame of the apparatus was designed to provide a stable foundation on which the three principle sub-systems could be mounted. The major focus in the design of the frame was to enhance the reliability and validity of the data collected during the image acquisition process. Key aspects considered were the deformation of the inspection table under typical loading conditions and overall vibration absorption characteristics of the frame. The finite element analysis package, *COMOSM*, was used to investigate the deflections of the frame and modifications to the design were made as required. A list of design parameters used in the design of the base frame for the inspection apparatus is presented below.

- Max. pallet size : 325 x 325 mm
- Max. pallet load = 20 kg
- Max. permissible deflection under max pallet load = 0.001 m
- Mass of the trolley track = 30 kg

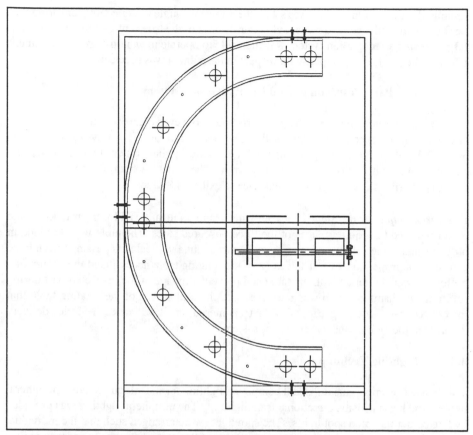

Figure 4.1. A side elevation of the inspection apparatus.

4.2 The Camera Positioning System

Any part recognition and inspection processes performed by the inspection apparatus would be dependent on the information contained in digital images of the six principle views of a part. The acquisition of these images required the development and design of a camera positioning system capable of positioning the digital camera at three principle positions (ref Fig 3.2); top dead centre (TDC), side dead centre (SDC) and bottom dead centre (BDC). This necessitated the design of an unique trolley track system in which the camera is rigidly mounted onto a trolley running in an I-beam type track of constant curvature (ref Fig 4.2). The trolley is positioned in the track using a chain drive system powered by a 12 V DC motor delivering 20 Nm of torque at 75 rpm, capable moving the camera from station to station (eg. TDC to BDC) in a traverse time of 5.4 seconds.

The trolley was designed to be flexible with respect to the type of digital camera used. The maximum permissible trolley load was set at 10 kg. The deflection of the camera focal axis had to be restricted to ensure that the variance in the focal point of the camera was minimised. An upper limit of 2° was set for the allowable angle of deflection, ensuring that the focal point of the camera remained within 17 mm of the true centre of the part (ref Fig 4.3). The image recognition system was then able to operate in an non-fuzzy environment in which the position of the part with respect to the camera was precisely known.

4.3. The Part Positioning And Manipulation System

The part positioning system. The pallet transport conveyor controlled and regulated the movement of the inspected parts through the inspection apparatus. The conveyor was driven by a 12 V DC motor running through a 2:1 reduction pulley system that presented the conveyor shaft with 40 Nm of torque at 35 rpm. The pallet transport conveyor was then able to perform a full traverse, from entrance to exit, in 4.8 seconds.

The part manipulation system. The rotary inspection table allowed for the rotation of the pallet required to acquire images of the three obscured sides. The table was designed in order to enable the rotation of the inspected part about a vertical axis passing through the centre of the inspection area. The design of the rotating table necessitated the design of a pallet that could interact with the table and prevent slippage of the inspected part during rotation. A chamfered step was machined into the top surface of the rotating table that introduced a self-locating capability that countered small variances in pallet delivery inherent in the pallet transport conveyor system.

4.4 Lighting Technique

The inspection apparatus utilised a compound lighting format which included peripheral lighting and high intensity, directional front lighting. The peripheral lighting was provided by four corner mounted spot lights. The spot lights were arranged such that the individual incident beams converged at the geometric centre of the inspection area in order to reduce the degree of shadow casting [4]. The high intensity front lighting was provided by a circular tube florescent lamp mounted on the camera positioning trolley. The camera was mounted concentrically (with respect to the circular tube) on the trolley. The directional capability of the high intensity front lighting system ensured a constant incident light at every camera position.

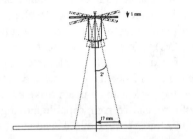

Figure 4.3. A schematic diagram illustrating the defection of the camera's focal axis resulting from a 1mm defection of the camera mounting plate.

5   Typical functioning of the Inspection Apparatus

This section details the typical functioning of the inspection apparatus, explicitly defining the interaction of the camera positioning system and the part manipulation system. A typical inspection routine would begin with the transportation of the inspected part to the rotary inspection table by the pallet transport conveyor. The part, at this stage, would then be positioned at the geometric centre of the *inspection envelope*. The inspection envelope is a complex geometric surface that defines all possible camera viewing positions with respect to the inspected part. The envelope is described mathematically in equation 5.1, where r is the radius of curvature, ϕ is the angle of rotation from the Y axis and φ is the angle of declination from the Z axis (ref Fig 5.1). The radius of curvature, r, is a constant defined by the geometry of the I-beam camera track. However, both the angle of rotation, ϕ, and the angle of declination, φ, are variable within the limits indicated in equation 5.1.

$$
\begin{aligned}
r &= const \\
0 &< \phi < 360 \\
0 &< \varphi < 90
\end{aligned}
\qquad \ldots 5.1
$$

The camera positioning system was capable of positioning the camera at any specified angle of declination, φ. Similarly the part manipulation system was capable of positioning the part at any specified angle of rotation, ϕ. The coordinated interaction of the camera positioning system and the part manipulation system could therefore position the camera at any viewing point within the inspection envelope as defined by equation 5.1.

6   The Apparatus Control System

The design of the inspection apparatus control system was focused by the identification of two principle objectives; the control system needed to be computer based and the control system had to ensure the smooth and coordinated functioning of the inspection apparatus. This necessitated the control of three drive motors and the collection of data from the respective three position feedback circuits.

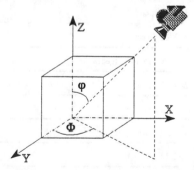

Figure 5.1. A figure defining the angles of declination, φ, and rotation, Φ, for an arbitrary camera viewing point.

The Camera Positioning System. In order to ensure the required degree of precision in the positioning of the camera, a two speed traverse system was adopted. For the last 50 mm on approach and the first 50 mm on exit from any of the three stations, the systems DC motor, MTR1, was driven with half its rated voltage (ie 6 Vdc). Outside of these limits the motor was driven with its full rated voltage (12 Vdc). Through this voltage switching, a rudimentary ramping of the motor speed was achieved and consequently the degree of overshoot in the trolley positioning was reduced. Position feedback information was gathered using optical sensor based electronic circuitry incorporated on the camera track.

The Part Positioning System. The pallet transport conveyor system was driven by a DC motor, MTR2, and the position of the pallet along the conveyor was determined by a limit switch feedback circuit. The rudimentary speed ramping system used in the camera positioning system was again adopted in the part positioning system to improve the accuracy and precision of the pallet positioning.

The Part Manipulation System. Once the inspected part had been positioned in the plane of the camera, part rotation was still required to present the obscured faces of the inspected part to the camera. The part manipulation system used a 12 Vdc DC motor, MTR3, to rotate the inspection table. Angular position of the table was monitored using a rotary optical encoder mounted on the shaft of the motor. The feedback signal, PRF1, was used to monitor the angular position of the table.

A further set of control signals was required to control the Inspection apparatus's pneumatic circuits, responsible for raising and lowering either the mass station or the inspection table. Two control signals, PNM1 and PNM2, were used to activate the pneumatic circuits. Table 5.1 summarises the control signals and feedback signals used to control the inspection apparatus.

CONTROL AND FEEDBACK SIGNAL CHART		
SYSTEM	CONTROL ACTION	SIGNAL
Camera Positioning System	MTR1 speed 1	CPC1
	MTR1 speed 2	CPC2
	MTR1 direction	CPC3
	Camera position feedback	CPF1
Part Positioning System	MTR2 speed 1	PPC1
	MTR2 speed 2	PPC2
	Pallet position feedback	PTF1
Part Manipulation System	MTR3 speed 1	PMC1
	Pallet Rotation feedback	PRF1
Pneumatic Circuits	Activate circuit 1	PNM1

CONTROL AND FEEDBACK SIGNAL CHART		
SYSTEM	CONTROL ACTION	SIGNAL
	Activate circuit 2	PNM2

Table 5.1. Control signal and feedback signal chart listing all the signals used in the Inspection apparatus control system. A definition of the 4 bit code words used in the table follows:

MTR	:	DC motor number
CPC	:	Camera positioning system Control signal
CPF	:	Camera positioning system Feedback signal
PPC	:	Part positioning system control signal
PTF	:	Pallet transport conveyor feedback signal
PMC	:	Pallet manipulation system control signal
PRF	:	Pallet rotation feedback signal
PNM	:	Pneumatic circuit control signal

The control system has been coded using Microsoft Visual C++ , version 4.1. Communication between the processor and the Inspection apparatus was achieved using the standard Input / Output card via D-type 25 pin parallel port. The DC motors were controlled from the computer using an application specific driver circuit [6]. The circuit was designed to use control signals sourced from the parallel port to switch relays in the driver circuit to change voltage supply and polarity to the motor. The driver circuit enabled the computer to select direction and perform crude, two step speed control.

7   Conclusion

An automated visual inspection apparatus was developed using cost effective, off-the-shelf technologies based on mechatronic principles. The inspection apparatus was developed as a computer based technology and therefore was suitable for integration into a CIM environment. The inspection apparatus was incorporated into a low cost CIM system, still under development at the University of Natal, Durban, and has enhanced the effectiveness of that system.

Specifically designed camera positioning and part manipulation systems were developed that enabled the inspection apparatus to position the single digital camera at known view points within the apparatus' inspection envelope. The control resolution for both the camera positioning system and the part manipulation system was such that both the angle of declination, φ, and the angle of rotation, Φ, could be incremented in steps of $5°$. The computer based control system was able to smoothly coordinate the functioning of the two systems, accurately and precisely positioning the camera at specified view points within the limits imposed by the control resolution of the motors.

Through this interaction of the camera positioning system and part manipulation system, the inspection apparatus was suitable for use as a test bed for the implementation of view planning algorithms and inspection by multiple view techniques. Further research is being conducted in order to further investigate the development of reliable and accurate 3D models of inspected parts by implementing view planning algorithms. Research will include investigations into different methods of archiving such data and sharing it with the CIM environment.

8 References

1   T.S. Newman, and A.K. Jain, A Survey of Automated Visual Inspection, *Computer Vision and Image Understanding*, Vol. 61, No. 2, March 1995, pp 231-262

2   M.P. Groover, Automation, Production Systems, and Computer Integrated Manufacturing, *Prentice Hall International*, 1987.

3   W. Brent Seales and O.D. Faugeras, Building Three Dimensional Object Models From Image Sequences, *Computer vision and Image Understanding*, Vol. 61, No. 3, May 1995, pp 308-324

4   L. Van Gool, P. Wounbacq and A. Oosterlinck, Intelligent Robotic Vision Systems, *Intelligent Robotic Vision Systems* (S.G. Tzafestas, Ed.), Dekker, New York, 1991, pp 457-507

5   A.D. Marshall, The Automatic Inspection Of Machined Parts Using Three-dimensional Range And Model-based Matching Techniques, *Ph.D. dissertation*, Department of Computing Mathematics, University of Wales, College of Cardiff, 1989.

6   G. Kaplan, Industrial Electronics, *IEEE Spectrum*, Vol. 31, No. 1, 1994, pp 74-76

Non - Destructive Testing for Thin - Film Magnetic Rigid Disks

A. Wichainchai, P. Nabokin, S. Rukvichai, and Ch. Puprichitkul*

Magnetic Research Lab. , Applied Physics Department, Faculty of Science,
King Mongkut's Institute of Technology Ladkrabang, Bangkok 10520, Thailand.

*Failure analysis Department, Seagate Teparuk Thailand Ltd. Co., Thailand

Abstract.   An inexpensive non-destructive testing system was designed and built to support magnetic characterization of longitudinal recording thin-film rigid disks for hard disk drive application. The instrument's principle is magneto-optic Kerr-effect measurements of the magnetic film M-H hysteresis loop. These measurements being combining with ordinary testing by reading/writing head provide the data on rigid disk quality and the sources of read/writing failure. The system operates in AC mode with magnetic field amplitude up to 4 kOe and rise/fall times more than 1 ms. The magnetic field parallel to the plane of disk, is produced by a pair of "big" magnetic heads (small electromagnets) supplied with triangular current pulses. The testing program, disk positioning and data processing are computer controlled. The main system's features are: 2% - relative magnetization resolution, 1% and 2% - repeatability and accuracy of coercive force H_c determination for H_c values (500 - 3500) Oe, comparing with commercial systems' (1 - 5)% absolute accuracy for H_c.

1 Introduction

Magnetization - field, M-H hysteresis loop is the most important characteristic curve for magnetic recording thin-film media, providing the major parameters of the media: coercive force, remanent magnetization, squareness' S and S^* [1]. Magnetic recording has reached to the density of 1 Gbits/inch^2 and is expected to grow to higher densities [2]. Writing difficulties due to the high coercivity of disks and the effect of thermal relaxation for high density longitudinal recording media result in vital necessity of testing the thin-film disks magnetic parameters: coercivity, magnetization and S^*. Non-destructive measurements of M-H loops are needed both at thin- film disks manufacturing and the hard disk components testing for assembling and read/write failure analysis.

Some instruments are available now for non-destructive M-H measurements. For rigid disks of any diameters the most interesting commercial proposals are of Digital Measurements Systems, Inc. [3] and Innovative Instrumentation Incorporated [4]. The first is the magneto-optic Kerr effect utilizing machine operating in DC mode (3 seconds field sweep time) with high power 4 inches pole pieces electromagnet producing magnetic field up to 13 kOe in the 5.25 inches gap. The second is the magnetic head instrument utilizing scaling principle : the recording a strip of oppositely directed magnetization by head with head gap and head - media separation four orders of magnitude bigger that of an actual recording head, then examining the transition region at the ends of this strip by rotating the disk at low speed (typically 12 rpm) past a sensor. Both systems have fine technical specifications but they are very expensive.

We will describe here AC-mode operating system which we have built for non-destructive testing of rigid disks. The setup is based on the magneto-optic Kerr-effect examining of the film magnetization at checked point (region) of the disk under remagnetizing of this region with AC magnetic field produced by symmetrical pair of big magnetic heads.

2 Technique

Schematic diagram of the test set is shown on Fig.1. Planopolarized laser's light beam is splittered by a beamsplitter onto two : one - to the tested sample, then through absorbing filter (or polarizer, for equalizing light beams intensities) to #1 photodetector, another - to #2 photodetector directly. The photodetectors signals are amplified by differential amplifier with subtracting of low frequency additions due to the laser noise and the system mechanical vibrations (splitting of the beam and differential detection doesn't principle, it depends on : 1) desirable ΔM resolution, 2) how large is the laser noise and 3) how well, about vibrations, the system is constructed and placed). The amplifier's output signal proportional to the sample's magnetization in the region of the light beam spot is fed into Y input of a digital oscilloscope operating in average XY mode with X input signal, from magnetic field sensor, proportional to the magnetic field acting on the sample at the checked point. The averaged M - H loop is stored by digital oscilloscope and / or computer processed or copied by a printer.

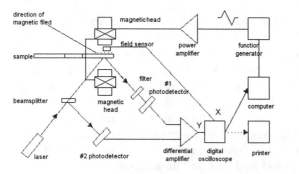

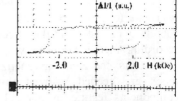

Fig. 1. Schematic diagram of the test set

Fig. 2. Transverse Kerr-effect
M-H loop recording of rigid disk

Transverse Kerr effect is used for M-H curves recordings : the change of the light beam intensity reflected by magnetic surface is proportional to the component of magnetization perpendicular to the plane of the light incidence. The incident angles $\theta \geq 50°$. Light beam spot at the reflecting disk surface is elliptical in shape with main axes about 1.5 mm x 1.0 mm. Longitudinal Kerr-effect geometry can also be used in our system.

The most serious problem for nondestructive magneto-optic testing is the high amplitude magnetic field creation. The problem is solved for DC-mode systems by use of big electromagnet with placing of the testing rigid disk in the inter-poles gap [3]. Such Kerr-effect system for longitudinal recording disks is needed in very high power and high stable programmable DC source for the electromagnet (it is not the ordinary problem to create (5 - 10)kOe in the gap 10cm) and is needed in good dampers for mechanical vibrations' insulation (because single M - H loop is recorded, without averaging). Conventional recording magnetic heads which are used for data writing can not be used with optical testing systems because of their extremely small gaps and heights of flights (about 0.3 μm and 0.03 μm respectively).

We have built symmetrical system of big magnetic heads $\equiv$ small electromagnets, with soft steel (transformer sheets) cores of about 20 mm^2 cross section area, 2 mm head gap and 4 mm inter-heads poles separation. The coils of the heads system, 4 Ohms total resistance in series, are supplied with alternating current from function generator through 300 W power audio amplifier. Magnetic fields up to 4 kOe in amplitude, triangular pulses with rise/fall times (10-1) ms, were produced by the system at the coils currents less than 10 A in amplitude. Repetition rates about $(100 \times \text{pulse width})^{-1}$ were used to operate without the coils overheating.

A small sensitive coil located at the inter-poles gap, with linear integration circuit, is used as AC magnetic field sensor. Correction for the checked point, which is the light beam spot at the surface of the disk, was obtained during calibrating procedure with additional sensitive coil located at the checked point without disk and calibrated by use of standard disks. The magnetic field homogeneity in the operating volume, being specified better than 2 % , was checked by high precision moving of the additional coil through operating volume (steps : 10 μm in Z direction, perpendicular to the disk plane, and 100 -150 μm in X-Y directions).

3 Results

An example of M-H loop recording for rigid disk with Co-Cr-Pt magnetic layer (300 $\overset{\circ}{A}$ thickness), covered by diamond like carbon film (200 $\overset{\circ}{A}$ thickness) and lubricant film (25 $\overset{\circ}{A}$ thickness) is shown on Fig. 2. The recording was made in transverse KE geometry, $\theta = 50°$, H parallel to the easy axis , one light beam optical scheme, with laser diode light source ($\lambda = 670$ nm, beam power 2 mW, low frequency noise $\simeq 0.5$ %), field rise time 1.5 ms, 256 sequence averaging mode, at ordinary third floor room with undamped equipment, without additional antivibration efforts. The total change of the reflected light intensity corresponding to the magnetization reversal (from $- M_s$ to $+ M_s$) is $2 \Delta I / I_{refl} = 2 \times 10^{-3}$, M_s is the saturation magnetization.

The main systems specifications (for the two-beam optical scheme) are as following :

$\Delta M / M_s$ resolution   2%
coercive force : repeatability   1%
            accuracy   2%   relative to standard disk
magnetic field amplitude   up to 4 kOe
test time :   1 sec for 1 point
            30 sec for disk change
measurement area   1.0mm x 1.5mm
disks diameters   90 , 65mm (larger and smaller sizes possible)
capability for measurements   both H ll. easy axis and H $\perp$ easy axis
determined parameters $M_{remanent} / M_s$, coercive force, S*

The system has similar to [3], [4] magnetometers features (for example, 2% accuracy of coercive force determination compared to 1% [3] and <5% [4]) with total cost of the equipment which the setup consists of about one order less.

References

1.   J. E. Manson , Recording measurements, in *Magnetic Recording Handbook: Technology and Applications*, edited by C. D. Mee and E. Daniel, Mc Graw-Hill, Inc.(1990), pp. 396-399.
2.   D. N. Lambeth, E. M. Velue, G. H. Bellesis, L. L. Lee, and D. E. Laughlin, Media for 10 Gb/in.^{2} hard disk storage: Issues and status, *J. Appl. Phys.* **79**, 8 (1996) 4496-4501.
3.   Kerr - Effect Magnetometer System, Model 660 L - KEM, *Rev. Sci. Instr.* **63**, 4 (1992) p 2347.
4.   R. M. Josephs, Remanent Moment Magnetometer, *US Patent* # 5 136 239 (1993).

Digital Filtering of Images Using Compact Optical Processors

Yu-hua Li, Yi-mo Zhang, Wen-yao Liu and Hong Hui

The Faculty of Precision Instruments and Photoelectrons Engineering

TianJin University, TianJin 300072,P.R.China

Tel: (086)022 27404459, Fax: (086) 022 27404547

Abstract. Based on the optical 4-f system, designing the compact automatic 2-D target classification implementation. The length of the system can be reduced to about 1/10 of original system length by using two Fourier transform telephoto lenses and an adding phase compensating lens in a 4-f optical system. The focal length of the phase compensating lens is calculated using optical theory. Lastly, the result of a computer simulation with four - kinds of aircraft classification is given.

1 Introduction

Automatic on-line 2-D multi-target classification, which is an important branch of pattern recognition, has been researched in many practical application fields for nearly two decades[1-4]. However, a satisfactory result has not been obtained yet. Its application is limited by four main factors: 1) the precision of classification; 2) the abilities of error-tolerance; 3) the requirement of real-time classification; 4) the size of system. The techniques of traditional Computer target classification can meet the factors of (1) and (4) [5-6], but its consuming time will be much longer than our expecting, especially when the shapes of the target are very complicated. Compared with Computer 2-D data processing, optical image processing has many advantages due to optical features, such as high speed, amass-parallel processing model, large capacity etc.. Therefore, Combining them with the hybrid opto-electronic hybrid processors have been proposed. It will have a future in industry automation, missile guidance, fingerprint identification, and security applications etc[7-8].

The initial efforts and developments are directed at using it for large space-bandwidth product systems, with emphasis on the large information-handling capacities of such systems[9-10]. Its typical structure is shown in Fig.1. It is based on an optical 4-f system that is used for optical information processing. In Fig.1 f is the focal length of the Fourier transforming lenses. The outside object is collected by CCD and as an input pattern f(x,y) is placed in plane P_1. The 2-D Fourier transform F(u,v) of f(x,y) is taken by the Fourier lens L_1 and is presented in plane P_2. The matched spatial filter, which represents the 2-D Fourier transform G*(u,v) of a test pattern g(x,y), is placed in plane P_2. The transmitted light through plane P_2 consists of the product F(u,v)G*(u,v) and is Fourier transformed by the lens L_3. The correlation of f(x,y) and g(x,y) is then formed in plane P_3, and the outside object will be classificated — the basic method of optical pattern recognition.

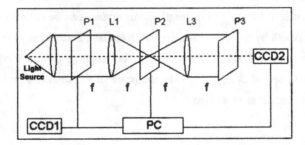

Fig.1 The typical structure of opto-electric hybrid 2-D target classification system

This optical system has a physical length equal 4f, and this physical length is often unacceptably large in its many practical application fields. Meanwhile, these opto-electric hybrid processors required high quality optical components with system stability requirements that are best met only in a laboratory environment. These requirements limit the development of hybrid opto-electronic element in its application fields.

In this paper, we propose the new-style optical construction of real-time compact optical processors to be used for automatic 2-D target classification. It is based on an optical 4-f system, and in order to reduce the system length, a five-lens optical system is designed. As a result, total optical system length will be compressed about 10 times. This system will serve as a test bed to further determine applications for which processors could be useful.

1248

2. The Theory of Compact Optical Processors Designing

2.1 Fourier Transform Basic Properties of Lenses and Optical 4-f System

It is very useful that a 2-D Fourier transformation may be obtained from a positive lens[11-13]. The simple optical system is shown in Fig.2.   In Fig.2, P_1 as a planar light source, P_2 as output plane of Fourier transformation lens.   Let us assume that the complex light field at P_1 is $f(x,y)$.   Then the complex light distribution at P_2 $g(\alpha,\beta)$ may be determined by means of Huygen's principle:

$$g(\alpha,\beta) = C \iint_{S_1} f(x,y)\exp(ikr)dxdy \tag{1}$$

where S_1 denotes the P_1 surface integral, C is an arbitrary complex constant, and

$$r = \left[L^2 + (\alpha - x)^2 + (\beta - y)^2 \right]^{1/2} \approx L + \frac{1}{2L}\left[(\alpha - x)^2 + (\beta - y)^2 \right] \tag{2}$$

At last, we can obtain:

$$g(\alpha,\beta) = C_1 \exp\left[-i\frac{\pi}{\lambda f}\left(\frac{1-v}{v} \right)(\alpha^2 + \beta^2) \right] \times \iint_{S_1} f(x,y)\exp\left[-i\frac{2\pi}{\lambda f}(\alpha x + \beta y) \right]dxdy \tag{3}$$

where $v = f/L$, C_1 is another arbitrary complex constant.

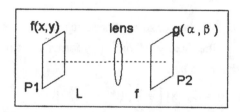

Fig.2 Geometry for determination of the optical Fourier transformation

From equation (3), clearly, except for a spatial quadratic phase variation, $g(\alpha,\beta)$ is the Fourier transform of $f(x,y)$. In fact, the quadratic phase factor vanishes if $L = f$. Furthermore, it can easily be shown that a quadratic phase factor also result if the signal plane P1 is placed behind the lens, and the quadratic phase factor vanishes too if $f = L$.   It is a typical optical 4-f system.

2.2 The Phase Transformation of Thin Lenses and the Theory of Designing Shortened Optical 4-f System

Optical 4-f system has a physical length 4f, and the sizes of the system seem very long for its application. Therefore, one of the important issues is the miniaturization of the dimension of optical processors to meet the requirement of available space in a particular application. Meanwhile, the sizes of the 2-D Fourier transform of an input object that is produced optically must exactly match the dimension of the 2-D Fourier transform which is stored on spatial filter. The focal length f of a Fourier lens must satisfy:

$$f = \frac{N_2 d_1 d_2}{\lambda} \tag{4}$$

where d_1 and d_2 are the pixel spacing of the two planes, N_2 is the number of pixels in the second plane, and λ is the wavelength used.

In the equation (4), the focal length f has been fixed. The choice of lenses and input, output plane spacing has a significant effect on processor's size. The consequence is that the exact 2-D Fourier transform is performed by the lens but is multiplied by a quadratic phase front according to Fourier transform basic properties of lens. Thus, another method of designing shortened processors should be considered.

We know, lenses are made of glass or some other transparent material. When a light ray passing lens, accord the theory of optical information processing, the phase transformation can be shown to be:

$$T(x, y) = C_2 \exp\left[\, i\frac{2\pi}{\lambda f}(x^2 + y^2)\,\right] \tag{5}$$

where C_2 is a complex constant.

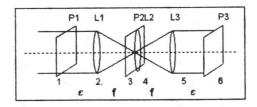

Fig.3 The structure of optical 4-f system with adding an phase compensating lens

We consider reduction sizes of the optical processors. Therefore, a new-style optical 4-f system which adding a phase compensating lens has been designed , as shown in Fig.3. In Fig.3, L_2 is an adding phase compensating lens, it immediately adjacent to behind of plane P_2. At first, let us assume that Fourier lenses L_1 and L_3 are same, its focal length is f, and the focal length of phase compensating lens L_2 is f_2. According to the theory of optical information processing, we can obtain:

$$g(x_6,y_6) = C\left\{ \left\{ \left\{ \left\{ [f(x_1,y_1)*h(x_1,y_1)]T(x_2,y_2)\right\}*h(x_2,y_2)\right\}F(x_3,y_3)T(x_4,y_4)\right\}*h(x_4,y_4)\right\} \times \right.$$
$$T(x_5,y_5)\left. \right\}*h(x_5,y_5) \qquad (6)$$

where $f(x_1,y_1)$ and $g(x_6,y_6)$ correspond to the complex light field at input plane P_1 and output plane P_3, $h(x,y)$ is the spatial impulse response, and $T(x,y)$ is the phase transformation of the lens. C is an arbitrary complex constant, and ' * ' denotes the spatial convolution.

In the equation, let us assume that all lenses and planes are so thin that can be negligible in their thickness, and the spatial filter $G^*(u,v)=1$. Then equation (6) may be written in the integral form:

$$g(x_6,y_6) = \iint\limits_{S1}\iint\limits_{S2}\iint\limits_{S4}\left[\iint\limits_{S5} \exp(i\frac{k}{2}\Delta)dx_5 dy_5 \right] f(x_1,y_1)dx_4 dy_4 ... dx_1 dy_1 \qquad (7)$$

where, $\iint\limits_{S1}...\iint\limits_{S5}$ denote the surface integrals in plane S_1, S_2, S_4, and S_5, and

$$\Delta = \{\frac{1}{\varepsilon}[(x_2-x_1)^2+(y_2-y_1)^2] + \frac{1}{f}[(x_4-x_2)^2+(y_4-y_2)^2] + \frac{1}{f}[(x_5-x_4)^2+(y_5-y_4)^2]$$
$$+\frac{1}{\varepsilon}[(x_6-x_5)^2+(y_6-y_5)^2] - \frac{1}{f}[(x_2^2+y_2^2)+(x_5^2+y_5^2)] - \frac{1}{f_2}(x_4^2+y_4^2)\} \qquad (8)$$

Through calculating, we can get, when $\varepsilon \neq f$, the additional quadratic phase fronts of optical system will remove too using a phase compensating lens L_2, and its focal length of L_2 satisfies equation (9)

$$f_2 = \frac{f^2}{2(f-\varepsilon)} \qquad (9)$$

where ε is the distance between plane P_1 and lens L_1 (or L_3 and P_3).

When $\varepsilon = 0$, $f_2 = f/2$. In this condition, the plane P_1 and plane P_2 are immediately adjacent Fourier lenses L_1 and L_3. So that, the size of total optical system is compressed 2 times, and its physical length only equal to 2f.

2.3 The Telephoto Fourier Lens and Compact System Designing

Applying the basic theory of geometrical optics[14-15], we know that a telephoto Fourier lens can reduce its working distance. The telephoto lens is consisted by two lenses (a positive lens L'$_1$ and a negative lens L'$_2$), and the working distance L' of telephoto lens (the distance between first lens and the focal plane of the telephoto lens) is less than its focal length f'. We define that k'= L' / f' is its compressing ratio, and $k' \leq 1$ for the telephone lens. Therefore, we can farther shorten the size of optical system in condition of keeping the focal length of Fourier lens invariability. We can get (10) using the Principe of geometrical optics

$$\frac{1}{f'} = \frac{1}{f_1'} + \frac{1}{f_2'} - d' \frac{1}{f_1' f_2'}$$

$$x_H' = f' \frac{d'}{f_2'} \tag{10}$$

where f'$_1$ and f'$_2$ are separately the focal length of L'$_1$ and L'$_2$, and d' is the distance between them, X'$_H$ is the distance between object main plane and lens L'$_1$, and f' is the focal length of telephoto lens.

As a result, we design a new-style compact optical processor, it is shown in Fig.4. Comparing with Fig.3, we replace two single Fourier lenses (L$_1$ and L$_3$) in optical system with two combinations of the positive and negative lenses (L$_1$ and L$_4$, L$_3$ and L$_5$) in order to shorten size of the 4-f optical system. In Fig.4, an input pattern f(x,y) is placed in plane$_1$ (LCD) (Liquid Crystal Display), and the matched spatial filter which represents the 2-D Fourier transform G*(u,v) of a test pattern g(x,y) is placed in plane$_2$ (LCLV) (Liquid Crystal Light Valve). This one acts as a filter by displaying a stored image that has been transformed electronically. The correlation result of f(x,y) and g(x,y) is then formed in plane P$_3$, where correlation peaks appear, the images match.

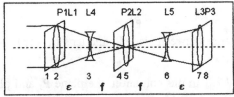

Fig.4 The farther structures of compact optical 4-f system with the telephoto Fourier lenses

In Fig.4, we design that the lens L_1 and L_3, and L_4 and L_5 are same, i.e. $f_1 = f_3$, $f_4 = f_5$, and the optical system is symmetry about the axis of the P_2 and L_2. One of the reasons for this is to reduce cost of designing and manufacturing lenses, and another, to reduce some image aberration in this optical system.

Keeping this in mind, we calculate the focal length f_2 of the phase compensating lens L_2 when the plane P_1 and plane P_3 are immediately adjacent lenses L_1 and L_5. $\varepsilon = X_H^{(1)} = X'_H^{(2)}$, applying equation (9) and (10), we get,

$$f_2 = \frac{f \times f_4}{2(f_4 - d)} \tag{11}$$

where, f is total focal length of combination lenses L_1 and L_4 (or L_3 and L_5), f_2 is focal length of lens L_4 (or L_5), and d is the distance between lenses L_1 and L_4 (or L_3 and L_5).

2.4 The Example of Compact Optical Processors Designing

In input plane P_1, we selected LCD is PVG161501PYN model LCD. The panels have 150 × 160 pixels, and each measures about 270 × 270um with a center-to-center spacing of about 300 × 300 um. In the spatial filter plane P_2, we selected LCLV is 40 lines / mm, the number of pixels in each row of the filter $N_2 = 128$. The model of the diode laser is TOLD9412(S), its wavelength is 0.65 um. Through calculating, the focal length of Fourier lens f = 1477 mm, and 4 f = 5908 mm and 2 f =2954 mm. It will be very large and unacceptable length.

We design compact optical processors according to above method. We select: 1) the focal length of the positive lens L_1 and L_3 is 170 mm, $f_1 = f_3 = 170$mm; 2) the focal length of the negative lens L_4 and L_5 is -20mm, $f_4 = f_5 = -20$mm; 3) the distance between L_1 and L_4 (or L_3 and L_5) is 152.3mm, i.e. d=152.3mm. Through calculation, the total focal length of Fourier lens that is combined by two lenses L_1 and L_4 (or L_3 and L_5) is 1477 mm, and the focal length of the phase compensating lens L_2 is 9.87 mm, i.e. f_2=9.87mm according equal (11). The compressing ratio is k = 1/5. Thus, the total length of the compact 4-f optical system that we design is only 609 mm, and the system length is compressed about 10 times.

3 Simulation Result of Using Compact Optical 4-f System in Target Classification

As an example which the compact 4-f optical system application, taking four classes of aircraft(bomber, rocket, airliner and fighter as shown in Fig. 5) as the targets to be classificated, we have taken the bomber as test pattern. In the match spatial composite filter is BPOF (binary phase-only filter)[16-17]. The result of a computer simulation is shown in Fig.6.

4 Summary

A scaled shortened processor with a total system length equal to about f/3 is proposed, and the system length is compressed about 10 times. In future, it will be small enough to fit the PCI (Peripheral Component Interconnect) slot of a personal computer with processing up 30 flames per second. It will be acted a part of computer and accelerated image processing speed of the computer.

5 Acknowledgment

This work was supported by the research fund of the National Nature Science Foundation of P.R.China and Post-Doctoral Foundation of P. R. China.

References

1. D.Casasent, "High capacity pattern recognition associative processors," N.N. Vol.5, 687-698(1992)

2. Qing Tang, "Multiple-object detection with a chirp-encoded joint transform correlator," App. Opt., Vol.32,No.26, 5079(1993)

3. J.A.Davis, "Compact optical correlator design," App.Opt. Vol.28, No.1, 10(1989)

4. D.L.Flannery, "Real-time coherent correlator using binary magnetooptic spatial light modulators at input and Fourier planes,"App. Opt., Vol.25, No.4, 466(1986)

5. A.J.Katz, "Generating image filters for target recognition by genetic learning," IEEE Trans. PAMI, Vol.16, No.9, 906(1994)

6. T.Poggio, "Computational vision and regularization theory," Nature, No.317, 314(1985)

7. Bahram Javidi, "Position-invariant two-dimensional image correlation using a one-dimensional space integrating optical processor : application to security verification", Opt. Eng. Vol.35, No.9, 2479(1996)

8. Thomas J.Grycewicz, "Experimental comparison of binary joint transform correlators used for fingerprint identification," Opt. Eng., Vol.35, No.9, 2519(1996)

9. Juris Upatnieks, "Portable real-time coherent optical correlator," App. Opt., Vol.22, No.18, 2798(1983)

10. X.J. Lu, "Basic parameters for miniature optical correlators employing spatial light modulators," Opt. Eng. Vol.35, No.2, 429(1996)

11. Francis T.S.Yu, 《 Optical Information Processing 》, A Whiley-Interscience Publication, 1983

12. Guo-guang Mu, Yuan-ling Zhan, 《Optics》, The People Educational Publication, P.R.China, 1978

13. Zhi-jiang Wang, 《 The handbook of Optical Technology 》 ,The Mechanic Publication of China, 1987

14. R. Kingslake, 《 Lens Design Fundamentals 》 ACADEMIC PRESS, INC. 1978

15. S.G.Lipson & H.Lipson, 《 Optical Physics 》 , Second Edition, Cambridge University Press, 1981

16. Danny Roberge, Yunlong Sheng, "Optical real-time correlator for implementation of phase-only composite filters," Opt. Eng. Vol. 35, No. 9, 2541(1996).

17. - B.V.K.Vijaya Kumar, "Tutorial survey of composite filter designs for optical correlators,"App. Opt. Vol. 31, No. 23, 4773 (1992).

Fig.5(a) The four kinds of aircraft
(fighter, rocket, bomber, airliner)

Fig.5(b) The test pattern (bomber)

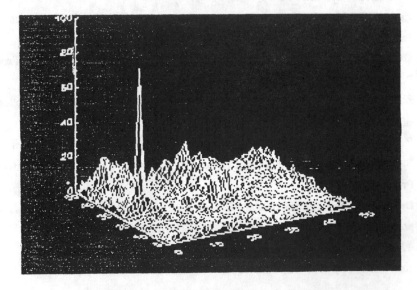

Fig.6 The result of computer simulation on BPOF correlation intensity pattern

Robotics and AGV Systems

An Easy-to-Use Robotic Welding System for Shipyards[1]

Marcelo H. Ang Jr.*, Lin Wei*, and Lim Ser Yong*

Centre for Intelligent Products and Manufacturing Systems
Department of Mechanical and Production Engineering
National University of Singapore 119260

*Gintic Institute of Manufacturing Technology
71 Nanyang Drive
Singapore 638075

Abstract. Welding is a fundamental task in ship building and repair. The welding process involves joining stiffeners with large steel plates which may be flat or curved, and fastening webs, T-bars, and brackets to the steel plates. Harsh working conditions coupled with the complexity of the welding patterns represent a major challenge for designing robotic welding systems. Welding webs, T-bars, brackets and sections with curvature is not currently feasible using automated systems. In this paper we describe a robotic welding system designed welding stiffeners, webs, T-bars, brackets, and flat steel panels for shipyards. The central component of the system is a robotic manipulator that can be programmed in a *walkthrough* mode. The welder programs the robot by guiding it once through the actual welding process as though he is doing it himself. Critical positions and orientations are recorded during teaching and the robot thereafter does the unpleasant welding job. Welding is further improved with arc-sensing technology. An intelligent welding subsystem is incorporated to automatically select the welding and other parameters. We also present an easy-to-use interface for the welding operators. We describe the details of the robotic welding system for shipyards and highlight system integration aspects.

1. Introduction

Welding is fundamental task in shipyards and marine/offshore companies. It is used in building and repairing structures. The welding process involves joining plates together and joining stiffeners with large steel plates which may be flat or curved, and fastening webs, T-bars, and brackets to the steel plates. In ship building, and more so in ship repair, welding is done in restricted and confined spaces with a very harsh environment. These environmental factors coupled with the complexity of the welding patterns represent a major challenge for designing robotic welding systems.

Welding steel plates together and fastening the first set of stiffeners can be done almost fully automatically together with the handling and transport of plates and stiffeners. These operations are shown in Fig. 1. Automatic and semi-automatic welding systems that exist today are confined to relatively simple welding operations.

[1] Presented at the International Conference on Computer Integrated Manufacturing, 21-24 October, 1997. Project supported by NSTB Grant 17/3/16 and Far East Levingston Shipbuilding Ltd. For more information, contact Marcelo H. Ang Jr. (mpeangh@nus.sg). The support of the Singapore National Science and Technology Board through research grant NSTB/17/3/16 is gratefully acknowledged.

These operations include straight welds (Fig. 1) or smoothly-curved weld paths wherein a flexible track can be laid to guide the welding head. The challenge is the more complicated welding paths as shown in Fig. 2. Fig. 2 shows the welding paths (in bold) required to weld "cross-stiffeners" onto the plates in Fig. 1. Welding the second set of stiffeners ("cross-stiffeners) or webs is done manually due to the complexity of the welding paths required. The cross-stiffeners run perpendicular to the first set of stiffeners. Fig. 2 shows the motions required.

The welding operations require horizontal and vertical motions as well as rounding motions around the corner. The size of the plates and stiffeners may vary. For example, the stiffeners can be as high as 4 m. Currently, the process is done manually with the operator having to stop and start the welding. This affects the quality of the weld. A robotic solution is found the be the most feasible approach to complicated welding paths, as shown in Fig. 2. With a robotic system, the starts and stops during welding operations can be minimised.

It can be observed that all shipyards involve welding operations with motions that shown in Fig. 2. The variation is in the size of the webs, stiffeners and plates. It should also be emphasised that unlike manufacturing environments, the parts and workpieces in shipyards are rarely standardised. Each part has a unique shape and dimension with some similarities. This makes traditional robotic programming of detailed teaching not feasible since the time spent in robotic teaching is not worth the effort since only one piece will be handled. The use of the robotic system cannot be justified. This emphasises the importance of having an **easy** teaching method, i.e., much easier than the actual welding.

We have developed a Ship Welding Robotic System (SWERS) to address this complicated welding. SWERS is currently in use by Keppel Far East Levingston Shibuilding in Singapore. In this paper, we describe the SWERS and highlight important issues. We discuss the current status and future plans for the SWERS.

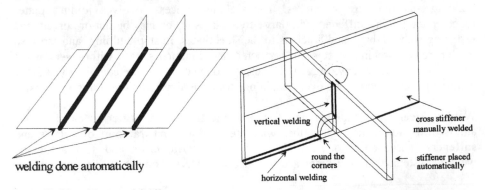

welding done automatically

vertical welding

round the corners

horizontal welding

cross stiffener manually welded

stiffener placed automatically

Fig. 1 Welding Plates and Stiffeners

Fig. 2: Manual Welding of Cross-Stiffeners

2. Methodology

Robotic welding has been recognized as the next step in technological advancement of shipyards[1]. There many commercially available robotic welding systems such as the robotic systems of Odense Steel Shipyard Ltd in Denmark, but almost all are not attractive to the Singaporean shipyards. The are many important reasons for this. Robotic welding system are very complicated to use, they require a robot programmer and/or application engineer which shipyards do not have. The workforce is welding operators and supervisors whose training is up to the high school level at best. Also, CAD data of plates, webs, stiffeners are not available. Furthermore, the workplace is very unstructured and locations of work pieces are always uncertain. CAD-based robotic solutions such as those in [2] are not suitable. Offline robot programming systems [3] are also not feasible because of the uncertainty and unstructured nature of the environment.

The need for improved productivity and quality is however crucial to the survival of shipyards. A very important requirement of our robotic solution is its ease of use, that is, it must be "idiot-proof" such that an unskilled or untrained worker can operate it out of his/her own "common sense". This is one of the most important design targets in our development of SWERS. Our approach is to develop a robotic system as a super tool for humans to use, rather than developing a completely autonomous system. The importance of human operators in man/machine teams has been recently recognized to be more useful that purely autonomous systems, especially so in welding [4].

Our approach to developing SWERS to exploit currently available technology and build on top of it to make it useable and attractive for shipyards to use. A very important consideration in deciding which robotic welding system to use is the "openness" of its architecture. The system must allow us low level access to its controller so that we add our own features to the system. This is crucial because we do not want the robotic system to restrict our ability to implement advanced algorithms to make the robotic system usable by shipyards. For example, we need to be able to give current and/or torque commands to each robot joint motor, thereby by-passing the robot controller.

Although many robotic systems claim to provide full access, almost all of them were built without the "openness" feature we require. They did not expect that the target customers would do R&D on a robot, and hence would require low level access. We have however found a robot manufacturer, REIS robotics, who is willing to work with us on this. They have provided a customised version of an EPROM in their robot controller that allows us a clean and multi-level access to their controllers, the lowest level of which is complete by-pass of their robot controller through software.

Another important criterion in our design is the use of industrially proven hardware for implementing our algorithms and for smooth integration into the robot controller. We have decided on the use of the industrially proven VME bus system, which is the same bus used by the robot controller. Our algorithms implementing the additional features run on a single board computer (the HyperSparc 125 MHz board) that plugs into the same VME bus of the robot controller. Our HyperSparc works "hand-in-hand" with the robot CPU, which is a 68030 based computer.

The SWERS has the following main features:

- A VMEbus-based open architecture robot controller.
- Off-the-shelf robot and welding equipment.
- A new robot teaching method called *Walk-Through Teaching* (WTT).
- Custom designed *Man-Machine Interface* (MMI).

3. SWERS

SWERS consists of a six-axis articulated welding robot (REIS SRV6) that is mounted upside down on a 3-axis gantry system. A schematic diagram of SWERS is illustrated in Fig. 3. Fig. 4 shows a picture of SWERS at FELS. The workspace of the robot is about 3 meters in diameter and the gantry moves the base of the robot in a work volume of 12 m x 12 m x 2 m. The gantry was designed to accommodate the 12m x 12 m panel size requirements in FELS. The whole gantry moves on a pair of tracks with length 16 m.

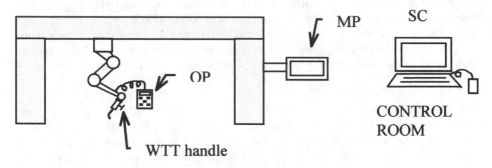

Fig. 3: Schematic Diagram of SWERS

Fig. 4: SWERS in the Shipyard Work Area

A key feature of the system is the walk-through-teaching (WTT) handle. An operator using the WTT handle specifies welding positions and motions. Advanced algorithms are employed in the robot controller to allow the operator to grab the WTT handle and move the robot as if the robot is operating in a zero-gravity environment. Parameters can be adjusted to change the dynamic response of the robot from the forces exerted by the operator. The robot can be made to appear very light of fast, or to behave as if it's underwater or in a very viscous environment.

Fig. 5 shows the various subsystems of SWERS and their components. The REIS robot controller is a 68030-based computer on a VME bus. We have installed a single board HyperSparc 125 MHz computer on the same VME bus to run our enhancements in conjunction with the 68030-based robot controller. The standard teach pendant of the REIS robot has been supplemented with the OP, MP, SC, and WTT handle to allow a welding operator to easily use the system without the need for technical knowledge on robot programming. The HyperSparc acts as the supervisory controller that allows high level and low level access to the robot controller.

3.1 Robot Subsystem

The Reis SRV6 robot used is a 6 degree-of-freedoms articulated configuration robot. It has a work envelop of 3m diameter and a repeatability of ±0.05mm. The main feature of the robot is the VME-based open architecture controller. The robot subsystem incorporates our walk-through algorithms and the robot controller. The welding controller sits on the same VME bus as the robot controller and information is shared between the robot and welding subsystems via shared memory. Communications between the HyperSparc and the robot and welding subsystems are via shared memory, while communications with the other five subsystems are via serial communications.

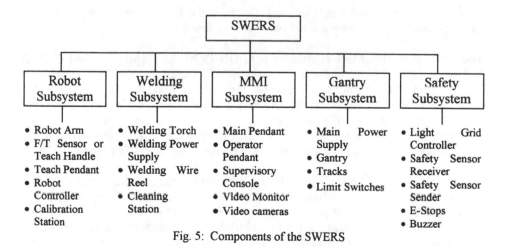

Fig. 5: Components of the SWERS

Mounted together with the robot are the welding accessories such as the wire reel drum, the wire feeder, a torch cleaning station. The welding torch and a WTT Handle (Walk-Through Teach Handle) are equipped at the end of the robot arm. The welding power source is placed on top of the gantry. The robot controller, Supervisory Console and the gantry controllers are housed in the air-conditioned Control Room at one end of the gantry. The control Room moves with the gantry.

3.2 Welding Subsystem

The welding system consists of a power source, a wire feed unit, a welding torch and a torch nozzle cleaning unit. It also comes with a built-in seam tracking sensor for following of welding path. It compensations for errors in welding taught paths due to workpiece dimensional errors (due to heating) or positioning errors.

Table 1 shows a list of important welding process parameters. We have done a thorough study of the welding process and arc sensor characteristics. Many of the welding process parameters are coupled with arc sensor parameters. This study together with the current welding code resulted in an intelligent database of values for the different welding process and arc sensor parameters. This database consists of recommended values for the welding process (Table 1) and arc sensor parameters. The user, however, is allowed to select and adjust parameter values before welding. Typically adjustments are done by a welding engineer in the control room; and welders need not adjust the values. The recommended values serves as the default values for the parameters if the user does not select or input values. An improved feature we are currently working on is the provision for online adjustment of welding current and voltage during welding. The welding position (vertical, overhang, or horizontal) can also be automatically determined from the taught positions.

3.3 Man-Machine Interface (MMI)

To facilitate the walk-through teaching and to make the system easier to operate, custom-designed MMI software and hardware are developed.

The hardware of the MMI consists of a MAIN PENDANT (MP), an OPERATOR PENDANT (OP, Fig. 6) and a SUPERVISORY CONSOLE (SC, Fig 7). Important welding and other positions are recorded through the operator pendant (OP). The main pendant (MP) consists of a touch screen LCD that allows the operator to move the gantry or robot base. Through the MP, the robot can be set to WTT mode or to welding mode. In the welding mode, the safety subsystem is activated and the robot does the actual welding using the positions previously taught in WTT mode.

The Supervisor Console (SC) provides a higher level user interface for the welding supervisor. In the SC, important welding parameters, such as welding voltage and current can be set or default parameters from an intelligent database can be used. This database contains important parameters ideal for different welding configurations. The database has been developed through a detailed welding process study and in consultation with current welding practices and codes. From the taught positions using the WTT handle, the system can automatically select the correct welding parameters. As in any intelligent system, the user can always over-ride these parameters through the SC.

The OP is attached on the robot near the welding torch. It is a control pendant for WTT only. The MP and the OP are located outside the gantry. They are for overall control of the SWERS and data/parameters entry.

The handle located at the welding torch is for WTT. The handle is designed under the considerations of ease in teaching and no interference during welding.

The SUPERVISORY CONSOLE is located inside the control room of the SWERS. It consists of a monitor, a keyboard and a mouse. Upon start off the SWERS, the user firstly uses the SC to select and adjust the welding parameters according to the workpiece he is working on.

A Graphical User Interface (GUI) is developed to facilitate the operation of the system. Fig. 7 illustrates an example of the menus and dialog boxes of the GUI. The GUI allows the user simply to click the mouse button to select the desired welding data such as leg length, wire type, welding mode and no. of passes .The selected data is matched with the default and optimized values of the welding parameters such as current, voltage, speed, oscillation etc. The user may adjust the values of the parameters if he/she so desired. The default values of the parameters are based on a welding study and are consistent with the welding specifications. The user can, however, add, delete, and change the records in the database. This ensures that the database grows and improves with experience, The user updates the data using the mouse and the keyboard.

Table 1: Important Welding Process Parameters

- Welding Mode (Position)
- Leg Length
- Wire Type
- No. Of Passes
- Welding Current
- Welding Voltage
- Welding Speed
- Stick out Length
- Oscillation Frequency
- Oscillation Amplitude
- Oscillation Hold Time

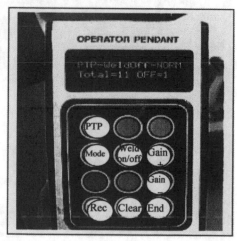

Fig. 6: Operator Pendant

The MP is a device for overall control of the whole welding system. It is the only device in the commands and controls movements of the system. Using the MP located near the control room at the side of the gantry, the user can move the gantry, transfer the robot on the gantry, clean the welding torch, and start the teach mode and welding mode. For safety purpose, it can only be operated when the OP is hanged up on the robot arm.

The MP consists of a touch screen terminal. The touch screen has the advantages of ease of use and programming flexibility. Fig. 8 illustrates some screen pages available in the MP. The screen displays the desirable buttons for the user to press at different states. It also displays the robot and welding status.

The OP is a hand-held device for the operator to perform walk-through teaching. It is activated by the MAIN PENDANT located remotely from the robot. The OP is connected on the robot at a position near the welding torch. To begin the teaching, the operator firstly removes the OP from the robot arm and a sensor detects that the OP is detached. After the teaching procedure, he/she attaches the OP back to the robot and all the buttons are then disabled. The operator then leaves the work envelope for the MP. The feature is to ensure that no operator is around the robot when it is moved by the MAIN PENDANT.

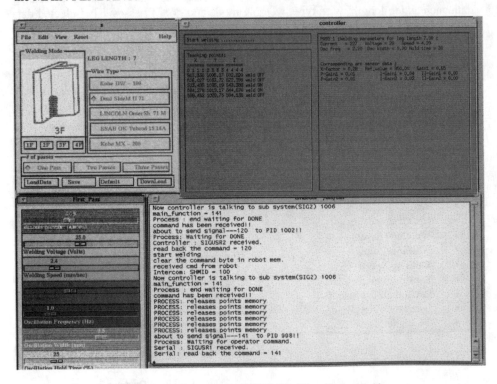

Fig.7 : Supervisory Console GUI (Welding Parameters)

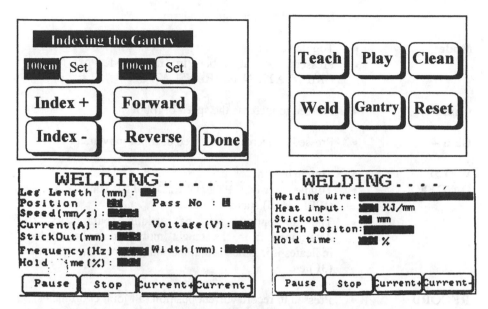

Fig 8: Sample MP (Touch Screen) Pages

As shown in Fig. 6, there are eight buttons and a 2x16 LCD on the OP. The LCD displays the current teaching state. The *PTP* button allows Point-to-Point for walk-through teaching. As the operator leads the torch to the point to be recorded, he/she can select the mode of robot movement by pressing the *Mode* button. Three modes can be select with the toggle button, namely the Normal, Rotate and Translation modes. The Normal Mode allows the operator to rotate and shift the position of the torch. This mode involves more computation time thus the torch is slower to follow the hand movement. On the other hand, the Rotate Mode and the Translate Mode decouple the two modes. It allows a easier movement either by rotating the torch about the TCP (Tool Center Point) Point or just moving the torch position. The *Gain* + and *Gain* - buttons increases or decreases the gain of the robot speed. The operator can adjust the gain value for the easiness of moving the robot.

There are two states, namely *Weld On* and *Weld Off*. The operator can teach the robot for the desired state by pressing the corresponding buttons as he/she tracing the robot path. The point just recorded can be deleted if he/she so desired. Table 2 lists the description of each buttons.

Table 2 *Description of the OP Buttons*

PTP	• Pressed to put robot in walk-through Point-to-Point mode • State after pressed: PTP (WELD OFF)

		• Enable: Walk-through' program
Mode	:	• Toggle button • Pressed to put robot in the mode of NORM, ORIENTATION, or TRANSLATION
Gain +		• Pressed to increase the speed of robot
Gain -		• Pressed to decrease the speed of robot movement
WELD ON/OFF	:	• Toggle button • Toggle between PTP (WELD OFF) state and PTP (WELD ON) states • The appropriate sets of welding parameters are determined from the taught motion data base which indicates if it is horizontal motion (2F) or vertical motion (3F) etc.
RECORD	:	• Pressed to record current position (all joint coordinates) • Pressed to record welding parameters under PTP (WELD ON) state. • Pressed to change welding path (for PTP state) and welding modes. • Used in the following states: PTP (WELD ON); PTP (WELD OFF)
CLEAR	:	• Pressed to CLEAR the previous pressed data entered.
END	:	• Pressed to get into IDLE state • States can be terminated: PTP (WELD ON); PTP (WELD OFF)
LCD display	:	• show the following - states [PTP(WELD ON); PTP(WELD OFF) - Number of points recorded

---

3.4 Gantry Subsystem

The 3-axis gantry is driven by a pair of AC induction motors along the flow of the panel (X drive). The robot is transferred across the panel by an AC servo motor (Y drive). Due to the height of the webs and limited size of the robot work envelope, the robot has to be raised up for the next welding zone. (Z drive). The communication between the robot controller and the gantry controller is through simple RS232 interface.

The robot controller is housed inside an air-conditioned control room. The room is mounted together with the welding source at one end of the gantry. Due to the long span of the gantry, the cables lines for controller and welding equipment are up to 30m. The drum for wire feed unit and the nozzle cleaning station are located on top of the gantry and moved together with the robot.

The gantry is to index the robot at a number of points along the webs, The locations of these points are determined such that they ensure the robot to cover the entire required welding area. The fine positioning of the arm compensates position errors contributed by the gantry.

The MP is a touch-screen panel and serves as the user interface for the operator to move the gantry and the robot mounting base. A video monitor is also available so that the operator can see the work area as seen from the top by the robot.

3.5 Safety System

Generally, the overall system is designed and constructed under standard safety guidelines of industrial robots [5]. Some of the safety features are as follows:

- Emergency Stops are available on the Reis Robot Controller, MAIN PENDANT and OPERATOR PENDANT.

- Software module of each subsystem includes *exception handling routines* which ensure that the software will not be aborted abnormally leaving the system in an unknown state.

- In "TEACH mode", the velocity of the robot is restricted at a safe speed and the robot cannot be moved on the gantry.

- The MAIN PENDANT is operational only when the TEACHING PENDANT and all other control stations are deactivated.

- A sensor is implemented to detect whether the OPERATOR PENDANT is detached from the robot.

- The program ensures that the robot is in up position when the robot moves in X or Y direction.

- Light curtain is installed for vertical safe-guarding of the robot work envelope.

4. Performance Evaluation

The SWERS have been certified and performance has been good. Certification includes extensive laboratory tests of welded specimens. These include hardness tests and micro-edge inspections, and macro-etch tests for cracks and porosity. Extensive experimental runs and demonstrations have been conducted. Typical runs involving welding paths that are U-shaped (100 mm down, 600mm right, 100mm up). The tests have shown that the quality of the weld by the SWERS is better than the manual FCAW and SMAW in terms of uniformity, shape and consistency. Macro-etching inspection and micro hardness tests have shown that the samples are free of porosity and cracks. For the cycle time, the SWERS also performed better than the manual processes (Table 3)

Table 3: Typical Cycle Times

Time	SWERS	FCAW	SMAW
Set-up	N/A	60 sec	30 sec
Teaching	150 sec	N/A	N/A
Welding	234 sec	306 sec	608 sec
Total	**384 sec**	**366 sec**	**638 sec**

The cycle time is expected to be even faster in actual workpieces where it has longer welding seams and repeated patterns. For such job, the PTP mode and "Teach-Weld-Weld" mode can be used so that the teaching time can be significantly reduced.

The flexibility of SWERS has also been demonstrated in pipe welding operations. Although not originally designed for this application, the general applicability of the walk-through teaching methods has made the application to pipe welding very straightforward with minimal effort.

5. Conclusions

We have described the SWERS that we have developed for Far East Levingston Shipbuilding Ltd. The system demonstrates an attractive and feasible approach to robotics applications wherein state-of-the-art technology is used and additional features are added to make the system attractive to industry. An important requirement is the "openness" of commercially available systems to allow further R&D work to improve and/or add features.

A unique feature of SWERS is the walk-through programming capability, which is very natural and easy to use. The difficulty and tedium involved in robot programming has always been a disadvantage in choosing a robotic solution. Walk-through programming addresses this problem and is applicable not only in robot welding but in general robot programming as well.

An increasing trend in robotic applications is the use of the robot as a "super-duper" tool for humans to use. This opposes the original thinking in which robots would replace humans. Indeed the SWERS demonstrates how the robot can be used as an effective tool. The concept of "de-skilling" is demonstrated by SWERS. The welding skill is delegated to the robot, while the programming and decision making tasks are retained by the human.

SWERS is continuously evolving through improved features. Current work hopes to improve the productivity of SWERS by incorporating additional features. These features include the welding of similar parts by teaching only reference positions, and the closer coordination of gantry and robot motion during welding.

Acknowledgements. The contributions of the following members of the project team have made this project a success. They include Ng Teck Chew from Gintic Institute of Manufacturing Technology; Fung Mok Wing, Patrick Yang, Danny Tan, and Zeng Xiao Ming from Singapore Productivity and Standards Board; Roy Lim Weng Chin, Lim Tow Koon, Wee Teck Guan, Lee Weng Kee and Sam Weng Kuan from Far East Levingston Shipbuilding Ltd; Ng Joon Leong from University of Western Australia, Yang Jiandong, Lee Kim Seng, Kang Yanjun, and Pubudu N. Pathirana from National University of Singapore. The support of the National Science and Technology Board (Grant NSTB/17/3/16), Gintic Institute of Manufacturing Technology, Singapore Productivity and Standards Board, and Keppel Far East Levingston Shiobuilding Ltd are also acknowledged.

References

[1] C E Skjolstrup and S Ostergaard, Shipbuilding using automated welding processes, *Welding International Review*, February 1994, pp. 213-217.

[2] T. Kangsanant, and R.G. Wang, CAD-based robotic welding system with enhanced intelligence, in *Proceedings of the Third International Conference on Computer Integrated Manufacturing*, 11-14 July 1995, World Scientific Publishing, Singapore.

[3] R.O. Buchal, et.al., Simulated off-line programming of welding robots, *International Journal of Robotics Research*, Vol 8, No 3, June 1989, pp. 31-43.

[4] Arc welding is still very much a craft, *ABB Robotics Review, No. 1, 1993*.

[5] *Singapore Standard CP53: 1990: The Safe Use of Industrial Robot.*

Fixturing 2D Concave Polygonal Objects Using a Three Fingered Flexible Fixturing System

H. Du and Prof. Grier C I Lin

Centre for Advanced Manufacturing Research
School of Engineering
University of South Australia
The Levels, SA 5095 Australia

Abstract. This paper presents the development of an algorithm for fixturing 2D concave polygonal objects using a three fingered flexible fixturing system. The three fingered flexible fixture system includes two computer numerically controlled (CNC) modules and an auxiliary mechanism. One module has two fingers which can be positioned along two circumferences of two circles, whereas the other module has one finger which can be adjusted along a slot. The auxiliary mechanism applies the clamping force to the object and the control of the clamping force can be achieved automatically.

The algorithm presented in this paper is the implementation of the concept of the maximal inscribed circle of a polygon, and it is an extension of the work conducted by the authors. The algorithm mainly consists of determining clamping points on the concave polygonal object and the optimal configuration of the three fingers in the fixturing system. The algorithm is implemented by a program written in C language for Windows. Examples of the application of the algorithm are described in this paper. The results show that the algorithm can guarantee a secured fixture in a trial with a concave polygonal object.

1 INTRODUCTION

A fixture is a device that locates and holds a part for machining, assembly, inspection and other processes in manufacturing. In order for a Flexible Manufacturing System (FMS) to be really flexible, every process needs to be as flexible as possible. It is thus desirable that a fixture is flexible to accommodate geometries and shapes of a variety of parts or assemblies in manufacturing.

Research has been conducted into the minimum number of fingers needed to immobilise an object in robot hand grasping; this has provided a new approach to flexible fixturing. Markenscoff et al [1] have established that four contact points suffice to immobilise a generic 2D object, and seven suffice to immobilise a generic 3D object. On the other hand, Czyzowicz et al [2] have shown that generic 2D and 3D polygonal objects can be immobilised respectively by three and four frictionless point contacts. There are two advantages to using the minimum number of fingers in a fixture. One advantage is that it may lead to more adaptable or more efficient fixturing techniques and fixturing planning algorithms. The other advantage is that the possibility of using fewer fixtures may lead to more efficient re-fixturing methods in machining process involving frequent repositioning of fixtures. Therefore, a need has occurred to implement the concept of the minimum number of fingers in fixturing for further development.

Based on the number of contact points for a planar object, a three fingered fixturing system is proposed by the authors and is currently being developed. Figure 1 shows a schematic diagram of the system, while Figure 2 shows a prototype of the system. The three fingered fixturing system consists of two computer numerically controlled (CNC) modules, one fixed and the other movable. The fixed module has two fingers positioned along two circumferences of two adjacent circles, while the movable module has one finger which can be adjusted along a slot. A detailed mechanical design of the system can be found from the paper published by Lin and Du [3].

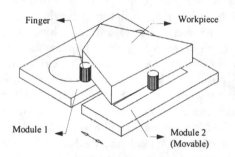 

Figure 1 A three fingered fixturing system   Figure 2 Prototype of the system

The motivation behind this research is that an algorithm needs to be developed for the three fingered flexible fixturing system. The authors have developed an algorithm for fixturing planar objects [4], but it can only deal with convex polygons. This paper presents the development of an algorithm for fixturing a 2D concave polygonal object, based on the concept of the maximal inscribed circle in a polygon. In comparison with the previous one, this algorithm is faster and more efficient. It will be tested in the three fingered fixturing system in the future. In this paper, after reviewing some previous research in fixturing algorithms, the proposed algorithm is presented and explained, followed by an example of its application. Finally, conclusions are given as to the efficiency of this algorithm.

2 LITERATURE REVIEW

In recent years considerable progress has been made on fixturing research, and substantial literature exists on automated fixture design and flexible fixturing, as reviewed by Hazen and Wright [5] and Trappey and Liu[6]. Two major geometric algorithms proposed by Brost and Goldberg, Wallack and Canny, along with the concept of the maximal inscribed circle, are discussed as follows.

Brost and Goldberg [7] have proposed an algorithm for fixturing prismatic polyhedral objects using a modular fixturing system, consisting of fixture elements and a translating clamp, all mounted on a fixture plate. Wallack and Canny [8] have also proposed an algorithm which is similar to that developed by Brost and Goldberg. Their algorithm synthesises a class of modular fixtures with four round locators to be in form closure on a split lattice that can be closed like a vice. Due to the characteristics of modular fixturing, these algorithms generate all the possible configurations for a polygonal workpiece. However, they can only be applied in modular fixturing.

The concept of the maximal inscribed circle has been proposed by many researchers to secure stable grasps for both a three fingered robot hand in robotics and fixturing system. Rimon and Burdick [9] have shown that three contact points, where a circle maximally inscribed meets the sides of a polygon, can determine a feasible equilibrium grasp, and thus these three points secure a stable fixturing configuration. Even though a few schemes for determining the maximal inscribed circle have been suggested by these authors, no details are given regarding their implementation. Czyzowicz et al [2] have explored the idea of using the maximal inscribed circle with a Voronoi diagram to immobilise a polytope. But similarly, no practical implementation has been given. The Voronoi diagram is a very useful tool in computational geometry to determine the maximal inscribed circle, but it is complicated and difficult to implement in fixturing. Preparata and Shamos [10] have proposed converting the nonlinear problem of the maximal inscribed circle into a linear program, but the implementation of the idea is not straightforward.

The algorithm previously developed by Lin and Du [4] was the implementation of the concept of the maximal inscribed circle in a polygon for fixturing in manufacturing. However, it was not very efficient and it can only deal with convex polygons.. The algorithm developed here is faster and more efficient.

3 THE ALGORITHM

The determination of the maximal inscribed circle in a polygon lies the principal part of the algorithm. Once the maximal inscribed circle is obtained, three clamping points for fixturing are chosen either on or close to tangential points of the circle maximally inscribed within the boundary of the object. These three tangential points secure the fixturing configuration. In this section, an overview of the algorithm is followed by an explanation.

3.1 Overview of the Algorithm

Figure 3 shows a flow diagram of the algorithm. Once the geometry of the data of a polygon is extracted from a CAD file, the algorithm checks whether it is the convex polygon or not. The algorithm then applies the downhill simplex method (Press et al [11]) to determine the maximal inscribed circle in the polygon. After the circle is found, the Newton-Raphson method (Faux and Pratt [12]) is used to obtain three tangential points where the maximal inscribed circle meets the sides of the polygon. Configuration of the three fingers in the three fingered system can then be determined from these three tangential points. Optionally, the fixturing configuration can be verified and simulated by using two software modules developed by Du and Lin [13]. The algorithm has been implemented in a program written in C language for Windows.

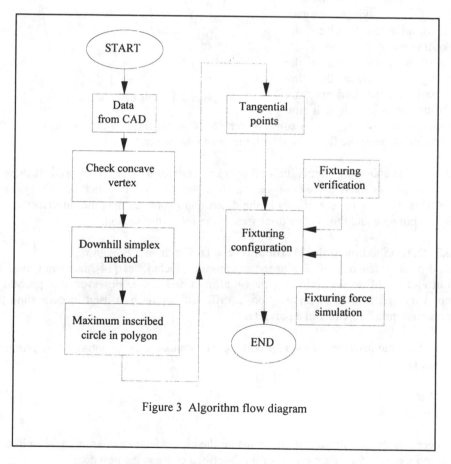

Figure 3 Algorithm flow diagram

3.2 Algorithm for Checking Concave Polygons

Vertices of a polygon are referred to by their indices, which increase counter clockwise. It is assumed that the first index of the polygon be known, and the linear constraint equations for all edges be determined.

The concave vertex can be found by using the following procedure:

1) Start from the first vertex of the polygon, for example, vertex 1 in Figure 4, obtain the middle point $P(x,y)$ of the line segment formed by the first vertex and the third vertex.

2) Check if the point is within the polygon, by comparing its position relative to the two edges of the second vertex. The lines formed by the two edges provide enough information to determine if the point is within the polygon.

3) If the point is within the polygon, replace the first vertex with the third one, and return to step 1. If it is not,

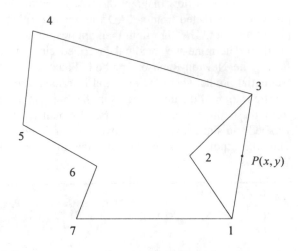

Figure 4 Finding the concave vertex

then the second vertex is a concave vertex. Next set flags to the two edges of the vertices. Replace the first vertex with the third one, return to step 1.

Repeat the above procedure, until all polygon vertices have been checked. If there is any vertex which is a concave vertex, then the polygon is concave. The concave vertices will provide constraints in the determination of the maximal inscribed circle in the polygon and this will be discussed in the following section.

3.3 Determination of the Maximal Inscribed Circle in a Polygon

The penalty function of nonlinear programming techniques [14], has been used by Lin and Du [4] to determine the maximal inscribed circle. However, this process is not very efficient, so in this section, a different approach is used to determine the maximal inscribed circle in a polygon.

The maximal inscribed circle in a polygon is formulated as an optimisation problem, listed as follows:

Maximise $\qquad\quad r$ $\hfill$ (1)

Subject to $\qquad\quad \mathbf{x} \in \mathbf{X}$

where $\mathbf{x}$ is the coordinates of the points on the circumference of an inscribed circle in the polygon, and r is the radius of the inscribed circle in the polygon.

The above unconstrained optimisation problem is solved by using the downhill simplex method for multiple dimensions [11]. The downhill simplex method requires only function evaluation, not derivatives. This is the main reason why it is used to optimise the maximal inscribed circle. The method can be briefly explained as follows:

The method must start with N+1 points, defining an initial simplex. N is the dimension of the optimisation problem. If one of these points is used as an initial point $\mathbf{P}_0$, then the other N points are

$$\mathbf{P}_i = \mathbf{P}_0 + \lambda \mathbf{e}_i \qquad (2)$$

where the $\mathbf{e}_i$'s are N unit vectors, and where λ is a constant which is an estimate of the problem's characteristic scale of length.

This method involves many steps, and most steps only move the point of the simplex where the function is largest through the opposite face of the simplex to a lower point. These steps are constructed to conserve the volume of the simplex. If this can be done, large steps will be taken to expand the simplex in one direction or another. When the simplex reaches a valley floor, the method contracts in the transverse direction and tries to moves rapidly down the valley. Where the simplex tries to pass through a very small hole, it contracts itself in all directions, pulling itself in around its lowest point. The optimised points have thus been found.

Having discussed the downhill simplex method, the procedure of the unconstrained optimisation is explained as follows. With a given point in the polygon as the centre of a circle, the maximal radius of the circle within the polygon is obtained by comparing the distances between the centre point and all edges of the polygon, and the distance between the central point and the concave vertex. The maximal radius is the minimum of all distances. As shown in Figure 5, for example, r_1 is the maximal radius of the feasible circle within the

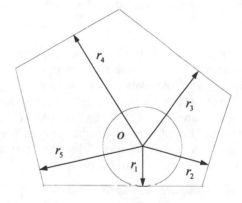

Figure 5 Maximal radius of an inscribed circle

polygon, with a given point o as the centre of a circle in the polygon.

Once the maximal radius of the circle has been obtained from a given point as the centre of the circle, the downhill simplex method generates the search direction of the maximal radius of the circles within the polygon. The initial points of the simplex can be generated by a random function to achieve the automation of calculation. A new centre point is then generated by the algorithm. This process repeats until the optimisation method converges to the maximal point. The final result of the optimisation is a maximal inscribed circle in the polygon.

An alternative geometric method for determining the maximal inscribed circle is to locate the maximal radius of the circle according to any three edges of a polygon. By searching all three edge combinations and the circles based on them, the maximal inscribed circle can be found. However, this method requires an exhaustive search of all possible combinations. If the number of edges of the polygon is large, the computation time will be extensive. By comparison, the proposed algorithm only searches the circle maximally inscribed in the polygon, with an initial estimation of the centres of the circles. Hence, it is efficient and its computation time is small.

The key issue in the above algorithm lies in determining the maximal inscribed circle in a polygon. Once the three clamping points are obtained, configuration of the three fingered system can be determined accordingly [4]. Optionally, the fixturing configuration can be verified and simulated by using the software modules developed by Du and Lin [13]. In the next section, an example is given of an application of the algorithm.

4 EXAMPLE OF APPLICATION OF THE ALGORITHM

In this section, a concave polygonal object is used as an example to show the effectiveness of the algorithm and the flexibility of the three fingered system developed.

A program of implementation of the algorithm has been written in C language for Windows. An input interface has been developed to read the CAD data in DXF file format. The CAD data for a concave polygonal object with five edges is generated by a CAD package as an input for the algorithm. The coordinates of the vertices of the polygon are (0,0), (20, 90), (100, 140), (140, 50) and (200, 0). Figure 6 shows the maximal inscribed circle in the concave polygon and the results of the calculation. Figure 7 shows the results of the three tangential points where the maximal inscribed circle meets the sides of the polygon.

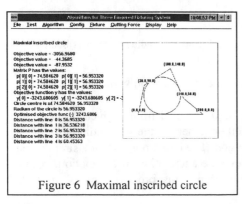

Figure 6 Maximal inscribed circle

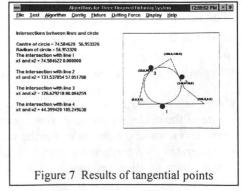

Figure 7 Results of tangential points

After the three fixturing points having been obtained, the fixturing verification module can be used to verify whether the fixturing configuration is valid. The results in Figure 8 show that the fixturing configuration is valid for this example.

The configuration of the three fingers in the three fingered system can then be obtained from the three clamping points. Figure 9 shows one of the fixturing configurations of the three fingers and the results of the calculation. The results include the coordinates of the vertices of the object and the three fingers, and the distance between the two modules. The results of the fixturing configuration are used to configure the three fingers in the three fingered system. It shows that the algorithm can guarantee a secured fixturing configuration for the three fingered fixturing system in a machining process.

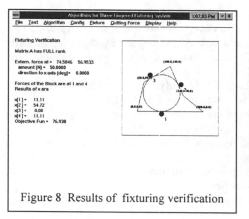

Figure 8 Results of fixturing verification

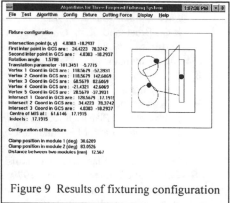

Figure 9 Results of fixturing configuration

5 CONCLUSIONS

An algorithm has been developed for fixturing concave polygonal objects by using the concept of the maximal inscribed circle for a three fingered fixturing system. It is an extension of the research conducted previously by Lin and Du [4]. The difference between them is that this algorithm can deal with concave polygonal objects, and is more general than the previous one. In addition, by considering the distance between the centre point and the concave vertices of a polygon, the algorithm tests the suitability of the concave vertices as clamping points.

The results of an application of the algorithm show that the algorithm guarantees a secured fixture in a trial with a concave planar polygonal part. The results also show that the algorithm is faster and more efficient than the previous one. The algorithm will be tested in the future in a real manufacturing operation for the three fingered fixturing system.

6 REFERENCES

1. X. Markenscoff, L. Ni, and C.H., Papadimitriou, The Geometry of Grasping, *International Journal of Robotics Research*, **9**, 1(1990) 61-74.
2. J. Czyzowicz, I. Stojmenovic, and J. Urate, Immobilising a Polytope, *Lecture notes in Computer Science*, **519** (1991) 214-227.
3. G.C.I. Lin and H. Du, On Developing a Flexible Fixture for Planar Objects, Proc of the International Conf. on Manufacturing Automation, Hong Kong, 28 - 30 April 1997, **2** (1997) 629-634.
4. G.C.I. Lin and H. Du, Fixturing Planar Objects Using the Concept of the Maximal Inscribed Circle, Australasian Conference on Manufacturing Technology, New Zealand, February 12-13, (1997).
5. F.B. Hazen and P.K. Wright, Workholding Automation: Innovations in Analysis, Design, and Planning, *Manufacturing Review*, American Society of Mechanical Engineers, **3**, 4 (1990) 224-237.
6. J.C. Trappey and C.R. Liu, Literature survey of fixture-design automation, *International Journal of Advanced Manufacturing Technology*, **5**, 3 (1990) 240-255.
7. R.C. Brost and K.Y. Goldberg, A Complete Algorithm for Synthesizing Modular Fixtures for Polygonal Parts, IEEE International Conference on Robotics and Automation, (1994) 535-542.
8. A. Wallack and J.F. Canny, Planning for Modular and Hybrid Fixtures, IEEE International Conference on Robotics and Automation, (1994) 520-527.
9. E. Rimon and J. Burdick, New Bounds on the Number of Frictionless Fingers Required to Immobilise Planar Objects, *Journal of Robotic Systems*, **12**, 6 (1995) 433-451.
10. F.P. Preparata and M.I. Shamos, *Computational Geometry: an Introduction*, Springer-Verlag (1988)
11. W. Press, B. Flannery, S.Teukolsky and W. Vetterling, *Numerical Recipes in C*, Cambridge University Press, (1979) 305-309.
12. I.D. Faux and M.J. Pratt, *Computational Geometry for Design and Manufacture*, Halsted Press, (1979) 297-300.
13. H. Du, and G.C. I. Lin, Clamp Location Optimisation in Fixturing Configuration, Proceedings of the 12th International Conference on CAD/CAM Robotics and Factories of the Future, Middlesex, UK, (1996) 1054-1060.
14. Bazaraa, M.S. and Shetty, C. M. 1979, *Nonlinear Programming: Theory and Algorithms*, John Wiley & Sons.

Study of Self-Learning Control for Special Shape Robots

Sui Qing   Xie Ming

School of Mechanical & Production Engineering
Nanyang Technological University, Singapore 639798

Abstract. In this paper, self-learning control for special shape robots is studied thoroughly by a case that a special shape robot (i.e., having the shape of a Greek cross "+"), which has a perception system for sensing environment, passes through a crack whose width is less than the width of the robot body. This means that the robot cannot absolutely pass through the narrow crack unless it knows how to take advantage of its self special shape intelligently. The objective of this work is to investigate how to automatically generate a trajectory for the robot based on its own behavior. The advantage of self-learning control for special shape robots is demonstrated and verified from principle and experiment views.

1 Introduction

With the development of robotic technique, special shape robots and associated control technique are studied widely recently. The special shape robots are useful systems in the military or defense industry. Their applications can also be found in manufacturing industry or practical life for the purpose of performing some particular tasks (e.g., snake robots exam and repair pipelines). In this paper, a self-learning control approach for special shape robots is studied. The study case considered is a cross mobile robot (i.e., having the shape of a Greek cross "+"), which has a perception system for sensing environment, passing through a crack which width is less than the robot body's width or length. This means that the robot cannot pass through the narrow crack unless it knows how to take advantage of its self special shape intelligently. Normally, it is difficult to plan their trajectories and program them in advance for the special shape robots like the cross mobile robot. It should be admitted that the case to be studied here is not an easy problem, so choosing such a case is without lose of generality for studying behavior-based cognitive control for special shape robots. The objective of our work is to investigate how to automatically generate a trajectory for the robot based on its own behavior.

In the paper we first discuss the state and action space of the special mobile robot, and give some useful definitions and conclusions. Then we investigate a self-learning algorithm using reinforcement learning idea [1][2][3](more precisely, Q-learning) for the special robot. The presented algorithm, however, is somewhat different from the traditional learning method in that 1) in the traditional algorithm the space for choosing a random action is fixed, now we change this static space to a dynamic space so that its size reduces gradually with the training time increasing; 2) For the states whose corresponding Q values are never updated before, we modify the mapping from state space to action space so that the Q values of the most similar states will updated when no unique action for those states exists. Finally, some simulation results for the cross mobile robot are given in order to prove the validity and efficiency of the developed self-learning control strategy for special shape robots.

2 The State and Action Space of the Mobile Robot

In fact, teaching a robot to perform a task essentially reduces to learning a mapping from perceived state space to desired action space. The preferred mapping is one that maximizes the robot's performance at the particular task. In this section, we will discuss the state and action space in accordance with our problem considered in this paper, and give some useful definitions and conclusions.

2.1 Action space
First let's give the following definitions and conditions for an action:

Definition 1. Supposing that T_1 and T_2 are two motion modes for a moving robot, if their step lengths are equal but their motion directions are opposite (e.g., translate along positive and negative x axes or rotate in a clockwise and counterclockwise directions), then the T_1 and T_2 are called a pair of symmetric motion modes;

Definition 2. The action is invalid if it includes two adjacent steps which are symmetric.

Condition 1. The robot has n modes for each motion step, i.e., T_1, T_2, ..., T_n (n is an even number). Where T_1 and $T_{n/2+1}$, T_2 and $T_{n/2+2}$, ..., $T_{n/2}$ and T_n are symmetric respectively.

Condition 2. Each action of the robot includes N motion steps;

According to the above definitions and conditions, we have the following conclusion:

Conclusion. If an action is denoted as $A_1 ... A_i A_j ... A_N$ ($N \geq 2$, $j=i+1$), where $A_i A_j$ ($1 \leq i, j \leq N$) corresponds to the robot motion modes $T_u T_v$ ($1 \leq u, v \leq n$), and $|u-v| = n/2$, then this action is invalid.

Let's illustrate the above conclusion by giving an example. Suppose that a robot moves in 2D space, more specifically, it moves along positive X, negative X, positive Y, negative Y axes of its own coordinate system, or rotates about the origin of the coordinate system in the clockwise and counterclockwise directions. And again suppose that each action includes 2 steps, then we have 36 different types of actions totally. In fact, however, there are only 30 reasonable actions in the action space after reducing 6 invalid actions according to the conditions and conclusion given above.

2.2 State space

Assume that the robot is equipped with proximity sensors so that the robot can detect if some obstacles exist around. The state of the mobile robot in a given environment at a time is determined by the following factors 1) the robot self orientation, i.e., relationship between the robot coordinate system and world coordinate system; 2) the goal orientation, i.e., the direction from the robot coordinate origin to the goal point; 3) the environment information which can be detected by the robot sensors. Now we give the following general conclusion.

Conclusion.

IF

1) A robot self orientation is represented by N_s quadrants approximately (detected by one sensor), and any self orientations of the robot which belong to $[(2p/N_s)\cdot(i-1),(2p/N_s)\cdot i]$ $(i=1,2,3,..., N_s)$ are regarded as identical; 2) the direction of the goal is represented by N_g quadrants approximately (detected by one sensor), and any directions of the goal which are in $[(2p/N_g)\cdot(i-1),(2p/N_g)\cdot i]$ $(i=1,2,3,..., N_g)$ are regarded as identical; 3) the environment information around the robot is detected by n sensors, i.e., the environment states N_e is represented with n bits ($N_e = 2^n$)

THEN

States in the state space are $N_s \cdot N_g \cdot N_e$.

Now let's give an example. Suppose a robot self orientation is represented with 8 quadrants (N_s=8). The goal direction is represented with 8 quadrants (N_g=8). The robot is equipped with 4 sensors which are used to detect the information of the environment. Therefore, the all states in state space are: $N_s \cdot N_g \cdot N_e = 8\cdot 8\cdot 2^4$.

3 The Learning Algorithm

Here we wouldn't plan to describe the basics of reinforcement learning due to the space limitation. For a review of reinforcement learning in general, see [2][3][4]. In the light of the problem concerned in this paper, we give the following specific algorithm using reinforcement learning idea for the special shape robot cognitive control . This algorithm is somewhat different from the traditional learning method in that 1) in the traditional reinforcement algorithm the space for choosing a random action is fixed, now we change this static space to the dynamic space so that its size reduces gradually with the training time increasing; 2) For those states whose corresponding Q values are never updated before, we modify the mapping from state space to action space so that the Q values of the most similar states will updated when no unique action for those states exists.

THE LEARNING ALGORITHM
1) Define Step: $X\_STEP, Y\_STEP, \Theta\_STEP$
2) Initialize $Q[s][a]$ whose initial entries are 0;
3) For ()

      (1) Call function *WORLD()* to observe the current world state s

          WORLD () {

               Read the sensors on the robot to get the state of the robot in the environment.

          }

      (2) If *rand()* $\in$ *(0, rd)* (Let $rd \in (0,1)$)

          Choose an action that maximizes $Q(s,a)$

    Else

          Choose a random action

      (3) Call function *ACTION()* to carry out the action a

          ACTION() {

               Actuate the robot and check collision.

          }

      (4) Call function *WORLD()* to observe the next state s' and *CRITERION()*, let the reward $r = F_r(J)$ and $E(s') = \max_a(s',a)$

          CRITERION() {

$$J = \alpha_1 \cdot d_1 + \alpha_2 \cdot d_2 + \alpha_3 \cdot d_3$$

              where d_1 : the distance from the robot to the goal;

                          d_2 : the angle from the self orientation of robot to the desired orientation;

                          d_3 : the complexity of the environment;

                          $\alpha_1, \alpha_2, \alpha_3$ are weight factors and $\alpha_1 + \alpha_2 + \alpha_3 = 1$.

          }

      (5) Choose parameter β and γ, and update $Q(s, a)$

$$Q(s,a) = Q(s,a) + \beta[r + \gamma E(s') - Q[s,a]]$$

End For()

4 Simulation

In order to testify efficiency and feasibility of the self-learning algorithm for the control of the special shape mobile robot, we simulated some typical cases. Due to the limit of space, we present here only one case, i.e., a mobile robot with cross shape passes through a narrow crack. The width of the crack is less than width and length of the robot body. This means that the robot cannot absolutely pass through the arrow crack unless it takes advantage of its self special shape intelligently.

Simulation description:

Mobile robot: cross shape; the width is Lr. (the length is equal to the width).

Environment: one narrow crack formed by several broken lines; the width of the
                  crack is Le.

Goal: at the lower left corner of the environment.

Conditions: 1) the width of the crack is less than the width of robot body, i.e.,
                  Le<Lr.

                  2) the robot can move along X axis and y axis of its own
                  coordinate system, or rotate in clockwise and counterclockwise
                  directions.

                  3)   each action includes 2 steps.

From above description, we know that the robot body's length is greater than the
crack's width. So in order to pass through the crack, the robot has to rotate properly
near the crack, then translate, after that, rotate and translate again, maybe repeat such
procedures a lot of times. That means that no collision will happen only if the robot
understands how to take advantage of its own special shape. Fig.4.1---Fig.4.2 show
the results of training. Fig.4.1 is the training result for first time. Fig.4.2 is for 70th
time. From these Figures, we can see that the robot moved at random basically at the
beginning, and could pass through the crack very easily and smoothly at last.

5 Conclusion

In this paper, the technique of self-learning control for special shape robots is
studied thoroughly by a case. The main contribution includes some useful definitions
and conclusions for the state and action space, a self-learning algorithm for special
shape robots, and a lot of simulation results. In addition, the advantage of self-
learning control for special shape robots is demonstrated and verified from principle
and experiment views.

References

1.   Jonathan H. Connell, Sridhar Mahadevan, "Robot Learning", Kluwer Academic
     Publishers, 1993.
2.   Richard S. Sutton, "Learning to Predict by Method of Temporal Defferences", The
     Journal of Machine Learning, 3(1), 1988.
3.   Richard S. Sutton, "Reinforcement Learning", Kluwer Academic Publishers, 1992.
4.   Leslie Pack Kaelbling and Stanley J. Rosenschein, "Action and Planning in Embeded
     Agents" , Robotics and Automation, (6) 1990.
5.   Meystel, Alex, "Autonomous Mobile Robots Vehicles with Cognitive Control", Word
     Scientific Publishing Co. Pte. Ltd 1991.

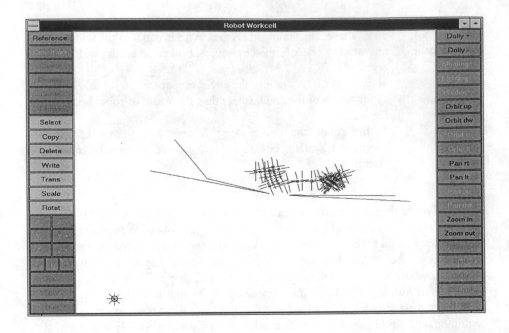

Fig.4.1 The training result for the first time

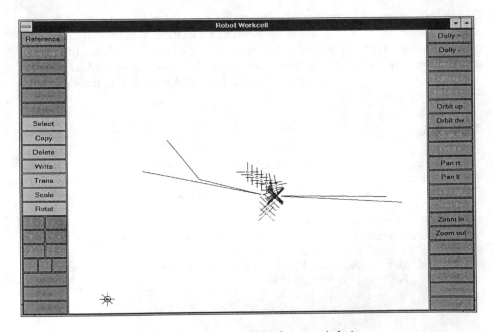

Fig.4.2 The training result for the seventieth time

DEVLOPMENT OF AN HIGHLY FLEXIBLE TWO-LEGGED WALKING DEVICE

PD Prof. h. c. Dr.-Ing. Josef Schlattmann

University of Paderborn, Germany

Abstract. The paper addresses the development of a static steady walking two legged totally modulized robot.

1   Mobile Robots

Mobile robots are often used in environments that are dangerous for humans, or where humans cannot reasonably be expected to work. In these environments they carry out, for example inspections, maintenance work or are used as observers in potentially dangerous situations. Typical applications are in irradiated areas in nuclear power stations, for manipulating tasks in laboratories or for sealing leaks in chemical plants as well as for fires in potentially explosive environments. These robots usually move using tracks or rollers. Only a few are constructed as walking robots and thus move around on feet. The construction of the movement unit is based on the conditions under which the robot has to fulfil its tasks /1/.

1.1   Robot Concepts

Wheel and track drives are suitable for movement on flat and uneven ground. However, application limits are set by the potential to overcome obstacles, to move on stairs or curbs and to manoeuvre in narrow, winding passages. Special constructions are suitable to overcome certain obstacles by having devices that enable them to place the tracks or driven wheels on steps or curbs and then to move onto them. However these devices considerably increase the machine's weight and are usually only suitable for mounting straight stairs. The size of these extra machine parts further reduces the machinc's ability to manocuvrc in narrow spaces.

Walking robots use another principle of movement. Movement in these machines is not realised by a rolling movement, but by moving the body relative to the feet. As this movement is limited by the length of the legs, it is necessary for the feet to change their position to enable it to travel long distances. Nature shows us many examples of solutions for this problem.

2   Walking Forms

When considering these examples from nature we should remember that machines at the moment, are not yet capable of dynamic movement over long distances, as, for example, a galloping horse would be. This is because of the machines relatively low power capacity compared to its body weight as well as the complexity of control necessary. "Dynamic" movement means that movement is realised in conditions in which it cannot be stopped because the existing conditions are not stable. In contrast we should name, for example, the stable form of movement of insects, which is characterised by always having at least three legs touching the floor at any time. They describe a surface on the ground, i.e. the positions where the legs touch the ground are not in one line /2/.

2.1 The Arangement of Legs Provides Stability

That a surface is exactly defined by three points, or, in other words, a body with three legs on a surface is stable, is the reason for animals to usually have at least four legs. Also, animals usually prefer stable forms of movement where at least three legs describe a surface on the ground while the other leg(s) move to a new position. The energy required to co-ordinate the leg movement increases with the number of legs, while at the same time the energy needed to keep up stability decreases. During the process of leg movement in a quadruped (horse, dog, cat, etc.) leg movement must be exactly co-ordinated. However, in a six-legged being (crabs, insects, etc.) there are a multitude of possible stable leg positions due to the 2x3 arrangement of the legs /3/. Thus, the fewer legs an animal has the more likely it is to get into an unstable state. For this reason the energy necessary to regulate stability must be greater. Only creatures with highly developed brains are capable of moving continually on two legs. The high amount of energy needed to stabilise movement is justified by optimal mobility and the ability to move in any direction from any position.

The information about the movement of creatures has been transferred to walking robots. Most of these robots have six legs and copy the movement of insects. These walking robots can be used on even surfaces and open terrain. With the relevant co-ordination it is even possible to overcome obstacles. The long body form is a distinct disadvantage as it prevents manoeuvring in narrow passages and movement up spiral stairs. Additionally, it requires a greater amount of energy to co-ordinate leg movement and to calculate the individual leg load necessary because the positioning of the feet in uneven terrain cannot be predicted.

Principally, two-legged robots have the same flexibility of movement as humans, however, due to the technical difficulties in controlling dynamic movement, robots are not yet able to make an unlimited amount of controlled steps. Continual improvements in sensors and information processing make definite improvements possible in the near future.

3   The New Robot Concept

At the University of Paderborn, a concept for a two-legged robot has been developed using a systematic process. This combines the advantages of a six-legged creature with the mobility of a two-legged one. The aim was to develop a modular robot that would be capable of offering maximal mobility with minimal control technology. Also, the weight should be low enough to make it suitable for use by mobile units of the fire brigade.

3.1   The Feet

The invention of a new foot form that enables a two-legged walking machine to be able to move in a stable way was the basis of the design. Each foot is shaped so that it forms a standing surface within which the machine's centre of gravity falls, and can be placed inside the area covered by the other foot. The foot shape enables the robot to stand stable on each foot and to move its body relative to the foot inside an area limited sideways by the expansion of the foot and forwards and backwards by the maximal extension of the leg. Furthermore, by continually transmitting both foot surfaces the robot can be moved in any direction. At no time during this movement does the robot become unstable. As the foot shape allows movement in any direction, it is necessary to have a leg construction that enables equal movement in any direction, i.e. it has a rotation-symmetrical working space and enables the foot to be positioned in any of six degrees of freedom.

Figure 1 shows by example of a "one-legged-structure" the stabile area. This area is limited by projection of the foot-area along the direction of gravity force. The area is limited by the stability border. Transgressing this border would cause the robot to be in an unstabilized mode.

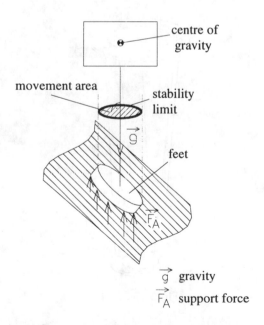

Fig. 1. project movement area with stability limit

Fig. 2. Form of the feet

3.2　　The Legs

All these criteria are covered by constructing the legs as hexapods. The hexapod, also known as the "Steward platform" shown in figure 3, has been used successfully for many years in the movement devices of flight simulators /4/. It consists of six legs each of which can be altered in length. The legs join two platforms so that by altering the length of the legs the position and orientation of the platforms to one another inside the construction limits can be varied in any way. Further characteristics are the high rigidity of the construction, the simple rotationally symmetrical construction and the high load capacity relative to the low weight of the hexapod. These characteristics

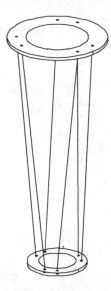

Fig. 3. Hexapod

are the reasons for the system having recently been used in the construction of various manipulating systems.

The construction of the leg as hexapod combined with the foot shape described above enable the robot's body to be moved inside the constructive limits in any direction in relation to the foot on the floor as well as the other foot. This enables the robot to perform manipulating tasks and walking operations. A robot of this type is capable of moving on slopes, straight and spiral staircases and, while moving and stationary, of performing exactly controlled activities.

The figure 4 shows the complete assembly. Table 1 is an overview of the important design parameters.

Table 1. Technical Dates of the Two-Legged-Robot

overall width	700 mm
max. overall height	1400 mm
min. overall height	900 mm
max. step length	400 mm
max. stroke length of a leg	500 mm
motion speed	150 mm/sec
weight of the mechanical unit	80 kg
max. balancing moment of the MVS	25 Nm
max. allowable tilting moment	80 Nm

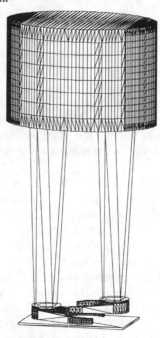

Fig. 4. Two-Legged-Robot

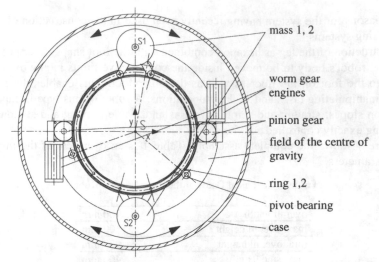

mass 1, 2

worm gear
engines

pinion gear

field of the centre of
gravity

ring 1,2

pivot bearing

case

Fig. 5. The Mass Displacement System

3.3   The Mass Displacement System

To extend the movement of the machine's centre of gravity outside the foot area a special device that enables the machine's centre of gravity to be moved within the construction limits is used. At a further stage of development this mass displacement system (*MassenVerlagerungsSystem MVS*) will also be used for dynamic impact stabilisation.

This function is fulfilled by two excentric co-axial masses that are capable of rotating on a ring guide and that are driven by a servo drive. The positioning of these masses can translate the common centre of gravity inside the shaded area shown in figure 5.

3.4   The Control System

Because of the modular construction the control tasks can be based on solutions for individual systems with low complexity. The co-ordination of these systems takes place through a higher ordered unit. The legs are completely identical in their construction and control systems. The control task for the legs is limited to position control of its linear units.

The desired position of the foot platforms at the end of the leg and thus the position and orientation of the foot surface are given by a separate system that is responsible for the route planning. Load cells contained in the linear unit of the leg give the route planner the required information about the foot load.

With the aid of the foot geometry and the measured foot load, the area can be calculated in which the body may be moved relative to the feet with a given acceleration without the system becoming unstable. If movement is planned outside this area, one leg is moved while the robot is stationary, in such a way to enable this foot to be newly positioned. Because the foot surfaces can be placed inside one another they form a redundant system in view of the stability of the body. As long as the foot surfaces do not need to be re-positioned the robot stands on one foot and the other foot is parallel to and above it. The stressed leg can be re-positioned by moving the appropriate linear units into the new position. Control of stability is guaranteed through the foot on which the machine is standing. If the calculated load moment is too high the moving foot is stopped and lowered. Because the dynamic force on the moving foot is relatively small compared to the static forces that act during a collision with an obstacle the moving foot is used to measure the load during movement. This measurement is used as an input quantity for collision recognition.

3.5 The Concept for Collision Recognition

A sudden change in the load is interpreted as a collision of the foot with an obstacle and leads to a stop in movement. When lowering the foot, the collision means that the foot has been placed on the surface. This starts an algorithm that, under consideration of the measured moment in the foot surface centre of gravity, places the foot squarely on the surface. If the foot collides during movement to a desired position the movement is stopped and then moved back a given distance in the opposite direction. After moving back, the foot is then lowered as described above.

4　Module Oriented Optimisation through a Structural Construction

The algorithm for calculating the position of the foot relative to the surface and the positioning of the foot as well as decision making in case of a collision or stability loss can be tested with simple test equipment. These factors can be optimised independently of each other. Errors in movement behaviour can be relatively easily allocated to the modules and corrected. The development of the control tasks can take place independent of the whole construction subject to the interface. After a short development period, a basic version can be implemented that can be immediately employed in a limited task area. It can be further developed by information gained while in operation and can be adapted to take on new tasks.

The advantage of the modular design of the construction offers itself to the development of the robot as a project in the field of simultaneous development at university. At the beginning of the project the development team were working on the following projects:

- construction of the feet
- development of a prototype of the "Stewart platform"
- design of a device to help the robot back on its feet, without outside help, after falling
- construction of the mass displacement system (MVS)
- development of an alternative solution for driving the "Stewart" platform with the aim of making the legs lighter and more dynamic.

The aim during the first stage was to simultaneously develop a prototype in a short amount of time that could be used as a functional model for the development of the modules used for planning routes, recognising collisions and measuring the foot forces. The prototype has been designed so that further development is possible by merely changing the individual parts. Basic units such as the MVS, feet and body remain the same. To a great extent, the information about the characteristics of this unit gained through the prototype can thus be transferred to the new models. The number of possible source errors can be lessened by optimisation. In a later development stage these components will be adapted to suit the specific requirements. Further details will be given in the oral presentation.

5 Conclusions

Future development is aimed at the optimization of control algorithm in order to improve system stability during locomotion.
A future focal point in research is developing strategies towards coping with obstacles. Development is enhanced by employing virtual reality (VR) and simulation of the mechanical components. The gained results are to be directly transfered towards the mechanical unit. This allows for a highly flexible layout of the controls, which can be swiftly adapted to changing environment data.

6 References

/1/   *Alexandre, P.; Preumont, A.* Walking Machines: A State of the Art in Europe. Zeitschriftenaufsatz: European Journal of Mechanical Engineering, Band 40, Heft 1, Seite 27-33, 1995

/2/   *Pfeiffer, F.; Cruse, H.* Bionik des Laufens - technische Umsetzung biologischen Wissens. Zeitschriftenaufsatz: Konstruktion, Band 46, 1994, Heft 7/8, Seite 261-266

/3/   *Ferrell,C.:* A Comparsion of three Insect - Inspired Locomotion Comtrollers. Zeitschriftenaufsatz: Robotics and Autonomous Systems, Band 16 (1995) Heft 2-4, Seite 135-139

/4/   *Fichter, E. F.* A Stewart Platform-Based Manipulator: General Theory and Practical Construktion. Zeitschriftenaufsatz: The International Journal of Robotics Research, Band 5, Heft 2 Sommer 1986. Massachusetts Institute of Technology.

Design of Active Links for Modular Robot[1]

Wei Tech Ang   and   Ming Xie
(techang@post1.com)   (mmxie@ntu.edu.sg)

School of Mechanical and Production Engineering
Nanyang Technological University, Singapore 639798

Abstract. This paper presents a new design concept of modular robot. Our design of modular robot features two distinctive characteristics (i) we do not separate links and joints, the modules are *active links*[2]; (ii) each module has two degrees-of-freedom: a rotational DOF and an angular DOF. The active links are serially connected to form a modular robot manipulator. Making use of the design and geometric advantages of the active links, we have also developed a closed-form geometric inverse kinematics method that is more efficient than the conventional analytical and numerical methods. This allows more computational resources to be allocated to the vision system that is computationally intensive so as to achieve real-time guidance.

1. Introduction

The conventional applications of robotic system involve manufacturing and assembly tasks. In recent years, much effort has been made in developing new field of non-manufacturing applications for robots. The perceivable trend of technology is moving towards the direction of service robot, i.e. a robot that can provide services for both human and machine [1,2,3]. Saturation in the industrial robot market, the increasing numbers of elderly people in most developed countries, the shortage of manpower in hospitals, the involvement of human in hazardous or harsh working environment etc. are some of the important factors that are providing the necessary background for the development of service robots.

Here, we are developing a general purpose vision guided mobile service robot for structured indoor environment. The robot is targeted to perform a large variety of tasks ranging from house-keeping to mail delivery to machine maintenance. In order for the robot to adapt to tasks of different nature, we want it to possess both software and hardware flexibility. The re-programmability nature of software systems has made software flexibility easily attainable. Hardware flexibility, on the other hand, can be achieved by adopting a modular design for our robot arm. *Modular robot* introduces a new dimension to hardware flexibility and ensure an individual global optimal arm geometry or degree-of-freedom for each task in hand.

Research in the area of modular robot design has been addressed by both the robotics system developers/ manufacturers (e.g. TOMMS by Toshiba Corp. [4]), as

---

[1] This project is supported by the Ministry of Education, Singapore under the grant RG72/96
[2] "Active" in the context of modular robotics refers to the module that is the motion provider.

well as academic institutions (e.g. the "Modular Robot System" by University of Stuttgart [5], the "Reconfigurable Modular Manipulator System (RMMS) by Carnegie Mellon University [6, 7] and a nameless modular robot system by University of Toronto [8] and RISC modular robotic system by Nanyang Technological University [9]). Some other literature in this area had presented pneumatic and electro-mechanical modular robots for small parts assembly [10, 11], for spot welding [12], and other material handling applications [13]. Modular robotics has also been of interest in the context of space stations as reported in [14].

The TOshiba Modular Manipulator System (TOMMS) in [4] consists of one type of joint module (active) and one type of link module (passive) with one degree-of-freedom and the control unit. The system allowed two types of configuration, namely the horizontal type and the vertical type. A single control software was developed such that it could control any configuration of assembled manipulators without special handling.

The Modular Robot System by University of Stuttgart [5] consists of a variety of rotational joints actuated by AC motors in conjunction with differential gears and links of square cross-section. The two modular units manufactured are : a two DOF "main axis" (large module) joint and a 2 DOF "wrist" (small module). Due to its weight the main axis can only be used at the base, it would have to be substantially reduced in weight (and torque) for use at a location such as the elbow of the arm.

The "Reconfigurable Modular Manipulator System" (RMMS) by CMU in [6,7] consists of "rotate" (roll) and "pivot" (pitch) type one DOF rotary joints actuated by DC motors in conjunction with harmonic-drive transmissions, and links of circular cross-section. Each module is self-contained with sensor interfaces, power amplifiers and control processors. The joint designs were accomplished by using V-band flanges. It was commented in [8] that this joint design were more suitable to modular robotics than those reported in [5].

The modular robot by University of Toronto in [8] consists of rotary joints (active with one DOF) of two sizes, prismatic joints, a Roll-Pitch-Yaw end-effector and passive connectors. To avoid difficulties and efficiently connecting a vast number of wires, a serial communication method was selected. The design of the connector made use of a 45° connection scheme which enable the arm to be reconfigured into in-plane or out-of-plane configuration.

The RISC modular robot system by Nanyang Technological University, Singapore in [9] once again consists of two types of one DOF active joint modules: revolute and prismatic, and two types of passive link modules with "connecting ports". The "connecting ports" were the places where the joint modules were attached, a cubic box link had six ports while a square prismatic links had ten.

The objective of our research work is to propose a new concept of modular robot that made use of our two DOF active links. The two main research interests in modular robotics have been mechanical design and motion control. The following sections describe the proposed design of the active links and the geometric inverse kinematics method.

2. Design of Modular Active Links

Our design of modular robot features two main differences as compared to those as reported in [4,5,6,7,8,9]:

(i) There is only one type of module, i.e. we do not separate links and joints. Therefore, the modules are *active links*.

(ii) Each module has two degrees-of-freedom: a rotational DOF with a limit-to-limit range of 360° and an angular DOF with a limit-to-limit range of 180° as shown in Fig. 1.

To cater for different mechanical stiffness and torque requirements, three sizes of modular active links are designed namely Type L, M and S. Every module houses two YASKAWA Σ series AC servo motors, one for each DOF. Harmonic Drive transmission is proposed in our design to provide the necessary torque because of its following features:

(i) High positional accuracy and repeatability, since 10% of the teeth are always meshed;

(ii) Low backlash, (≈1 arc minute) due to pre-loading and radial mesh of the teeth;

(iii) High efficiency (approximately 79 percent for 100:1 reduction ratio) because of the single stage speed reduction.

The design of a modular link is presented in Fig. 2. The modules are to be serially assembled to form a manipulator. The rotational DOF of a module is connected to the angular DOF of its predecessor. The rotational motor is housed in a hollow cylindrical motor housing, whose base flange is connecting to the base flange of the previous module. The motor output shaft is directly coupled to the Harmonic Drive, and the output spline of the Harmonic Drive is attached to a rotating plate which connects to the upper angular DOF section. For the angular DOF, the motor power is fed into the horizontally oriented Harmonic Drive via timing belt and pulleys. A output shaft extension is bolted to the output spline of the Harmonic Drive which drives the connector flange and hence the next module.

To ensure modularity and compatibility, (see Fig, 3) the dimensions of the connector and base flanges, the bolt holes size and positions are standardised throughout all types of modules. When Module i is removed, Module $i+1$ can be connected to Module $i-1$ without problem. The size of the hollow cylinder and the distance between the flange legs, however, are different in order to suit the different sizes of motors and Harmonics Drives. The slot in the connector flange enables the signal and power cables to run through without difficulty.

The central control unit (CCU) and power source are separated from the arm. In our design, the power and communication cables are run internally, the signal and power cables have to run through the modules to reach the motors and sensors. To avoid interference problem, the signal wires and power cables of each module are in separate strands. The strands of signal wires through a module are bundled into a single strand, and so do the power cables. Fig. 4 shows the distribution of the wires running internally to the four modules, two bundles of four strands run through the base, two bundles of three through the Module 1-2 interface, two bundles of two through Module 2-3 interface and two strands through Module 3-4 interface.

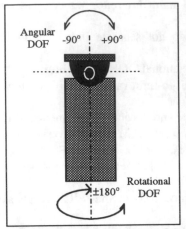

Fig. 1. The two DOF active link

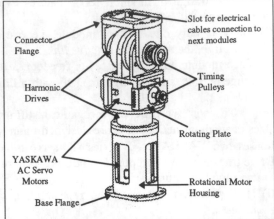

Fig. 2. Design of a modular link

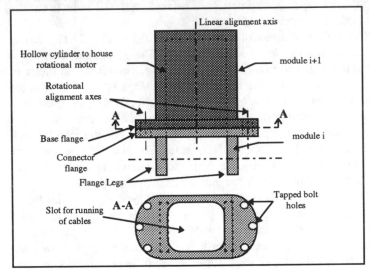

Fig.3. Design of connecting flanges

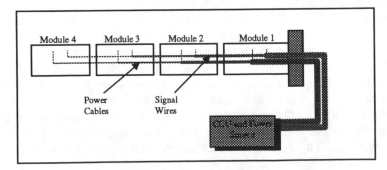

Fig. 4. Distribution of internal signal wires and power cables

Each module is self-contained with its wiring networks and communicate with its predecessor and successor through wire connectors at the base and connector flanges. In a module, there are male type wire connectors at the base flange, evenly spaced around the internal circumference of the hollow rotational motor housing. These male connectors are connected to the female type wire connectors fixed to the four sides of the rectangular slot at the connector flange of the predecessor module.

Fig. 5 explains the arrangement of wires within a module, the wires are constrained at various appropriate positions and enough slacks (additional length) are given to ensure limit-to-limit operations.

The physical and technical specifications of the modular active links are summarised in Table 1.

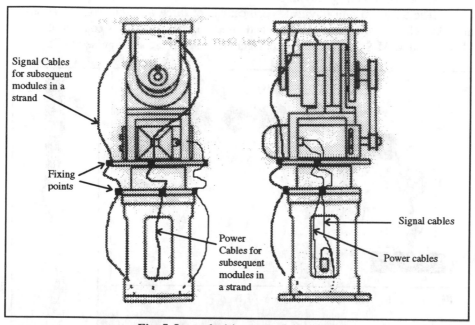

Fig. 5. Internal wiring network in module

Table 1. Technical specifications of modular active links

Item		Active Link Type L	Active Link Type M	Active Link Type S
Dimension *l* x d (mm)		340 x Ø140	310 x Ø130	250 x Ø110
Weight (kg)		4.4	3.5	2.0
Motor Power Rating (W)	rotational	200	100	30
	angular	100	100	30
Rated Output Torque (Nm)	rotational	60.7	25.3	7.5
	angular	40.5	30.3	9.6
Maximum Output Torque (Nm)	rotational	150.9	75.1	22.1
	angular	120.1	91.0	28.2
Rated Output Speed (rpm)	rotational	30	30	30
	angular	15	20	20

3. Modular Arm

3.1 Mechanical Design

Fig. 6 shows a 4-links manipulator made up by four serially connected active links. The degree-of-freedom of the manipulator is scaleable. If an application requires higher flexibility and mobility, additional modules can be added to provide more degrees-of- freedom. Conversely, the DOF of the manipulator can be scaled down when the application is simple. A number of configurations are possible with different combination of the three types of modules (i.e. Type L, M and S), depending on the requirement of the task in hand. Fig. 7 shows a modular arm with a 4-links M-M-S-S configuration.

The technical and physical specifications of three different configurations of modular robot arm (not including the end-effector), namely 4-links configurations L-M-M-S and M-M-S-S, and 3-links configuration L-M-S are summarised in Table 2.

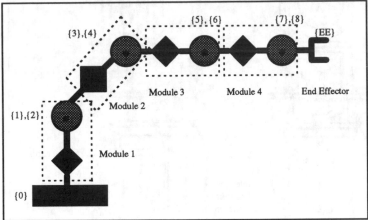

Fig. 6. Serially connected modular arm of 4

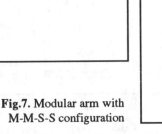

Fig.7. Modular arm with
M-M-S-S configuration

Table 2. Technical specifications of different configuration of modular arm

Item	Configurations		
	L-M-M-S	M-M-S-S	L-M-S
Design Payload (kg)	4.5	2.0	9.0
Tip Velocity (m/s)	2.73	2.54	1.76
Reach (m)	0.87	0.81	0.56
Total Weight (kg)	13.4	12.8	9.9
Overall length (m)	1.21	1.14	0.9

3.2 Control

3.2.1 System Hardware

The modular arm is to be guided by real-time vision system. The central control unit consists of a Pentium 200 PC with 32MB RAM, vision system (2 CCD cameras and a SMT303 frame grabber card), a TMS 320C81 DSP based image processing system by LSI and a NextMove PC 8-axes motion controller card.

3.2.2 The Geometric Inverse Kinematics Method

Instead of getting the inverse Jacobian matrix (or the Pseudoinverse matrix in cases where there is redundancy in the number of DOF) from pure mathematical manipulations by the conventional analytical [15] and numerical [16] methods, we are proposing a geometric inverse kinematics method that exploits the modular characteristic and the design advantages of the two DOF active links. The objective of the geometric inverse kinematics method is to achieve a good solution (not necessary the best solution) by injecting geometric insight into the analysis. Due to the constraint in the number of pages, only a brief description of the method will be discussed.

Consider a desired position P_0 in 3D space and orientation $V_0 = [\hat{n}_0 \quad \hat{s}_0 \quad \hat{a}_0]$ with respect to the base co-ordinate system {0} of the modular arm (refers to Fig.6 for assignment of co-ordinate frames). Refer to Fig. 8, now we offset Frame {0} in the direction of Z_0 such that Frame {0}, {1} and {2} coincides.

Let T_0 be the target position offset from P_0 in the direction of - a_0 with an magnitude equal to l_e the length of the end-effector, $T_0 = [\ ^0T_x \ ^0T_y \ ^0T_z \ ]^T = P_0 - l_e \, a_0$ (Refer to Fig.8). If we rotate Frame {0} by $\theta_1 = \tan^{-1}(\ ^0T_y / \ ^0T_x)$ about the Z_0 axis such that T_0 is housed in the X_1-Z_1 plane, we transform the problem into a 2-dimensional.

Once the problem is simplified to 2D, the inverse kinematics or joint angles for subsequent DOF can be solved by simple geometric manipulations. The solution in the case of 3-links configuration is trivial as shown in Fig. 9. The solution of a 4-links configuration, on other hand, can be further broken down into two different implementation schemes as shown in Fig. 10 depending on the distance from T_0 to {0}. Details in dealing with different orientations of the tooltip will not be discussed.

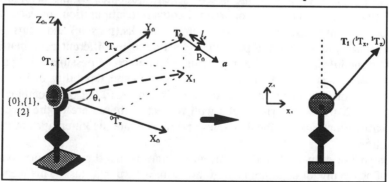

Fig. 8. Transformation into a 2D problem

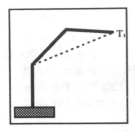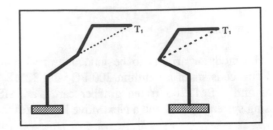

Fig.9. Inverse kinematics solution of 3-link configuration

Fig.10. Two implementation schemes of inverse kinematics solution of 4-links configuration

The geometric inverse kinematics method of a 4-links M-M-S-S configuration is formulated using Mathematica Ver.2.2, together with the closed-form polynomial method proposed by Raghavan et. al. [15] and the numerical method proposed by Kholsa [16]. The efficiencies of the three algorithms are evaluated and simulation results have shown that the geometric inverse kinematics algorithm consumes 7.1 and 20.6 times less CPU time than the numerical and the closed form polynomial algorithm respectively. This would allow more computational resources to be allocated to the vision system that is computationally intensive so as to achieve real-time guidance.

4. Modular Service Robot

The development of modular active links and modular arm is part of the effort to develop a general purpose vision guided service robot for structured indoor environment (Fig.11). The main components that made up our service robot consist of a mobile base of 3 DOF (2 translational DOF in the X-Y or ground plane and 1 rotational about the Z-axis), one or two modular robot arms (depending on the application) and a robot head with a pair of CCD cameras mounted on a modular neck. The computing system, image processing system and the motor amplifiers are all housed in the mobile base.

4.1 System Advantages

The fundamental aim of the design schemes adopted by the other modular robot developers is to achieve an optimum hardware configuration for each task in hand. Most of them had chosen to have a variety of links (passive) and joints (active) of different sizes, and hardware flexibility is achieved by different combinations of the stock of modules. The fact that there might be numerous possible configurations for a given task has exerted tremendous pressure on the flexibility of the control software. The software has to decide on the optimum DOF, the best combination of hardware, check the singularity points with respect to the nature of the given task, perform calibration, analyse the dynamics and re-evaluate the kinematics of the new configuration.

One major draw back of these design schemes is that they are built around the concept of achieving an optimum configuration for a particular task. This means that the optimal configuration for a grasping operation might not be able to tighten screws due to singularity problem. For a robot that is required to perform a variety of tasks,

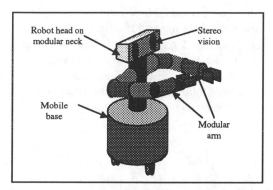

Fig. 11. A schematic view of the service robot

the engineers would have to reconfigure the robot for almost every task. This situation is definitely intolerable for our application as a general purpose service robot. This again comes back to the problem of limited hardware flexibility, which means that the fundamental problem has not yet been solved.

In addition, most researchers has been spending a lot of effort in achieving hardware flexibility and neglected system simplicity and efficiency which is utmost important for real-time control. In other words, they are heading towards the direction of developing a complicated system with ideal flexibility and adaptability in both software and hardware which requires a great deal of engineering effort and money.

The essences of our proposal are design simplicity, control efficiency and hardware flexibility through control strategy. Firstly, instead of *the optimum* configuration for a given task, our emphasis is placed on achieving a robot with *an optimal* DOF for a variety of tasks. With the additional 3 DOF from the mobile base, we are able to adopt a control strategy that is inspired by the way human being perform tasks. When the robot arm is unable to reach certain positions or orientations along the planned trajectory, the mobile base will re-position the robot so that the desired positions and orientations are within its reach and out of the singularity zones. In this way, nearly all positions and orientations in the workspace can be reached without difficulties. This is our proposed solution to "hardware flexibility".

Secondly, since there is only one type of module in our design, and all modules are serially connected, the tasks of dynamics analysis, forward kinematics and calibration of a new configuration is greatly simplified. Hence, much engineering effort and resources can be saved in the software development.

Last but not least, by taking the design and geometric advantage of the active links we are able to develop a simple and efficient inverse kinematics scheme. The significant reduction of CPU consumption time of the modified geometric method over both the conventional closed-form and numerical methods has made more computational resources available to the real-time visual servoing system.

5. Conclusion

We have presented our research work on the design and control of a new concept of modular robot arm. The modular robot arm is made up of serially connected active links with 2 DOF. Three sizes of active links have been design to meet different requirement of torque and material stiffness depending on the task in hand. Exploiting the design and geometric advantage of the active links, we have developed a simple and efficient geometric inverse kinematics scheme. The modular arm is to be guided by real-time vision system, and this new inverse kinematics scheme allows more computational resources to be allocated to the computational intensive real-time visual servoing.

References

[1] Hasegawa Y., "A New Wave of Japanese Robotization", *Proceedings of the 4th International Conference on Control, Automation, Robotics and Vision*, Singapore, Dec.1996, pp.1640-1645.

[2] Xie M., "Towards visual Intelligence of Service Robot", *Proceedings of the 4th International Conference on Control, Automation, Robotics and Vision*, Singapore, Dec 1996, pp. 2217-2221.

[3] Dario P., Guglielmenlli E., Genovese V. and Toro M., "Robot assistants: Applications and evolution", *Robotics and Autonomous Systems*, Vol.18, 1996, pp. 225-234.

[4] Matsumaru T., "Design and Control of the Modular Robot System: TOMMS", *IEEE International Conference on Robotics and Automation*, 1995, pp. 2125-2131.

[5] Wurst K.H., "The Conception and Construction of a Modular Manipulator System", *Proceedings of the International Symposium Industrial Robotics*, 1986, pp.37-44.

[6] Schmitz D., Khosla P. and Kanade T., "The CMU Reconfiguarble Modular Manipulator System", *Proceedings of the 19th International Symposium on Experimental Robotics*, Sydney, Australia. 6-10 Nov, 1988, pp. 473-488.

[7] Paredis C.J.J., Brown H.B., Casciola R.W., Moody J.E. and Khosla P.K., "A Rapidly Deployable Manipulator System", *In International Workshop on Some Critical Issues on Robotics*, Singapore, 1995, pp. 175-185.

[8] Cohen R., Lipton M.G., Dai M.Q. and Benhabib B., "Conceptual Design of a Modular Robot", *ASME J. Mechanical Design*, 114: pp. 117-125, Mar 1992.

[9] Chen I-M., "On Optimal Configuration of Modular Reconfigurable Robots", *Proceedings of the 4th International Conference on Control, Automation, Robotics and Vision*, Singapore, Dec.96, pp.1855-1859.

[10]Gruzdev, Korytko and Yurevich, "Modular Electro-Mechanical Industrial Robots", *Electrotekhnika*, Vol.55, No.4, 1984, pp. 4-7.

[11]Harrison, Weston, Moore and Thatcher, "Industrial Applications of Pneumatic Servo-Controlled Modular Robots", *Proceedings of the 1st National Conference on Production Research (UK)*, 1986, pp.229-236.

[12]Smith and Crazes, "Modularity in Robots-Technical Aspects and Applications", *Proceedings of the International Conference on Robotics and Automotive Industry (UK)*, 1982, pp.115-122.

[13]Muck and Mammern, "Modular Mechanical Engineering", *Proceedings of the International Conference on Advances in Manufacturing*, 1984, pp. 271-282.

[14]Jenkins L.M., "Telerobotic Work System-Space Robotics Applications", *IEEE International Conference on Robotics and Automation*, 1986, pp. 804-806.

[15]Raghavan M. and Roth B., "Inverse Kinematics of the General 6R Manipulator and Related Linkages", *Transactions of the ASME*, Vol. 115, 1993.

[16]Khosla P.K. and Neuman C.P., "An Algorithm for Seam Tracking Applications", *The International Journal of Robotics Research*, Vol.4, No.1, MIT, 1985.

[17]Xie M., "Robotic Hand-Eye Co-ordination: Solutions with uncalibrated stereo cameras", *Journal of The Institution of Engineers, Singapore*, Vol.37, No.1, 1997.

An Adaptive Tracking Controller for Robots Including Motor Dynamics

Yin Guang and Song Qing

School of EEE, Nanyang Tech Univ, Singapore

Abstract. This paper presents a direct adaptive tracking control scheme for robotic applications with motor dynamics. It is proved that the tracking error is semi-globally asymptotically stable and all parameters are bounded. Simulation results are provided to illustrate the performance.

1 Introduction

The main focus of this paper is to develop an observer-based adaptive control scheme which considers both robotic manipulator and motor dynamics without using the measurement of acceleration. We attack the same problem posed in [1], that is, the tracking control of rigid-link electrically-driven robots with the same assumption regarding actuator parameters as in [2], but use a different approach, *i.e.* the adaptive observer-based control algorithm. We first construct a velocity estimator and then design a stable-feedback embedded controller which uses only position measurements. The voltage input control signal,therefore, does not depend on the acceleration with the estimated velocity variable. This adaptive control law is designed to achieve semi-globally asymptotical stability of the link position tracking error in the face of parameter uncertainties. The result is an extension of the work of Slotine and Li [3], adding the non-negligible motor dynamics. In this way, we combine the approaches of embedded control as in [1] with the adaptive observer-based control concept to guarantee semi-globally asymptotical stability.

2 Preliminaries

Assume that the robot under consideration is driven by armature-controlled DC motors with voltages as inputs. The model of a robot manipulator including motor dynamics and some physical properties of the robot is known, and can be used in the control synthesis procedure. Following [4], a model of the robot manipulator, including the dynamics of DC motors, can be expressed as:

$$D(q)\ddot{q} + C(q, \dot{q})\dot{q} + g(q) = K_t I \tag{1}$$
$$L\dot{I} + RI + K_b\dot{q} = u \tag{2}$$

where $q \in \Re^n$ represent link angles; $D(q) \in \Re^{n \times n}$ is the link inertia matrix, $C(q, \dot{q})\dot{q}$ represents Coriolis and centrifugal force; $g(q)$ represents gravitation terms. $K_t \in \Re^{n \times n}$ is a diagonal positive definite matrix. The vectors u and $I \in \Re^n$ represent the armature input voltage and current. $R, L, K_b \in \Re^{n \times n}$ are diagonal positive definite matrices of the resistance, inductance and back EMF of the motor respectively.

In this study $\|x\|$ denotes the Euclidean norm of an arbitrary vector $x \in \Re^n$ [5]. This norm induces a norm $\| \cdot \|_i$ on the space $L(\Re^n, \Re^n)$ of linear transformation such that for $A \in L(\Re^n, \Re^n)$, we define:

$$\|A\|_i := \sqrt{\lambda_M(A^T A)} \tag{3}$$

where $\lambda_M(A^T A)$ is the largest eigenvalue of the (positive semi-definite) symmetric transformation $A^T A$. Finally, assuming that $\|A(t)\|_i$ is bounded, we define:

$$\alpha_m := \sup_{x \in \Re^n} \|A(t)\|_i, \quad \alpha_M := \inf_{x \in \Re^n} \|A(t)\|_i \tag{4}$$

Property 1 — Inertia [6]: The inertia matrix $D(q)$, defined in (1), is symmetric, positive definite and uniformly bounded by the following condition:

$$\lambda_m \|x\|^2 \leq x^T D(q)x \leq \lambda_M \|x\|^2 \tag{5}$$

It is easy to see then that $D^{-1}(q)$ is also uniformly bounded since:

$$\frac{1}{\lambda_M}\|x\|^2 \leq x^T D^{-1}(q)x \leq \frac{1}{\lambda_m}\|x\|^2 \tag{6}$$

where λ_m and λ_M are positive scalar constants and $x \in \Re^n$ is an arbitrary vector. It should be noted that the above bounds only hold for revolute robots.

Property 2 — Skew Symmetry [6]: The inertia matrix $D(q)$ and the Coriolis and centrifugal matrix $C(q, \dot{q})$ are not independent, and the matrix $\dot{D} - 2C$ is skew symmetric by selecting a suitable definition of the matrix $C(q, \dot{q})$. However, the following quadratic form is always zero regardless of the choice of $C(q, \dot{q})$.

$$x^T(\dot{D}(q) - 2C(q, \dot{q}))x = 0 \tag{7}$$

for an arbitrary vector $x \in \Re^n$.

Property 3 — Coriolis and centrifugal Matrix: Given arbitrary vector v_1 and v_2, there exists:

$$C(q, v_1)v_2 = C(q, v_2)v_1 \tag{8}$$

Property 4 — Boundedness on Coriolis and centrifugal Matrix: The norm of $C(q, \dot{q})$ satisfied the inequality:

$$\|C(q, \dot{q})\| \leq k_c\|\dot{q}\| \tag{9}$$

where k_c is a constant.

Property 5 — Linear in Parameters: The manipulator dynamic equation (1) is linear in its parameters:

$$D(q)\ddot{q} + C(q, \dot{q})\dot{q} + g(q) = Y(q, \dot{q}, \ddot{q})\theta \tag{10}$$
$$D(x_1)\phi_1 + C(x_1, \phi_2)\phi_3 + g(x_1) = Y(x_1, \phi_1, \phi_2, \phi_3)\theta \tag{11}$$

where $\theta \in \Re^p$ is a vector of unknown constant parameters and $\phi_1, \phi_2, \phi_3 \in \Re^n$. Matrix $Y \in \Re^{n \times p}$ is usually known as the manipulator regressor [6].

In this paper, the parameter θ is assumed to be unknown, while the structure of the regressor matrix Y is assumed to be available. The target is to design a control law to ensure asymptotic tracking of the desired trajectory q_d. Let $q_d(t) \in C^3$ denote a desired link trajectory which is continuously differentiable up to the third order, and the desired trajectory and its first two time derivatives are bounded. The $3n$-dimensional state-space Equation (1) and (3) for the model of the robot manipulator and motor are given by:

$$\dot{x}_1 = x_2 \tag{12}$$
$$\dot{x}_2 = D^{-1}(x_1)[-C(x_1, x_2)x_2 - g(x_1) + K_t x_c] \tag{13}$$
$$\dot{x}_c = L^{-1}(-Rx_c - K_b x_2 + u) \tag{14}$$

where $x_1 := q$, $x_2 := \dot{q}$ and $x_c := I$. The starting point is to define a tracking error as the following:

$$e = x_1 - q_d \tag{15}$$
$$x_r = \dot{q}_d - (x_1 - q_d) \tag{16}$$
$$\dot{x}_r = \ddot{q}_d - (x_2 - \dot{q}_d) \tag{17}$$
$$s = x_2 - x_r = \dot{e} + e \tag{18}$$

Now, with the definitions:

$$\dot{\hat{x}}_r = \ddot{q}_d - (\hat{x}_2 - \dot{q}_d) = \dot{x}_r + \tilde{x}_2 \tag{19}$$

$$\hat{s} = \hat{x}_2 - x_r = s - \tilde{x}_2 \tag{20}$$

where $\hat{x}_2$ determined later is the estimate of x_2 and $\tilde{x}_2 = x_2 - \hat{x}_2$ is the observation error vector. Now, we can rewrite (13) in terms of Equation (18) to yield:

$$
\begin{aligned}
D(x_1)\dot{s} &= D(x_1)[\dot{x}_2 - \ddot{q}_d + (x_2 - \dot{q}_d)] \\
&= -C(x_1, x_2)x_2 - g(x_1) + K_t x_c - D(x_1)\dot{\hat{x}}_r + D(x_1)\tilde{x}_2 \\
&= -C(x_1, x_2)s - D(x_1)\dot{\hat{x}}_r - C(x_1, \hat{x}_2)x_r - g(x_1) + K_t \tau \\
&\quad -K_t \eta + D(x_1)\tilde{x}_2 - C(x_1, x_r)\tilde{x}_2
\end{aligned}
\tag{21}
$$

where $\tau \in \Re^n$ is an embedded control input vector to be defined. This embedded control signal can be viewed as an adaptive control variable that specifies a desired link torque. As shown later, this control signal τ is actually embedded inside the overall control strategy which is designed at the voltage input level for each motor. This control strategy can be best described in terms of the perturbation term:

$$\eta = \tau - x_c \tag{22}$$

If η equals to zero, we can see from (21) that the desired control torque is effectively applied to each joint of the robot. In general, η will not be equal to zero, so we must design our overall control strategy to make η "small" enough so that we can still achieve the desired stability result of the link position tracking error. To do this, the dynamic characteristics of the perturbation term is needed. Therefore, we differentiate equation (22) to yield:

$$\dot{\eta} = \dot{\tau} - \dot{x}_c \tag{23}$$

And multiply (23) by L to give

$$L\dot{\eta} = L(\dot{\tau} - \dot{x}_c) = L\dot{\tau} + Rx_c + K_b x_2 - u \tag{24}$$

where the vector $u \in \Re^n$ is the final control signal that determines the input voltage for each motor.

It is reasonable to assume that the matrices K_t, L, R, and K_b are accurately known , because they are basically motor related parameters. Hence the final control signal u can be described from the torque signal τ.

Now, we select the control torque τ as:

$$\tau = K_t^{-1}[\hat{D}(x_1)\dot{\hat{x}}_r + \hat{C}(x_1, \hat{x}_2)x_r + \hat{g}(x_1) - K_D \hat{s}] \tag{25}$$

where K_D is a positive definite matrix and define:

$$d_m = \lambda_m(K_D) \quad \text{and} \quad d_M = \lambda_M(K_D) \tag{26}$$

Therefore, the final voltage input is given by:

$$u = L\dot{\tau} + Rx_c + K_b x_2 - K_t^T(s + \tilde{x}_2) \tag{27}$$

Hence, using (25) we obtain:

$$
\begin{aligned}
D(x_1)\dot{s} &= -C(x_1, x_2)s + Y_k\tilde{\theta} - K_D\hat{s} \\
&\quad -K_t\eta + D(x_1)\tilde{x}_2 - C(x_1, x_r)\tilde{x}_2 \\
&= -C(x_1, x_2)s + Y_k\tilde{\theta} - K_D s - K_t\eta \\
&\quad +[D(x_1) - C(x_1, x_r) + K_D]\tilde{x}_2
\end{aligned} \tag{28}
$$

where:

$$Y_k\tilde{\theta} = Y(x_1, \hat{x}_2, x_r, \dot{\hat{x}}_r)\tilde{\theta} = \tilde{D}(x_1)\dot{\hat{x}}_r + \tilde{C}(x_1, \hat{x}_2)x_r + \tilde{g}(x_1) \tag{29}$$

$$\tilde{D}(x_1) = \hat{D}(x_1) - D(x_1) \tag{30}$$

$$\tilde{C}(x_1, \hat{x}_2) = \hat{C}(x_1, \hat{x}_2) - C(x_1, \hat{x}_2) \tag{31}$$

$$\tilde{g}(x_1) = \hat{g}(x_1) - g(x_1) \tag{32}$$

The following Lemma is useful for the stability analysis:

Lemma 2.1. *Consider the function $f(\cdot, \cdot) : \Re \times \Re \to \Re$*

$$f(w_1, w_2) = 2\alpha_1 w_1 w_2 - \alpha_2 w_1^2, \quad w_1, w_2 \in \Re^+ \tag{33}$$

where α_1 and α_2 are positive constants, then we have:

$$f(w_1, w_2) \leq \frac{\alpha_1^2}{\alpha_2} w_2^2 \tag{34}$$

3   Adaptive Velocity Estimator

In order to remove the acceleration measurements in the final control signal, we design a velocity estimator to reconstruct the velocity state. In addition, we use these estimated states together with the adaptive control signal to get semi-globally asymptotical stability.

Based on the above model, the proposed velocity estimator is described by:

$$\dot{\hat{x}}_1 = \hat{x}_2 + \tilde{x}_1 \tag{35}$$

$$\hat{x}_2 = z + k\tilde{x}_1 \tag{36}$$

$$\dot{z} = \dot{\hat{x}}_r + k\tilde{x}_1 \tag{37}$$

where $\hat{x}_1$ and $\hat{x}_2$ are the estimates of x_1 and x_2 respectively, and $\tilde{x}_1 = x_1 - \hat{x}_1$ and $\tilde{x}_2 = x_2 - \hat{x}_2$ are the observation error vectors. And z is an auxiliary variable, and k is a positive constant.

Comparing equations (12) and (35), (13), (36) and (37), we can derive the following error equation:

$$\dot{\tilde{x}}_1 = \dot{x}_1 - \dot{\hat{x}}_1 = x_2 - \hat{x}_2 - \tilde{x}_1 = \tilde{x}_2 - \tilde{x}_1 \tag{38}$$

From this we can obtain:

$$
\begin{aligned}
\dot{\tilde{x}}_2 &= \dot{x}_2 - \dot{\hat{x}}_2 = \dot{x}_2 - \dot{z} - k\dot{\tilde{x}}_1 = \dot{x}_2 - \dot{\hat{x}}_r - k\tilde{x}_2 \\
&= D^{-1}(x_1)[-C(x_1,x_2)x_2 - g(x_1) + K_t\tau - K_t\eta - D(x_1)\dot{\hat{x}}_r] - k\tilde{x}_2
\end{aligned}
\tag{39}
$$

In addition, using the embedded control law (25), we get:

$$
\begin{aligned}
D(x_1)\dot{\tilde{x}}_2 &= -C(x_1,x_2)x_2 - g(x_1) + K_t\tau \\
&\quad -K_t\eta + D(x_1)\dot{\hat{x}}_r - kD(x_1)\tilde{x}_2 \\
&= -C(x_1,x_2)\tilde{x}_2 - C(x_1,x_2)\hat{x}_2 - g(x_1) + \hat{D}(x_1)\dot{\hat{x}}_r - K_t\eta \\
&\quad +\hat{C}(x_1,\hat{x}_2)x_r + \hat{g}(x_1) - K_D\hat{s} - D(x_1)\dot{\hat{x}}_r - kD(x_1)\tilde{x}_2 \\
&= -C(x_1,x_2)\tilde{x}_2 + Y_k\tilde{\theta} - C(x_1,\hat{x}_2)s - K_D\hat{s} \\
&\quad -K_t\eta - kD(x_1)\tilde{x}_2 \\
&= -C(x_1,x_2)\tilde{x}_2 + Y_k\tilde{\theta} - C(x_1,\hat{x}_2)s - K_Ds \\
&\quad -K_t\eta - [kD(x_1) - K_D]\tilde{x}_2
\end{aligned}
\tag{40}
$$

To simplify the stability analysis, k is now explicitly defined in terms of bounding coefficients and a nonlinear damping gain [7] as the following:

$$k = \frac{1}{\lambda_m}(d_m + d_M + \sigma_1^2\frac{k_0}{d_m} + \sigma_2^2\frac{k_0}{d_m}) \tag{41}$$

where k_0 is a positive, nonlinear damping gain, and σ_1, σ_2 are positive "bounding" constants which will be determined later.

The parameter update law is provided by:

$$\dot{\hat{\theta}} = -\Gamma^{-1}Y_k^T(s + \tilde{x}_2) \tag{42}$$

where the $\Gamma \in \Re^{p \times p}$ is a diagonal positive definite matrix

4   Stability Analysis

We now show that observation errors converge to zero if the initial error is within a compact set. In other words the equilibrium point is Lyapunov semi-global stable. To simplify the proof of the following theorem, we define $x_s := [s^T \tilde{x}_2^T]^T$, $x_f := [s^T \tilde{x}_2^T \tilde{\theta}^T \eta^T]^T$, and $F := diag\{D(x_1) D(x_1) \Gamma L\}$.

Theorem 4.1. *Assume that for any $t \geq 0$, $\|\dot{q}(t)\| \leq k_q$, and for the control signal in (25) and (27) as well as adaptive law given by (42), if the control gain k_0 of (41) is chosen such that:*

$$k_0 > 1 \qquad (43)$$

Then

$$\lim_{t \to \infty} e = \lim_{t \to \infty} \dot{e} = \lim_{t \to \infty} \tilde{x}_2 = 0 \qquad (44)$$

Moreover, a region of attraction is given by:

$$B = \left\{ x_f \in \Re^{3n+p}, \|x_f\| < \sqrt{\frac{\varrho_m}{\varrho_M}(k_0 - 1)} \right\} \qquad (45)$$

where ϱ_m and ϱ_M are positive constants defined as $\varrho_m = min\{\lambda_m, \lambda_m(L), \lambda_m(\Gamma)\}$, $\varrho_M = max\{\lambda_M, \lambda_M(L), \lambda_M(\Gamma)\}$.

Proof: Omitted due to the page limit and available upon request.

5 Simulation

The performance of the adaptive controller and velocity estimator have been applied to the simulation plant, referred to a two-link robot arm with DC motor. The dynamic model is described by (1), (2) and a parameterization scheme of this robot is given in [3]:

$$\theta_1 \ddot{x}_{11} + (\theta_3 c_{21} + \theta_4 s_{21})\ddot{x}_{12} - \theta_3 s_{21}\dot{x}_{12}^2 + \theta_4 c_{21}\dot{x}_{12}^2 = k_{t1} x_{c1} \qquad (46)$$
$$(\theta_3 c_{21} + \theta_4 s_{21})\ddot{x}_{11} + \theta_2 \ddot{x}_{12} + \theta_3 s_{21}\dot{x}_{11}^2 - \theta_4 c_{21}\dot{x}_{11}^2 = k_{t2} x_{c2} \qquad (47)$$

where $x_1 = [x_{11}^T \ x_{12}^T]$, $\theta = [\theta_1 \ \theta_2 \ \theta_3 \ \theta_4]$ $c_{21} = \cos(x_{12} - x_{11})$ and $s_{21} = \sin(x_{12} - x_{11})$. It is clearly linear in terms of the four parameter $\theta_1, \theta_2, \theta_3$ and θ_4, which are related to the physical parameter of the links with the load treated as part of the second link. Defining the component of matrix C as:

$$C(1,1) = C(2,2) = 0 \qquad (48)$$
$$C(1,2) = (\theta_4 c_{21} - \theta_3 s_{21})\dot{x}_{12} \qquad (49)$$
$$C(2,1) = (\theta_3 c_{21} - \theta_4 c_{21})\dot{x}_{11} \qquad (50)$$

the skew-symmetry of the $\dot{D} - 2C$ can also be satisfied.

6   Conclusion

In this paper, we have proposed an adaptive control law with a velocity estimator for the robot link. Using this estimated speed, the controller is capable of tracking a given desired trajectory with guaranteed stability The dynamic part of the controller consists of a nonlinear velocity estimator which guaranteed semiglobal uniformly asymptotic stability of the error dynamics.

References

1. M. M. Bridges, D. M. Dawson, and X. Gao, "Adaptive control of rigid-link electrically-driven robots," Proc. IEEE Conference on Decision and Control, vol. 1, pp. 159–165, 1993.

2. S. S. Ge and I. Postlethwaite, "Nonlinear adaptive control of robots including motor dynamics," Proc. American Control Conference, pp. 1423–1427, 1993.

3. J.-J. E. Slotine and W. Li, "Adaptive manipulator control: A case study," IEEE Transactions on Automatic Control, vol. 33, pp. 995–1003, 1988.

4. M. W. Spong and M. Vidyasagar, Robot Dynamics and Control, New York: John Wiley & Sons, 1989.

5. M. Vidyasagar, Nonlinear System Analysis, New Jersey: Prentice Hall Inc., 1978.

6. R. Ortega and M. W. Spong, "Adaptive motion control of rigid robot: A tutorial," Automatica, vol. 25, pp. 877–888, 1989.

7. T. Turg, D. M. Dawson, and P. Vedagarbha, "A redesigned dcal controller without velocity measurements: Theory and experimentation," Proc. IEEE Conference on Decision and Control, vol. 1, pp. 824–828, 1994.

8. S. Sastry and M. Bodson, Adaptive Control, New York: Prentice Hall, 1979.

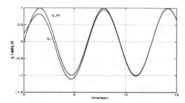

Figure 1: Output tracking q_1 and its desired trajectory q_{1d} for first link.

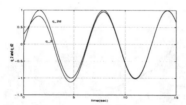

Figure 2: Output tracking q_2 and its desired trajectory q_{2d} for second link.

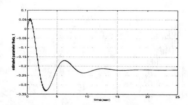

Figure 3: The estimated parameter value $\hat{\theta}_1$.

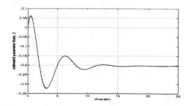

Figure 4: The estimated parameter value $\hat{\theta}_2$.

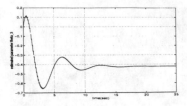

Figure 5: The estimated parameter value $\hat{\theta}_3$.

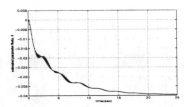

Figure 6: The estimated parameter value $\hat{\theta}_4$.

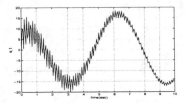

Figure 7: Control signal u_1 of the first link.

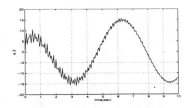

Figure 8: Control signal u_2 of the second link.

Development of A New Industrial Robot

Hong-You Lee and W. Wälischmiller

Hans Wälischmiller GmbH, Tele Robot Engineering
Klingleweg 8, D-88709 Meersburg, Germany
E-Mail: hongyou@mee.hwm.com and ww@mee.hwm.com

Abstract. This paper presents a new general purpose and modular industrial robot, trade named TELBOT, developed by Hans Wälischmiller GmbH. The robot has six revolution joints. A unique drive system of the robot contains all of the electronics, cables, motors, reduction gears and encoders. The transmission of motions of the arm are translated from the unique drive system by sets of concentric tubes that make up to the links and sets of bevel gears that make up the joints. There is no electrical parts in the arm itself. This allows unrestricted rotation of each joint. That means, all the revolution joints can rotate over 360 degrees continuously. The arm itself is relatively compact and lightweight. The inverse kinematics problem of this robot has still up to 16 solutions, which correspond to the different robot configurations. The Jacobian matrix of the robot is very simple when it is formulated with respect to a selected special coordinate system, so that it is possible to be inverted symbolically. Even more, the singularities of the robot can be identified easily, because the determinant of the Jacobian is very concise. The robot has been used for steam generator maintenance in Canada and for decommissioning glove boxes in Japan.

1   Introduction

A general purpose and modular industrial robot, trade named TELBOT, has been developed by Hans Wälischmiller GmbH (Wälischmiller and Frager, 1995). The TELBOT robot has been successfully used for cleaning steam generator tubes at the primary side in CANADA (Bains, Majarais, and Scott, 1995). The robot has been also used for decommissioning glove boxes in Japan. The robot has six degrees of freedom and a unique drive system (Fig. 1). All the motors, measuring system and the transmission gear box of the manipulator are located in the robot base. There is not any electrical cables from the base to the end-effector. The transmission of the motion from the motors to the end-effector is realized via concentric tubes and bevel gears, so that all the revolution joints can rotate over 360 degrees continuously. Very recently, Prof. Jacques M. Hervé of Ecole Centrale Paris suggested the first author a reference (Kersten, 1977) with a conceptual design of the similar manipulator kinematics. Because of the novel design and motion transmission, the end-effector or the tool can reach its work position in any orientation. Furthermore, due to there is no electrical cable in the whole arm, the manipulator can be easily protected from the dirty environment, and also some of the parts can be easily changed if necessary.

The present paper will focus on the novel mechanical design, the inverse kinematics, singularity identification and the application in hazardous environments.

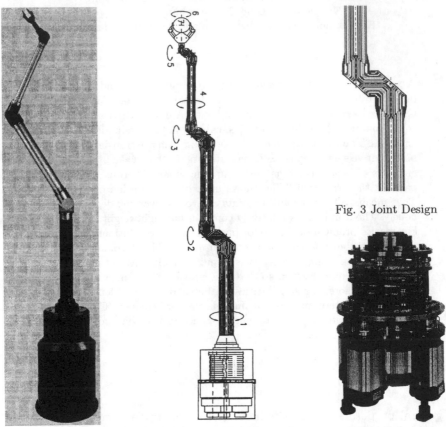

Fig. 3 Joint Design

Fig. 1 TELBOT Robot   Fig. 2 Transmission Principle   Fig. 4 Robot Base

2   Mechanical Design

The TELBOT robot has six degrees of freedom and a unique drive system that allows unrestricted motion of each joint (Fig. 1 and Fig. 2). The rotation of all the revolution joints and the opening and closing movement of the gripper are transmitted with the concentric tubes in the arm and bevel gears in the joints. The mechanical system possesses many novel properties and advantages:

• **Unique Drive System**

A unique drive system is located at the robot base. All of the motors, electronics and cables are contained within the base (Fig. 4). There is no electrical parts in the arm itself.

- **Motion Transmission**

The transmission of motions of the arm is translated from the unique drive system by sets of concentric tubes that make up the links and sets of bevel gears that make up the joints (Fig. 2 and Fig. 3).

- **Kinematic Skeleton**

The first three joint axes of the TELBOT robot are in a standard configuration with one vertical and two parallel horizontal axes. The three wrist joint axes have an offset between the fourth and sixth joint axis, so that all the three wrist joints can also be rotated over 360 degrees continuously (Fig. 5).

- **Modular Design**

The modular design of the gear block allows for adaptation to various transmission ratios and moments. Tubes and joints with different sizes are designed for meeting the requirements of different payloads. The modular parts can be quickly changed when it is needed.

- **Unrestricted Joint Rotation**

The novel mechanical system allows unrestricted rotation of each joint. All the revolution joints can rotate over 360 degrees continuously.

- **Compact Arm and Lightweight**

The arm itself is relatively compact and lightweight. This facilitates increasing the workspace and improving the dynamic properties.

3 Geometrical Modeling

A kinematic diagram of the TELBOT manipulator showing all link parameters is given in Fig. 5. The directions of the joint axes are sequentially labeled with the unit vectors $\vec{s}_i$, $i = 1, 2, \ldots, 6$. The directions of the common normals between two successive joint axes $\vec{s}_i$ and $\vec{s}_j$ are labeled with the unit vectors $\vec{a}_{ij}$, $ij = 12, 23, \ldots, 56$, and the lengths of the common normals (link lengths) with a_{ij}. The mutual perpendicular distances between pairs of successive links $\vec{a}_{ij}$ and $\vec{a}_{jk}$ along the joint axis $\vec{s}_j$ are denoted by S_j, $j = 1, 2, \ldots, 6$, and are called joint offsets. The joint angles θ_i and the twist angles α_{ij} are measured by right-hand rotations about $\vec{s}_i$ and $\vec{a}_{ij}$, respectively.

A robot base frame is located with $\vec{z}$ axis coaxial with $\vec{s}_1$, the axis of the first joint. The position of the end-effector is specified in the robot base frame by $\vec{R}_H$ while the orientation is defined by the pair of orthogonal unit vectors $\vec{a}_{67}$ and $\vec{s}_7$.

It is important to note that only the joint angles (θ_j) are unknown quantities; all other Denavit-Hartenberg parameters, i.e., twist angles (α_{ij}), offsets (S_j), and link lengths (a_{ij}) have constant values that are listed in Table 1.

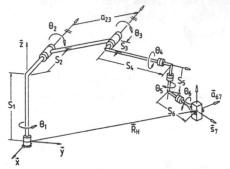

Fig. 5 Robot Kinematic Skeleton

Table 1: Link Parameters

i	S_i	$a_{i(i+1)}$	θ_i	$\alpha_{i(i+1)}$
1	S_1	0.0	θ_1	$\pi/2$
2	S_2	a_{23}	θ_2	0.0
3	S_3	0.0	θ_3	$\pi/2$
4	S_4	0.0	θ_4	$\pi/2$
5	S_5	0.0	θ_5	$\pi/2$
6	S_6		θ_6	

4 Inverse Kinematics

It was shown by Lee (Li) and Liang (1988) and Lee (Li) and Reinholtz (1996), that there are up to 16 solutions of the inverse kinematics for robots with 6R joints. The TELBOT robot with non-zero offset of the 5th joint has no closed-form solution of the inverse kinematics either, although it still has special geometry. It will be shown that the inverse kinematics problem of the TELBOT manipulator can be reduced to a 16th degree polynomial in one unknown and five linear equations in the other five unknown joint angles. The univariate polynomial is derived from only three constraint equations. Once the Polynomial is solved for the unknown joint angle numerically, all the other five unknown joint angles are determined by solving linear equations. The real solutions correspond to the different robot configurations. The optimal configuration for a given task and limited workspace can be chosen once all the possible solutions are found. Because all the joint rotations are transmitted by concentric tubes in the arm and concentric bevel gears in the joints, the joint angles must be then converted in the rotation angles of the tubes and finally rotation of the motors.

It is well known that a 16th degree polynomial for the general 6R manipulator can be derived from 14 closure equations (Lee and Liang, 1988), while the derivation of such a polynomial for the TELBOT robot, because of its special geometry, requires only three closure equations. The basic approach is to disconnect the manipulator into two subchains at two joints and then formulate scalar products of the vectors representing the joints and links for both subchains. These scalar equations are then solved simultaneously to yield a single polynomial equation in one unknown joint angle. Once the polynomial is obtained and solved for the joint angle, all the remaining joint variables follow from linear equations.

4.1 Univariate Polynomial

The 16th degree polynomial in one unknown is derived by using the following three equations:

$$[\vec{s}_4 \cdot \vec{s}_1]_A = [\vec{s}_4 \cdot \vec{s}_1]_B \tag{1}$$

$$\left[\vec{R} \cdot (\vec{s}_1 \times \vec{s}_4)\right]_A = \left[\vec{R} \cdot (\vec{s}_1 \times \vec{s}_4)\right]_B \tag{2}$$

$$\left[\vec{R} \cdot \vec{R}\right]_A = \left[\vec{R} \cdot \vec{R}\right]_B \tag{3}$$

where

$$\left[\vec{R}\right]_A = [(S_2 + S_3)\vec{s}_2 + a_{23}\vec{a}_{23}] \tag{4}$$

$$\left[\vec{R}\right]_B = -\left[S_4\vec{s}_4 + S_5\vec{s}_5 + S_6\vec{s}_6 - (\vec{R}_H - S_1\vec{s}_1)\right] \tag{5}$$

The three closure equations are derived by disconnecting the TELBOT manipulator at joints 1 and 4. This divides the manipulator into two subchains A and B and the vector sum of subchain A equals the sum of subchain B. Note that the subscripts A and B on a bracketed quantity indicates that all vectors within the brackets must be expressed using the vectors and rotations associated with that subchain.

Although it may not be immediately apparent, these three scalar equations contain only θ_5, θ_6 and $(\theta_2 + \theta_3)$, as unknowns. It should, therefore, be possible to solve this system. A key to efficient formulation is to express vectors in the most convenient coordinate frame before the scalar and triple scalar product operations are performed. A change in reference frame is easily accomplished by using standard joint and twist rotation matrices.

The given end-effector position vector and the orientation matrix are defined in the robot base frame as follows:

$$\vec{R}_{HO} = \vec{R}_H - S_1\vec{s}_1 = \begin{bmatrix} r_x \\ r_y \\ r_z - S_1 \end{bmatrix} \quad M_H = \begin{bmatrix} \vec{a}_{67} \\ \vec{s}_7 \times \vec{a}_{67} \\ \vec{s}_7 \end{bmatrix} = \begin{bmatrix} h_{00} & h_{10} & h_{20} \\ h_{01} & h_{11} & h_{21} \\ h_{02} & h_{12} & h_{22} \end{bmatrix} \tag{6}$$

The three eqs. (1)-(3) can be evaluated and may be written as:

$$c_5 h_{22} + s_5(-h_{20}c_6 + h_{21}s_6) = \cos(\theta_2 + \theta_3) \tag{7}$$

$$c_5\left[c_6(h_{20}S_5) - s_6(h_{21}S_5) + (-h_{12}r_x + h_{02}r_y)\right] +$$
$$s_5\left[c_6(h_{10}r_x - h_{00}r_y - h_{21}S_6) + s_6(-h_{11}r_x + h_{01}r_y - h_{20}S_6) + (h_{22}S_5)\right]$$
$$= (S_2 + S_3)\sin(\theta_2 + \theta_3) \tag{8}$$

$$c_5 S_4(H_z - S_6) + s_5(-c_6 S_4 H_x + s_6 S_4 H_y) - c_6 S_5 H_y + s_6 S_5 H_x +$$
$$\left[(1/2)(S_4^2 + S_5^2 + S_6^2 + R^2 - (S_2 + S_3)^2 - a_{23}^2) - S_6 H_z\right] = 0 \tag{9}$$

where S_i and a_{23} are constant parameters defined in the Table 1, $c_i = \cos\theta_i$, $s_i = \sin\theta_i$ (the abbreviations c_i and s_i will be used throughout this paper), and

$$[H_x, \ H_y, \ H_z]^T = M_H \vec{R}_{HO} \tag{10}$$

$$R^2 = r_x^2 + r_y^2 + (r_z - S_1)^2 \tag{11}$$

The three closure equations (7)-(9), which contain all the information needed to derive the 16th degree polynomial, now can be written in the following form:

$$F_1(\theta_5, \theta_6) = \cos(\theta_2 + \theta_3) \tag{12}$$
$$F_2(\theta_5, \theta_6) = (S_2 + S_3)\sin(\theta_2 + \theta_3) \tag{13}$$
$$F_3(\theta_5, \theta_6) = 0 \tag{14}$$

Where the functions $F_1(\theta_5, \theta_6)$, $F_2(\theta_5, \theta_6)$ and $F_3(\theta_5, \theta_6)$ are linear combinations of the terms $(c_5 c_6, c_5 s_6, c_5, s_5 c_6, s_5 s_6, s_5, c_6, s_6)$.

Eliminating $(\theta_2 + \theta_3)$ from Eqs. (12) and (13) gives

$$F_1^2 + [F_2/(S_2 + S_3)]^2 = 1 \tag{15}$$

The next step is to eliminate θ_5 (or θ_6) from Eqs. (14) and (15). For convenience, the two equations can be expressed as polynomials in $x_5 = \tan(\theta_5/2)$ and $x_6 = \tan(\theta_6/2)$ as follows:

$$A_0 + A_1 x_5 + A_2 x_5^2 = 0 \tag{16}$$
$$B_0 + B_1 x_5 + B_2 x_5^2 + B_3 x_5^3 + B_4 x_5^4 = 0 \tag{17}$$

where A_0, A_1 and A_2 are quadratic polynomials in x_6, and B_0, B_1, B_2, B_3 and B_4 are quartic polynomials in x_6.

Eliminating x_5 from the two eqs. (16) and (17) yields the desired 16th degree univariate polynomial in x_6 with constant coefficients e_i:

$$\sum_{i=0}^{16} e_i x_6^i = 0 \tag{18}$$

4.2 Determination of All the Joint Variables

Once the desired hand position and orientation (6) is specified, all e_i coefficients of Eq. (18) can be calculated. The roots of the polynomial give the values of θ_6 which correspond 16 possible configurations of the manipulator. Each value of θ_6 will give a unique value of x_5 by solving the two equations (16) and (17). For each pair of θ_5 and θ_6, a value of $(\theta_2 + \theta_3)$ can be easily obtained from equations (12) and (13). Then for each set of $(\theta_2 + \theta_3)$, θ_5 and θ_6, the joint angles θ_1 and θ_4 can be determined from the following two pairs of equations:

$$[\vec{s}_1 \cdot \vec{x}]_A = [\vec{s}_1 \cdot \vec{x}]_B \tag{19}$$
$$[\vec{s}_1 \cdot \vec{y}]_A = [\vec{s}_1 \cdot \vec{y}]_B \tag{20}$$

$$[\vec{s}_4 \cdot \vec{a}_{45}]_A = [\vec{s}_4 \cdot \vec{a}_{45}]_B \tag{21}$$
$$[\vec{s}_4 \cdot \vec{s}_5 \times \vec{a}_{45}]_A = [\vec{s}_4 \cdot \vec{s}_5 \times \vec{a}_{45}]_B \tag{22}$$

The equations (19)-(22) can be formulated so that they are linear in terms of sines and cosines of θ_1 and θ_4 respectively, thus can be solved for directly.

The joint angles θ_3 can be easily determined from the following equations, respectively:

$$\left[\vec{R} \cdot \vec{a}_{34}\right]_A = \left[\vec{R} \cdot \vec{a}_{34}\right]_B \tag{23}$$

$$\left[\vec{R} \cdot (\vec{s}_3 \times \vec{a}_{34})\right]_A = \left[\vec{R} \cdot (\vec{s}_3 \times \vec{a}_{34})\right]_B \tag{24}$$

The two equations (23) and (24) are linear in terms of sines and cosines of θ_3. If θ_3 is found, the joint angle θ_2 can be determined using the value of $(\theta_2 + \theta_3)$.

It should be noted that if $(\theta_2 + \theta_3)$ near to zero or 180 degrees, the proposed equations (19) and (22) cannot be used. The joint angles θ_1 and θ_4 must be calculated by using some other closure equations.

5 Singularity Identification

5.1 Jacobian Matrix

It is well known that the Jacobian matrix of a 6R robot is defined by the following matrix:

$$Jacobian = \begin{bmatrix} \vec{s}_1 & \vec{s}_2 & \vec{s}_3 & \vec{s}_4 & \vec{s}_5 & \vec{s}_6 \\ \vec{R}_{P1} \times \vec{s}_1 & \vec{R}_{P2} \times \vec{s}_2 & \vec{R}_{P3} \times \vec{s}_3 & \vec{R}_{P4} \times \vec{s}_4 & \vec{R}_{P5} \times \vec{s}_5 & \vec{R}_{P6} \times \vec{s}_6 \end{bmatrix} \tag{25}$$

where $\vec{R}_{Pi}$ with $i = 1, 2, 3, 4, 5, 6$ are the position vectors of the point P to an arbitrary point of the axis of i_{th} joint. If the point P is selected at the cross point of the axes of joints 3 and 4, for example, the six position vectors $\vec{R}_{Pi}$ can be defined as following:

$$\vec{R}_{P1} = -(S_2 + S_3)\vec{s}_3 - a_{23}\vec{a}_{23} \tag{26}$$
$$\vec{R}_{P2} = -S_3\vec{s}_3 - a_{23}\vec{a}_{23} \tag{27}$$
$$\vec{R}_{P3} = \vec{0} \tag{28}$$
$$\vec{R}_{P4} = \vec{0} \tag{29}$$
$$\vec{R}_{P5} = S_4\vec{s}_4 \tag{30}$$
$$\vec{R}_{P6} = S_4\vec{s}_4 + S_5\vec{s}_5 \tag{31}$$

Evaluating the Jacobian matrix with respect to the coordinate system $(\vec{a}_{34}, \vec{s}_3, \vec{s}_4)$ which is attached to the joints 3 and 4, results in the following:

$$Jacobian = \begin{bmatrix} s_{23} & 0 & 0 & 0 & s_4 & c_4 s_5 \\ 0 & 1 & 1 & 0 & -c_4 & s_4 s_5 \\ -c_{23} & 0 & 0 & 1 & 0 & -c_5 \\ c_{23}(S_2 + S_3) & s_3 a_{23} & 0 & 0 & c_4 S_4 & c_4 c_5 S_5 - s_4 s_5 S_4 \\ -c_2 a_{23} & 0 & 0 & 0 & s_4 S_4 & s_4 c_5 S_5 + c_4 s_5 S_4 \\ s_{23}(S_2 + S_3) & -c_3 a_{23} & 0 & 0 & 0 & s_5 S_5 \end{bmatrix} \tag{32}$$

Note that $c_{23} = cos(\theta_2 + \theta_3)$ and $s_{23} = sin(\theta_2 + \theta_3)$.

The lengths of the axes S_2, a_{23}, S_3, S_4 and S_5 are scaled in the Jacobian matrix so that the maximum value of S_2, a_{23} S_3, S_4 and S_5 is equal to 1. This reduces the influence of the dimension of lengths and also makes the Jacobian matrix dimensionless. Actually, all the lengths and the end-effector position $\vec{R}_H$ in the controller algorithm are divided by the maximum value of the original values of S_2, a_{23}, S_3, S_4 and S_5.

5.2 Determinant of the Jacobian Matrix and Singularities

The determinant of the Jacobian matrix of the robot is evaluated

$$D_{jacobi} = c_2 s_4 c_5 S_5 [c_3 c_4 a_{23} + s_4 (S_2 + S_3)] - s_5 (s_{23} S_4 + c_2 a_{23})(c_3 S_4 - s_3 s_4 S_5)$$

$$(33)$$

Obviously, the robot is in singularity configurations when the joint angles meet one of the following conditions:

1. $\theta_2 = \pm 90$ degrees, and at the same time, $\theta_3 = \pm 90$ degrees, it means that the first joint axis $\vec{s}_1$ is parallel to the upper arm $\vec{a}_{23}$, and at the same time, the upper arm $\vec{a}_{23}$ to the fourth joint (lower arm) axis $\vec{s}_4$;

2. $\theta_3 = \pm 90$ degrees, and at the same time, $\theta_4 = 0$ or 180 degrees, the upper arm $\vec{a}_{23}$ is parallel to the fourth joint axis $\vec{s}_4$ and at the same time, the third joint axis $\vec{s}_3$ to the fifth joint axis $\vec{s}_5$;

3. $\theta_4 = 0$ or 180 degrees, and at the same time, $\theta_5 = 0$ or 180 degrees, the third joint axis $\vec{s}_3$ is parallel to the fifth joint axis $\vec{s}_5$, and at the same time, the fourth joint axis $\vec{s}_4$ to the sixth joint axis $\vec{s}_6$;

4. $\theta_2 = \pm 90$ degrees, and at the same time, $\theta_5 = 0$ or 180 degrees, the first joint axis $\vec{s}_1$ is parallel to the upper arm $\vec{a}_{23}$, and at the same time, the fourth joint axis $\vec{s}_4$ to the sixth joint axis $\vec{s}_6$.

Finally, it should be noted that the Jacobian matrix, which is expressed with respect to the coordinate system $(\vec{a}_{34}, \vec{s}_3, \vec{s}_4)$ attached to the joints 3 and 4, can be inverted symbolically, and even more the inverse matrix is not very complex. This greatly facilitates the analysis of velocity and statics.

6 Application

An important application of the TELBOT system is for nuclear steam generator maintenance. Fig. 7 shows the steam generator primary head tube cleaning system block diagram. The steam generator contains a shell-and-tube type heat exchanger with thousands of tubes. The plurality of the tubes is encased in a vertical cylindrical pressure vessel with a hemispherical lower end (bowl). The tube ends are inserted through a plate (tube-sheet) and welded and forms a planar cap over the bowl. The tubes are bent into an upside down "U" shape. These tubes must be inspected and repaired from within the bowl quarter spheres. The steam generator bowl is a high radiation area. Some robotic systems were

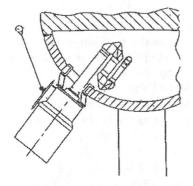 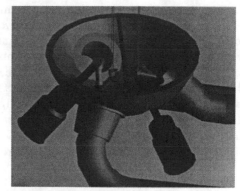

Fig. 6: Man Way Mounting Position   Fig. 7: Steam Generator Maintenance

developed and used for cleaning the tubes, for example, COBRA robotic system developed by B&WNT (Tidwell and Glass, et al, 1991). An advantage by using TELBOT system is that the motors, gear block and all the electrical cables are OUTSIDE of the bowl (Fig. 6 and Fig. 7). The robot arm parts which are inside the bowl can be very easily protected from dirty radioactive materials. Some arm parts can be quickly changed if necessary, because of the modular design. This can greatly reduce the cost of nuclear steam generator maintenance.

Once the manipulators were installed at the man-ways, the operator could initiate the calibration program. An initializing routine was used to calibrate the manipulator base coordinates with respect to the overall tube-sheet reference coordinates. The manipulator automatically moved to a pre-programmed location at the tube-sheet underneath a tie-rod location which was convenient to use for calibration as its position was known and was visually different from the tubes. They were also well spaced. For this initial calibration the stand-off distance of the tool to the tube-sheet was greater than the working distance. This allowed for any misplacement of the manipulators during the manual setting up operation. In order to ensure the manipulators moved in a plane parallel to the tube-sheet, the tool had a metallic "finger" that was used to touch the tube-sheet in three places each time the "finger" touched the tube-sheet a circuit was closed and the operator was alerted that the manipulator was in contact with the tube-sheet. From the three locations the manipulator base frame was calculated through the software and referenced to the tube-sheet plane. The manipulator could now be operated in straight line interpolation mode with respect to the tube-sheet. This made the tele-operation of the manipulator much easier than using the individual joint control. Once the manipulator was in this mode it was referenced to the actual tubes. Again the manipulators were initiated to move to the tie-rod locations. The manipulator was then tele-operated so that the center of the tube adjacent to the tie rod location was in the center of the field of view of the cameras. This was simply carried out using a set of cross hairs over-layed on the monitors. Once this was achieved for two or more tie rod locations the manipulator was referenced with respect to the tubes. During the cleaning process this type of calibration was repeated several times.

This means that the robot base frame is determined with respect to the world frame (attached to the bowl) in two steps. First, only the plane of tube-sheet is calibrated to the robot base frame, and then the position of the tubes with respect to the robot base frame. This greatly simplified the process of the calibration of the robot base frame with respect to the world frame, because there are not three or more sharp points in the bowl whose position is known exactly and could be touched by tool tip of the robot. The problem to plug the finger into the tubes is that the depth of the tool finger in the tube is very difficult to be exactly measured by using normal camera. Therefore, the plane of the tube-sheet is first determined , and then the position of tubes.

Another important application of the TELBOT robot is for decommissioning glove boxes in Japan. The robot is used for dismantling boxes in radioactive environment and for loosing screws on the boxes and then to cut the boxes. The novel properties of the robot, such as no limit continuous rotation of the tool, greatly facilitate reducing the working time and the cost.

7 Conclusions

The TELBOT system is a universal industrial robotic system. Its novel kinematic structure, control system and simulation system allow itself to be able to perform complicated tasks in hazardous environments and in manufacture automation. The TELBOT manipulator without a doubt has proven itself in the field and its revolutionary design coupled with its flexibility and proven reliability will impact the advanced manipulator field for years to come.

References

1. Bains, N., Majarais, B. and Scott, D.A., 1995, "Cleaning of Steam Generator Tubes at the Primary Side Using a General Purpose Manipulator", ANS conference, San Francisco, October.
2. Kersten, L., 1977, "The Lemma Concept: A New Manipulator", International Journal of Mechanism and Machine Theory, Vol. 12, pp.77-84.
3. Lee (Li), H.-Y. and Liang, C.-G., 1988, "Displacement Analysis of the General Spatial 7-Link 7R Mechanism", International Journal of Mechanism and Machine Theory, Vol. 23, pp.219-226.
4. Lee (Li), H.-Y. and Reinholtz, C.F., 1996, "Inverse Kinematics of Serial-Chain Manipulators", ASME Journal of Mechanical Design Vol.118 (September), pp.396-404.
5. Tidwell, P. H., Glass, S. W. and et al, 1991, "COBRA - Design and Development of a Manipulator for Nuclear Steam Generator Maintenance", Proceedings of the 2nd National Applied Mechanisms and Robotics Conference, USA, Vol. 1, pp. IVB1-1 to IVB1-10.
6. Wälischmiller, W. and Frager, O., 1995, "TELBOT - A Modular Tele-Robot System for Hostile Environments with Unlimited Revolutionary Joints and Motors, Drive Sensors and Cables in the Base", Proceedings of the ANS 6th Topical Meeting on Robotics and Remote Systems, Monterey, California, USA, Vol. 1, pp.274-280.

Adaptive Gait Planning for Multi-Legged Robots Using a Stability Compensation Method

Wenjie Chen, K. H. Low and S. H. Yeo

School of Mechanical and Production Engineering
Nanyang Technological University
Republic of Singapore 639798

Email: m94013h61 | mkhlow | myeosh@ntuvax.ntu.ac.sg

Abstract.   Adaptive gait planning is an important consideration in the development of the control system for any multi-legged robot applied to rough terrain. The problem of adaptive gait generation is to find a sequence of suitable foothold on rough terrain so that the legged system maintains static stability and walking continuity. Due to the limit of static stability, deadlock situation during finding suitable foothold may occur if terrain contains a large amount of forbidden zone.   This paper develops an improved method for adaptive gait planning through active compensation of stability margin, through the adjustment of center of gravity (CG) in longitudinal and lateral axes. Simulation results show that the presented method provides the legged machine with a much better terrain adaptivity and deadlock-avoidance ability.

Key words: *Multi-legged Robot, Gait Planning, Motion Planning.*

1 Introduction

The problem of gait planning in multi-legged system can be generally formulated as how to coordinate the motion of leg and body to make the machine traverse over a particular terrain with static stability manner from the starting position to the given objective location along a trajectory specified in advance [6]. The task of an adaptive gait planning in control includes determining the optimal schedule for the lifting and placing of the legs, and finding suitable foothold as supporting point(s) for transfer leg(s) through the whole trajectory [3, 6]. To achieve an effective adaptive gait, some limits, namely, terrain condition, kinematic (or geometry) constraint and static stability, must be taken into account in planning [4, 10].

A considerable amount of prior work has been devoted to the study of adaptive gait over past decades [1-4, 6, 9]. Representative of the these works is *free gait* algorithm which was first recognized and formalized by Kugushev and Jaroshevskij [3] and subsequently improved and developed for a hexapod robot by McGhee and Iswandhi [6]. The general strategy of free gait is to find a sequence of support points so that there is an overlap of the *existence segment* of each support state with that of the proceeding support state. To reduce the complexity of the problem, the algorithm determines the support pattern sequence only one stage forward rather then over the whole trajectory. However, such an approach may lead to a situation of deadlock if the terrain contains a large number forbidden zones. For a four-legged robot, due to fewer choices about selecting its steps, free gait algorithm may become more susceptible to deadlock situation. Hirose [2] presented a hierarchical algorithm to overcome the deadlock problem of quadruped robot. In his algorithm, the selection of new support points was conducted through the three reflexes from sensors, which trend to restrict the search area of new foothold. More recently, an improved free gait algorithm based on the approach of graph search over whole trajectory [9] was also presented to avoid deadlock.

Careful examination of the methodology of all algorithms for free gait shows that the trade-off between stability and adaptability is critical for successful generation of free gait. Deadlock will occur if a contradiction appears between stability and adaptability. It is clear that to enforce or compensate the minimum value of stability margin would decrease the likelihood of deadlock [1, 5, 8].

In this paper, a method to improve adaptive gait planning through the compensation of stability margin is proposed. The compensation of stability margin here is implemented by using CG adjustment in longitudinal and lateral axes. The significant new feature of this method is that it can provide machine a much greater adaptivity over rough terrain with a reasonable desired walking speed. The remainder of the paper is organized as follows. The scheme of compensation of stability margin for a quadruped prototype is described in section 2. Some necessary terminology about adaptive gait is introduced in section 3. The method to improve adaptive gait planning under consideration of stability margin compensation is addressed in section 4 and the strategy of CG adjustment is also included in this part. The simulation result is given in section 5 to show the promising behavior of the proposed method. Finally, a summary and the conclusions are made in section 6.

2 Description of stability margin compensation for a quadruped robot

For the study of adaptive locomotion of legged system, a quadruped robot, referred as NTU-Q1, is at present under construction at the Nanyang Technological University. A preliminary simulation prototype of NTU-Q1 is shown in Fig. 1. When completed, NTU-Q1 will be a self-contained

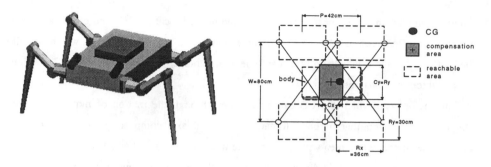

Fig. 1. Computer model of NTU-Q1

Fig. 2. The compensation area of NTU-Q1

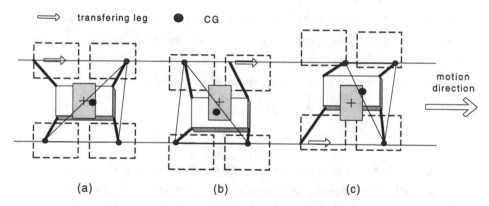

(a)　　　　　　　　(b)　　　　　　　　(c)

Fig. 3. Top view of CG adjustment of NTU-Q1 in some support pattern to maintain static stability. (a) CG adjusted forward only in longitudinal direction (b) CG is adjusted only in lateral direction (c) CG is adjustment in both longitudinal and lateral directions.

walking robot intended for material handling on irregular terrain environment. Each of its four legs will has three independent rotary joints arranged in an arthropod configuration. Two terrain scanners located in the front of body will be used to provide terrain preview data which can then be processed by the control computer to predict suitable foothold position. To achieve superior terrain adaptivity, a device for the compensation of stability has also been incorporated into the design of this machine.

The stability margin compensation scheme adopted for NTU-Q1 involved CG adjustment relative to body frame along longitudinal axis and body frame translation in lateral axis. Such an arrangement will allow the machine to enhance the minimum value of stability margin, while independent of machine's locomotion in longitudinal direction. To adjust CG position in longitudinal axis effectively, a moveable weight which may consist of batteries and control parts is considered in the design of the machine. The moveable weight can move forward and backward along the longitudinal axis so as to shift CG to the desired location. On the other hand, due to arthropod configuration of each leg, the machine possesses a large lateral stroke which will allow the body frame to translate laterally in a sufficient large range. The adjustment of CG position in lateral direction is thus obtained.

The region with which the CG can be adjusted through the motion of movable weight in longitudinal axis and that of body frame along lateral axis composes a *compensation area*. Clearly, the compensation area will be a rectangle if the body is not allowed to rotate. The length of compensation area in lateral direction depends on the position of support legs in lateral stroke, while the length in longitudinal direction is determined by the maximum distance covered by moveable weight. The size of compensation area may affect the extent of compensation to the stability margin. A reasonable size for compensation area is needed to maintain machine's static stability on any support pattern permitted by the reach of support leg. For NTU-Q1, if legs are assumed to have only longitudinal motion and all support points on the same side are on central line of lateral stroke, then the required compensation area is determined by (see Fig. 2)

$$C_y = R_y$$
$$C_x = R_x - \frac{C_y}{W} P.$$

(1)

According to the parameter value given in Fig. 2, the adjusting distance of CG in longitudinal axis, C_x, is about 20.2 cm. Some preliminary simulations have shown that such an adjusting distance of CG is acceptable in the mechanism design aspect.

As long as appropriately adjusting CG position in this compensation area, the machine will be able to keep its static stability at any instance. Figure 3 shows several examples of how to adjust CG position in compensation area to maintain the static stability of machine.

3 Terminology

The *support state* [7] of a K-legged locomotion system is actually the state of all legs touching or lifting the ground. A support state is called *F-state* if its all legs are supporting leg, otherwise, called *T-state*. The shifting of support state from one to another is caused by the occurring of at least one event. An *event* [3] of a gait is the placement or lift of a leg. An entire gait always consists of a finite number of support states corresponding to the events occurring. To take the sequence of a support state into account, an integer number n is used for each support state according to the time order. If the location of a support state in the series formed by all support states is n, then this support state will be denoted as S(n). The time length occupied by a support state is called the *state period*. A state period associated with S(n) can be denoted as $\nabla t(n)$. The moment of shifting from S(n-1) to S(n) is denoted as $t^-(n)$, likewise, the moment from S(n) to S(n+1) as $t^+(n)$, both of which are measured from the origin time.

Three frame-systems, namely, *global-frame*, *body-frame* and *trace-frame* are used to describe the machine's motion. The global frame is fixed on the ground, used to defined the motion of body-frame. The body frame is attached on the body chassis and used to depict leg motion relative to body. While trace-frame, which is fixed on the compensation area, is used to describe CG adjustment relative to compensation area. During the period of a support state, the body of machine may move or rest. A *body motion state* [11] is defined as 1 if the body has moved, and as 0 if the body has not moved. The *locomotion state* of a machine in any time is possibly one of four cases, (T, 0), (T,1), (F,1), and (F,0). If the locomotion state of S(n) is (F,0), such a S(n) is called *transitional state* which will be used in our research for CG adjustment in some cases. The *foot condition* of a state is the position of the support feet respect to the body-frame, which can also be presented by the *kinematic margins* [6] of the feet. Kinematic margin of a leg is the length of a vector opposite motion direction from current support point to the intersection with the reachable area boundary. In this paper, $K_i(n)$ is used to denote the kinematic margin of leg i in S(n). It is clear that foot condition in different instance of this state will be different if body motion state is 1 in a support state. The foot condition at the moment of the beginning of a support state is called *initial foot condition*, which determines the locomotion characteristics of the support state.

Both body speed and swing speed of transfer leg have direct effect on the shifting of support states. To consider these effect, two parameters $\beta(n)$ and $\alpha(n)$, also called β-factor and α-factor respectively, are defined by

$$\beta(n) = \frac{v_s}{v_s + v_b(n)}, \qquad \alpha(n) = \frac{v_b(n)}{v_s + v_b(n)} \tag{2}$$

where v_s is swing speed which is given by system specification, $v_b(n)$ is body speed in S(n). Swing speed here is assumed a constant specified in advance while body speed may be changed in different support state. For a given task, it is often useful to specify a desired β-factor in advance, this desired β-factor can be denoted by β. Since it is assumed $v_b(n) \geq 0$, therefore $v_s + v_b(n) \geq 0$. The following relation about $\beta(n)$ and $\alpha(n)$ are always hold in a support state

$$1 \geq \beta(n) > 0, \quad 1 > \alpha(n) \geq 0, \quad \alpha(n) + \beta(n) = 1. \tag{3}$$

It can be proven that, if the sequence of support state is periodic, $\beta(n)$-factor in quantity will equate to the duty factor defined in periodic gait.

The following terminology is also helpful for understanding of this paper. The *support pattern* [7] associated with a given support state is the convex hull (minimum area convex polygon) of the point set in a horizontal plane which contains the vertical projections of the feet of all supporting legs. *Stroke of a leg* [11], denoted by R_x, is a length of reachable area of this leg in the motion direction. *Foothold* [6] is a permitted cell which lies in the reachable area of a swing leg.

4 Improved adaptive gait planning

The problem of how to use the concept of stability compensation to plan the adaptive gait for NTU-Q1 to travel over rough terrain will be addressed in this part, which includes the rule of selection of transfer leg and locomotion state, the way to select foothold and the strategy of CG adjustment.

4.1 Rules of selecting transfer leg and support state

The selection of swing leg is based on the initial foot condition, and selecting process follows the rules:

Rule 1: the leg with minimum kinematic margin will always first be selected as transfer leg.

Rule 2: a leg can't be selected twice successively

Rule 3: if there are two or more legs with minimum kinematic margin, the leg that, when lifted, machine will have the largest stability margin should be selected as swing leg.

Rule 4: if above rule can't still determine swing leg, then the sequence of swing leg, 1-4-2-3, should be used, which possess largest stability margin in period gait [7].

Rule 1, also referred as minimum kinematic margin principle in [1], means that if only one leg has minimum kinematic margin, then this leg will be selected as transfer leg. Rules 2 through 4 give the way of how to determine transfer leg when there are two or more legs with minimum kinematic margin.

Having determined the transfer leg, it is still needed to consider whether this swing leg is lifted immediately, which result in a T-state, or this leg is lifted after a period, which result in an F-state. The way for generation of support state here is given by following rules.

Rule 5: if support state tends to be static unstable when selected leg is lifted, then an F-state should be used.

Such an F-state is actually a transitional state. During the period of this transitional state, CG of the machine will be adjusted to a suitable position so as to guaranty the selected swing leg can be lifted.

Rule 6: if the minimum value of kinematic margin among all legs is zero, then a T-state should be used.

For the case which minimum kinematic margin is greater than zero, both T-state and F-state can be used for machine locomotion. However, it is often desired to suitably use the characteristics of two types of support state for the purpose of keeping $\beta(n)$ of desired β so as to make the motion of machine more "smooth". If the kinematic margins of all legs at instance $t'(n)$ are arranged in order from the smallest to the biggest and denoted by $K(1,n)$, $K(2,n)$, $K(3,n)$ and $K(4,n)$, following rule can be used for the above purpose.

Rule 7: If $K(2,n) > \dfrac{[R_x - K(1,n)]\alpha(n)}{\beta(n)}$ and $K(1, n) > 0$, then an F-state should be used, otherwise a T-state should be used.

As will be seen in part 6, above rules can generally result in machine achieving a better locomotion behavior when walking over rough terrain.

4.2 Foothold Selecting

The underlying logic for foothold selection here is to find a region in which all feasible footholds satisfy the constrains requirement. After the region is determined, the suitable foothold will be selected from this region according to some optimum rules.

4.2.1 Geometry constraint

Geometry constraint requires that all feet of the machine must be placed within respective reachable area. To determine the foothold region satisfying the geometry constraint, the maximum distance covered by the transfer leg should be determined. Clearly, if without the motion of body, the maximum distance covered by swing leg in S(n), $C_g(n)$, is R_x-$K(1,n)$, where R_x is the stroke of the swing leg. $C_g(n)$ may increase if having the aid of body motion. Denote the maximum distance body can translate in S(n) as $D(n)$, this distance, in geometry, should be equal to the minimum kinematic margin of support legs of S(n), i.e., $D(n)=K(2, n)$. Therefore, with the aid of body motion the maximum distance the swing leg can cover should be $C_g(n) = R_x$-$K(1, n)$+$K(2, n)$. Notice that $K(1, n)$ and $K(2,n)$ should be measured at the instance $t'(n)$.

4.2.2 Kinematic constraint

When the legged system walk at a desired β specified in advance, the following cases may appear: 1) Swing leg reaches the boundary of its reachable area, while the support legs don't. 2) Support leg with kinematic margin $K(2,n)$ reaches the bounder of its reachable area, while transfer leg doesn't. The time through which transfer leg reaches its boundary is $(R_x$-$K(1,n))/v_s$ and the time body covers the distance of $D(n)$ is $D(n)/v_b$, which is also the time that the support leg with $K(2, n)$ reaches its boundary. It is obvious that if $D(n)/v_b > (R_x$-$K(1, n))/v_s$, then transfer leg reaches boundary earlier than support leg, thus first case appears; if $D(n)/v_b < (R_y$-$K(1, n))/v_s$, then the support leg reaches boundary earlier than the transfer leg, second case appear; if $D(n)/v_b = (R_y$-$K(1, n))/v_s$, both of transfer leg and support leg will reach their boundary at the same time. In any case, to satisfy the kinematic constraint, the maximum distance the swing leg can cover in S(n) is determined in the form

$$C_k(n) = \min\{(R_x - K(1,n) / \beta(n), D(n) / (1 - \beta(n))\} \tag{4}$$

Obviously, $C_k(n) \le C_g(n)$. if appropriately adjusting the value of $\beta(n)$, $C_k(n)$ will be able to get its maximum value $C_g(n)$. It can be found in equation (4) that if let $(R_x - K(1,n)) / \beta(n) = D(n) / (1 - \beta(n))$,

$$\text{or } \beta(n) = \beta_m(n) = \frac{K(2,n)}{R_x - K(1,n) + K(2,n)}, \tag{5}$$

then $C_k(n)$ will be equal to $C_g(n)$, i.e., R_y-$K(1, n)$+$K(2,n)$. At this point both the support leg and the swing leg will reach their boundary simultaneously. It means that as long as $\beta(n)$ is adjusted to $\beta_m(n)$ in S(n), the machine will achieve maximum $C_k(n)$.

4.2.3 Stability constraint

Static stability requires the CG of the machine located within support polygon or the stability margin of the machine positive at any time. Since the backward motion of machine is forbidden, only the *front stability margin* (S_l) needs to be considered in foothold selection. It has been proven that the front stability margin in the period of a support state is monotone function and the minimum value occurs at the end of that support state [4]. It implies that if the front stability margin of S(n) is S_l, then under the consideration of static stability, the maximum distance body can translate in S(n) will be S_l. Thus the maximum distance transfer leg can cover under the condition of the stability constraint, $C_s(n)$, can be determined using the same way as that in section 4.2.2 in the form

$$C_s(n) = \min\{(R_x - K(1,n)) / \beta(n), S_l / (1 - \beta(n))\}. \tag{6}$$

Comparing $C_s(n)$ with $C_k(n)$, the maximum distance that transfer leg can cover under kinematic constrain and stability constrain, $C(n)$, should be the lesser one which can be expressed as

$$C(n) = \min\{(R_x - K(1,n)) / \beta(n), \min(S_l, D(n)) / (1 - \beta(n))\}. \tag{7}$$

Likewise, the value $\beta_m(n)$ can also be obtained by

$$\beta_m(n) = \frac{\min(S_l, D(n))}{R_x - K(1,n) + \min(S_l, D(n))}. \tag{8}$$

If without the compensation of stability margin, S_l will passively depend on the support polygon. It is possible that S_l is very small in some support states if terrain is very rough. At that case, $C(n)$ may decrease even become zero, thus terrain adaptivity of machine will be reduced, the deadlock case may occur. If there is compensation of stability margin, S_l will be independent on the support polygon. The value of S_t can be increased through adjusting CG position within the compensation area.

$C(n)$ defines a range (or region) in which any a feasible foothold can be selected as a support point for next state $S(n+1)$. Such a defined range is also called *foot range*.

4.2.4 Foothold Selection

In general, we select suitable foothold based on the such a rule: the feasible foothold wit maximum distance from last support point will always be selected as next support point for transfer leg. It means if forbidden zone totally locates inside of foot range, then the foothold wit a distance of $C(n)$ will be selected as support point, if the forbidden zone crosses the from boundary of foot range, then the foothold nearest to forbidden zone will be selected as suppo point. This rule tends to make the swing leg get maximum ability of obstacle avoidance.

Once the support point is selected based on the foot range, the real distance covered by body can be finally determined. If the distance from the last support point to selected foothold is $c(n)$, then the state-period of $S(n)$ will be

$$\nabla t(n) = c(n) / (v_s + v_b(n)) = c(n)\beta(n) / v_s. \tag{9}$$

The real distance covered by body is $d(n)=c(n)\alpha(n)$ or $c(n)(1-\beta(n))$, and the kinematic margin of the transfer leg in $S(n+1)$ is $c(n)\beta(n)$.

After all parameters of $S(n)$ are determined, the beginning time, $t'(n+1)$, can be determined by t' $(n+1)= t'(n)+ \nabla t(n)$ and the kinematic margin of all leg of $S(n+1)$ at the instance $t'(n+1)$, i.e., state foot condition of $S(n+1)$, is also obtained by.

$$K_i(n+1) = K_i(n)-d(n) \qquad \text{if leg } i \text{ is support leg in } S(n)$$

$$K_i(n+1) = K(1,n)+c(n)\beta(n) \quad \text{if leg } i \text{ is transfer leg in } S(n). \tag{10}$$

4.3 Strategy of CG adjustment

Static stability of a $S(n)$ is heavily associated with its support pattern. To clearly address the strategy of CG adjustment, represent the support pattern at the beginning of $S(n)$ as $P'(n)$ and the support pattern at the end of $S(n)$ as $P^+(n)$. Here $P'(n)$ and $P^+(n)$ also represent the area covered by them when observed from body frame. Denote the location of CG projection relative to the trace frame at instance $t'(n)$ as $g'(n)$, at $t^+(n)$ as $g^+(n)$, and assume the support state at $t'(n)$ is stable. In addition, denote the intersection area of support pattern $P^+(n)$ and $P'(n+1)$ as A, the intersection area between $P^+(n)$ and compensation area, $\Re$, as B, and the intersection area between $P'(n+1)$ and $\Re$ as C, the relationship of which can be written in mathematics by

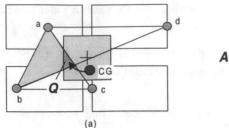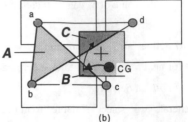

(a)   (b)

Fig. 4. Schematic diagram for the strategy of CG adjustment, where a-b-c is support pattern at $t^+(n)$ and a-b-d is support pattern at $t^-(n+1)$. (a) the case in which $Q \neq 0$, (b) the case in which $Q = 0$.

$$A = P^+(n) \cap P^-(n+1)$$

$$B = P^+(n) \cap \Re \qquad\qquad (11)$$

$$C = P^-(n+1) \cap \Re.$$

It is clear that if $g^+(n) \in B$, then support state S(n) is stable at $t^+(n)$, if $g^+(n) \in C$, then S(n+1) at instance $t^-(n+1)$ is stable, if $g^+(n) \in A$, then the shifting from S(n) to S(n+1) is stable. Therefore to guaranty static stability of S(n) and the shifting from S(n) to S(n+1), the CG location at $t^+(n)$ should satisfy

$$g^+(n) \in Q = A \cap B \cap C = P^+(n) \cap P^+(n+1) \cap \Re \qquad\qquad (12)$$

where Q is the intersection area among A, B and C. Since we have assumed S(n) at $t^-(n)$ stable, if $g^-(n) \in Q$ and let $g^+(n) = g^-(n)$, then CG can satisfy stability requirement during period of S(n) without any adjustment. However, if $g^-(n) \notin Q$, CG location must be adjusted to $g^+(n)$, a new position satisfying equation (12).

It should be noted that, in the terrain containing a large amount of forbidden zone, it is possible that $Q = 0$. In that case, it is necessary to insert a *transitional support state* S([n]), Since S([n]) is with full support legs, the support pattern P([n]) covers $P^+(n)$, $P^-(n+1)$ and $\Re$. Thus we can get

$$g^+(n) = g^-([n])$$
$$\qquad = P^-([n]) \cap P^+(n) \cap \Re = P^+(n) \cap \Re$$
$$g^+([n]) = g^-(n+1) \qquad\qquad (13)$$
$$\qquad = P^+([n]) \cap P^-(n+1) \cap \Re = P^-(n+1) \cap \Re.$$

It means that if $Q = 0$, CG should be first moved to a position $g^+(n)$ within B, then through transitional state S([n]) to final position $g^+([n])$ within C. The state period of S([n]) should be determined according to the moving speed of movable weight.

The strategy of CG adjustment can be summered as follows. First look for the area Q, B and C. Second, if $Q = 0$, judge whether $g^-(n)$ is located inside the interior of Q, If true, then let $g^+(n) = g^-$

(n), that means no any adjustment is needed; otherwise find a point as $g^+(n)$ in Q, and then move CG from $g^-(n)$ to $g^+(n)$. Third, if $Q = 0$, then judge whether $g^-(n)$ is inside B, if true, $g^+(n) = g^-(n)$, otherwise, find a point as $g^+(n)$ within B, move CG from $g^-(n)$ to $g^+(n)$, and then add a transitional state following S(n), find a point in C as $g^-(n+1)$, move CG from $g^+(n)$ to $g^-(n+1)$ during transitional state. Obviously, $g^+(n)$ and/or $g^-(n+1)$ may be not unique. Figure 4 shows the schematic picture of CG adjustment strategy.

5 Simulation Results

The program for adaptive gait generation has been written in Maple V Language (release 3) and implemented on a PC486 computer. The animation is conducted using UNIX system and displayed on SUN-station through a commercial software package DADS (release 8), which also helps to do dynamic and kinematic analysis of the prototype. The same simulating model as that of NTU-Q1 shown in Figure 1 is used and only the locomotion with zero crab angle is considered in the simulation. This program allows the user to set up the terrain data, select desired β-factor and/or vary initial foot condition according to requirement to test the terrain adaptivity of prototype.

Figure 5(a)-(j) shows ten representative phases of a simulation experiment with swing speed of transfer leg, $v_s = 25$ cm/s, and desired β-factor $\beta = 0.8$. Here, the transfer leg is indicated with an arrow near that leg. It can be seen from Figure 5(c), (e), and (f) that the negative stability margin or deadlock situation would occur at time t_1 (11.9s), t_2(15.3s) and t_3(19.5s) if without compensation of stability margin. The trajectories of CG adjustment relative to float-frame in longitudinal and lateral direction are shown in Figure 6.

Many different terrain conditions have been investigated with the program. It is noteworthy that as long as the maximum width of forbidden zone is less than stroke of reachable area, no deadlock cases have been found even although the machine traverses such a continuous forbidden zone as ditch. It is also been found that if reducing the size of forbidden zone or increasing the value of β-factor in the program, the continuity of body motion will be improved. If the terrain becomes perfect, the program will automatically produce a periodic gait for locomotion of the machine. These features are highly desired in the control of system.

6 Conclusions

This paper has presented an alternative method for the improving of the terrain adaptivity of a multi-legged machine traveling over rough terrain. It is done through 2-D CG adjustment intending to compensate stability margin according to the terrain condition. This method determines foothold based on the current support pattern rather then over the whole trajectory.

Due to the aid of stability compensation, such determined foothold is capable of maintaining the machine static stability during current support state and the shifting from current support state to the proceeding support state. Therefore this way can easily help machine to avoid deadlock case unless the forbidden zone is larger than its reachable area. Some simulation works have been implemented to test the effectiveness of the proposed method.

The concept of stability margin compensation has been applied to the design of the quadruped robot NTU-Q1. Since the moveable weight provides a way to control stability margin actively, it is believed that NTU-Q1 will possess a larger degree to accommodate rough terrain during locomotion.

It should be pointed out that the trajectory of CG adjustment is not unique. Although a feasible CG trajectory has been given in this paper, apparently, it may not be the most optimum one. Further research is therefore necessary in finding the strategy to achieve the most optimum trajectory of CG adjustment based on the given criteria such as minimum adjusting distance and smoothness.

References

1. Y. Ding, and E. Scharf, "Deadlock avoidance for a quadruped with free gait," Proc. of IEEE Conf. Automation and Robotics, (1994), pp. 143-148.

2. S. Hirose et al., "A study of design and control of a quadruped walking vehicle," Int. J. of Robotics Res., vol. 3, no. 2, (1984), pp. 113-133.

3. E. I. Kugushev and V. S. Jaroshevskij, "Problems of selecting a gait for an integrated locomotion robot," Proc. Fourth Int. Conf. Artificial Intelligence Tbilisi, Georgian SSR, USSR, (1975), pp. 789-793.

4. Y. J. Lee and Z. Bien, "A hierarchical strategy for planning crab gaits of a quadruped walking robot," Robotica (1994) vol. 12, pp. 23-31.

5. D. A. Messuri and C. A. klein, "Automatic body regulation for maintaining stability of a legged vehicle during rough-terrain locomotion," IEEE Journal of Robotics and Automation. vol. RA-1, no. 3, Sept. (1985), pp. 132-141.

6. R. B. McGhee and G. I. Iswandhi, "Adaptive locomotion of a multilegged robot over rough terrain," IEEE Trans. System, Man and Cybernetics, vol. SMC-9, no. 4, (April 1979), pp. 176-182.

7. R. B. McGhee and A. A. Frank, "On the stability properties of quadruped creeping gaits," Math. Biosci., 3, 331-351, (1968).

8. P. V. Nagy and S. Desa, "Energy-based stability measures for reliable locomotion of statically stable walkers: theory and application," The Int. Journal of Robotics Research, vol. 13, no. 3, (June 1994), pp. 272-287.

9. P. K. Pal and K. Jayarajan, "Generation of free gait--a graph search approach," IEEE Tran. Robotics. Autom. vol. 7, no. 3, (June 1991), pp. 299-305.

10. J. Stoner and R. H. Davis, "A gait simulator for a quadruped walking robot," Robotica (1992) vol. 10, pp. 57-64.

11. S. M. Song and K. J Waldron, *Machines That Walk: The Adaptive Suspension Vehicle*, The MIT Press, Cambridge, MA, (1989).

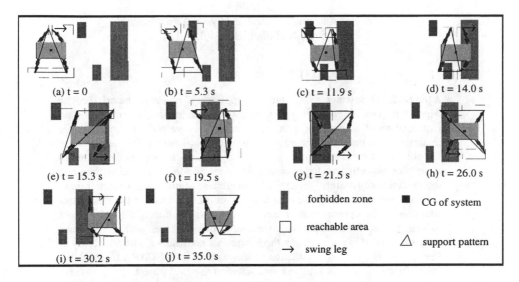

Fig. 5. Some representative phases when simulation model of NTUQ-1 walks over a typical rough terrain

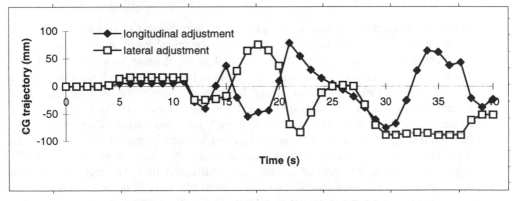

Fig. 6. CG adjustment trajectory in longitudinal and lateral direction

Application of Pareto-Based Multiobjectives Genetic Algorithm in Minimum Time Motion Planning

Ang Mei Choo

B.Sc (Maths.), M.Sc (Eng. in Control Systems)
Dept. of Industrial Computing, Faculty of Information, Science and Technology
National University of Malaysia, 43600 Bangi, Malaysia.

Dr. A.M.S. Zalzala

BEng, PhD, MIEEE, AMIEE
Dept. of Automatic Control and Systems Engineering
University of Sheffield, S1 3JD, UK

Abstract. This paper looks into the application of Pareto-based Genetic Algorithm (GA) in obtaining the minimum time motion planning for an industrial robot. A common practice in multiobjective optimisation GAs for minimum time motion planning is to apply classical aggregation approach to the objective formulation while in this study, Pareto-ranking method is used. A suitable objective vector is organised to include the total travel time and the two joint constraints: velocity and acceleration limits. The objective is to obtain a optimal motion with minimum travel time and within the kinematics limitations. The optimisation process involves first producing a fixed number of joint displacements using the genetic operators, and then scaling the travel time such that it is not violating the kinematics constraints. The feasibility of this method is shown by simulation results with an RTX SCARA robot. Cubic spline functions are used in the construction of the joint trajectory.

1. Introduction

Minimum time motion planning for a industrial manipulator is an important subject in the area of robotics. However, most multiobjective optimisation GAs used in minimum time motion planning apply classical aggregation approach to the objective formulation (see for e.g., [8, 9, 11]). This paper is intended to develop a Pareto-based multiobjective genetic algorithm, which has recently being proposed [4, 5], for the minimum time motion planning of an RTX robot (with six joints). The travel distance and kinematics characteristics of each joint are varied. Thus, to obtain a feasible minimum motion time, the combination of the kinematics characteristics for their joints such as velocities, accelerations and jerks have to be optimised. The objective vector in this project includes the total travel time, the criticality to joint velocity and acceleration limits. The GA motion planning is using the cubic spline theory [7] to construct the joint trajectories. The method allows manipulator motion time, velocity, acceleration and jerk to be scaled such that the kinematics constraints are met. Note that the path planning is carried out in joint space and only kinematics constraints (joint velocities, accelerations and jerks) are considered in this studies. This paper is organised as follows: Section 2 explains the application of cubic spline joint trajectories and the time scaling method. Section 3 incorporates the parameters obtained in Section 2, i.e. optimal motion time, the criticality to velocity and acceleration limits, into the objective function formulation. Section 4 gives the Pareto-based GA motion planning. Section 5 shows the results when applying this method. Section 6 gives the discussions and conclusions.

2. Cubic Spline Joint Trajectories and Time Scaling Method

Each joint trajectory is fitted to a number of joint displacements at a sequence of time instants by using piecewise cubic polynomials. Let $t_1 < t_2 < t_3 < t_4 < ... < t_{n-2} < t_{n-1} < t_n$ be an ordered time sequence and the position (or knot) of the jth joint at time $t = t_i$ is $\theta_{ji}(t_i)$. Thus, the vector of knots for the jth joint along the path is given as $[\theta_{j1}(t_1), \theta_{j2}(t_1),...,\theta_{jn}(t_n)]$. The interval time is defined as $h_i = t_{i+1} - t_i$ $(i = 1, 2,..., n-1)$ whereas the velocity and acceleration of joint j at knot i are denoted as v_{ji} and w_{ji} respectively. The cubic spline joint trajectories allow the joint velocities, accelerations and jerks to be evaluated at every instant of motion time [7]. Every initial motion time will be tested such that it is optimal and the motion will not violate the kinematics constraints. The intervals must be scaled up or down depending on the scaling factor in equation (4). The knot velocities, accelerations and jerks should be compared with their own limits to obtain the time-optimal path satisfying the joint constraints. Let the absolute values of the jth joint velocity, acceleration and jerk limits denoted as VC_j, WC_j, and JC_j respectively. The scaling factor λ can be obtained as follows:

$$\lambda_1 = \max_j \left[\max_{i = 1,...,n} \left\{ |v_{ji}|, |v_{ji}(t_i)| \right\} / VC_j \right] \tag{1}$$

where $\bar{t}_i$ satisfies $w_{ji}(\bar{t}_i) = 0$ and is in the interval $[t_i, t_{i+1}]$,

$$\lambda_2 = \max_j \left[\max_i |w_{ji}| / WC_j \right] \tag{2}$$

$$\lambda_3 = \max_j \left[\max_i \left| \frac{w_{j,i+1} - w_{ji}}{h_i} \right| / JC_j \right] \tag{3}$$

$$\text{and } \lambda = \max\left(\lambda_1, \sqrt[2]{\lambda_2}, \sqrt[3]{\lambda_3}\right). \tag{4}$$

If the time interval h_i is replaced by λh_i for $i = 1, 2,..., n-1$, then the velocity, acceleration and jerk will be replaced by factors of $1/\lambda, 1/\lambda^2, 1/\lambda^3$, respectively. These changes assure the satisfaction of constraints on velocities, accelerations and jerks [7].

3. Objective Formulation

The objective vector following from the definition of Pareto optimality [13] would look like this:

$$\text{Minimise: } (\sum_{i=1}^{n-1} h_i, \ 1-\lambda_1, \ 1-\lambda_2) \tag{5}$$

$$\text{subjects to constraints: } \lambda_3 \leq 1,$$

where h_i is the time interval i, $1-\lambda_1$ is the criticality to the velocity constraints, $1-\lambda_2$ is the criticality to acceleration constraints. The parameters: λ_1, λ_2 and λ_3 are computed by equations (1), (2) and (3) respectively. Criticality is a measurement of a trajectory on how close it is to the joints' velocity and acceleration limits.

4. Pareto-based GA Motion Planning

The schematic diagram for GA minimum time motion planning is illustrated as follows:

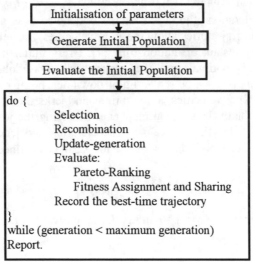

4.1 Generate Initial Population

From the knots vector described in Section 2, the path is encoded directly as string of floating point to be used by the GA as $[\theta_{11}, \theta_{12}, ..., \theta_{1n}; \theta_{21}, \theta_{22}, ..., \theta_{2n}; ...; \theta_{m1}, \theta_{m2}, ..., \theta_{mn}]$. The knot values of each joint are generated randomly using a transition scheme. The joint is restricted to move in only 5 directions from any knot. To create the initial population, two trajectories, one begins from start position and the other one begins from end position will be generated. The trajectories start from the end position will have higher tendency to move downward whereas the other one will have higher possibility to move upward. A valid trajectory will be created once they meet at a particular point. The travel time between intermediate knots is initially set to their lower bounds for each joint, according to the formula as follows:

$$(h_1, h_2, h_3, \ldots, h_{n-1}) = \left(\max_j \frac{|\theta_{j2} - \theta_{j1}|}{VC_j}, \max_j \frac{|\theta_{j3} - \theta_{j2}|}{VC_j}, \ldots, \max_j \frac{|\theta_{j,n} - \theta_{j,n-1}|}{VC_j} \right) \tag{6}$$

and then the time scaling method will convert the time into a feasible one so that the trajectories will not exceed the limits on all the six joints

4.2 Evaluate The Population

All trajectories are ranked based on its total travel time, the criticality to the joints' velocity and acceleration limits (see Section 3) based on Pareto ranking. According to the procedure proposed by [4], an individual in a population can be ranked by counting the number of individuals that dominate it. When all individuals are ranked, the fitness values will be assigned to them according to their rank. In this paper, the fitness assignment is done by interpolating a linear function [12] from the best individual (rank = 0) to the worst individual (rank<maxpop). Following that, same rank individuals will receive the same fitness values by averaging the total values assigned to them. Fitness sharing uses a sharing parameter h to control the extend of sharing, which is a measurement of the maximum distance between individuals that could form niches. It is computed as follows [4]:

$$h = [8C_n^{-1}(n+4)(2\sqrt{n})^n / N]^{1/(n+4)} \tag{7}$$

where n is the number of decision variables, and N is the population size. It is also known as smoothing parameter for the Epanechnikov kernel [10]. Each trajectory's share count is set to zero initially, and then it is incremented by a certain amount for every trajectory in the population, including the trajectory itself. Raw fitness value is then recalculated by dividing the values with the total share count. The share count can be obtained by using the Epanechnikov kernel [4]. However, the calculation of Epanechnikov kernel function will require the inverse of the sample covariance matrix and also the determinant of the covariance matrix [10]. To avoid the problem of matrix singularity, an approximation is implemented and the share count function will look like the following [6]:

$$sh(d^*) = \begin{cases} 1 - (d^* / h)^2 & \text{if } d^* / h < 1 \\ 0 & \text{otherwise} \end{cases} \tag{8}$$

where d^* is computed such that each decision variable x_i is weighted by its variance. The variance is calculated as follows:

$$\sigma_i^2 = \frac{N\left(\sum_{j=1}^{N} x_{ij}^2\right) - \left(\sum_{i=1}^{N} x_{ij}\right)^2}{N(N-1)} \tag{9}$$

where N is the number of observations or in other word the population size and d^* is computed as:

$$\left(d^*\right)^2 = \sum_{i=1}^{n} \frac{(x_i - y_i)^2}{\sigma_i^2}. \tag{10}$$

4.3 Selection Scheme And Mating Restriction

Selection scheme is a process to determine the number of trials a particular individual is chosen for reproduction. The selection technique adopted in this project is based on stochastic universal sampling (SUS) introduced by [1]. This method uses a single spin and N equally spaced pointers, where N is the number of population size. Arbitrary recombining pairs of trajectories may produce new offspring that do not represent any niche [3]. It is therefore desirable to reduce crossover between trajectories of different niches. A simple mating restriction scheme is implemented by setting the mating parameter to be equalled to the sharing parameter [3]. The mating parameter is actually the measurement of maximum distance between individuals (trajectories) that allow them to be paired for recombination. Thus, if an individual (trajectory) is within a distance of mating parameter, then a mating companion is found and mating can be performed, otherwise another individual (trajectory) is tried.

4.4 Genetic Recombination

The selected trajectories will be paired up for crossover or recombination subject to their mating distance (i.e. mating restriction) and cross-over probability. A robot trajectory consists of joint angles that may produce large position jump in the offspring strings after conventional crossover. To tackle this problem, a customised genetic operator named as path redistribution and relaxation operator is used [9]. The technique involves fitting cubic splines onto the offspring's knots with each time interval set to one. The path length is then computed as the Euclidean distance between the start and end knots along the splines. Each joint knots are then 'redistributed' evenly over these splines at equal intervals. The paths are then relaxed by moving each knots with a small step towards the point that will bisect the line between its neighbouring knots. Single point cross-over is applied in this GA motion planning. A special purpose mutation operator [9] named as injection is implemented where a fixed number of new trajectories are injected into the population and randomly replace selected trajectory. The number of new trajectories is kept low but it has the effect of preventing premature convergence and creating new search space. After the recombination, the parent trajectory motion time is passed on to the child trajectory as the initial time. The child path together with the motion time will be tested by time scaling method to obtain an optimal motion which will not violate the kinematics constraints. The resulted optimal time and the criticality values to joint velocity and acceleration limits will be used to access the performance of the child trajectory.

5. Results

The multiobjective optimisation GA incorporating all the techniques described in the sections above is implemented. In this simulation, the robot is assumed at rest initially, and comes to a full stop at the end of the time interval. In other words, at the initial time $t = t_1$ and the terminal time $t = t_n$, the joint velocity, and joint acceleration are given as: $v_{j1} = 0 = v_{jn}$, and $w_{j1} = 0 = w_{jn}$. The initial and final configurations of the path planning are shown in Table 1. The limits of the velocities, accelerations and jerks are given in Table 2 [2]. The path redistribution-relaxation operator is experimented with different population size, and 0.9 crossover probability [9]. The injection rate is 2.5% of the population size. The results for 100 generations with different population size are shown in Table 3. The motion profiles for 200 population size are given in Figures 1-4.

Table 1. Initial and final configurations for the path planning of RTX robot.

	Column (m)	Shoulder (rad)	Elbow (rad)	Yaw (rad)	Pitch (rad)	Roll (rad)
Initial Configuration	0.4	$-\pi/6$	$-\pi/3$	$-\pi/2$	0	$-\pi/4$
Final Configuration	0.8	$\pi/6$	$\pi/3$	$\pi/2$	$-\pi/6$	$\pi/4$

Table 2. Velocity, acceleration and jerk constraints for RTX robot.

	Column	Shoulder	Elbow	Yaw	Pitch	Roll
Velocity	0.1116	0.1654	1.2092	1.9715	1.3780	1.2412
Acceleration	1.7755	6.2018	14.081	31.055	28.063	26.180
Jerk	297.59	894.67	3718.9	3377.6	3933.1	4172.7

Note: The zed velocity, acceleration, and jerk are in m/s, m/s², and m/s³, respectively. The other joint angle velocities, accelerations and jerks are in rad/s, rad/s², and rad/s³, respectively.

Table 3. Results From GA Minimum Time Motion Planning

Population Size	Minimum Time(sec)
100	3.9530
200	3.9335
300	3.9049

6. Discussions And Conclusions

Pareto-based multiobjective GA involves ranking process of non-dominated solution to generate the joints trajectories using cubic spline function can produce a optimum path which results in minimum time. This result of minimum motion time improves as the population size increases. However the processing time in obtaining results also increases. Therefore an appropriate population size has to be selected to obtain a reasonably minimum motion time with acceptable length of processing time.

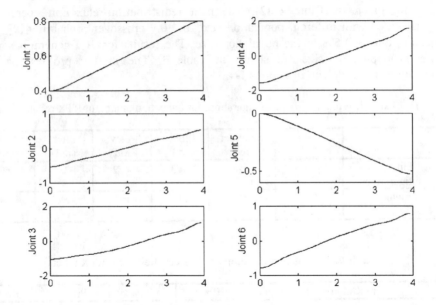

Fig. 1. Position vs. Time

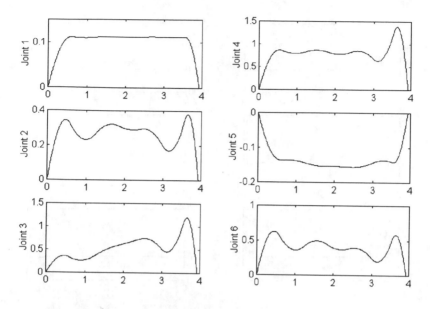

Fig. 2. Velocity vs. Time

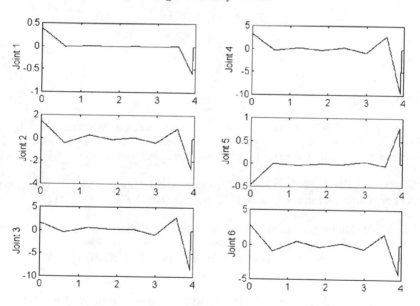

Fig. 3. Acceleration vs. Time

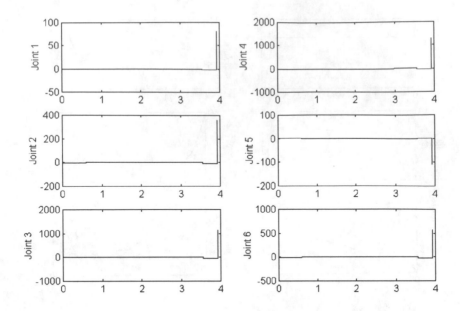

Fig. 4. Jerk vs. Time

References

1. Baker, J. E. (1987). Reducing bias and inefficiency in the selection algorithm, *Proceedings of the Second International Conference on Genetic Algorithms*, pp. 14-21 Grefenstette, J. J. ed., Lawrence Erlbaum Associates, Publishers.

2. Cao, B., and Dodds, G. I. (1994). Time-Optimal and smooth joint path generation for robot manipulators. *Int. Conf. CONTROL '94*, (Coventry, UK), pp. 1122-1127.

3. Deb, K and Goldberg, D. E. (1989). An Investigation of Niche and Species Formation in Genetic Function Optimization. *Genetic algorithms: Proceedings of the Third International Conference on Genetic Algorithms*, pp. 42-50. Schaffer, J. D. ed., Morgan Kaufmann.

4. Fonseca, C. M. and Fleming, P. J. (1995a). Multiobjective genetic algorithms made easy: selection, sharing, and mating restriction. *First International Conference on 'Genetic Algorithms in Engineering Systems: Innovations and application', GALESIA '95*, (Sheffield, UK, 1995), pp. 45-52.

5. Fonseca, C. M. and Fleming, P. J. (1995b). An Overview of evolutionary algorithms in multiobjective optimisation. *Evolutionary Computation.*, **3**, No. 1, pp. 1-16.

6. Fonseca, C. M. (1996). Personal communication.

7. Lin, C.-S., Chang, P.-R. and Luh, J. Y. S. (1983). Formulation and optimisation of cubic polynomial trajectories for industrial robots. *IEEE Trans. Automatic Control*, **AC-28**, No. 12, pp. 1066-1073.

8. Rana, A. S. and Zalzala, A. M. S. *An Evolutionary Algorithm for the Collision Free Motion of Multi-Arm Robots*. Research Report 570, Depart. of ACSE, University of Sheffield, UK, April, 1995.

9. Rana, A. S. and Zalzala, A. M. S. (1996). *Minimum Time Motion Planning of the RTX Robot Using an Evolutionary Algorithm*. Research Report 620, Depart. of ACSE, University of Sheffield, UK, April 1996.

10. Silverman, B. W. (1986). *Density Estimation for Statistics and Data Analysis*. London: Chapman and Hall.

11. Wang, Q. and Zalzala, A. M. S.(1995). *Genetic Optimisation and Experimentation for the PUMA 560 Manipulator*. Research Report 567, Depart. of ACSE, University of Sheffield, UK, April 1995.

12. Whitley, D. (1989). The GENITOR algorithm and selection pressure: why rank-based allocation of reproductive trials is best. *Genetic algorithms: Proceedings of the Third International Conference on Genetic Algorithms*, pp. 116-121. Schaffer, J. D. ed., Morgan Kaufmann.

13. Yacov, Y. H., Tarvainen, K, Shima, T. and Thadathil, J. (1990). *Hierarchical Multiobjective Analysis of Large-Scale Systems*. New York: Hemisphere Publishing Corporation.

NC Programming

A Curvilinear Grid Generation Technique Applied to Tool-Path Planning of a Five-Axis Milling Machine

D. Batanov*, E.Bohez*, S.S. Makhanov**,
K. Sonthipaumpoon** and M. Tabucanon*

*School of Advanced Technologies,
Asian Institute of Technology, Bangkok 12120, Thailand
**Faculty of Information Technology,
King Mongkut's Institute of Technology, Bangkok 10520, Thailand

Abstract. We propose a new algorithm to correct tool-paths of a five-axis milling machine based on elliptic grid generation. Our weighting function is represented in terms of errors related to nonlinear kinematics of the milling machine. Numerical experiments reveal that the proposed technique provides a significant increase in the accuracy of milling.

1 Introduction

Tool-path planning and optimization of a five-axis milling machine has received considerable attention [1-9]. Recent results have displayed a number of sophisticated methods to optimize a "zigzag", spiral or window frame pattern[1-3]. These methods are often combined with techniques dealing with geometric complexity of a work-piece [4,5]. There exists also a variety of off-line techniques to generate a non-uniform tool-path, for instance, methods based on the Voronoi diagram [6] or on neural network modeling[5]. However, a robust algorithm to generate such complicated patterns is still an open problem. On the other hand, the "structured" zigzag pattern is simple and robust and therefore it has been used in conventional CAD/CAM systems.

Adaptive grid generation techniques constitute an essential tool in various CFD and CAD/CAM applications. (See[10,11] for fairly comprehensive reviews on this topic).

In this paper, we present a technique to optimize the zigzag tool-path based on an adaptive grid refinement. We generate the path in a curvilinear "computational" space adapted to milling errors expressed in terms of some global spatial estimates. The resulting curvilinear grid determines the zigzag tool motion adapted to the error variations.

Finally, numerical experiments demonstrate a significant increase in the accuracy of milling.

2 Mathematical formulation

Let $S \equiv S(u,v) = (x(u,v), y(u,v), z(u,v))$ be a machined surface, where u and v are independent variables. Consider a set of points $\{(u,v)_{i,j}\}$, $0 \le i \le N_\xi$, $0 \le j \le N_\eta$, being a discrete analogy of a mapping from the "computational region" $\Delta = \{0 \le \xi \le N_\xi, 0 \le \eta \le N_\eta,\}$ into the "physical region" $D = \{0 \le u \le 1, 0 \le v \le 1\}$, i.e. the points $(u,v)_{i,j}$ belong to some curvilinear structured grid .

Next, consider the tool-paths $T_{i+0.5,j}(\xi)$, $T_{i,j+0.5}(\eta)$ of a five-axis milling machine between the points $(u,v)_{i,j}$, $(u,v)_{i+1,j}$ and $(u,v)_{i,j}$, $(u,v)_{i,j+1}$, respectively, and let us define the following surfaces:

$$T_\xi(\xi,\eta) = L_\eta(T_{i+0.5,j}, T_{i+0.5,j+1}), \qquad if \ \ i \le \xi \le i+1 \ \ and \ \ j \le \eta \le j+1,$$

$$T_\eta(\xi,\eta) = L_\xi(T_{i,j+0.5}, T_{i+1,j+0.5}), \qquad if \ \ i \le \xi \le i+1 \ \ and \ \ j \le \eta \le j+1,$$

where L_τ corresponds to a linear approximation in the τ-direction, $\tau = \xi$ or $\tau = \eta$.

A quality criterion of a global tool path is formulated in terms of the spatial estimate

$$\omega_\tau = |S(\xi,\eta) - T_\tau(\xi,\eta)|,$$

corresponding to the zigzag motion in the τ-direction.

Let $N_\xi \times N_\eta$ be an arbitrary but fixed number of grid points. We call "an optimal tool path in the τ-direction", a solution of the following optimization problem:

$$\underset{(u,v)_{i,j}}{\mathrm{minimize}} \ E_\tau, \ i = 1, \ldots, N_\xi, j = 1, \ldots, N_\eta,$$

where

$$E_\tau = \underset{\xi,\eta}{\mathrm{max}}\{\omega_\tau\}.$$

Note that the above estimate could be modified in order to involve some additional technological effects such as machining techniques, tool inclination, etc.[8]. However, such modifications do not affect the main results presented in this paper.

Furthermore, observe that the tool path depends on a prescribed policy to calculate the orientation of the tool. Such a policy can be formulated in terms of some supplementary relations between components of the normal vector and the rotation angles.

Therefore, in this section we formulate the problem of tool-path planning in terms of a grid adaptation. As a matter of fact, we have to redistribute a given number of points belonging to a structured curvilinear grid in order to find the best approximation with regard to the maximum-norm. However, the tool path is not linear with respect to ξ or η. Moreover, the derivatives $\partial T_\tau / \partial \xi$, $\partial T_\tau / \partial \eta$ could be discontinuous. Hence, prior error estimates, for instance [12,14], can not be used to optimize the grid.

3 Non-linear kinematics of a five-axis milling machine

In order to calculate the tool-path between successive positions, we first transform part-surface coordinates into machine coordinates. Secondly, the rotation angles and the corresponding machine coordinates are assumed to change linearly between the prescribed points. Finally, we invoke an inverse kinematic equation technique[3], [8]. Observe that inverse kinematics depend upon the structure of a machine. For the machine configuration presented in this paper (Fig.1) the resulting kinematics equations, involving two rotations and three translations, are given by

$$M=B(A(P+R)+T)+C,$$

where M, P, R, T, C are respectively the machine, the part-surface, the rotary table, the tilt table and the cutter center coordinates. A, B are respectively the matrices of rotation about the primary and the secondary axes.

The rotation angles a and b are given by

$$a=\arctan(i/j), \quad b=\arctan((i^2+j^2)^{1/2}/k),$$

where (i, j, k) denotes the normal vector.

The five variables (X, Y, Z, a, b) (where $(X, Y, Z)\equiv M$) provide reference inputs for servo-controllers of the milling machine.

Remarks

A procedure to compute the rotation angles becomes "ill-conditioned" in the regions where $|i + j|< \varepsilon$, or $|k|< \varepsilon$ (ε is a small number), for instance for "flat" surfaces[15]. As a matter of fact, small variations of the normal vector in the above mentioned regions can lead to sharp variations of the rotation angles and to numerical instabilities. Therefore, in order to provide a "continuous" change of the rotation angles, special interpolation procedures are implemented in the regions where the normal vector degenerates. Note also that if the a-angle "jumps" from a small value to 360^0 or vice versa, then the sequence of the angles should be "smoothed" in order to minimize the difference between the successive values.

4 Grid generation technique

Next, we propose to apply a modification of an elliptic grid generator [12,14], [16]. A grid generation problem is formulated in terms of Euler equations for a linear combination of functionals characterizing some desirable properties of the grid, namely

$$I = I_s + \lambda I_v,$$

where

$$I_s = \iint_\Delta [(\nabla \xi)^2 + (\nabla \eta)^2] d\xi d\eta,$$

$$I_v = \iint_\Delta J\omega_\tau d\xi d\eta,$$

and where J is the Jacobian of the mapping. λ is a calibration parameter. Observe that I_s is related to the smoothness of the grid (Winslow-functional[17]), whereas I_v is related to an adaptation of the grid to regions of large milling errors. Fig.2 illustrates a grid adapted to an elliptical region.

In order to compute E_τ for a fixed λ, we implement the following iterative algorithm:

1) Generate the initial grid G.
2) Calculate $\omega_{\tau,i,j}\equiv\omega_\tau(\xi_i, \eta_j)$ and E_τ on G.
3) Scale and smooth the grid-function $\omega_{\tau,i,j}$ [16].

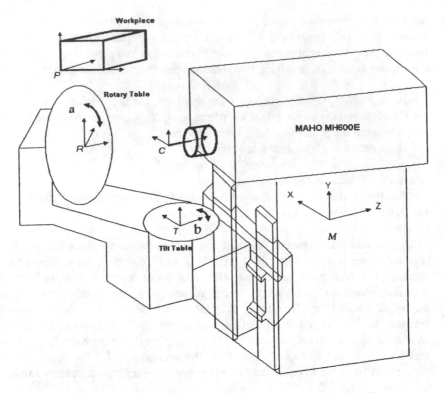

Fig.1 Five-axis milling machine with rotary axes on the table

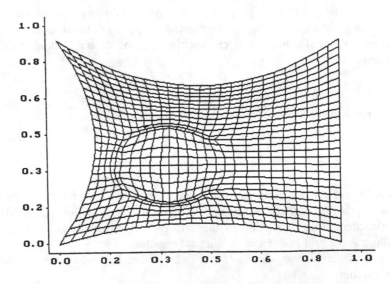

Fig.2 Grid adapted to an ellipse-shaped zone

4) Generate the adapted grid G^{new}, i.e. solve numerically the corresponding Euler equations.

5) Calculate $\omega_{\tau,i,j}^{new}$, E_{τ}^{new} on G^{new}.

6) If $|E_{\tau}^{new} - E_{\tau}| > \delta$ (where δ is some prescribed value) then $E_{\tau} := E_{\tau}^{new}$, $\omega_{\tau,i,j} := \omega_{\tau,i,j}^{new}$ return to step 3.

Note that there exist various numerical strategies to increase the convergence and the stability of the grid generation algorithm(see, for instance, [14]).

Finally, in order to find an "optimal grid", we apply an appropriate modification of the standard one-dimensional search[18]. But observe that the proposed procedure only searches a local minimum. The existence of a global minimum is still an open problem.

5 Application

A large series of surfaces has been tested by means of the proposed algorithm. Our techniques provide a significant increase in accuracy of about 40%.

A typical testing surface is depicted in Fig.3. The corresponding error distribution on the grid 25x25 and the zones of large milling errors are depicted in Fig.4. The grid adapted by means of the weighting function ω_{τ}, is shown in Fig.5. Clearly, the grid adaptation corresponds to error variations. The convergence of our proposed algorithm requires 4 iterations, for a prescribed $\delta=0.01$. The resulting λ-value is 180.5. Table 1, presenting errors for the sequence of nested rectangular and curvilinear grids, demonstrates that an increase of grid-points in 144 times leads to an accuracy increase of about 140 times for the uniform grid and of about 235 times for the adaptive grid.

Conclusions

A new algorithm to correct tool-paths of a 5-axis milling machine based on elliptic grid adaptation has been proposed and tested. Numerical experiments reveal that the proposed technique provides a significant increase in the accuracy of milling.

Finally, we remark that the proposed approach is in particular suitable to be embedded in a conventional CAD/CAM software in so far as the algorithm deals with a curvilinear version of the standard zigzag pattern. Such modifications only require an additional mapping obtained by means of the grid adaptation procedures.

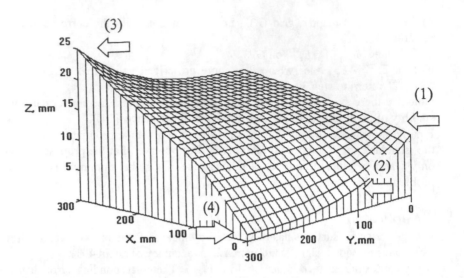

Fig.3 Machined surface, (1),(2),(3),(4) regions of large milling errors

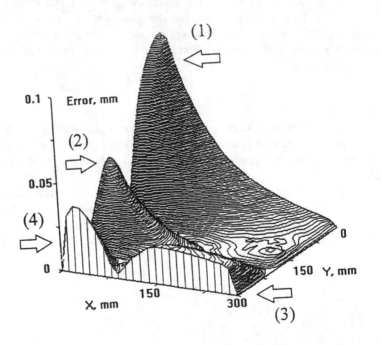

Fig.4 Spatial error distribution

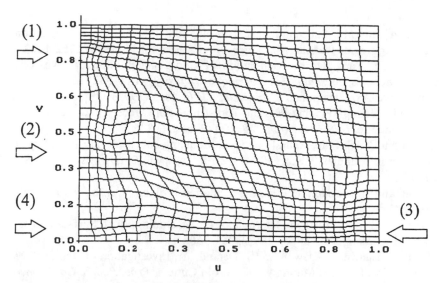

Fig.5 Grid adapted to milling errors

Table 1. Error on uniform and adaptive grids

Number of grid-points	Step for uniform grids, mm	Max. error on uniform grids, mm	Max. error on adaptive grids, mm
100	30.00000	0.14233	0.09538
400	15.00000	0.05692	0.02845
1600	7.50000	0.01594	0.00715
3600	3.75000	0.00752	0.00480
6400	1.87500	0.00421	0.00268
10000	0.93750	0.00161	0.00096
14400	0.46875	0.00099	0.00041

References

1.M.I. Kamien and L.Li, Subcontracting, Coordination, Flexibility and Production Smoothing in Aggregate Planning, *Management Science*, **36**, 11 (1990) 1352-1363.

2. J.Chou and D. Yang, On the Generation of Coordinate Motion of Five-Axis CNC/CMM Machines, *Journal of Engineering for Industry*, **114** (1992) 15-22.

3. Y. Koren, Five-Axis Interpolators, *Annals of CIPR*, **44**, 1 (1995) 379-382.

4. K. Kim and J. Jeong , Finding Feasible Tool-Approach Directions for Sculptured Surface Manufacture, *IIE Transactions*, **28**, 10 (1996) 829-836.

5. S.-H. Suh and Y.-S. Shin, Neural Network Modeling for Tool Path Planning of the Rough Cut in Complex Pocket Milling, *Journal of Manufacturing Systems*, **15**,5 (1996) 295-324.

6. R. Ahmadi and H. Matsuo, The Line Segmentation Problem, *Operation Research*, **39**, 1 (1991) 42-55.

7. R. Lin and Y. Koren, Real Time Interpolator for Machining Ruled Surfaces, *Proc. ASME Annual Meeting*, DSC-**55**, 2, (1994), 951-960.

8. E.A. Gani, B.Lauwer, P.Klewais and J. Detand, An Investigation of the Surface Characteristics in Five-Axis Milling with Thoroid Cutters, *Proc. Pacific Conference on Manufacturing* (1994) 53-60.

9. Y. Huang and J.H. Oliver, Integrated Simulation, Error Assessment, and Tool Path Correction for Five-Axis NC Milling, *Journal of Manufacturing*, **14**,5 (1995) 331-344.

10. V.D. Liseikin, The construction of structured adaptive grids-a review, *Journal of Computational Mathematics and Mathematical Physics*, **36**, 1 (1996) 1-32.

11. J.F. Thompson and N.P. Weatherill, Aspects of Numerical Grid Generation: Current Science and Art, *AIAA-93-3539-CP* (1993) 1029-1070.

12. S.A. Ivanenko, Construction of Curvilinear Meshes for Computing Flows in Basins, in *Modern Problems of Computational Aerodynamics*, CRC Press, London (1992), pp.211-250.

14. S.A. Ivanenko, Adaptive Grids and Grids on Surfaces, *Journal of Computational Mathematics and Mathematical Physics*, **33**, 9 (1993) 1179-1194.

15. M. Daniel, A Note on Degenerate Normal Vectors, *Computer Aided Geometric Design*, 12 (1995) 857-860.

16. J.U. Brackbill and J.S. Saltzman, Adaptive Zoning for Singular Problems in Two Dimensions, *Journal of Computational Physics*, **46**, 3 (1982) 342-368.

17. A.M Winslow, Numerical Solution of Quasilinear Poisson Equation on Non-Uniform Triangle Mesh, *Journal of Computational Physics*, **1**, 2 (1967) 149-172.

18. W. T. Vetterling, S.A. Teukolsky, W. H. Press and B.P. Flannery, *Numerical Recipes in Pascal*, Cambridge, University Press (1990).

Intelligent, Distributed NC Systems

Sally M. Chan
Terence L. Lammers

Boeing Commercial Airplane Group - Information Systems
P.O. Box 3707, MS 6C-MT
Seattle, WA, USA 98124
{sally.m.chan, terence.l.lammers}@boeing.com

Abstract   This paper presents the integration of Artificial Intelligence Technologies and Object-Oriented Technology into a domain engineering process, Object-Oriented Domain Engineering (OODE). This process is demonstrated in the Numerical Control (NC) domain. Specifically, the paper discusses how the integration of four techniques can be used to produce intelligent, distributed NC applications. These four techniques are: Knowledge Acquisition (KA), the Knowledge Based Systems (KBS) approach, Object-Oriented Analysis and Design (OOA&D), and architecture based on a distributed, heterogeneous object bus.

The first part of the paper discusses Knowledge Acquisition methods, and how these methods capture requirements and expert knowledge for KBS development and for OOA&D. Here the paper describes what a KBS is, its purpose, benefits, and application to the NC domain. The second part of the paper talks about the application of Object-Oriented Analysis and Design to NC for domain engineering. This section describes the OODE process and deliverables in the NC domain, and discusses how these support the rapid development of intelligent distributed NC applications. The third part of the paper presents an overview of a preliminary heterogeneous, distributed object architecture for the NC domain. This architecture then could be used as the framework for developing intelligent, distributed NC applications. Finally, the paper presents the future direction of AI and OOT as seen by the authors.

1. Introduction

This paper discusses the integration of Artificial Intelligence (AI) Technologies and Object-Oriented Technology (OOT)[2][3][6] to achieve rapid development of intelligent Numerical Control (NC) systems based on a distributed architecture. The AI Technologies discussed are Knowledge Acquisition (KA)[5] and Knowledge Based Systems (KBS). The paper shows the use of KA to build a KBS application. In addition it shows that KA is needed to extend and complement Object-Oriented Analysis and Design (OOA&D) for Object-Oriented Domain Engineering (OODE)[1]. Both AI and OOT are important to form an NC framework[4].

Communication by applications across heterogeneous environments (such as the one created by integrating AI and OOT) is enabled by an object bus architecture with its Object Request Broker (ORB). This architecture encourages reuse, open frameworks, and standard object interfaces, all of which make Rapid Application Development (RAD) possible. Rapidly developing applications by extending a framework defines a new software development paradigm.

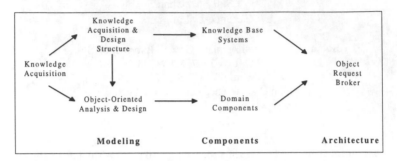

Figure 1. Information Map

Figure 1 is an information map of the content of this paper. It shows the use of KA by both KBS development and OOA&D. It shows how the products of the two approaches, KBS and domain components, are integrated by the object bus architecture, represented in the diagram by the Object Request Broker.

2. Artificial Intelligence Technologies

The goal of Artificial Intelligence Technologies is to develop computer systems which can mimic human intelligence. In order to mimic human intelligence, knowledge must be acquired and stored. This section discusses how Knowledge Acquisition (KA) and Knowledge Based Systems (KBS) accomplish this.

2.1 Knowledge Acquisition (KA)

The Knowledge Engineers (KEs) use the KA process to gather a domain expert's complex problem solving knowledge to develop a KBS. Knowledge Acquisition and Design Structure (KADS) is a methodology widely used by KEs. KADS approaches KBS development as a modeling activity. It uses cognitive models and Problem Solving Templates (PSTs) to illustrate the reasoning pattern domain experts follow when solving complex business problems. KADS consists of four modeling steps: the Process Model, the Conceptual Model, the Problem Solving Template, and the Strategic Model. The Process Model structures the behavior used to solve problems and complete a process. The Conceptual Model structures the information needed to solve problems and perform a process. The PST illustrates the reasoning patterns inherent in business and system processes. The Strategic Model identifies the high level workflow and core processes of the system to define the architecture of the KBS.

Figure 2 illustrates the application of KADS modeling techniques to the NC domain to develop a manufacturing KBS. The models in the Figure represent preliminary work which has not been implemented in any specific application.

KADS models are highly abstracted descriptions of typical problem solving tasks. They are easy to understand because their cognitive structure represents high level collaborations and because they identify clear boundaries and linkages for business processes. KADS models help KEs identify the information necessary for KBS development effectively, therefore shortening the system development cycle.

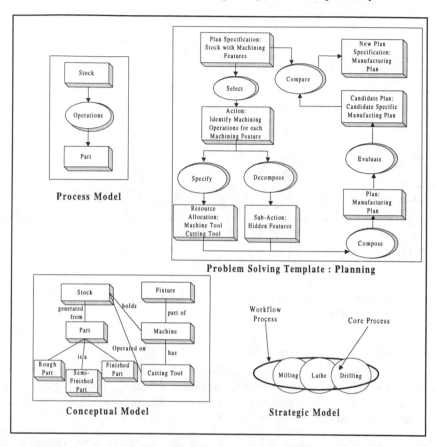

Figure 2. Examples of KADS Models for Knowledge Acquisition

In KBS development, the modeling focus is on "how" the domain expert solves a complex business problem; in OOA&D, the modeling focus is on "what" the interactions between objects are in the system. Despite these different focuses, KA can benefit OOA&D as a process for gathering systems requirements. Before any application is developed, the requirements need to be understood, collected, and documented. In OOA&D, user requirements are defined in Use Cases, which are textual descriptions of an actor's interaction with the application. From the Use Case, one can identify the system or domain objects, their collaborations, relationships, and subsystems. Applying KADS models increase understanding of the problem and this enhances the application of OOA&D. For example, the Process Model can graphically capture user requirements and interaction in Use Cases. The Problem Solving Templates (PSTs) add understanding of how something is done, and the Process Model and Conceptual Model aid the abstraction of application-level Use Cases into Domain

Use Cases for framework development. Figure 3 shows the integration of KA models and OOA&D models; the un-shaded boxes represent the KA models.

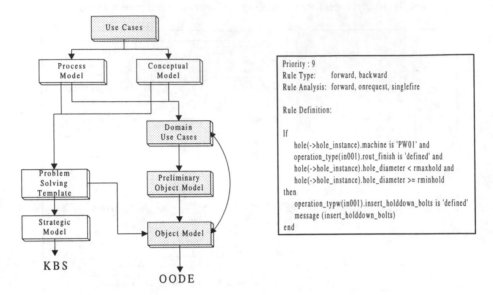

Figure 4 box text:

Priority : 9
Rule Type: forward, backward
Rule Analysis: forward, onrequest, singlefire

Rule Definition:

If
    hole(->hole_instance).machine is 'PW01' and
    operation_type(in001).rout_finish is 'defined' and
    hole(->hole_instance).hole_diameter < rmaxhold and
    hole(->hole_instance).hole_diameter >= rminhold
then
    operation_typw(in001).insert_holddown_bolts is 'defined'
    message (insert_holddown_bolts)
end

Figure 3. Integration of KADS and OO Models

Figure 4. Sample KBS Rule

2.2 Knowledge Based Systems (KBS)

KA is an input to KBS development. The KBS processes subject matter expert's knowledge for solving complex problems within a domain. It is designed to solve diagnostic, monitoring, configuration, planning, design, or scheduling problems. A KBS consists of an Inference Engine and a Knowledge Base.

The Knowledge Base is usually implemented in rules, such as IF_THEN rules, meta rules, and/or methods. Figure 4 shows an example of a manufacturing rule in a KBS. Knowledge Engineers collect these rules in KA sessions from subject matter experts. The Inference Engine processes the rules in the Knowledge Base by forward or backward chaining, agendas, truth maintenance, confidence factors, conflict resolution, and by using daemons. Other functionality like explanation and pattern matching are additional capabilities which facilitate KBS development. A KBS can provide realistic conclusions to actual problems as well as the steps and materials involved in the reasoning process.

Using a KBS adds to an application the complex decision making capability which usually requires human expertise. The KBS Inference Engine separates the decision-making logic from the process control logic. It does so by not embedding business decision rules within application code, as in conventional programming. Separation accommodates ever-changing business requirements and provides maintainability. NC is a good area for KBS application because NC programming requires experience and knowledge.

3. Object-Oriented Domain Engineering

3.1 Background

A manufacturing KBS is only one component of the NC domain framework. To develop complex systems such as NC applications, it is crucial to also apply OOT to maximize reuse, provide plug-and-play capability, and enable RAD. Object-Oriented Domain Engineering (OODE) supports these goals by creating a framework of collaborating components. OODE extends the techniques and concepts of Object-Oriented methodologies and applies them to framework development.

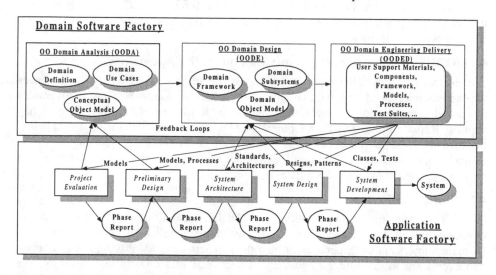

Figure 5. Application and Domain Engineering Dual Lifecycles

OODE is one of the two lifecycles in a dual lifecycle software development method. The OODE lifecycle consists of three phases: Object-Oriented Domain Analysis (OODA), Object-Oriented Domain Design (OODD), and Object-Oriented Domain Engineering Delivery (OODED). These phases run parallel to the Application Engineering (AE) lifecycle. Figure 5 shows the interrelations, including feedback loops, between the OODE and AE lifecycles. The major OODE end user deliverables are a Domain Object Model, a Domain Framework, and User Support Materials.

3.2 Object-Oriented Domain Engineering Lifecycle Phases

Object-Oriented Domain Analysis (OODA) is the first phase of OODE. It has three key deliverables: 1) the Domain Definition; 2) the Domain Use Cases; and 3) the Conceptual Object Model. The Definition focuses and constrains the OODE process. The Domain Use Cases capture the domain requirements, and validate the completeness of the Model. The Conceptual Object Model captures the domain classes, their relationships, and their responsibilities. The Conceptual Object Model and Domain Use Cases span multiple domain applications and the domain through time.

The second phase of OODE is OODD. Its deliverables are: 1) the Domain Framework; 2) the Domain Subsystems; and 3) the Domain Object Model. These are elaborated from the OODA deliverables. Once the Domain Object Model is complete, a make or buy study can determine the best approach for implementing the framework.

OODED is the third phase of OODE. The major risk OODE faces is that once a framework is delivered, projects will not use it. From the application project's point of view, the framework could be viewed as an unknown entity. This critical issue is addressed by the OODE Delivery Phase which supplies training, documentation, and support to transition the framework into the application project. The major deliverables in the OODE delivery phase are the Test Suites and User Support Materials. Test Suites are derived from the Domain Use Cases. Thus the Domain Use Cases play a role in the OODE process from modeling through testing, ensuring the robustness and correctness of the framework.

4. OODE Deliverables for the Numerical Control (NC) Domain

The previous section of the paper presented the OODE process. This section describes the deliverables produced by that process for the NC domain. The deliverables discussed in this section are preliminary and meant to illustrate OODE concepts. They support Rapid Application Development (RAD) in the NC domain or any domain, because they are in place before an application development project begins. The project only needs to customize the deliverables for its special needs and then can rapidly assemble them into an application.

4.1 Domain Definition

After a domain has been identified for OODE, the first step is to define it. Combinations of events and actors within the domain define the domain boundary. Domain subject matter experts contribute their expertise to identifying domain actors and events and to drawing the boundary. Experience with OODE has shown that the key to good domain definitions is effective Knowledge Acquisition techniques. Thus KADS Process Models complement the Domain Definition Model. Figure 6 is a sample domain boundary diagram, showing an NC Programmer as an initiating actor.

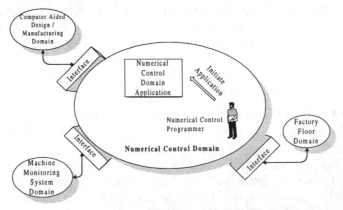

Figure 6. Numerical Control Domain Boundary

4.2 Domain Use Cases

Domain Use Cases capture abstract business processes which span an entire domain. The KADS Process and Conceptual Models make the process of abstracting application-specific Use Cases to domain level Use Cases easier. They are applicable to all the applications in the domain. It is the goal of the Domain Use Cases to not only capture what exists in a domain, but to describe "to be" processes as well. "To be" processes are defined by reengineering the existing processes in light of future business requirements and technological advancements. By capturing these processes OODE fulfills its responsibility for having components ready when they are needed by an application project, even a future project.

A Domain Use Case contains the actors, the intent of the Use Case, assumptions which underlie the Use Case, and the results achieved by it. The actor initiates the Use Case and it is carried forward to completion by the interaction of the system and the actor. System responses to actor events are mapped to domain classes which provide them. From the Domain Use Case the domain's classes are identified and later formalized in the Domain models.

4.3 Domain Models

Domain Use Cases capture essential business processes, and the Domain Models build on these to define generic domain classes and their relationships. The OODE process produces and elaborates three models: the static object model, the dynamic object model, and the subsystem model. These three make up the Domain Object Model.

The static object model presents the domain classes, their inheritance and composition relationships, responsibilities, services, and attributes. Figure 7 portrays only one NC subsystem from the entire NC static object model. The model in Figure 7 is preliminary and does not represent a framework which has yet been implemented.

Class collaboration relationships are modeled in the OODE dynamic domain class model and the subsystem models. They indicate how the domain objects interact over time and demonstrate whether the class responsibilities have been well distributed.

4.4 Domain Subsystems

A domain subsystem is both a container for the classes which compose it, and a controller for their activities. It presents a public interface, or contract, to the world. The contract in Figure 8 is the half circle on the right edge of the subsystem boundary. To fulfill its contract the subsystem sequences the services of its internal objects. This sequence is indicated by the numbers on the lines. The lines themselves represent service requests from the subsystem to its members to support its contract.

The domain subsystem models continue the process of elaborating the object models, but their primary purpose is to aggregate classes into larger entities. Subsystems increase the granularity of the domain components, as seen in Figure 8, where the NC Strategy Planner Subsystem aggregates a number of classes. This is a goal of OODE: to enable Rapid Application Development by reducing the number of components needed to assemble an application.

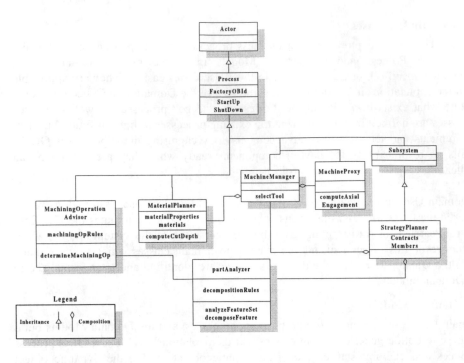

Figure 7. NC Strategy Planner Subsystem Static Object Model

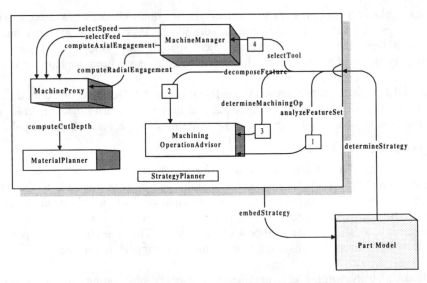

Figure 8. NC Strategy Planner Subsystem

5. Distributed Object Architecture

This part of the paper presents a preliminary heterogeneous distributed object architecture for the NC domain. The architecture is the culmination of the first two phases of the OODE process. Figure 9 presents the architecture in terms of subsystems and an object bus. These are the building blocks for application development.

The architecture is heterogeneous because components built with a variety of languages and for different platforms can be integrated into it. It is distributed because components execute on independent hosts and communicate via an object bus. The Inter Bus Gateway manages communication between different object bus standards.

The architecture is the foundation of Rapid Application Development, because components plug quickly into the object bus with minimal effort through standard interfaces such as the CORBA Interface Definition Language (IDL). OODE supports this approach because components are specifically designed for assembly and plugability.

5.1 Benefits

Domain architectures are beneficial because they enable comparisons among domains. Common, reusable components which exist in more than one domain are easily identified. This leads to cross-domain reuse and greater return on investment.

Distributed architectures make the best use of available resources. For example, in Figure 9 the StrategyPlanner is seen hosted on a separate machine, because it is computationally intensive. Other components are located on one host because they need fewer resources or in order to communicate locally. By distributing components, data movement on the network is localized and therefore minimized overall.

Heterogeneous architectures maximize the reuse of existing applications and components. Applications need not be rewritten, but only made compatible with a standard object bus interface.

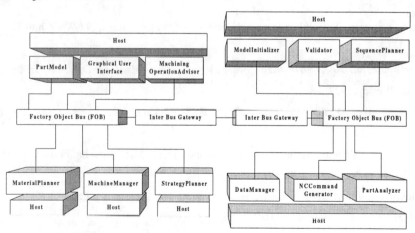

Figure 9. Distributed, Heterogeneous OO Numerical Control Architecture

6. Conclusion

This paper concludes that the integration of Artificial Intelligence (AI) and Object-Oriented Technology (OOT) will lead to the rapid development of intelligent, distributed Numerical Control applications. AI and OOT complement each other. KA methods are strong where OOT is weak. Their integration leads to better framework design. Frameworks in turn enable the rapid development of applications.

At the architectural level, the object bus allows domain subsystems to communicate with each other through standard interfaces. Heterogeneous components such as a KBS and domain components are integrated by a distributed object architecture. These lead to rapid assembly of applications and software reuse.

7. Future Direction of OO Technology and Artificial Intelligence

This paper describes how OOT and AI complement each other. In the future there will be a more complete merger of the two. The State Object is the most common type in the current practice of OOT. State Objects change state but are not aware it. Aware Objects know their state and make use of it to perform their responsibilities. At this point AI begins to be deeply integrated into OOT. Intelligent Objects have integrated Knowledge Based Systems which give them the capability to perform complex, unstructured services. Artificial Life Objects are intelligent and have the ability to learn and evolve. This means that Artificial Life NC systems can rapidly adapt to new parts, new tools, new materials, and new business requirements. Living Systems are systems of Artificial Life Objects. These Systems demonstrate behavior which emerges from the interaction of their members. Because its objects evolve, the entire System evolves to respond to changes in its environment. Some of the enabling AI technologies which contribute to this are machining learning, genetic programming, neural nets, and adaptive programming.

References

[1] Chan, S., Lammers, T. , *Object-Oriented Domain Engineering*, Reuse'97 Workshop Tutorial, 1997

[2] Jacobson, I. *et al.*, *Object-Oriented Software Engineering - A Use Case Driven Approach*, Addison Wesley, 1992

[3] Orfali, R., Harkey, D., Edwards, J., *The Essential Distributed Objects Survival Guide*, Katherine Schowalter, 1996

[4] Schmid, H. A., "Creating Applications from Components: A Manufacturing Framework Design", IEEE Software, November 1996, Vol. 13, No. 6, pp. 67-75

[5] Tansley, D., and Hayball, C., *Knowledge-Based Systems Analysis and Design: A KADS Developer's Handbook*. Prentice Hall, 1993

[6] Wirfs-Brock, R., Wilkerson, B., and Wiener, L., *Designing Object-Oriented Software*, Prentice Hall, 1990

Automation Components

DEVELOPMENT OF AN AUTOMATIC AND FLEXIBLE FEEDING SYSTEM FOR DELICATE CYLINDRICAL PARTS WITH ASYMMETRICAL FEATURE

Patrick S.K.Chua
School of Mechanical and Production Engineering
Nanyang Technological University
Nanyang Avenue
Singapore 639798

ABSTRACT

This paper concerns the development of a flexible and automatic feeding system which is capable of feeding cylindrical parts which are delicate or powdery in nature and possess asymmetrical features such as a groove near to one end. Controlled by a programmable logic controller and electropneumatics, the system has active orientating capability. System evaluation results showed that the average jam rate is below 10 % and the percentage of correctly orientated parts is above 90 %. With fine tuning, the system could become a very useful feeder for industry in the future.

1   INTRODUCTION

In industries where assembly costs form a large portion of the production costs, manufacturers are increasingly looking towards assembly automation to improve efficiency as well as productivity. Special purpose automatic assembly machines are however only economically feasible in very large volume production and they are usually dedicated only to that particular product, thus lacking flexibility.

Recently, there is an increase in demand for customised products with special features and options to meet the particular needs of the customer. To meet this market needs, manufacturers must be able to reduce manufacturing lead times, accommodate for product changeovers and part design changes in order to stay competitive. It is envisaged that flexibility of the assembly systems is the key to survival in the competitive manufacturing market. Therefore there is now a growing interest in the development of a flexible parts presentation system to achieve a total flexible manufacturing system

Conventional vibratory feeders commonly seen in industry today have the following shortcomings or limitations in scope:

(a)   They are usually dedicated types and therefore unsuitable for productions in which there are frequent batch changes or in which the product life cycle is short.

(b)   They are also not suitable for fragile, delicate or powdery parts because of the vibration of the feeder.

(c)   The orientating devices used in the vibratory bowl feeders are usually of the passive type, i.e. by rejecting parts which travel on the track in the undesired orientation. Thus there is a reduction in efficiency.

(d) They are incapable of feeding and orientating cylindrical parts which have unsymmetrical features at the circumference e.g. a groove near one end as shown in Fig. 1.

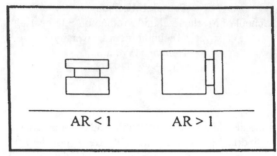

Fig. 1 :Asymmetrical Parts

Moreover, existing designs of non-vibratory feeders have similar shortcomings as mentioned in items (a), (c) and (d) above. These limitations in scope of existing feeding systems call for a flexible, non-vibratory feeding system with active re-orientating devices capable of orientating cylindrical parts even if these parts have asymmetrical features such as a groove near one end.

2   OBJECTIVE AND SCOPE

The work involved the conceptualisation, experimental evaluation of the conceptualised design, design, construction and evaluation of a suitable feeding system for feeding and orientating fragile asymmetrical cylindrical parts of various aspect ratios.

3   LITERATURE REVIEW

When it comes to the feeding of delicate powdery parts, vibratory feeders are certainly unsuitable for use because cracking and dust accummulation can occur as a result of the rigorous vibration. Many non-vibratory feeders have been developed over the years but the objective of their developments was as a low cost alternative to the more expensive vibratory bowl feeders and not because delicate parts were to be fed. Several such non-vibratory feeders can be found in Boothroyd et al [1, 2], Denhamer [3] and Chua and Lim [4].

Another reason for developing non-vibratory feeders was flexibility which necessitated the use of feature recognition sensors such as infra-red fibre-optic sensors and cameras. These feature recognition devices require that the surface that the part is travelling be non-vibratory in nature inorder to ensure reliability in identification of the orientation of the moving part. Examples of such flexibile non-vibratory feeders can be found in Pherson et al [5] and Chua and Sung [6].

The works of Pherson et al [5] and Chua and Sung [6] concentrated on the feeding of non-rotational parts with asymmetrical feature such as a blind hole at one corner of a rectangular part. No work was carried out on the feeding of rotational parts with an asymmetrical feature such as a groove near to one end. The reason for this lack of work on rotational part is probably due to the additional difficulty of the part's ability to roll freely. To be able to convey it without vibratory means while at the same time orientating it to the desired orientation for pick-up by the manipulator is not an easy task.

4   DESIGN SPECIFICATIONS AND REQUIREMENTS

The feeding system to be developed should be able to feed cylindrical parts as follows:

Part Diameter (mm)	Aspect Ratio (AR)
13 - 25	0.5 - 0.8
	1.2 - 1.6

Table 1 : Cylindrical Parts of Interest

The part has a groove of 2 mm (width and depth) near to one end as shown in Fig. 1. The feed rate should not be less than 10 parts per minute.

5   CONCEPTUAL DESIGN

The components required of the feeding system is shown in Fig. 2.

Fig. 2 : Sub-systems of the Feeding System

Various conceptual designs were generated and the final design that emerged as the most suitable one is as shown in Fig. 3.

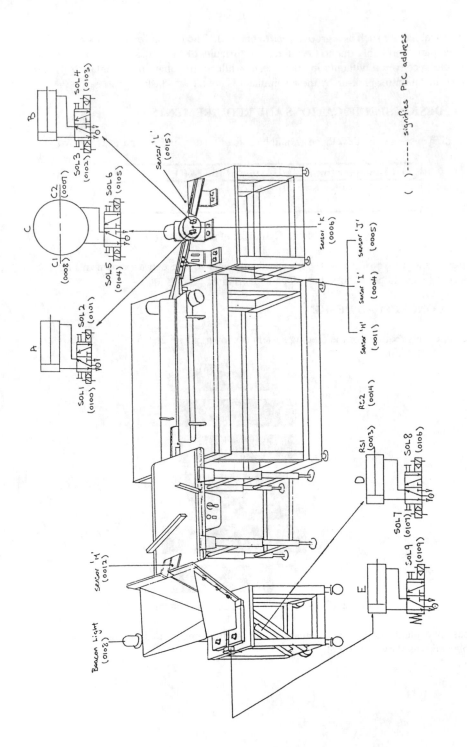

Fig. 3 : The Final Design of the Feeding System

( )------ signifies PLC address

DESCRIPTION OF FEEDING SYSTEM DEVELOPED

6.1 Loading Module

The loading module consists of a hopper for feeding the cylindrical parts. It is flexible in the sense that minimum adjustment/alignment is made when there is a change of part size. Fig. 4 illustrates the fabricated design of the hopper unit.

The mechanism of the hopper basically uses the reciprocating action of a pneumatic linear actuator. When the actuator is activated, the piston rod extends and pushes those parts that fall on the track up to the gate. The opening gap of the gate can be adjusted to cater for any change in part size so as to control the feedrate.

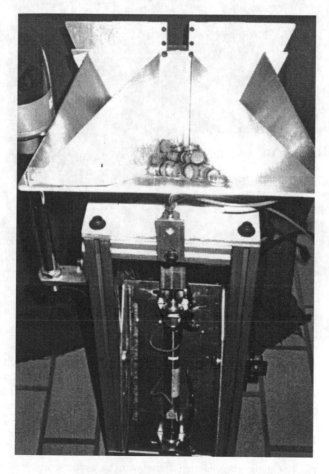

Fig.4 : Front View of Loading Module (Hopper)

6.2   Singularising Module

The singularising module consists of a conveyor belt system which comes after the hopper unit and its requirement is to separate those parts that come close to each other before proceeding to the next station. Figure 5 illustrates the fabricated design of the singularising conveyor unit.

The unit consists of a rough conveyor belt which gives a good grip of the parts (without rolling off) after having been deflected from the deflector blades.

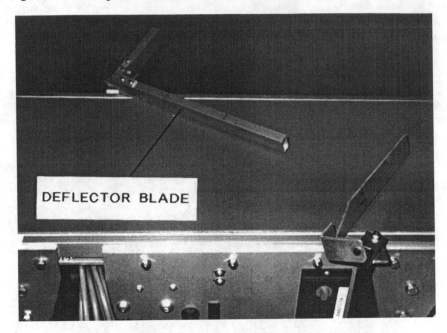

Fig. 5 : Singularising Conveyor Unit

6.3   V-Belt Orientator

The V-belt orientator (Fig. 6) serves to orientate cylindrical parts of different aspect ratios (length/diameter). It consists of two inclined conveyor belts of different roughness and moving at two different speeds. Parts travelling on the conveyor are orientated till they reach a stable position by the means of rotation caused by the differential speed of the moving V-belt. The tilting angle of the V-belt was experimented and analysed for various aspect ratios of the parts and it was found that the optimum inclination is 30° [7].

Evaluation of the orientator showed that parts with an aspect ratio less than one will stabilize in resting aspect 1, whereas parts with an aspect ratio greater than one will stabilize in resting aspect 2 (Fig. 7).

Fig. 6 : V-belt Orientator

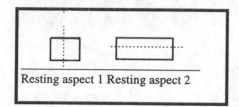

Resting aspect 1 Resting aspect 2

Fig. 7 : Resting Aspect of Parts

Parts with an aspect ratio not equal to one do not have any orientation problems. All of them managed to be orientated within the length of the V-belt. Whereas, parts that have aspect ratio of one or very close to one do not have one stable position. They usually assume the position of either resting aspect 1 or 2.

6.4   Transfer Mechanism

The transfer mechanism basically consists of thin aluminium plate and is installed just after the V-belt unit to transfer the orientated parts to the next station. Fig. 8 illustrates the fabricated design of the transfer mechanism.

From the figure, the two metal strips located by the side are to induce a springing effect to any part that comes in contact so as to assist the part to drop down the slope of the transfer mechanism smoothly. They can be adjusted in the gap opening to suit the different part sizes. The slope of the transfer mechanism is set between 25° to 30°. To avoid any part getting stuck, the entering edge of the transfer mechanism before the V-belt is chamfered.

ALUM.STRIPS

Fig. 8 : Transfer Mechanism

6.5 Unloading Module

The unloading module basically consists of upper delivery chute, holding device (re-orientator) and lower delivery chute. The holding device will re-orientate any part that is in an undesired orientation to the desired orientation. In other words, the principle of active orientation is employed. Fig. 9 illustrates the fabricated unloading unit.

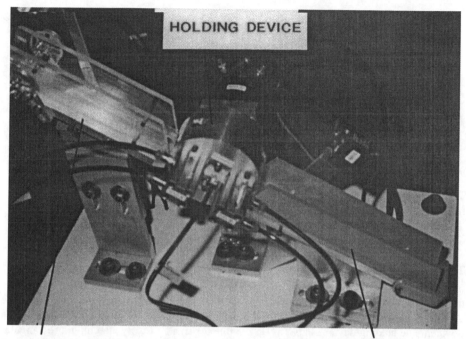

Upper Delivery Chute   *Lower Delivery Chute*

Fig. 9 : Unloading Unit

6.5.1   Upper Delivery Chute

The upper delivery chute is tilted at an angle of 30° with respect to the horizontal after evaluating the smooth sliding of parts at several angles and the height is made adjustable to provide flexibility. The whole section here is important because this is the area where the orientated parts are sensed for the desired orientation. The desired orientation for a part having an aspect ratio less than one is with the groove at the top whereas for a part having an aspect ratio more than one is the groove at the bottom (Fig. 10).

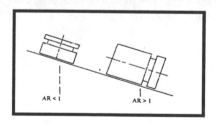

Fig. 10 : Desired Orientation for Range of Aspect Ratio

As for wrong orientation, the parts will be sensed by fine fibre-optic sensors. A pattern of on/off signals depending on the stage of the part is created as parts move past the strategically located sensors. This technique is relatively low in cost compared to the expensive vision system. Fig. 11 illustrates the positions of the sensors for wrong orientations.

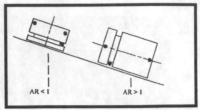

Fig 11 : Position of Sensors

Through-beam sensors are used for this application to detect grooves. The sensors shown in the figure above are all in the 'on' stage to activate the holding device.

6.5.2   Holding Device

The holding device re-orientates any part that is in an undesired orientation to the desired orientation through rotation of 180°. A rotary pneumatic actuator is installed to operate the rotation. Speed of rotation can be controlled through the throttle valves attached on the actuator. Linear compact pneumatic actuators are used to act as stoppers in the holding device. The holding device is also made flexible for small parts by attaching a rectangular slab to close up the gap.

6.5.3   Lower Delivery Chute

The lower delivery chute is also tilted at an angle of 30° with respect to the horizontal. This is the area where all completed parts in their desired orientation are collected. A photoelectric sensor is installed at the upstream end of the chute to alert the operator that excess parts collected along the chute are not being removed for packing.

6.6   Overall System

The individual modules of the fabricated designs are integrated together to form a full flexible feeding system as illustrated in Fig. 12. It is designed to operate continuously with minimal human intervention. Control of the whole system is achieved using programmable logic controller and electropneumatics.

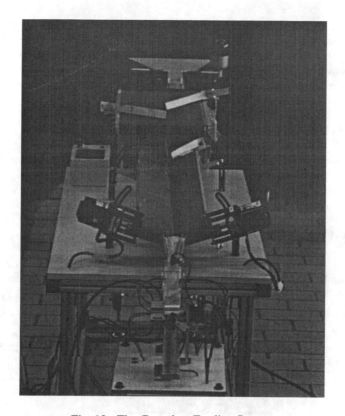

Fig. 12 : The Complete Feeding System

6.6.1 Operation Sequence of the New System

The whole system now consists of the feeder, singularising conveyor, V-belt orientator, re-orientator, chute and the controller unit.

First, the gate opening gap of the hopper, V-belt tilting angle, sensors mounting position and gap opening of the holding device have to be set for the particular part to be processed. The start button can be initiated to start the sequence.

Once there is a part/s ejected out from the hopper (detectable by parts sensor), the hopper will pause for a couple of seconds before the next action begins so as to prevent parts congestion at the next station. Parts arriving at the singularising conveyor will be separated by the deflector blades before proceeding to the V-belt conveyor. The V-belt will orientate the parts as they travel along the belt after which they will slide down the transfer mechansim to the upper delivery chute. Here, the sensors will detect the position of the groove to check if the part is in the correct orientation. Correctly orientated parts will be allowed to slide towards the lower delivery chute for collection.

If the part is in the wrong orientation, the stopper pins will be activated. They will stop the part in the holding device as it arrives from the chute. Once the part is in the holding device, the rotary actuator will rotate to turn the part to its correct orientation. Once this active re-orientation has been achieved, the stopper pins will retract to allow the part to be collected at the lower delivery chute. The rotary actuator will then rotate back to its original position and waits for the next part.

7   SYSTEM EVALUATION AND DISCUSSION

Evaluations were carried out to test the workability of the new system developed. They are :

(a)   Sub-evaluation at the transfer mechanism
(b)   Evaluation on the performance of the whole system for AR<1 and AR>1.

7.1   Sub-Evaluation at the Transfer Mechanism

An evaluation was done to analyse the orientation of the parts on the transfer mechanism after emerging from the V-belt. Fig. 13 illustrates the results obtained for four categories of parts.

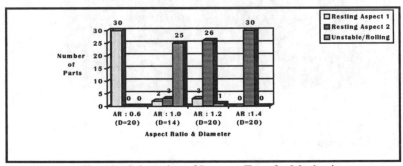

Fig. 13 : Orientation of Parts on Transfer Mechanism

It is observed that parts with an 'AR' less than or greater than 1 maintain the orientation from the V-belt well on the transfer mechanism. Whereas parts with an 'AR' very close to 1 do not have a stable orientation.

7.2   Evaluation Results for AR<1

Parts with a diameter of 20 mm but with two different aspect ratios (0.6and 1.2), were run to test on the efficiency of the feeding system for an input size of 30 parts. The parameter settings were as follows :

Speed of Singularising Conveyor	Speed of V-belt
85 mm/s	Smoother belt : 65 mm/s Rougher belt : 110 mm/s

Table 2 : Parameter Settings

Data was collected on parts completed per minute at the last station. A total of 6 runs were conducted for the same part size. The results are illustrated in Figs. 14 and 15 respectively.

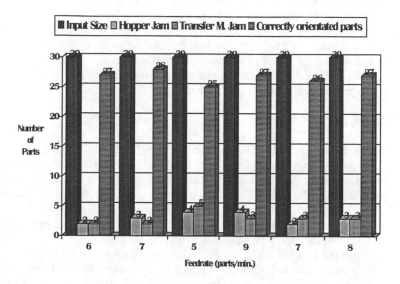

Fig. 14 : Feeder Performance for AR<1

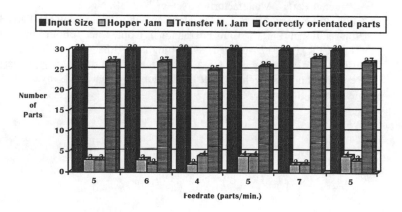

Fig. 15 : Feeder Performance for AR>1

The results showed that for both aspect ratios, the average percentage of correctly orientated parts is above 90 % and that jam rates are typically in the region of 10 %. Such results are indeed encouraging and indicate that the performance of the feeding system can be improved through fine tuning.

8   CONCLUSION

A flexible non-vibratory feeder with active orientating capability for fragile or powdery cylinderical parts with asymmetrical feature has been developed. Delicate and/or powdery cylindrical parts (including those with asymmetrical features) with aspect ratios ranging from 0.5 - 0.8 and 1.2 - 1.6 can be fed with this feeding system. It represents the first of its kind in providing for such feeding needs. Employing active orientation technique, it has a high feeding efficiency due to the absence of rejection. Fine tuning will certainly improve the performance of the feeding system.

REFERENCES

[1]   Boothroyd, G., Poli, C.R.and Murch, L.E., Handbook of Feeding and Orienting Techniques for Small Parts, Dept. of Mechanical Engineering, University of Massachusetts, Amherst, Massachusetts.

[2]   Boothroyd, G., Poli, C.R.and Murch, L.E., Automatic Assembly, Marcel Dekker Inc., USA, 1982.

[3]   Denhamer, H.E., Interordering : A New Method of Component Orientation, Elsevier Scientific Publishing Co. , 1980.

[4]   Chua, P.S.K. and Lim,L.E.N., "Development of an Automated Feeding, Orientating and Joint-Straightening System for Desk-Top Pen Holder", 10th International Conference on Production Research (10th ICPR), 14 - 17 August 1989, Nottingham, U.K.

[5]   Pherson, D., Boothroyd G. and Dewhurst, P., "Programmable Feeder for Non-Rotational Parts", Manufacturing System, Vol. 13, No. 2.

[6]   Chua, P.S.K. and Sung, E.,"Development of a Programmable Parts Feeder for Non-rotational Parts", 10th International Conference on Production Research (10th ICPR), 14 - 17 August 1989,Nottingham, U.K.

[7]   Justin Quak, Development Of A Flexible Feeder For Cylindrical Parts, Internal Report, School of Mechanical & Production Engineering, Nanyang Technological University, Singapore, 1993

Parametric Parts Feeder in Assembly Workcell for Mass Customization

Banna G. Rao

Dept. of Industrial Engineering and Engineering Management
Hong Kong University of Science and Technology
Clearwater Bay, Kowloon, Hong Kong

Abstract. Assembly automation is well established in the manufacturing industry, where economic advantages are proven in mass production. However, modern industries are following the concept of mass customization to meet business competition. The main challenge is to produce an increasing variety of products without any significant trade-off in production costs or lead time. Conventional mass production tooling in the assembly work cells are not flexible enough to handle the product variations and the present flexible assembly tooling is not giving better throughput rates. This research is focused on the issue of redesigning the tooling in assembly workcells to impart flexibility, while retaining the capability of high throughput rates. In this paper we have presented the detailed steps when redesigning a conventional parts feeder into a parametric parts feeder.

Key words: Mass customization, Assembly workcell, Parametric parts feeder

1 Introduction

1.1 Literature review

Modern industries are adopting the concept of mass customization in both manufacturing and service sectors to meet today's business competition. Mass customization is a new way of viewing business competition, one that identifies and fulfills customer demands without sacrificing efficiency, effectiveness and cost benefits [Pine 93]. The main idea of mass customization is the production and distribution of customized goods and services on a mass basis. The mass production environment is premised on high volume production in less variety at lower costs, but in the customized manufacturing environment, products are being produced in small batches and in a wide variety. Mass customization is aimed at uniting these two opposing strategies. In order to meet these challenges, today's business is looking for new approaches to combine mass production efficiency with customized production flexibility and variety. Research efforts were made to reduce the product design time and thereby achieve a short product realization cycle. Some researchers have applied axiomatic design methods to formulate a product family architecture (PFA) which allows for optimization of commonality in product design and process selection [Tseng 96]. However, there is also need for better tooling in manufacturing to suit mass customization requirements. Modular and flexible tooling is already in use in machining automation but is not well developed in assembly automation. Many attempts have been made to increase the flexibility of the tooling in the assembly workcells. Modular Orienting Devices (MODs) were developed to replace the fixed orienting devices on the conventional vibratory bowl feeders [Lim 93]. MODs offer greater flexibility in sequencing the orienting devices on the bowl feeder for the given features of the part to be fed. Recently modular and parametric orienting devices have been designed to further enhance the flexibility of the dedicated vibratory bowl feeders [Joneja 97]. These devices were designed as a subset of a Modular and Parametric Assembly Tool Set (MPATS) that is being developed by our group. The main idea is to impart modularity and parametric variability in the dedicated assembly tooling. This research explores the methods needed to redesign some special conventional parts feeders into modular and/or parametric parts feeders.

1.2 Problem Description

Automated assembly work cells are typically equipped with parts feeders, assembly fixtures and assembling tools. Redesigning these assembly tools to suit the mass customization needs is the main objective of our research, however in this paper our focus is limited to a subset of assembly tooling namely parts feeders. Feeding parts in the correct attitude/orientation and at a desired feed rate is an important requirement in automated assembly operations. Parts feeders are the devices used to segregate the parts and feed them into assembly machines. Existing parts feeders are designed for the particular dimensions of the part to be fed, and hence the same feeders can not be extended to feed the same type of part with varying dimensions. This inability has become an obstacle to meet diverse customer needs in a mass customization environment. We propose an approach to redesign these conventional feeders into modular and/or parametric feeding and orienting devices. Parts feeders in general can be classified as vibratory feeders or non vibratory feeders. The selection of parts feeder depends on the type of part to be fed. Conventional vibratory bowl feeders have been redesigned to equip with modular orienting devices [Lim 93] and further they have been modified and designed with modular and parametric orienting devices [Joneja 97]. However the same ideas have not been applied on non vibratory parts feeders. The utility of these feeders can be greatly extended by redesigning them into modular and/or parametric feeders. This paper presents the detailed steps involved when redesigning a Rotary Centerboard Hopper Feeder into a parametric parts feeder.

2. Approach

2.1 Redesign considerations

A good redesign should result in a feeder that performs its original function efficiently and economically within the imposed constraints.
1. New designs should provide the flexibility for manipulating all the critical operating parameters independently.
2. New functional modules/devices in the redesigned feeder should be easily reconfigured and also easily assembled and disassembled.

3. New modules/devices should fit together seemlessly and should not obstruct the flow of the parts during feeding.

Original designs of the non vibratory feeders were extensively developed and reported in a catalogue [Boothroyd 82]. For each feeder, catalogue contains drawings with description, general design data, details of the range of applications and operating speeds. The redesigning process can be approached in one of the following ways:

- Development of existing designs by introducing modifications to its functional modules/devices to suit the new design requirements. This type of work represents a large proportion of the design effort in industry.
- Generation of new designs from scratch by taking part features into consideration. It often requires a solution for problems which have not been encountered before and could require the addition of new resources to the organization. The introduction of a totally new design usually involves a large amount of experimental and prototype work [Farag 89].

We have followed the first alternative when developing the new designs for the parts feeders by modifying the existing designs into modular and/or parametric designs. All major steps of the redesign process are depicted in the flow chart shown below.

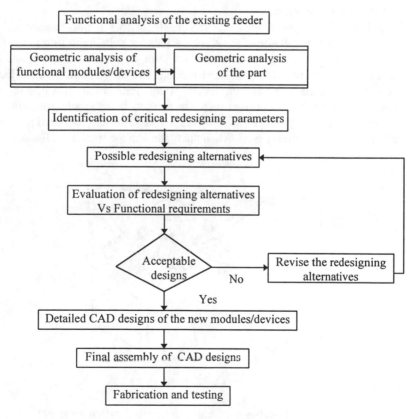

Fig. 1. Major steps of redesign process

3. Implementation

3.1 Example:

The Rotary centerboard hopper feeder is a non vibratory parts feeder. It is basically used to feed U - shaped parts to automated assembly machines. The rotary centerboard hopper feeder mainly consists of the following parts: a hopper, a hopper groove and a rotary bladed wheel. The hopper is a conical bowl designed to carry the parts when placed at random as a heap. It acts as a housing for the parts during agitation by the bladed wheel. It collects the rejected parts and returns them to the heap. It has a slot on its inclined wall to make a path for wheel movement. Rubber or cork may be placed along the inside wall of the hopper to reduce noise and damage to parts. A groove in the hopper bottom has dimensions such that the parts in the hopper may be accepted in the desired orientation only. The groove acts like an inclined track, allowing the parts to slide down and retaining them at the bottom. The upper end of the groove is curved so that the parts can be recirculated back into the hopper. The rotary bladed wheel rotates across the hopper and has blades whose dimensions are tailor-made to suit the parts to be lifted. The edges of the blades should be profiled to collect parts in the desired attitude and lift them clear of the bulk of the parts.

3.1.1 Principle of operation

The rotary wheel agitates the parts causing them to fall into the shaped groove in different orientations. Some of these orientations are suitable to fit into the groove and others are not. Once the blade sweeps up through the slot some of the unsuitable orientations are brought to suitable orientations and the remaining are rejected by the blade. Finally parts that fall with desired orientations are accepted by the groove for feeding. As the wheel continues to rotate further, these parts are collected by the blades and raised clear of the batch of parts. These correctly oriented parts slide along the blade onto the delivery track when the blade and the track are aligned.

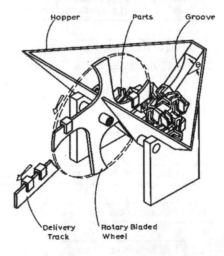

Fig. 2. Rotary centerboard hopper feeder [Boothroyd 91]

3.2 Redesigning process

3.2.1 Functional analysis of the feeder
The functional mechanism of the rotary centerboard hopper feeder performs the following basic operations:
1.   Collection of parts as a heap at the bottom of the feeder.
2.   Continuous agitation of these parts and allowing them to orient in correct attitude.
3.   Accepting the parts that are in correct orientation and rejecting the rest into the heap to make further agitation until they attain the desired orientation.
4.   Smooth conveyance of these parts to the feed track without disturbing their orientation.

In order to perform these operations we can identify the following components in the original design:
*   A Hopper to house the parts and direct the parts always towards the zone of agitation which further helps to orient the parts into right attitude.
*   A Hopper groove to accept the parts only in correct orientation.
*   A Bladed wheel that aids in both agitating the parts continuously until they attain correct orientation and also conveying the correctly oriented parts to the feed track without disturbing their orientation.

3.2.2 Geometric analysis of the part
Geometry of the part is important in two aspects in the redesign process:
1.   *To determine the possible orientations of the part*: Since the parts are filled into the feeder as a heap, it is the function of the orienting device to reject all but one orientation of the part. Hence a knowledge of the possible orientations of the part (fig. 3a) is important when designing the orienting device to maximize the occurrence of correct orientation. The orienting device restricts the resting poses of the part and attempts to orient every part in the desired attitude. This is an essential requirement to increase the feed rate of the feeder.
2.   *To determine the geometric constraints on the redesign:* The geometry (fig. 3b) of a typical U-shaped part can be represented by its parameters such as length of the part (L), height of the part (H), width of the part (W) and thickness of the part (T). The part has a central slot whose width and height are dependent on the thickness of the part. Among all the possible orientations shown in the fig.3a, the orientation 'g' is suitable for feeding and hence it dictates the parameters to be considered to fit the parts into the hopper groove in this orientation. Therefore the variations in width and height of the part play a vital role when redesigning the hopper groove. Since correctly oriented parts exactly sit on the blade to convey them smoothly onto the feed track, the variations in the dimensions of the central slot of the part are important to decide the redesign parameters of the blade cross section. The performance of the feeder is measured as the number of parts delivered per blade (N_p) per unit time. The number of parts that can be conveyed per blade may vary depending on the availability of space in the groove. The variations in the length of the part affect the available space for the correctly oriented parts to fit in the hopper groove. Therefore the variations in the part length alter the length of the groove at the hopper bottom.

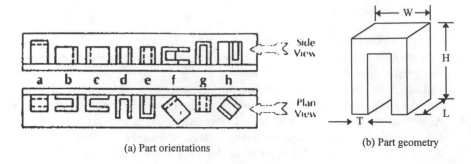

(a) Part orientations

(b) Part geometry

Fig. 3. Part geometry and orientations [Boothroyd 82]

3.2.3 Geometric analysis of the modules/devices

The geometry of the hopper cross section is very important for collecting the parts as a heap at the bottom of the hopper during agitation. Since it has a reducing cross section towards the bottom, parts can easily fall either into the hopper groove in correct orientation for acceptance or rejoin the heap. The dimensions of the hopper generally depend on the volume of the parts to be handled and the diameter of the bladed wheel. The important parameter of the hopper is the hopper wall inclination (θ) which helps the correctly oriented parts to slide down the groove and line up at the bottom of the hopper. If the parts are not correctly oriented, the inclined wall helps them to join the heap at the bottom to make further attempts. The geometry of the hopper groove depends on the parameters of the part in the correct orientation. In order to accept the correctly oriented parts into the groove, the width (W_g) and the height (H_g) of the groove should match the variations in the width and height of the part. The number of correctly oriented parts that can line-up in the groove depends on the length of the part and length of the groove (L_g). Hence 'L_g' should be enough to handle the variations in the part length to maintain optimum number of parts in the groove. The diameter of the wheel and the cross section parameters of the blade are important in the geometry of the bladed wheel. Wheel diameter (d) is specified in terms of part width (W) as the ratio of 'd/W' [Boothroyd 82] and hence the wheel diameter can be decided according to the variations in the width of the blade. In order to convey the correctly oriented parts onto the feed track without disturbing their attitude, there should be an exact fit between the central slot parameters of the part and the cross section parameters of the blade. Hence the width (W_b) and the depth (D_b) of the blade cross section are to be in agreement with the central slot parameters of the part when redesigning the blade.

3.2.4 Critical redesigning parameters

From the geometric analysis of functional modules/devices and the part, the following parameters are identified as the critical parameters for the redesign process: hopper wall inclination (θ), width of the hopper groove (W_g), height of the hopper groove (H_g), length of the hopper groove (L_g), width of the blade (W_b), depth of the blade (D_b), and diameter of the bladed wheel (d).

3.2.5 Solution alternatives and evaluation

We have come up with the following design as an alternative solution to parameterize the feeder modules/devices. The new design has a main body upon which all the other components can be assembled. The body itself is mounted on a base with an angular tilt adjustment. The hopper design is modified so that it has two halves, allowing for the groove width to be manipulated independently. The hopper wall inclination can be varied using the angular tilt mechanism on the base. The hopper groove is designed as a separate device with independent height adjustment and can be easily assembled to the hopper. The original design of the bladed wheel has been modified into two halves with a hollow expandable cover between the halves. The expandable cover is made up of flexible material and is fastened to the two blade halves. This cover fills the gap between the two halves and can prevent the parts from jamming. The parameterization of the wheel diameter is a difficult task because of its circular cross section. Therefore the ratio of 'd/W' as specified by Boothroyd *et al.* is used and the diameter is set to the maximum possible diameter for maximum width of the part that can be handled by the feeder. As mentioned in the geometric analysis of the part, the width (W_b) and the depth (D_b) of the blade are dependent on variations in the central slot dimensions of the part. Since we have set the wheel diameter suitable for the maximum part width, the corresponding depth of the blade can be fixed to handle the maximum part height. Since the manipulation of the parameters in the new design is done by adjusting the moving parts in the devices, the range of parameterization in the linear parameters is less than twice the dimension of the original part. For example, the height of the hopper groove (H_g) can be manipulated by the two sliding sections of the groove as shown in the simplified figure 4. However the maximum movement is restricted by the dimensions of the sections.

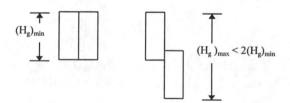

Fig. 4. Hopper groove height adjustment

3.2.6 Detailed designs & Final assembly

Detailed designs of all the components in the functional modules/devices are developed using the Pro/Engineer CAD package. A typical U- shaped part is selected and the designs of the feeder functional modules/devices are developed to suit a range of variations in dimensions of the part. These new designs are shown in the figures 5a & b. The final assembly of these modules/devices is developed to realize the CAD model of the assembled design and are shown in the figure 6.

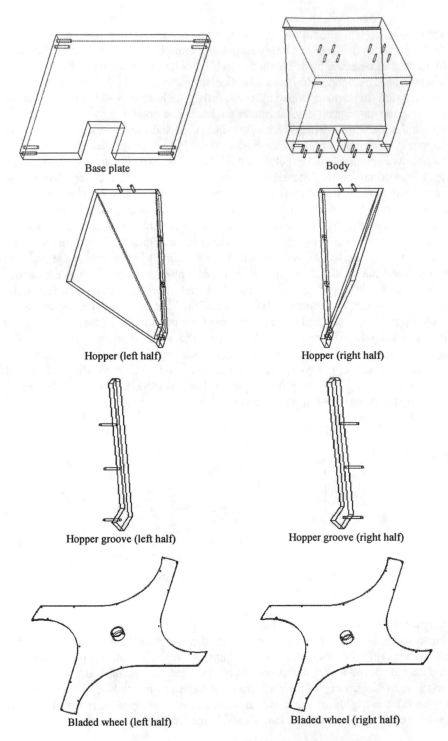

Fig. 5a. New designs of the feeder components

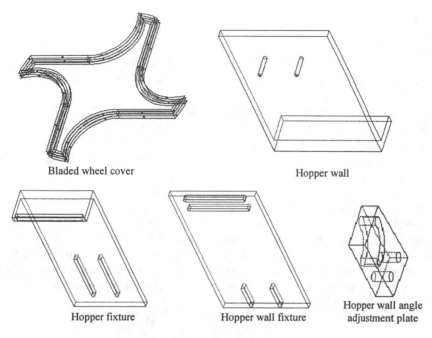

Bladed wheel cover

Hopper wall

Hopper fixture

Hopper wall fixture

Hopper wall angle
adjustment plate

Fig. 5b. New designs of the feeder components

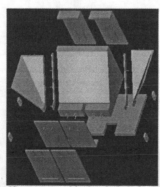

Hopper assembly (exploded view)

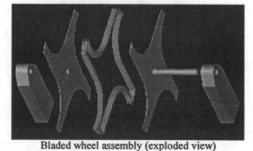

Bladed wheel assembly (exploded view)

Fig. 6. Final assembly of the parametric parts feeder

3.2.7 Fabrication and testing

Since the designs are available, the devices can be made and tested using the U-shaped part for which it is designed. A set of U-shaped parts with varying dimensions can be used to test the feeding characteristics of the feeder. After analyzing the performance of the feeder, we can fine tune the device designs for better results.

Conclusions. In order to meet the needs of the mass customization in a manufacturing environment, an approach has been proposed to redesign the tooling in the assembly work cell. We have focused our attention on a subset of hardware namely parts feeders in assembly work cell. Our approach is based on adopting the existing parts feeder designs and performing modifications to develop the new designs into modular and/or parametric parts feeder. Detailed steps of the redesigning process are explained using a rotary centerboard hopper feeder as an example. Though the new designs meet the basic functions of the feeder, there is enough scope for refinement. However our basic aim in this paper is to emphasize the need for modular and/or parametric tools in the assembly workcell which suit the needs of mass customization. One limitation to the validity of these designs is that we have not made the prototype and tested it to examine the feeder's performance characteristics. Hence our future work will focus in this direction.

4 References

1. Boothroyd, G., Poli, R. C., and Murch, L. E., Handbook of Feeding and Orienting Techniques for Small Parts, Dept. of Mechanical Engineering, University of Massachusetts, 1982.
2. Boothroyd, G., Assembly Automation and Product Design, Marcel Dekker, 1991.
3. Pine II, B. J., Mass customization: The New Frontier in Business Competition, Harvard Business School Press, Boston, 1993.
4. Farag, M. M., Selection of Materials and Manufacturing Processes for Engineering Design, Prentice Hall International (UK) Ltd., 1989.
5. Lim, L. E. N., Ngoi, B. K. A., Lee., S. S. G., Lye., S. W., and Tan., P. S., "Flexible Vibratory Bowl Feeding Using Modular Orienting Devices", Journal of The Institution of Engineers, Singapore, vol. 33, no. 4, June 1993.
6. Joneja, A., and Lee, N., "MPATS: A Modular, Parametric Assembly Tool Set for Mass Customization", Technical Report in Department of Industrial Engineering and Engineering Management, Hong Kong University of Science and Technology, Hong Kong, 1997.
7. Tseng, M. M., and Jiao, J., "Design for Mass Customization", CIRP Annals, vol. 45, no 1, 1996.
8. Pro/Engineer on-line manual, Version 15.

Advanced Mechatronic System for Turbine Blade Manufacturing and Repair

U. Berger, R. Janssen, E. Brinksmeier

Institute for Material Science (IWT), Division of Manufacturing Technologies
Badgasteiner Str. 3, 28359 Bremen, Germany

Abstract: Aspects of advanced technology for individual jet engine components repair are discussed. The technological solutions lie in the domain of cutting path planning and machining of difficult-to-cut materials like titanium alloys. The paper presents an integrated process chain for multi-axis high speed milling of TiAl6V4. The main development items are based on a simulation and verification system and a milling reference model. The milling reference model is established by milling experiments performed in order to enable safe machining processes and supports the machine operators in selection of cutting data for repair processes on turbine blades. The results of these research developments are applied to an European project in the field of Intelligent Equipment and Control.

1 Introduction

Jet engine components are an important field of modern aircraft technology. Especially the manufacturing of turbine blades and vanes due to safety and functional features will lead to higher engine efficiency, reduced fuel consumption and noise emission. On the other side, are turbine blades requiring regular and certified repair and maintenance cycles. The problem related to the manufacturing and repair processes is twofold:

- Due to optimised aerodynamic features modern turbine blades and vanes combine complex free form surface geometry and high finishing accuracy. Thus the machining processes require geometric flexibility and ensurance of precise cutting paths.
- Use of materials with high specific strength and heat resistance, mainly difficult-to-machine materials such as titanium alloys like TiAl6V4. The low heat conductivity and high chemical reactivity causes increased tool wear, high cutting temperatures and strong adhesion between tools and chips. Increasing the cutting speed leads to high thermal load of tool and workpiece, which affects the material microstructure. This transformation has to be avoided due to the functional behaviour of titanium parts under loading.

Titanium alloys are used mainly for the stationary vanes and the rotating blades in the "cold" compressor stages of a turbine. Since original manufacturing of turbine blades and vanes is done by contour milling under mainly automated manufacturing conditions, the repair processes are done by manual operation. The repair of turbine blades starts with the removal of the damaged area and replacement by raw block-material in a welding process. Finally the welded material has to be machined back due to the original shape. This operation is highly time and cost consuming and has a lack of sufficient process monitoring in view of traceability demands.

Automation of the machining process has not progressed significantly, due to the great difficulties encountered when trying to control specific aspects of a component's necessary repair characteristics:

- follow a precise path and adjust certain varying part parameters
- program the machining path to follow the complex geometric part surface in the repair sector.

Due to their exposure in the jet engine environment, airfoils can vary from other airfoils or new ones. It is therefore difficult to automate the machining processes required, because a unique repair process with its own parameters is necessary for each airfoil. Fig. 1 shows a typical turbine blade before repair takes place.

An advanced machining system based on a multi-axis high-speed milling center enables adaptive reshaping of blades with individual geometries due to the functional requirements. The RTD work performed so far to reach these goals are:

- Conceptualisation of a technology model combining geometrical and high speed milling features.
- Development of a milling reference model, based on analogous field test experiments, for selection of tool and coolant supply and setting of machining parameters. The milling reference model leads to a manufacturing database, which can be connected directly to the machine tool controller.
- Laboratory set-up of a 4-axis high speed milling center including tool and pallet changer and special fixtures such as experimental field test demonstrator, equipped with force/torque measurement sensors for in-process monitoring and evaluation of machining results.
- Set-up of a 3D shape measurement sensor system for integration in the machining platform, based on coded light and phase shifting principles.

The investigations and verification experiments performed so far contribute to increasing of process stability and tool life, improvement of surface roughness and avoidance of tensile residual stresses. The reference model implemented in the manufacturing demonstrator supports the staff involved in repair work preparation and the machine operators in selection of tools, coolant supply and machining parameters. The throughput time and machining costs can be substantially reduced. Furthermore, the part quality will be ensured at the required level.

Fig. 1. Jet engine components in repair status

2 Process Chain for Repairing Jet Engine Components

The machining of individual jet engine components as turbine blades and vanes require different technological issues than existing ones deliver. Applied systems should combine adaption features, easy and transparent programming and shorten the throughput cycles due to cost and delivery time demands.

To set up a multi-axis milling cell in view of reverse engineering and quality inspection, has been developed a process chain from geometry design to quality control of machined parts. The process chain is shown in Fig. 2 and consists mainly of two loops.

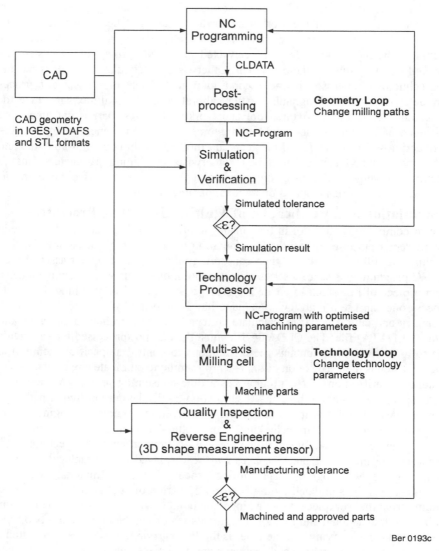

Ber 0193c

Fig. 2. Process chain for multi-axis milling cell

In the first loop a multi-axis NC programming system generates the milling path for repairing a turbine blade based on its geometry description. The geometry of the blade is constructed by a CAD system and then transferred into the NC programming system via neutral data exchange formats (e.g. IGES and VDAFS). The tool paths generated by the NC programming system describes the tool location of a pre-selected tool in CLDATA format. Then, a multi-axis postprocessor produces the necessary machine control codes for a specific NC-machine. These machine control codes are checked for their correctness by a realistic NC simulation system. In the simulation only geometric aspects of machining processes are considered. These include the detection of collisions in the machining environment, the visualization of material removal of the workpiece, handling of fixtures and jigs as well as the numerical verification of the workpiece surface quality. If deviations between the original CAD geometry and the simulated geometry exceed a specified machining tolerance, the NC program must be corrected in the NC programming system and checked again by the simulation system. Otherwise, the simulation result is put into the technology processor in the second loop that determines necessary process parameters for the NC programme. After calibration of the multi-axis machine tool center and set-up of the fixture, tool and workpiece, a test part is machined. The machined part is finally measured and a geometric model is generated by a reverse engineering software [1], [2]. If the geometry model is within the allowable tolerance zone, the real part can be machined. Otherwise, machining parameters must be adaptively changed. The machining technology processor is based on a milling reference model, which contains operational parameter settings.

3 Simulation and Verification of Multi-Axis Milling Processes

Due to complex geometry of turbine blades multi-axis milling machines are needed for the repair purpose. However, the generation of error-free NC programs for these machines is still a difficult and time-consuming task despite computer-assisted tools. An NC program consists of a series of cutter tool movements which remove material from a piece of raw stock to create a prototype part, mould or stamping die. In this process one can use a high level language like APT or a CAD system to define the geometry and cutter sequence, calculate cutter offsets and produce a cutter location data (CLDATA) file. The CLDATA file must then be postprocessed into a machine control data file which contains the instructions to control a specific machine tool. The next step is the verification of the NC programme to eliminate any errors.

Generally, verification of NC programs is a time-consuming process. Modern CAD systems provide a limited form of NC simulation by displaying machining paths with respect to workpiece. But it is difficult and often even impossible to imagine how the machine axes move just from looking at the cutter location data. This complicated process can be significantly simplified by using computer simulation technology. In principle, the motion of machine axes can be displayed on a graphical screen by compiling the NC programs. The milling process (removing material away from workpiece) can be simulated by the Boolean operation of subtracting the tool swept volume from the workpiece. From a kinematic point of view, a NC-machine is mainly built up of translational and rotational segments which can be divided into two types: one is workpiece carrying and the other is the tool carrying. These are represented by two kinematic chains, whose end segments are workpiece and tool respectively. Both chains start with the machine base as shown in Fig. 3.

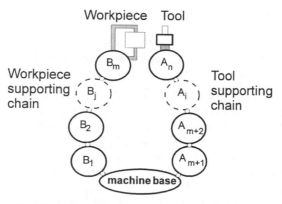

Fig. 3. Kinematic configuration of NC-machine

To describe the kinematic relationship between the segments the coordinate frames are embedded in each segment [3]. Hence, the kinematic configuration of these two chains can be represented in the model:

$$\begin{cases} T_m = B_1 * B_2 * ... * B_m & \text{for workpiece} \\ T_n = A_{m+1} * A_{m+2} * ... * A_n & \text{for tool} \end{cases} \tag{1}$$

In this model the (4x4) homogeneous transformation matrices T_n and T_m define the relative position and orientation of the tool with respect to the machine base respectively. Conceptually, the matrices A_i and B_j describe the spatial relationships between coordinate frames fixed relative to each of the two segments in the kinematic chains. These transformation matrices can be expressed as:

$$A_i = A(q_i) = \begin{cases} \text{Rot}(q_i) & \text{for rotat. sgmts} \\ \text{Trans}(q_i) & \text{for trans. sgmts} \end{cases} \tag{2}$$

where the motion axis is defined by the origin and motion direction. The input parameters to this model are the machine axis moving coordinate q_i. According to this kinematic model the transformation from the workpiece coordinate system to the machine coordinate system can also be easily derived for postprocessor development.

Since an NC-machine consists mainly of a set of segments, one can model the geometry of these segments separately. The geometric construction of the segments is done interactively by combining simple geometric primitives. The primitives currently implemented in the system are box, cylinder,sphere, cone and torus. They are geometrically defined by a few parameters, e.g. a box can be specified by length, height, width and a corner point. A solid modelling programme based on boundary representation is used to create and manipulate these primitives. To simplify implementation of the Boolean set operations the solid modelling programme is designed to handle only planar faced objects. It is, however, easily possible to approximate analytical surfaces as multiply-connected planar faces. For user convenience, the system supports also basic geometric operations of the object manipulations in the modelling space, such as translation, rotation and scaling etc. Once the segments have been geometrically created, they can be assigned to the corresponding machine components to build the whole machine geometry. This begins with the machine base component and completes at the tool and workpiece carrying components respectively. Each segment is specified with machine axis type, moving axis and working range. In order to build the simulation environment easily and quickly a library with a variety of geometric segments is also provided. A number of milling machines have been modelled by the system in [4].

4 Milling Reference Model and Manufacturing Experiments

The repair process for turbine components requires high demands on dimensional accuracy, surface roughness and material microstructure. To close the gap between simulation and verification of multi-axis milling processes, there will be established several milling reference processes according to material specifications in an experimental environment. This reference processes gain the required machining parameters; tool and workpiece behaviour and transformation of material properties under real machining conditions. Further experiments were performed in regard to increasing process stability and tool life, improvement of surface roughness and avoidance of tensile residual stresses. The reference processes will lead to an overall milling reference model, which supports both, the work preparation staff and the machine operators in choosing of machining parameters. The titanium alloy TiAl6V4 is in aerospace industrial use to a great extent and was also chosen for the first machining experiments. Titanium alloys perform high specific strength and heat resistance (strength-to-density ratio), but in any cases they are difficult to machine. The low thermal conductivity and high chemical reactivity of titanium alloys causes increased tool wear, high cutting temperatures and strong adhesion between tool and part surface [5],[6].

Accordingly, practical milling investigations on the titanium alloy TiAl6V4 were carried out on a multi-axis CNC-machining center. For this a representative series of uncoated and coated end mills of carbide metal and high speed steel were selected for the experiments. Experimental research focusses on cutting tool performance (e.g. breakage of cutting edges under high speed milling conditions and wear mechanisms while using different lubricants and cooling supply strategies). To characterize the performances of different tools and parameters, the cutting forces, the tool wear, the surface quality and the chip forming were analyzed.

The effect of the cutting speed on the tool life travel of uncoated HSS tools is shown in Fig. 4. It can be established that an increasing cutting speed reduces the tool life of HSS tools rapidly. An increase of the cutting speed to 90 m/min shortens the tool life down to 0,50 m. A further increase to 100 m/min is almost impossible. In this case, the tool failed after a milling path of 0,20 m. It can be established that milling processes under high speed conditions with cutting speeds over 100 m per minute are impossible with the HSS tools. For additional investigations, the experiments were extended with carbide tools. Compared with HSS tools the carbide tools have shown clearly better performance under high speed cutting conditions. Increasing cutting speeds also lead to reduced tool life travel, but compared with HSS tools higher cutting speed can be achieved. The effect of cooling lubrication on the tool wear is shown in Fig. 5. It illustrates the progress of the width of wear land VB_{max} in relation to the tool life travel. Generally, the correct use of coolants during machining operations greatly extents tool life. Especially in the case of milling the use of minimum quantity lubrication is suitable to reduce the tool wear [7]. In this case a special device atomizes the coolant and supplies it with very low flow rates.

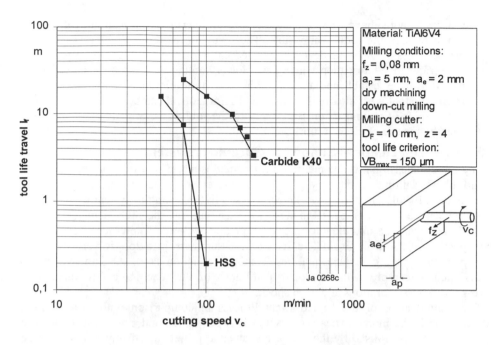

Fig. 4. Relation between tool life travel and cutting speed

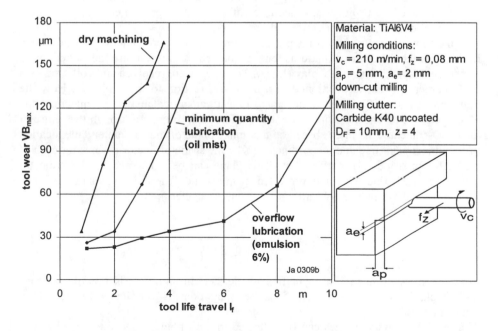

Fig. 5. Effect of cooling lubrication on the tool wear

The investigations have shown that the use of overflow lubrication with a 6% emulsion reduces the tool wear visible. The reduction of the heat generated by the process reduces also the tendency of welded chips on the tool surface. The use of minimum quantity lubrication which has a lower cooling effect is also shown in the figure. It can be established that also the use of minimum quantity lubrication decreases the tool wear clearly. For the use of HSS tools, it can be established that the tendency of the titanium to pressure weld to the tool flank can almost completely avoided by the use of oil mist lubrication.

5 Three Dimensional Shape Measurement System

The three dimensional shape measurement system is based on Coded Light Approach and phase shifting techniques [8]. The measurement process results are largely influenced by tolerances (e.g. workpiece, jig and fixtures and the system itself). These tolerances can be compensated for, by scanning of the 3D-geometry of the workpieces and the manufacturing environment as well as for intermediate and final inspection of the workpiece. The selected measurement principle is very robust in respect to changing surface qualities (coating, reflection etc.). Also the system is designed to comply with the rough conditions of the machining environment in respect to mist, vibrations and even shock.

The main feature of the Coded Light Process in comparison with Triangulation processes is the mathematical connection between space and time in a topometric scene, which is realised by the use of a sequential projection of n grey coded stripe patterns in the selected field of view.

The stripe patterns enable a differentiation of 2^n variable projection directions x_p, which can be securely identified by a characteristic bright/dark sequence and the specific code of the projection. A sequence of $n = 7$ stripe pattern enables therefore a differentiation of $2^7 = 128$ directions of projection.

Fig. 6a shows the physical principle of topometric measurement by use of Coded Light Process. The stripe patterns on the object of interest are generated by a transparent Liquid Crystal Display (LCD). The images are received and collected by a CCD camera, which is positioned in a defined direction to the projector. Now the point coordinates of a three dimensional object can be calculated by matching the plane generated from the stripe pattern and a line with direction through the centre of the CCD-camera. This line is based on the image plane of the camera and the location of the object point in camera coordinates {K}. The stripe projector defines for each of the different stripes a specific plane. This plane can be calculated by the projector centre and each line, which is projected in the scene. The existing planes are then according to the constraint (λ and μ are current parameters):

$$P_E = \lambda \begin{pmatrix} 0 \\ 1 \\ 0 \end{pmatrix} + \mu \begin{pmatrix} x_p \\ 0 \\ 1 \end{pmatrix} \tag{3}$$

For each image point can be a specific code defined through binarisation of the image and check, for which of the projections the image point is illuminated (value = 1) or not (value = 0).

The available binary images can be collected to a bit plane stack. Each pixel of this stack contains for the specific image point a n-ranging bit-sequence, which describes the projector direction to this scenery point. This code defines the projection direction x_p and the constraint of the plane. Therefore the measurement result of this process is singular and absolute. With the bit plane stack and the defined location of camera and projector the three dimensional coordinates of the object can be calculated. The main advantage of the Coded Light Process in comparison with other techniques is the fast image processing of topometric data. This is related to the fact, that regarding to the resolution of the CCD-camera only a small number of patterns has to be projected in the scene. Additionally, the analysis of the bit plane stack as well as the calculation of three dimensional coordinates can be done by simple image analysis operations. So a whole topometric scene can be measured within seconds. Also the Coded Light Process serves advantages at changing illumination conditions and different surface reflexions in comparison to other techniques.

Due to inaccuracies in scanning different views of the workpiece, points in the overlapping areas do not exactly occupy the same positions. This causes problems for constructing a solid model from the acquired data. Then, the data often includes points of environment elements such as fixtures etc. These points must be eliminated, otherwise the reconstructed model would be incorrect or invalid. Therefore, there has been developed a special-purpose algorithm overcoming the above mentioned difficulties. Fig. 6b shows a modelling example for a turbine vane. The model is scanned in six different views and obtained contains 23152 points.

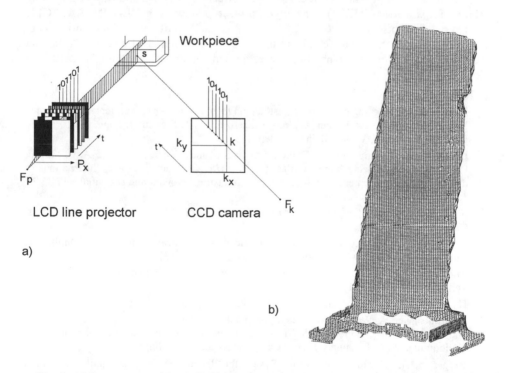

Fig. 6. Measurement of the Coded Light Approach and modelling example for a turbine vane (length appr. 150 mm)

6 Summary

The basic concept of the described developments is the integrated process chain for multi-axis milling. Due to the requirements in machining of individual jet engine components as turbine blades and vanes specific attention is paid to high-speed milling aspects. The geometrical loop concerning cutting path planning and adaptation is based on a mathematical model for a generic machine tool system. Features like collision avoidance, jig and fixture modelling and universal postprocessing are included. For quality inspection and reverse engineering purposes a 3D shape measurement system based on Coded Light Approach and phase shifting techniques is installed. The presented results of the carried out investigations show an overview of the possibilities for influencing the milling process of titanium alloys. Due to the investigations it can be established that higher cutting speeds on the machining of TiAl6V4 can be realized. The appropriate selection of tool materials, cutting parameters and coolant lubrication constitutes the basis for a safe machining process. Further investigations will be performed to achieve a wide range of machining data for suitable milling conditions to ensure increasing process stability. The pursued parameter database supports the machine operators involved in repair work preparation and selection of cutting data.

7 Acknowledgement

The presented paper is based on an RTD project funded by the the European Union in the ESPRIT Program with the title: Advanced Mechatronics Technology For Turbine Blades Repair (AMATEUR). The other project partners are ZENON S.A. (GR), Hellenic Aerospace Industry (GR), APS Gesellschaft für Automatisierung, Prozeßsteuerung, Schweißtechnik m.b.H. (DE), Bremen Institute of Industrial Technology and Applied Work Science at the University of Bremen (DE), Staubli France S.A. (FR).

8 References

1. U. Berger, Entwicklung eines Sensorgestützten Planungs- und Programmiersystems für den Industrierobotereinsatz in der Unikat- Einzel- und Kleinserienfertigung. Dissertation, Universität Bremen, Verlag Mainz, Aachen, 1995.

2. U. Berger, X. Sheng and A. Walter, Simulation and Verification of 5-Axis Milling Processes for Repairing Turbine Blades, *SURFAIR XI*, 11ème Journées Internationales d'Etude sur les Traitements de Surfaces dans l'Industrie Aéronautique et Spatiale, 12/13/14 Juin 1996, Cannes.

3. R. P. Paul, Robot Manipulators: Mathematics, Programming and Control, MIT Press, Cambridge, MA, 1981.

4. X. Sheng, Solid Modelling Based Geometric and Graphical Simulation of Multi-axis Milling Processes, PhD Dissertation, University of Bremen, 1992.

5. M. Eckstein et.al., Schaftfräsen von Titanlegierungen mit hohen Schnittgeschwindig-keiten, *VDI-Z* **133** (1991) 12, 28-34

6. H. Schulz, High-Speed Machining, Carl Hanser Verlag, Munich, 1996

7. E. Brinksmeier; A. Walter: Examples of the Employment of Reduced Amount of Coolant and Dry Machining, *Tribology - Solving Friction and Wear Problems*, 10th International Colloquium, 09. - 11. January 1996, Technische Akademie Esslingen.

8. Berger and A. Schmidt, Active Vision System for Planning and Programming of Industrial Robots in One-of-a-kind Manufacturing. *Proc. SPIE*, Automated 3D and 2D Vision, Casasent, D.P.; Hall, E.L.(eds.), VOL. 2588, SPIE Bellingham, WA, U.S.A., 1995.

The Development of an Intelligent Flexible and Programmable Vibratory Bowl Feeder Incorporating Neural Network

Patrick Soon-Keong Chua, Meng-LeongTay and Siang-Kok Sim

School of Mechanical and Production Engineering
Nanyang Technological University
Nanyang Avenue
Singapore 639798

Abstract This paper is concerned with the development of a vibratory bowl feeder which is not only flexible and programmable but also intelligent with neural network incorporated. Controlled by a microcomputer and driven by electropneumatics and mini-motors, the feeding system is capable of identifying the orientation of non-rotational parts with external and internal (e.g. blind holes of different depths and diameters) features as well as having the ability to actively re-orientate the parts so as to feed them in the desired orientation. The system developed was evaluated and the test results showed that with fine tuning, the feeding system would be an invaluable asset to industry.

1 Introduction

A parts feeder can be defined as a device that receives randomly oriented parts at its input and delivers these parts in the desired orientation at its output. The vibratory bowl feeder is a very common parts feeder in industry. The industry has accepted the use of the vibratory bowl feeder for parts feeding for many decades now and it appears that the industry is unlikely to turn away from it in the forseeable future. The reasons arc:

(a)   Its compactness.
(b)   Its proven track record in its ability and reliability to convey parts smoothly along the vibrating track.

However, two of its main draw backs are its mainly passive orientating system and its inflexibility in feeding different parts with the same feeder. In today's highly competitive manufacturing environment, the ability to rapidly adjust to product changes due to changing market demand is a key factor in survival. To make automated workpiece orientation feasible in batch production, it requires the feeders to be flexible, reliable, able to accommodate a wide range of parts and deliver them at the required feeding rate.

A programmable parts feeding system should have the flexibility to orientate most of the parts from one or more part families, with down time to changeover from one part to the next not requiring more than a few minutes. Attempts at flexible assembly system such as G.M.s' Programmable Universal Machine for Assembly (PUMA), Westinghouse's Adaptable Programmable Assembly System (APAS), and Olivetti's Programmable Assembly System (SIGMAS) have emphasized the need for such flexible feeding devices. Salmon and d'Auria [1] states that".......the lack of flexible feeding devices causes the main limits of programmable assembly". Beecher [2] states that "Whether or not programmable assembly devices succeed could very well be determined by the success or failure of parts feeding and orienting mechanisms".

2 Objective

The objective of this work is to develop a computer-controlled, flexible and intelligent programmable vibratory bowl feeder with the following salient features:
(a) Ability to reliably identify the orientation of moving parts on the vibrating track of the bowl feeder
(b) Application of neural network in the feature recognition process
(c) Ability to flip inverted moving parts travelling on the vibrating track
(d) Ability to perform active re-orientation of parts to the desired orientation
(e) Ability to reject parts that are faulty or unrecognisable

3 Literature Review

Several researchers such as Maul and Goodrich [3] had attempted to transform the dedicated vibratory bowl feeder into a flexible and programmable feeding system. They converted the vibratory bowl feeder into a flexible and manually programmable feeder which could be used for batch as well as mass production. The orientating devices such as wiper, pressure break, narrow track, slotted track, etc. of the bowl feeder were made adjustable by means of lead screws and dials were used to record the positions of the tools which are located along the bowl track near to the exit. The feeder could be programmed by placing a part in the desired orientation at each tool whose position was then adjusted and recorded.

Other attempts had also been made to increase the versatility of automatic parts feeders. One approach was to restrict the feeder's ability to only parts singularisation, leaving parts recognition to sensors at the exit of the feeder and the subsequent orientation by a manipulator [4]. This unfortunately slowed down the manipulator. The ideal situation was to have the parts already correctly orientated and waiting for the manipulator at the pick-up point.

Suzuki and Kohno [5] had developed a multi-parts feeding system using a multi-storey bowl feeder, separating unit and chutes and a parts holding/reversing unit in conjunction with a camera. The bowl feeder singularised the different types of thin parts in either one of two possible orientations via the chute to the holding/reversing unit where its orientation was identified by the camera system. Flipping then took place if the part was inverted. It could then be picked up by the manipulator for assembly. An important feature of their system was that the bowls did not need to be configured to perform part orientation and hence were easily adapted to accommodate different although limited part types.

Similar developments using the vibratory bowl feeder and camera system for image recognition were carried out by Cronshaw [6], Uno [7] and Warnecke et al [8]. The image recognition system was first taught to recognise the various possible input orientations of a part in the teach mode and the results stored in a computer as reference signatures. In the run mode, the image acquired was compared to the reference data for identification of orientation.

The above camera vision techniques, although achieving the objective of providing flexibility and programmability, were unable to detect the orientation of parts with only blind features and differentiate between blind features of different geometry. In an attempt to avoid the problems associated with camera vision systems (e.g. high costs, lighting, other environmental problems), Park [9] used fibre optic sensors to create a pattern of on/off signals as parts moved past the strategically located sensors. This silhouette detection technique would still be unable to detect blind features on a part.

4 Description of the Vibratory Bowl Feeder System Developed

The vibratory bowl feeder is designed to be flexible, programmable and intelligent. Instead of welded passive orienting devices such as wiper blades and permanently constructed track width, mini-motor controlled wiper blades and adjustable track width are incorporated into the system to make it programmable to suit parts of different sizes. As shown in Figure 1, the vibratory bowl feeder consists of nine specially designed stations along its track for feeding of non-rotational parts. These stations are controlled by a micro-computer.

4.1 Wiper Blade Station
Basically, this is a passive orienting device commonly used in the vibratory bowl feeder to reject or wipe off components that are stacked on top of one another and also those components that are higher than the set height limit of the wiper blade (i.e. specimen resting on its side). In this feeder, in order to achieve flexibility of such device, the wiper blade was attached to a ballscrew assembly driven by a miniature stepper motor, mounted vertically above the bowl track. This allows the wiper blade to be set at any desired height controlled by the computer.

4.2 Programmable Track Width Station
This is also a passive orienting device commonly used in the vibratory bow feeder to ensure that all the components would travel in single file longitudinally on the track, rejecting components that are travelling abreast to another part. In this feeder, the programmability of the track width is achieved by constructing a hinged wall pair that guides the components into a narrowing track. This hinged wall pair is also attached to a ballscrew assembly driven by a miniature stepper motor and therefore allows the narrow track to be set at any desired width controlled by the computer.

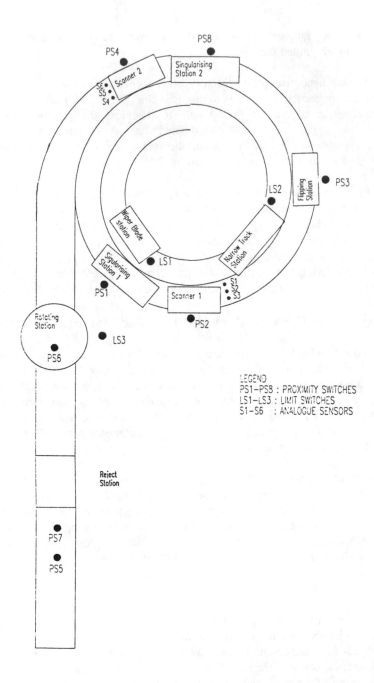

Figure 1. Schematic Layout of the Vibratory Bowl Feeder

4.4 First Scanning Station
This station serves to determine the orientation of a feeding component with an array of analog fibre optic sensors (scanners -- OMRON E3XA). The technique of sensing will

be discussed in section 5. When the orientation of the feeding component has been determined, the computer would decide on one of the following:

(a)   no action if component is in the desired orientation or 180° reversed.

(b)   request flipping operation if the component is upside-down.

(c)   activated scanning station #2 if a poor scan or identification is encountered.

4.5 Flipping Station

This station is designed to flip the component which is identified to be upside-down by the first scanning station. The orientating device consists of a 90° vee-track, an air-jet and a fibre-optic sensor. The flipping operation is activated upon request by the computer, after determining that the component is upside-down. Therefore, as the feeding component arrived and is sensed by the fibre-optic sensor, the air-jet would blow the component to rest on the other side of the vee-track, and thereby flipping it 180°.

4.6 Second Scanning Station (Scanner No. 2)

This station would make a second identification to determine the orientation of the feeding component. This station uses the multi-beam sensor which is modulated for ambient light rejection. In the same way, the computer would decide the appropriate actions to be taken after scanning such as:

(a)   send "accept" signal when component is in desired orientation.

(b)   request rotating operation if the component is 180° reversed.

(c)   reject the component if a poor scan or identification is encountered

(d)   reject an unflipped upside-down component.

In addition, as a precautionary measure, it would compare the scanning result with that of the first scanning station and would also attempt to use the successful first scan result whenever possible if a poor scan or identification is encountered.

4.7 Reject Station

This stage would reject any component upon sensing it unless an "accept" signal is received. This is to ensure that only parts determined to be in the desired orientation will enter the discharge chute. The reject station consists of a fibre-optic sensor and a diverting chute. Upon sensing of feeding parts by the fibre-optic sensor, the chute would open to divert the components out of the track. This reject operation would be inhibited when an "accept" signal is received from the computer.

4.8 Rotation Station

This is the last station of the whole system and is used to re-orientate a component that is 180° reversed. It consists of a 180° pneumatic rotary actuator which rotates a short chute by 180°, two pneumatic actuators acting as stoppers at the entrance and exit of the chute and two fibre-optic sensors as part sensor. The overall sequencing of the rotation operation is controlled using a relay logic circuit.

The computer would signal the relay logic circuit to perform a rotation operation when required. The exit of the orientator would first be closed by the pneumatic actuator to catch the component. Upon sensing of feeding component in the orientator, the entrance would be closed by another pneumatic actuator and the rotary actuator would perform a 180° rotation before releasing the component out from the orientator. The feeding component would then slide out and activate another fibre-optic sensor to signal "end of cycle" and return the orientator to its initial condition. If the component is in the desired orientation, no signal would be generated by the computer and it would be allowed to slide through the station as it moves down the chute.

5 Pattern Recognition of Bowl Feeder System

A pattern/feature is a description of an object. It contains discriminatory information and characteristics pertaining to that object. A good pattern will be one that is as unique to the object as possible. Hence, pattern recognition is defined as the categorisation of input data or information into the appropriate pre-defined classes of information via the extraction of significant attributes or features from a vast amount of details.

In general, there are three main phases in pattern recognition. The first is the data acquisition phase, whereby a pattern representation for the objects has to be found. The pattern is normally digitised by a vision system. Digitisation is the conversion of the image of a scene or picture into discrete/digital form. The second phase is the data preprocessing and feature extraction phase, whereby the image is being processed to extract salient features. Lastly, the classifier which is in the form of a set of decision functions is applied to the extracted feature. This is the recognition stage where the input pattern will be recognised. The most commonly used techniques for pattern recognition can be categorised as follows:

(a)   Template matching
(b)   Structural and syntactic techniques.
(c)   Statistical pattern recognition techniques
(d)   Neural network.

The first three methods rely on pre-stored patterns (e.g. template patterns, ordered composition of simple subpatterns or primitive patterns, or features extracted from patterns) as basis for comparison with input pattern to perform the recognition task. These methods are not easily amendable to changes in pattern configurations that one would expect in large variety small batches of parts required in today's dynamic manufacturing environment. By virtue of its ability to learn and hence adapt to changing patterns/features of different batches of parts required in part orientation system, neural network is chosen for the task of feature recognition of parts. Furthermore neural network is less susceptible to variation in lighting conditions which might result in noisy or incomplete digitised input after pre-processing of images of the vision system.

5.1 Feature Recognition using Neural Network
The feature recognition ability of the bowl feeder system is achieved by four modules, namely:
(a)   Part singularising station.
(b)   Collection of scan signatures.
(c)   Preprocessing and image extraction.
(d)   Neural network module.

5.2 Singularising Station

This station is used to control the flow of part into the scanner. The singularising station which consists of a hinge, a pneumatic cylinder, and a fiber optic proximity sensor, will release one part at a time for scanning. The computer will first send a signal to singulate or release a part. The cylinder will then retract and a part will move forward until the part reach the proximity switch. The proximity switch will then send a signal to cause the cylinder to extend again and stop the subsequent part from moving forward.

5.3 Collection of Scan Signatures

The scanners consist of three analog fiber optical sensors arranged in a row across the track. The line image sensing technique employed in this system. Three signatures of the part profile will be captured by the three analogue sensors in the scanner block after the part has passed through it.

For full analogue conversion, it was found that the pacer rate of the data acquisition card must not be set above 100 kHz. A memory for an array of 600 floating point numbers was allocated for the purpose of storing the raw pattern. Since each row in the raw pattern set has three readings, the scan signature must not exceed 200 rows (200 × 3 = 600). Hence, the pacer was set at around 1-2 kHz to get around 100 scan rows of readings. A sample of the scan signature at pacer setting of 1.6 kHz and bowl speed of 8 is given in Figure 2. The collected signature has to be preprocessed before it is passed into the input of the neural network.

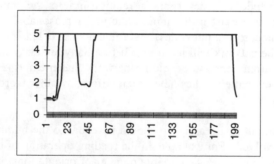

Figure 2: Sample of the scan signature

5.4 Preprocessing and Image Extraction

The collected signatures will be passed through a preprocessing module to extract the input pattern file as input into the neural network module. For example, if the input nodes of the network is specified as 30, then the preprocessing module will compress the raw pattern collected into 3 × 10 numerical values to be input to the network. The raw pattern collected usually consists of over 100 rows of values or 3 × (over 100 rows) of numerical values (See Figure 3).

5.5 Neural Network Module in Bowl Feeder System.

A control program has been developed to operate the training and operation of the neural network in this feeder system through a user interface menu. A brief description of the operation of the neural network module is as follows:

5.5.1 Configuring the Neural Network

The architecture of the feedforward backpropagation neural network [10] in terms of the number of input nodes, number of hidden nodes and number of output nodes can be re-configured through the settings in the configuration file. In general, the selection of the number of input nodes would depend on image resolution selected in relation to the part feature that could be discriminated.

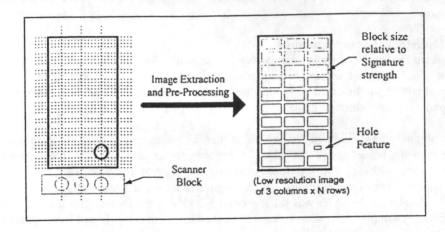

Figure 3. Preprocessing and Image Extraction

The selection of hidden nodes bears a relationship to the number of statistically significant factors that exist in the input pattern. A reasonable value to start with is given by ((input nodes + output nodes)$^{1/2}$ + a few nodes) [11]. If too few of the hidden nodes are used, the network will not train. If too many are chosen, the network beside taking forever to train, tends to develop a network that memorise everything and will not recognise a new pattern it has never seen before, in other words, the network does not generalize.

The number of output nodes assigned should be equal to the number of classifications or orientations involved. For example, in the feeding operation of the mixed specimens of A and B (shown in Figure 4), possible number of orientations of the parts are seven and hence the number of output nodes is set to seven when configuring the network.

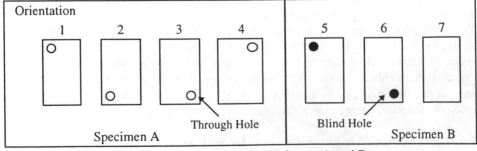

Figure 4. Orientations of Specimens A and B

5.5.2 Training the Neural Network

Two methods can be used in the training of the neural network. One method is to complete both weight adjustments of feedforward and backward activities in a single pass for each pattern presentation. The error is propagated back and the connecting weights are adjusted after each pattern is presented to the network. This is known as online training. The other method is to accumulate the changes of weights for all training sets in a batch, and then propagate the error back based on the grand total change in weight. This is called batch or epoch training.

The advantage of online training is that less memory is needed during training when compared with batch training. However, the disadvantage is that the network may 'unlearn' what it has learned in the previous pattern when it is trained by the present pattern. The batch training method is adopted for training the network in this project.

In this experiment, the neural network will be trained using the following training pattern sets collected:

> For scanning station #1 - S1SET1.PAT, S1SET2.PAT, S1SET3.PAT.
> For scanning station #2 - S2SET1.PAT, S2SET2.PAT, S2SET3.PAT.

The raw scan signatures of S1SET1.PAT collected are plotted and a sample of which is shown in Figure 2. The training pattern sets consist of all the seven orientations of the mixed specimens A and B. The neural network will be trained to recognise the seven orientations of the two specimens as shown in Figure 4. The training of the neural network consists of the preparation of the training configuration file (TRAIN.CFG), the preparation of the training pattern files (*.PAT) as inputs to the neural network module. After the training, weight files obtained are then used to operate the feeding action.

6 Testing the Accuracy of the Feature Recognition

6.1 Flipping Station Performance: Test on Recognition Ability of Scanner #1

The experimental procedure to test the scanner 1 (OMRON EX3A) recognition accuracy on the seven orientations of specimens A and B is described as follows:

Step 1 : The parts are arranged in orientation 1 on the bowl track.

Step 2 : Change the classification action of the TEST.CFG file to 1 0 0 0 0 0 0 0 0 0. This will cause the flipping station to flip part in orientation 1 and perform no action on the other orientations.

Step 3 : Record the number of parts in orientation 1 not flipped by the flipping station.

Step 4 : Go to Steps 1 to 3 but with the parts arranged in orientation 2 on the track and the classification action of the TEST.CFG file changed to 0 1 0 0 0 0 0 0 0 0. This procedure is repeated for all the other orientations (i.e. 3 to 7).

The results of the test are summarised in Table 1 below.

Table 1. Test Results on Flipping Station

Orientation	1	2	3	4	5	6	7	Average error (%)
Unflipped parts /100 parts	12	3	26	5	50	52	13	23

From Table 1 above, it can be observed that various orientations of specimen A (with through holes) gives a better set of results than that of specimen B (with Blind holes) Orientations 2 and 4 of specimen have better results when compared with the rest of the orientations. For the recognition of the blind holes specimens (orientations 5 and 6), the error is approximately 51%. This implies that either the sensors cannot differentiate the depth feature well or the neural network was not trained properly to recognise the depth variation.

6.2 Rotation and Reject Stations : Test on Recognition Ability of Scanner # 2

Similar to the flipping station test, the experimental procedure for the test on rotation and reject stations is as follows:

Step 1 : Arrange the parts in orientation 1 on the track.

Step 2 : Change the classification to 2 0 0 0 0 0 0 0 0 0 in the TEST.CFG file. This will cause the parts in orientation 1 to be rotated.

Step 3 : Record the error.

The above steps are repeated for the rest of the orientations with the appropriate classifications settings. The results of the performance of rotation and diverting chute stations are summarised in Table 2.

From Table 2, it is noted that the recognition ability of scanner 2 is very good.

Table 2. Test Results on Rotation/Diverting Chute Stations

Orientations	1	2	3	4	5	6	7	Average % error
No of mistakes per 100 parts	0	0	0	0	2	0	0	0.03

6.3 Testing on the Accuracy of the Overall Feeder System

The experimental procedure is as follows:

Step 1: Mix an equal number of specimen A and specimen B (20 from each group) and place the mixed parts in the bowl.

Step 2: Set the classification of the TEST.CFG file (0 1 2 1 3 3 3 0 0 0) such that the system orientates parts to orientation 1 at its output.

Step 3: Start the feeding operation of the feeder system.

Step 4: Record the number of parts in wrong orientations collected in the delivery chute and the number of specimen A rejected.

The results of the test is summarised as followed :

Table 3: Test results of the Overall Feeder System

No. of parts in wrong orientations at the delivery chute per 100 parts processed by the system.	0
No. of specimen A rejected by the system per 100 parts processed by the system	6

From the results in Tables 1 and 2, it is observed that there is a significant difference in recognition accuracy between scanner 1 and scanner 2. Scanner 1 has an accuracy of approximately 77% and scanner 2 an accuracy of approximately 99.97 %. Since both the scanners use the same neural network, the different in the accuracy is most likely due to the scanning ability of the sensors used.

From observations on the scan signatures collected from both stations, it is noticed that scanner 1 signatures are more ambiguous than scanner 2 signatures. This is caused by the inferior recognition ability of scanner 1 when compared with scanner 2. From the data collected in Table 3, out of 100 parts processed by the system, 6 parts of specimen A is rejected by the system. However, from Table 2, the scanner 2 can recognise specimen A almost perfectly, thus the rejected specimen A are probably due to the unflipped parts caused by the errors of scanner 1 and flipping station.

7 Discussion

From the results of the experiments described in Section 6, the main factor that impairs the recognition capability of the system is the sensor's reliability at scanner 1. Although various parameters of the neural network has an effect on its classification ability, the performance of the network is dependent on the input pattern used for training. Since real patterns are used for the training, the reliability of the sensor has a direct effect on the performance of the neural network. Hence, by replacing the sensors at scanner 1 with better sensors (e.g. multi-beam), the recognition ability can be enhanced because the latter is modulated for ambient light rejection..

Another problem of the feature recognition may be caused by irregular movement of part during the scanning process. This may cause the scanner to read in a pattern that is quite different from that used in the training set. Although the neural network can be trained to recognise as many patterns as possible by including all those patterns near the 'boundaries', it may be too time consuming to collect all the possible 'boundaries' pattern.

Another problem observed is that the sensors used in the scanner is not spaced close enough. This could cause some important feature on the part to be missed by the sensors. The present data acquisition card allows up to 16 single ended or 8 differential analogue inputs. Hence additional analogue sensor can be added to the scanner for a bigger and closer scan.

The classification ability of the feedforward backpropagation neural network is dependent on the number of different features of the similar parts used during training. Should different parts with different features be required in the manufacturing process, there is a need to re-train the neural network each time a different batch of parts are used. For the intelligent bowl feeder to be highly flexible, one approach is to used the Adaptive Resonance Theory (ART) neural network [10, 12]. ART is a two-layer nearest-neighbour classifier that stores an arbitrary number of spatial patterns using competitive learning. The neural network architecture is chosen because it updates its memory with any new input pattern without corrupting the existing patterns. The training algorithm conducts a search through the encoded patterns associated with each class of components trying to find a sufficiently close match with the input pattern. If no class exists, a new class of pattern is learned.

8 Conclusion

A flexible, programmable and intelligent vibratory bowl feeder with neural network incorporated has been developed. It represents one of the first attempts, if not the first, to perform feature identificaton of the feeding parts on the vibrating track as well as using neural network. It converts the vibratory bowl into a more versatile parts feeding equipment and also retains the compactness of the vibratory bowl. Evaluation results have been encouraging and with further fine tuning, the feeding system will be an important productive asset to industry.

References

[1]   Salmon, M. and d'Auria, Programmable Assembly System, *Computer Vision and Sensor-based Robots*, Plenum Press, London, 1979.

[2]   Beecher, R.C., "PUMA: Programmable Universal Machine for Assembly", *Computer Vision and Sensor-based Robots*, Plenum Press, London, 1979.

[3]   Maul, G.P. and Goodrich, J.L., "Programmable Parts Feeder", *Manufacturing Systems*, Vol. 13, No. 2, pp. 75 - 81.

[4]   Parham, R.J., "Automatic Parts Handling Systems", *Proceedings, of the 2nd International Conference on Assembly Automation*, U.K., May 18 - 21, 1981, pp.161 - 166.

[5]   Suzuki, T. and Kohno, M., "The Flexible Parts Feeder which Helps a Robot Assemble Automatically", *Assembly Automation*, Vol. 1 No. 2, Feb. 1981, pp. 86 - 92.

[6]   Cronshaw, A.J., "A Practical Vision System for Use with Bowl Feeders", *Assembly*, Sept. 1990, pp. 265 - 274.

[7]   Uno, M., "Multiple Parts Assembly Robot Station with Visual Sensor", *5th International Conference on Assembly Automation*, 22 -24 May 1984, Paris, France, pp. 55 - 64.

[8]   Warnecke, H.J., Schraft, R.D., Lindner, H., Schmid, S. and Sieger, E., "Components and Modules for Automatic Assembly", *Proceedings of the Automation-in-Manufacturing'91* (AiM'91) International Conference, 1991, Singapore.

[9]   Park, I. O., "Development of a Programmable Bowl Feeder Using Fiber Optic Sensor", *10th International Conference on Assembly Automation*, Japan, 1989, pp. 191 - 198.

[10]   Fausett, L., *Fundamentals of Neural Networks*: Architecture, Algorithms, and Applications, Prentice Hall International, 1994.

[11]   Russell C. Eberhart and Roy W. Dobbins. *Neural Network PC Tools.-A practical guide.* Academic Press, 1990.

[12]   Vellani, M and Dagli, C. H., Artificial Neural Network Approach in Printed Circuit Board Assembly, *Journal of Intelligent Manufacturing*, 1993, Vol. 4, No. 4, 109, 119, Chapman & Hall.

Enhanced Intelligent Control for Strip Thickness in a Rolling Mill

Shaowei Wang and A. Kiet Tieu

Department of Mechanical Engineering, University of Wollongong, NSW 2522, Australia

Abstract This article addresses two enhanced intelligent control schemes for rolling process. A mathematical model is formulated to simulate a rolling mill thickness control process; then two intelligent control schemes, an integral neural control and an enhanced fuzzy control are developed for the process. The integral neural control is composed of an inverse neural network controller, a PID controller and a process modelling neural network. The enhanced fuzzy control consists of a fuzzy inference controller and an adaptive neurofuzzy model parallel to the rolling process. The simulation results show that the integral neural control could significantly improve the strip thickness, and more excellent performance is obtained with the enhanced fuzzy control.

1   Introduction

The conventional PID control could achieve good performance for linear systems, but has a limitation when dealing with non-linear plants, whereas neural networks and fuzzy inference systems are finding increasing use in complex process control, due to their non-linear capabilities.

The application of neural networks for rolling process mainly focused on the parameter estimation[1-5]. The use of neural models for direct control had been investigated only by D. Sbarbaro-Hofer *et al* [6], in which static neural networks were created with data obtained by solving several equations of an analytical model of rolling mill. But the motivation of using neural networks to replace mathematical models (the equations of analytical models of the plants) is that neural network based controllers generally depend less on *a priori* knowledge of the process to be controlled but more on the input-output observations. And the input-output observations are the data which could be collected from the dynamic process, rather than those from the static equations. D. Sbarbaro-Hofer *et al* did not contribute a strategy of real time process application value. Fuzzy inference system has been successfully used for gain tuning in thickness control of rolling mill [7], but fuzzy application in rolling process has not been discussed widely in the literature.

In this article, a dynamic mathematical model for the rolling process is formulated which is represented by the connection of multiple transfer function blocks. Based on the simulation results of this dynamic model, a rolling process modelling neural network and an inverse neural network are developed, which, together with a PID controller, compose an integral neural control. Within the fuzzy control scheme, a fuzzy inference controller is implemented with rolling technology expertise, and based on the mathematical model simulation results, an adaptive neurofuzzy model is incorporated to enhance the control structure. The methodology proposed in this article could be verified by and applied to real time process.

2   Control Problem

The rolling process of a four-high mill could basically be described as shown in Figure 1, where f is rolling force, H strip input thickness, h output thickness, L and l the distances of input and output thickness sensors from rolling mill respectively.

Suppose S_0 is the roll gap with zero rolling force ($f = 0$), K is the rolling mill elastic modulus, the strip output thickness is given by

$$h = S_0 + f / K \tag{1}$$

where the rolling force is [8]

$$f = \sigma_c \left[\sqrt{\frac{DH\gamma}{2}} + \alpha \sqrt{\frac{fD}{E}} + \frac{\alpha^2 \mu Df(2-\gamma)}{2EH(1-\gamma)} \right] \tag{2}$$

where σ_c is compressive yield stress of the material being rolled(N / mm^2), D diameter of work rolls(mm), γ fractional reduction ($\gamma = \dfrac{H-h}{H}$), α dimensionless constant of value 1.08, E elastic modulus of the work rolls(N / mm^2), and μ effective coefficient of friction in the bite.

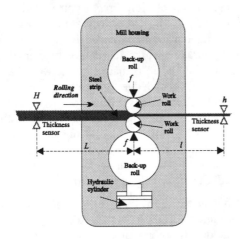

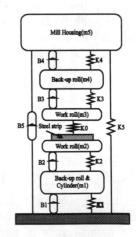

Fig. 1.   General sketch of a four-high mill rolling process

Fig. 2.   Mathematical model of a four-high rolling mill

The control objective is to maintain the output thickness as close as possible to the desired value h_{ref}. Equations 1 and 2 indicate that the input thickness variation ΔH will result in the output thickness deviation Δh, and their relationship is non-linear. The basic control concept is adjusting the work roll gap by ΔS to compensate the output thickness deviation Δh incurred by input thickness variation ΔH.

$$\Delta S = (M / K)\Delta H \tag{3}$$

where M is plastic modulus of steel strip, the value of which varies as to different input thickness values and different reduction values.

For this kind of non-linear process, the combination of pre-compensation control (feedforward control) and monitor control (feedback control) is conventionally a representative control scheme, as shown in Figure 3, in which the hydraulic valve and cylinder are omitted for simplicity. In order to increase the control performance to the maximum, it is necessary to set the parameters (e.g. M, K etc.) accurately at every operational point (corresponding to every different input thickness and reduction value) for pre-compensation control, and to obtain the accurate and prompt value of output thickness for feedback control. Unfortunately, it is difficult to obtain the values of M and K with sufficient accuracy, and the output thickness transport delay exists inevitably due to the distance between the output thickness sensor and the roll bite. As a result, an integral neural control and an enhanced fuzzy control are approached.

The control performance is indexed by the integral of the square error between output thickness reference value and measured value over an unit time, which is defined as

$$I = \frac{\int_{t=t_0}^{t_{max}} \left(h_{ref} - h\right) dt}{t_{max} - t_0} \tag{4}$$

The test input thickness profile is shown in Figure 4, and the desired output thickness is 3.5 mm.

3   Mathematical Model of Rolling Mill

In order to simulate the rolling process control and obtain data for neural network training and neurofuzzy model adapting, a mathematical model for the rolling mill is formulated.

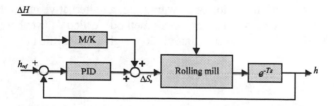

Fig. 3.   Feedback linear control and feed forward control for strip thickness

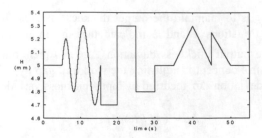

Fig. 4. Input thickness profile

Normally, a mechanical element of translational motion can be described as a mass-spring-damper system[9], in which the spring represents the elastic behaviour, and the viscous friction is represented by the damper. Therefore, the four-high rolling mill is modelled by a five-degree-of-freedom mass-spring-friction (damper) system, as shown in Figure 2, in which the roll force and hydraulic screw-down cylinder are considered as external forces to the system.

The force equation of the system is

$$\{[M]s^2 + [B]s + [K]\}[X] = [F]$$ (5)

where $[M]$ is mass matrix (kg), $[B]$ damping friction matrix (N/m/s), $[K]$ spring constant matrix (N/m), $[X]$ mass displacement vector (m) and $[F]$ external force vector (N). In block diagram, the Equation 5 is represented by connection of multiple transfer function blocks[10].

4   Integral Neural Control

As neural networks have great capabilities in solving non-linear problems by learning from the observations of input-output performance of the process[11], they are used to model the rolling process dynamics and the process inverse, with which the process modelling network and the inverse network are obtained.

There are several types of neural networks for non-linear system modelling, such as multiple-layer perceptron, radial basis function (RBF) network and back-propagation (BP) network. Here the most popular network BP network is used. To overcome the conventional BP network limitation of slow training, an alternative method to gradient descent, an approximation of Newton's method called Levenberg-Marquardt optimisation is used. The Levenberg-Marquardt weight update rule is

$$\Delta W = \left(J^T J + \lambda J\right)^{-1} J^T e$$ (6)

where ΔW is weight change matrix, J the Jacobian matrix of derivatives of each error to each weight, λ a scalar, and e an error vector. When the scalar is very large, the above expression approximates gradient descent, whereas when it is small the above expression becomes the Gauss-Newton method. The Gauss-Newton method is faster and more accurate near an error minimum, so the aim is to shift towards the Gauss-Newton method as quickly as possible. Thus, λ is decreased after each successful step and increased only when a step increases the error.

4.1 Feedforward Control with an Inverse Neural Network

With an inverse neural network, the feedforward control to compensate the output thickness deviation could be formulated as shown in Fig. 5, in which the hydraulic valve and cylinder are omitted for simplicity. The neural network for the rolling process inverse is trained with data from the simulation results of PID feedback control, taking input thickness and output thickness as inputs and roll gap as output.

4.2 Integral Neural Control

The result of feedforward control with an inverse neural network is much better than that of linear PID feedback control. But simulation results also indicate that the transient response of this control is still not quite satisfactory. Obviously that is the inevitable outcome of open loop control system, due to the dynamic behaviour of mill stand. In this sense, a feedback PID linear controller parallel to feedforward inverse neural network controller is implemented.

To overcome the output signal delay problem, a process modelling neural network is applied to estimate output thickness, which is trained with the data collected from PID control simulation results. The estimated output thickness is used to replace the delayed process output thickness as the feedback signal. The error between the process network output (after delayed) and rolling process output (only delayed value available) is also used as a feedback signal and passed to PID controller, taking into account the uncertainty of the process and inaccuracy of the process network.

The control structure is as shown in Figure 6, in which the hydraulic valve and cylinder are omitted. In this way the control performance is improved significantly. Figure 9b is the response of this control to the input given by Figure 4.

5 Enhanced Fuzzy Inference Control Structure

Fuzzy inference systems are able to successfully control non-linear processes[12] when the membership functions and rules are properly determined. For the rolling

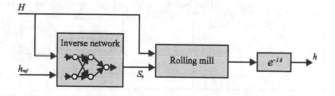

Fig. 5.   Feedforward control with an inverse neural network

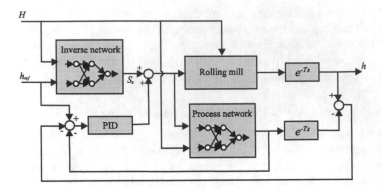

Fig. 6. Integral neural control

process, the deviation of output thickness Δh and its derivative $\dfrac{d\Delta h}{dt}$ are used as the inputs of the fuzzy controller.

To enhance the fuzzy control scheme, a rolling mill model is built by an adaptive neurofuzzy inference system to estimate the prompt process output thickness. The estimation result, instead of the delayed process output signal, is adopted as the input of fuzzy inference controller. And the error between the real process output (only delayed value available) and neurofuzzy model estimation (after delayed) is fed back to account for the real process uncertainty and neurofuzzy model inaccuracy. This enhanced fuzzy control structure is described as shown in Figure 7.

5.1 Fuzzy Inference Controller

The fuzzy inference controller has two input variables Δh and $\dfrac{d\Delta h}{dt}$, as stated above, and one output variable u, the control signal to hydraulic servo-valve.

The two inputs and one output are assigned different fuzzy sets as shown in Figure 8, based on rolling theory and rolling technology expertise. The linguistic variables are defined as follows:

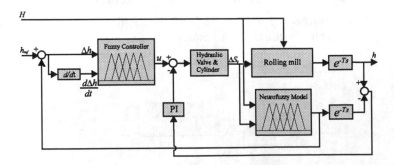

Fig. 7. The enhanced fuzzy control scheme

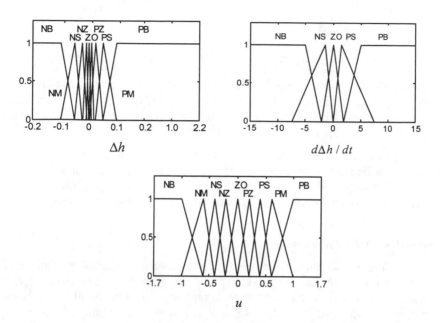

Fig. 8. Membership functions of fuzzy inference controller

PB: positive big;   PM: positive medium;   PS: positive small;

PZ: positive zero;   ZO: zero;   NZ: negative zero;

NS: negative small;   NM: negative medium;   NB: negative big.

With these fuzzy sets, 37 rules are used with Mamdani's fuzzy inference method[13].

5.2 Neurofuzzy Inference Modelling

The rolling process is modelled by a fuzzy inference system, and its parameters are turned by a back-propagation neural network structure based on the rolling process input-output data. The neural network structure minimises the fuzzy inference model output error. As the fuzzy inference model is optimised by neural learning, it is called adaptive neurofuzzy inference modelling.

For fuzzy process modelling, a more efficient fuzzy inference method, Sugeno method is employed[14]. Sugeno method[15] adopts singleton spikes to find the weighted average of a few data points, greatly simplifies the computation required to find the centroid of a two-dimensional shape, thus enhances the efficiency of the defuzzification process and speed up the adaptive learning.

For rolling process modelling, 6 membership functions are assigned to each input of the fuzzy inference system, S_0 and H. Therefore, there are 36 membership functions for the output of the fuzzy inference system, represented in the form of the following equation

$$u_{ij} = a_{ij}S_0 + b_{ij}H + c_{ij} \qquad\qquad i, j=1, 2 ,\ldots\ldots 6. \qquad\qquad (7)$$

where a_{ij}, b_{ij} and c_{ij} are parameters optimised by the neural network structure.

Generalised bell curve membership functions and triangular membership functions are separately employed for the fuzzy inputs. After training for 40 epochs, the neurofuzzy model with generalised bell curve membership functions estimates more accurate rolling process output than that with triangular membership functions.

The enhanced fuzzy control performs excellently, after several times of parameter regulation according the simulation results. The response of this control scheme indicates that the output thickness is almost perfect, as shown in Figure 9c, except for the several spikes, which are caused by the sudden changes of input thickness.

6   Results and Discussions

Figure 9 and Table 1 summarise the results of integral neural control and enhanced fuzzy control with PID linear control. The integral neural control improves the control performance greatly. This is due to the control philosophy that the non-linear relation between the input variables and output variable is compensated by the inverse neural network, and the PID controller is used to further minimise the error, especially the steady error.

Table 1. Comparison of the different control schemes

Control strategy	Index(t_0 =0)	Index(t_0 =5)
PID linear control	0.0082	6.2634e-005
Integral neural control	0.0024	2.1529e-006
Enhanced fuzzy control	0.0023	9.6746e-007

Note: the index is defined by Equation 4 as t_{max} =55.

More excellent performance obtained with enhanced fuzzy control is due to the fact that fuzzy inference controller is a non-linear controller, which could seize on the non-linear characteristics of the rolling process by itself. It is worth noting that the output thickness steady error could be controlled within a very small limit by these two enhanced intelligent control schemes. For the input thickness given by Figure 4, the output thickness error (except the spikes caused by sudden input changes) could be controlled within ± 0.0005 mm with the integral neural control, and the enhanced fuzzy control reduces the output thickness tolerance to $\pm^{0.0003}_{0.0002}$ mm.

7   Conclusion

From this work, it has been shown that rolling process and its process inverse can be successfully modelled by neural networks and the rolling process could be modelled also by neurofuzzy inference system. The integral neural control with process modelling network and inverse network improves the rolling process performance significantly. The enhanced fuzzy control with a fuzzy inference controller and an adaptive neurofuzzy model parallel to rolling process can obtain excellent performance, better than integral neural control.

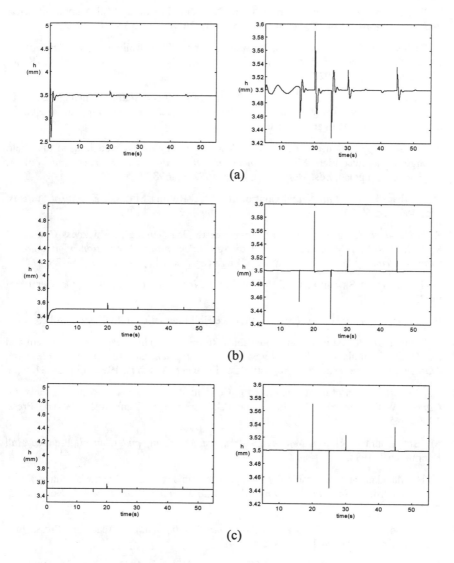

Fig. 9. Response of the rolling thickness control system with difference control schemes

(a) PID feedback control   (b) integral neural control   (c) enhanced fuzzy control

Note: The graphs in the right hand part are the magnification of those in left hand part
with starting time of 5th second

References

1. S. Wang and A. K. Tieu, Application of Neural Network to Width Estimation in Flat Rolling, *Proceedings of Second Biennial Australian Engineering Mathematics Conference '96*, The Institute of Engineer, Australia, 1996, pp.231-235.

2. N. F. Portnann, *et al*, Application of Neural Networks in Rolling Mill Automation, *Iron and Steel Engineer*, Vol. 72, February 1995, pp.33-36.

3. T. Fechner, *et al*, Adaptive Neural Network Filter for Steel Rolling, *Proceedings of the 1994 IEEE International Conference on Neural Networks*, Vol. 6, 1994, IEEE, Pitscataway, pp.3915-3920.

4. M. Roscheisen, *et al,* Neural Control for Rolling Mills: Incorporating Domain Theories to Overcome Data Deficiency, *Advances in Neural Information Processing*, Vol. 4, Hupson and Lippman, Eds, Morgan Kaufman, 1992, pp.659.

5. A. C. Tsoi, Application of Neural Network Methodology to the Modelling of the Yield Strength in a Steel Plate Mill, *Advances in Neural Information Processing*, Vol. 4, Hupson and Lippman, Eds, Morgan Kaufman, 1992, pp.698.

6. D. Sbarbaro-Hofer, *et al*, Neural Control of a Steel Rolling Mill, *IEEE Control Systems*, June 1993, pp.69-75.

7. K. Nose, *et al*, Application of Fuzzy Inference to Gain Tuning in Thickness Control of Multi-high Rolling Mills, *12th World Congress International Federation of Automation Control: Prints of Paper*, The Institute of Engineer, Australia, Vol. 6, pp.135-138.

8. W. L. Roberts, A Simplified Cold Rolling Model, *Iron and Steel Engineer*, October 1965, pp.75.

9. B. C. Kuo, *Automatic Control System*, Prentice Hall International, Inc. 1991, pp.128.

10. S. Wang, *et al*, A Mathematical Model for a Rolling Mill Thickness Control System and its Digital Simulation, *Proceedings of Second Biennial Australian Engineering Mathematics Conference '96*, The Institute of Engineer, Australia, 1996, pp.125-131.

11. A. G. Barto, Connectionist Learning for Control: An Overview, *Neural Networks for Control*, W.T. Miller, III, R. S. Sutton, and P.J. Werbos, Eds. Cambridge, MA: MIT Press, 1990. pp.5.

12. M. Brown and C. Harris, *Neurofuzzy Adaptive Modelling and Control*, Prentice Hall International, 1994.

13. E. H. Mamdani and S. Assilian, An Experiment in Linguistic Synthesis with a Fuzzy Logic Controller, *International Journal of Man-Machine Studies*, Vol.7, No.1, 1975, pp.1-13.

14. T. Terano, *et al*, *Fuzzy Systems Theory and its Application*, Academic Press, Inc., Harcourt Brace Jovanovich, Publishers, 1991.

15. M. Sugeno, *Industrial Applications of Fuzzy control*, Elsevier Science Pub. Co., 1985.

ADVANCED MANUFACTURING TECHNOLOGY

A Modular Head-Eye Platform for Real-time Active Vision [1]

Chee-Kong FONG[2]
jfong@cyberway.com.sg

Ming XIE
mmxie@ntu.edu.sg

Robotics Research Center
School of Mechanical & Production Engineering
Nanyang Technological University, Singapore 639798

Abstract. This paper describes the development of a high speed, controlled stereo head-eye platform which facilitates the rapid redirection of gaze in response to visual input. It details the mechanical device, which is based around geared DC motors, and describes the hardware aspects of the motion controller and vision system, which are implemented on a modular structured mobile robot. To arrive at an active head-eye relationship, two adaptations of head-eye coordination algorithms are implemented.

Our NTU Active Stereo Head (NASH) is a 6 DOF camera head. That includes independent auto-focus and independent control of pan for each of the two cameras. A combined set of pan and tilt motors are used for emulation of the neck movements. Servo control of all the axes is accomplished by a programmable multi-axis controller (PMAC) card.

To achieve real-time and robustness in our head-eye coodination problem, we adopted two methods. The first method involves using planar primitive, which requires to iteratively converge observed image towards that of the reference image. The second method is a closed-for solution, it uses two new images to determine where the position of the reference image has been taken.

1. Introduction

The ability to move one's head and eyes is imperative in order to perform even simple tasks. Imagine trying to catch a ball without moving either your head or eyes. At certain points during its flight, the ball may not even be in your field of view, so how are you expected to align yourself to catch it? Because animals and robot exist in a complex and dynamic world, one may argue that head and eye movements are necessary to function in the environment. In addition, one may also argue that moving one's head or eyes makes many visual task simpler. That is, movement can enhance the sensory acquisition process. For example, the ability to move our eyes allows us to track a moving object and thus we can predict it's position and/or velocity (such as in catching an object). Or, we can maintain the image of a moving object within our field of view to perform further processing. We can also move our eyes to view portions of the scene, to examine areas of interest or areas where we are uncertain of what we see.

---

[1] This research work is supported by Ministry of Education of Singapore (MOE) under the grant RG72/96.
[2] Author is a graduate student pursuing his Master degree.

In nature, heads are expected to be on shoulders and not on arms. This playing with words is actually referring to two different approaches in building head-eye systems. Some researchers are, rationally, interested in putting head-eye systems on a robot arms or have chosen other approaches [2,3,4]. NASH that is presented in this paper, on the contrary will be put on shoulders, that is a platform. Having a head and not just a pair of eyes on a robot arm (manipulator) is also more convenient for mounting on a moving platform. When talking about a 'head', the neck movements are to be included as part of the head. That is, having the three basic degrees of freedom (roll, pitch and yaw) in the manipulator will already do the job of the neck would do. In this sense, most of the head-eye systems implemented up to now are actually a pair of eyes and not head-eye systems.

Active vision research was initiated at the University of Pennsylvania in the early eighties [1]. This led to construction of robotic head-eye systems at a number of US universities including the University of Pennsylvania [5], Harvard University [6], and the University of Rochester [7,8]. These heads can be characterized as first generation heads, which to a large extent are based on home-made mechanisms etc. As active vision research has gained more wide-spread interest, there has also been an increased activity related to the construction of head-eye systems. Examples are those presented by Crowley [9], Prelove [10], TRISH [11], and KTH [12]. There seem to be two different approaches to the construction of head-eye systems, (1) general purpose, fast and high accuracy heads, e.g. the KTH head, and (2) heads which are flexible enough for a number of different tasks in computer vision, e.g. the Harvard head.

Under the Strategic Research Program: Vision and Control, Nanyang Technological University (NTU), there is also a considerable interest in the use of active vision to emulate an intelligent human behaviour, i.e. head-eye coordination. To enable further research works on robotic head-eye relationship, **NTU Active Stereo Head (NASH)** has been developed. At the same time, a customised mobile platform has been designed and it will be integrated with NASH to form a mobile head-eye system. To gain insight into the design and use of a head-eye system, the effort our research program was directed towards the construction of a head in category (2) as that mentioned in the later part of the above paragraph. Based on the experience from construction of such a prototype, subsequent new and improved designs would be specified and implemented. This paper presents the work carried out in order to construct the first prototype.

2. Requirements For a Head

Generally, a robotic head consists of a mechanical eye and a neck mechanism. There have been intensive studies on eye movements in human being. Less attention has been directed towards the neck movements and body movements. This might be the major source of confusion in the design of head-eye systems, because the eye-movements in present head-eye systems are actually a hybrid of eye and neck movements.

Eye movements are different from other movements that affect the quality of visual cues. This is true from the fact that the fovea centralis is occupied by images from the different parts of the scene, so, no geometrical distortion is generated in the

1430

image by normal eye movements. This is not the case with neck movements and other body movements. On the contrary, neck movements are supposed to change the perspective of the images and generate parallaxes, this is caused by the eccentric position of the spine. In a sense, body movements and neck movements are serving the same tasks, although there are still functional differences between them.

In the process of designing NASH, we have identified a number of different sets of degrees of freedom that are potentially of interest. There are grouped into (1) intrinsic parameters, (2) extrinsic parameters of individual cameras, and (3) head/neck parameters as shown below:

(1) Intrinsic camera parameters (eye configuration);
- Focus,
- Aperture,
- Zoom/Focal length.

(2) Extrinsic camera parameters (eye motion / position);
- Vergence and version.

(3) Head parameters (neck motion / position);
- Roll angle,
- Pitch angle,
- Yaw angle,
- Length of baseline - distance between cameras.

2.1 Intrinsic Camera Parameters

In the above list, though intrinsic camera parameters were mentioned, they are somehow not critical in our intended applications. Firstly, focusing can be provided by auto-focus Iris lens. Secondly, for the time being, NASH will only work in a structured environment with constant light conditions, thus camera aperture setting is not needed. And finally, since NASH will be mounted on a mobile platform, it is not crucial to have a zooming facility just to have a larger field of view.

2.2 Extrinsic Camera Parameters

By panning the cameras, it is possible to change the position where the optical axes intersect (the gaze or fixation point).

For change of gaze point, a symmetrical or independent control of the vergence angles θ_1 and θ_2 may be used. Most of the first generation head-eye systems used an arrangement where two cameras are coupled through use of a spindle to provide symmetric angles. In such a configuration, the gaze point may be changed only to a position along a line perpendicular to the baseline (line between the focal points). This is different in human vision, where the gaze point may move around in the plane without any motion by the neck. If the goal of active vision is to gain insight into biological vision, it is desirable to have independent vergence angle control for the two cameras.

In human vision, there is limited ability to perform independent tilt of the eyes. In general the use of the tilt will complicate the stereo reconstruction and it will not be considered, besides, tilt of the entire head can allow change of the gaze point to any position in the 3-D space much more appropriately and easily.

2.3 Head Parameters

As mentioned, the best possible accuracy for vergence stereo is achieved with symmetric camera vergence angles. To achieve symmetric camera vergence angles for any point in 3-D space, it is necessary to be able to perform panning for the head. Therefore the field of view is much larger and the specified accuracy will allow positioning of the point of interest close to the optical axis where a minimum of aberration is necessary. This consideration applies likewise for the tilt action of the head.

The human visual system baseline is about 6 cm,. This is much smaller than it will be possible to realize due to physical limitations (size of cameras, etc) for a binocular robotic head. If the cameras are to verge, then there has to be somewhere for the cameras and lens to move, thus the smallest realistic baseline has to be about 32 cm.

3. Head-Eye Coordination Algorithm

Having provided the mechanical structure for a head-eye system. The next step is to write the instruction sets in achieving a head-eye coordination solution, which sometimes also known as image based visual servoing or even mis-appropriately called hand-eye coordination.

Apparently, almost all the head-eye coordination algorithms proposed thus far used some form of convergent methods to servo the cameras to a position where the observed image is equal to the reference image.

At a particular position (called Reference View), the vision system can capture an image as reference (called Reference Image). Figure 1 shows one example that a camera is at a reference view position in which a reference image is taken.

In actual situation, the cameras may be at any position and orientation (as in Figure 2) other than the reference view position. Therefore, the observed image is different from the reference image.

Now, the problem is how to control the head-eye system which will bring the cameras to a position where the observed image is equal to the reference image? This problem has been extensively investigated so far.

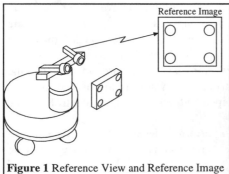

Figure 1 Reference View and Reference Image

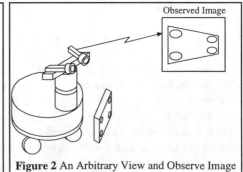

Figure 2 An Arbitrary View and Observe Image

There are many iterative approaches based on minimization process of a cost function [18], while there is only one deterministic method presented [20].

In the following sections, we will demonstrate two head-eye coordination solutions, one is a iterative method extracted from [19] while the second is a deterministic method developed by us.

3.1 Iterative Methods using Planar Primitive

Iterative approaches are based on the use of specially designed minimization process to generate motion control signal which guides head to a position where the image seen by eyes (the observation) looks like a reference image (the goal).

Planar primitive method was formed by Corke [19], in this method, it calculates a series of motions (translations and rotations) in order to iteratively converge the observed image towards the reference image (see Figure 3). The equations to work out the translation and rotation motions are as follows.

Translation:

$$Tx = k_1 * (x - x_c^r)$$
$$Ty = k_2 * (y - y_c^r)$$
$$Tz = k_3 * (L_0 - S_0)$$

Rotation:

$$\theta x = k_4 * [(S_0/(L_0-1)) + (L_2/(S_2-1))]$$
$$\theta y = k_5 * [(S_1/(L_1-1)) + (L_3/(S_3-1))]$$
$$\theta z = k_6 *\pm* a \cos(S_0*L_0)$$

where $k_{1,\ldots6}$ are control gains.

The main advantage iterative approaches has is that cameras do not have to be calibrated because 'self-calibration' will be achieved after the first few consecutive iterations.

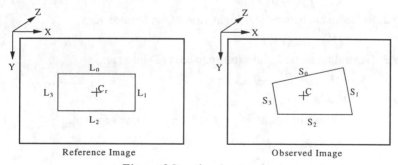

Reference Image   Observed Image

Figure 3 Iterative Approach

3.2 Deterministic Method

In our deterministic approach, the strategy for a closed-form solution is to estimate the position where the reference image has been taken. With two cameras and thus two known views, it is possible to (1) determine a third unknown view from 6 line segments by a combined least-squares estimation or (2) determine the third unknown view from 8 line segments by two separate least-squares estimations, where the first estimation is for rotation motion while the second one is for translation motion (see Figure 4).

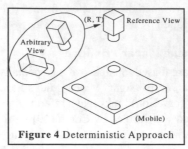

Figure 4 Deterministic Approach

The steps in arriving at the third unknown view from two known views can be demonstrated as follows:

1) Camera model: Perspective view.

$$x = X/Z$$
$$\text{and } y = Y/Z$$

2) Input views: A total of three views are needed, they are reference view, left view and right view each being represented by $V3$, $V2$ and $V1$ respectively.

3) Input primitives: From left and right views, extract at least 6 or 8 line segments.

4) **Step 1.** At reference view ($V3$),
   - A 2D line segment is described by,
   $$a_{v3}x_{v3} + b_{v3}y_{v3} + c_{v3} = 0$$
   - The corresponding projection plane is,
   $$a_{v3}X_{v3} + b_{v3}Y_{v3} + c_{v3}Z_{v3} = 0$$
   - If $S_{v3} = (a_{v3}, b_{v3}, c_{v3})$ and $P_{v3} = (X_{v3}, Y_{v3}, Z_{v3})$, then,
   $$S_{v3} * P_{v3}^{t} = 0 \qquad \ldots \ldots \ldots \quad (1)$$

5) **Step 2.** Relationship between left/right view and reference view,
$$P_{v3}^{t} = R_{v2v3} * P_{v2}^{t} + T_{v2v3} \qquad \ldots \ldots \ldots \quad (2)$$

6) **Step 3.** General equation 1, substitute equation (2) into (1),
$$S_{v3} * (R_{v2v3} * P_{v2}^{t} + T_{v2v3}) = 0 \qquad \ldots \ldots \ldots \quad (3)$$

7) **Step 4.** At left/right view,
   - From a pair of points $(P_{v2}^{'}, P_{v2}^{''})$ on a 3D line segment,
   $$S_{v3} * (R_{v2v3} * P_{v2}^{'t} + T_{v2v3}) = 0$$
   $$S_{v3} * (R_{v2v3} * P_{v2}^{''t} + T_{v2v3}) = 0$$
   - Subtract above two equations,
   $$S_{v3} * R_{v2v3} * (P_{v2}^{'t} - P_{v2}^{''t}) = 0$$

8) **Step 5.** General equation 2,
   define $\quad D_{v2} = (P_{v2}^{'} - P_{v2}^{''})/\| P_{v2}^{'} - P_{v2}^{''} \|$
   then $\quad S_{v3} * R_{v2v3} * D_{v2}^{t} = 0 \qquad \ldots \ldots \ldots \quad (4)$

The direction vector of a 3D line segment can be directly determined from the two known views [20]. It should be noted that equation (4) does not involve the unknown T_{v2v3}. Therefore, we can first estimate the rotation matrix and then the translation vector of the motion transformation between left/right view and reference view.

There are three independent unknown parameters in T_{v2v3}. Hence, a set of three line segments are enough to fully determine R_{v2v3} and T_{v2v3} from equations (3) and (4). In practice, it may be preferred to work with linear system and use Least-Squares estimation method. In order to do so, we can consider that there are eight unknown parameters in R_{v2v3} (factorize out, for instance, the parameter of row 3 and column 3 from R_{v2v3}, and recover it later by using the orthogonality constraint that the matrix R_{v2v3} has to meet). In this way, a set of eight line segments allow to establish a linear system for the estimation of the matrix R_{v2v3}. Once we know R_{v2v3}, T_{v2v3} can be estimated from a linear system with a set of three line segments only.

Contrary to iterative approach, our proposed deterministic approach needs the cameras to be calibrated. Since our approach uses 'one-move' from an unknown cameras position to a reference image position, therefore the perspectives of cameras have to be found.

4. Design of a Mobile Head-Eye Coordination Robot

Based on the considerations in Section 2, NASH was constructed. At the same time, a customized mobile platform was also designed to add locomotion to the head. The intended purpose for our robot is to serve as an experimental testbed for our future research works on active vision and also on autonomous robot. Mechanically, the robot is modular in design. Currently, it has 3 separate modules, and they are (1) the NASH module which consists of a pair of active 'eyeballs' (cameras), a neck (manipulator unit) and a baseline rig, this head module is mounted on top of a (2) command-center module (CCM) where it's backplane houses a Pentium computer running at 200 MHz, a DSP card and a motion controller card, in addition, the CCM also holds the battery units as well as all the servo motor amplifiers, below the CCM is the (3) mobile base module (MBM), this module has 3 independent wheels and each wheel has independent driving and steering actions. A schematic view of the mobile robot is shown in Figure 5.

4.1 Design of the NTU Active Stereo Head (NASH) Module

NASH is made up of a pair of eye units, the baseline rig, and a manipulator unit.

Lately, there has been a keen interest and rather novel concept in the design of robotic arm, that is modularity. Research works presented in [13,14,15,16,17] focused on complete modularity in which joints and links are classified as separate modules. In such a design, one joint provides one motion, therefore a range of different joints having different degree-of-freedoms is probable, so, when coupled serially together with the links, a customized manipulator is formed based on its intended application. Such an idea offers a higher flexibility in manipulator selection and configuration.

In the design of the neck, we work on similar concept, that is, the neck which is constructed like a manipulator, should be made up of modules. But in our design, we do not separate the link with the joints. In fact, we have integrated two joints (with the same or different degree-of-freedom) with a common link between them to form an 'active link'. Our active link is simpler in design and construction as compared to our counterparts. Two of such units are serially stacked together to form the neck. A schematic drawing on such an active link is depicted in Figure 6.

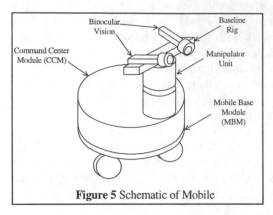

Figure 5 Schematic of Mobile

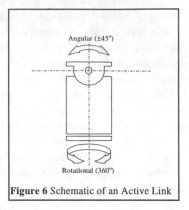

Figure 6 Schematic of an Active Link

The motion actuators in each active link include a pair of 100W MAXON motor (one for each DOF), each motor is coupled to a Harmonic Drive reduction gear via a set of timing pulley and belt.

To ensure compatibility, mounting hole dimensions on both the mating flange ends are the same. Electrical connectors of respective match are also located on each end of the flange. Each link has its own wiring network, and each network includes wires for its own motion actuators as well as wires for the neighbouring modules.

The technical specifications of an active link is summarized in Table 1.

The baseline rig provides the ability to change the length of the baseline dynamically. The rig containing the baseline mechanism represents also the connection between the eye units and the neck. This rig is purely for experimental reasons.

The only eye movement available on the eye unit is the panning action that allows both that cameras to verge and fixate at the point of interest.

4.2 Design of the Mobile Base Module (MBM)

To add mobility to NASH so that it can move around in a structured environment, a mobile base module (MBM) has to be designed. This module consists of a base plate with three wheels mounted underneath it. Each wheel has independent steering as well as driving, thus providing holonomic capability. Our mobile platform is capable of 3 DOF, two translational dof along the X-Y direction and the third is a rotational DOF about Z-axis.

Three wheels are mounted around and inside the circumference of a $\phi 160$mm base plate at an angle of $120°$ apart. Steering and driving actions from each wheel are realised by 2 separate sets of geared DC motor. Currently, the designed weight it can support is 115 kg. A schematic view of a wheel unit is shown in Figure 7.

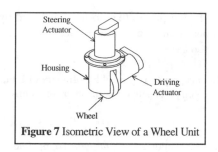

Figure 7 Isometric View of a Wheel Unit

As shown in the diagram, the wheel unit can be regarded as one independent unit. If future modification arises, the wheel can be removed easily and a new design can be fitted, without having to disturb the rest of the structure.

Technical specifications of the wheel unit can be summarized in Table 2.

Table 1 Active Link Specification Summary

Active Link Design Specification					
Active Link (Pan/Tilt)	1	Operating angular velocity	Pan	rpm	44
			Tilt	rpm	44
	2	Operating torque	Pan	Nm	0.12
			Tilt	Nm	4.12
	3	Operating power	Pan	W	0.55
			Tilt	W	18.98
	4	Acceleration	Steer	sec	1
			Drive	sec	1
DC Servo Motor (Electronic Commutation)	1	Assigned power rating		W	100
	2	Nominal voltage		Volt	30
	3	No load speed		rpm	4800
	4	Max. permissible speed		rpm	15000
	5	Designed operating speed	Pan	rpm	4400
			Tilt	rpm	4400
	6	Designed operating torque	Pan	Nm	0.22
			Tilt	Nm	0.22
Harmonic Drive Reduction Gear	1	Reduction ratio		-	1:100
	2	Max. input speed	Oil	rpm	14000
			Grease	rpm	5000
	3	Max. output torque		Nm	28
	4	Rated output torque at 2000rpm		Nm	7.8
	5	Designed input speed	Pan	rpm	4400
			Tilt	rpm	4400
	6	Designed output speed	Pan	rpm	44
			Tilt	rpm	44
	7	Rated output torque at 44rpm		Nm	27.84

Table 2 Wheel Unit Specification Summary

Wheel Unit Design Specification					
Wheel Unit	1	Operating angular velocity	Steer	rpm	44
			Drive	rpm	44
	2	Operating torque	Steer	Nm	1.2
			Drive	Nm	20.59
	3	Operating power	Steer	W	5.5
			Drive	W	94.92
	4	Acceleration	Steer	sec	0.3
			Drive	sec	0.3
DC Servo Motor (Electronic Commutation)	1	Assigned power rating		W	100
	2	Nominal voltage		Volt	30
	3	No load speed		rpm	4800
	4	Max. permissible speed		rpm	15000
	5	Designed operating speed	Steer	rpm	4400
			Drive	rpm	4400
	6	Designed operating torque	Steer	Nm	0.22
			Drive	Nm	0.22
Harmonic Drive Reduction Gear	1	Reduction ratio		-	1:100
	2	Max. input speed	Oil	rpm	14000
			Grease	rpm	5000
	3	Max. output torque		Nm	28
	4	Rated output torque at 2000rpm		Nm	7.8
	5	Designed input speed	Steer	rpm	4400
			Drive	rpm	4400
	6	Designed output speed	Steer	rpm	44
			Drive	rpm	44
	7	Rated output torque at 44rpm		Nm	27.84

5. Conclusion

We have presented a conceptual layout of a modular robot. The robot consisits of a stereo camera head system (NASH) mounted on a mobile platform. The mechanical parts and head-eye coordination algorithms were described in order to achieve active vision.

We presented a new concept of modular manipulator made up of a series of active links. Each active link has 2 DOF. By stacking two of the active links serially together, the neck of NASH is formed.

We have also presented two head-eye coordination approaches. The first is a iterative method using planar primitives, while the second method is a deterministic approach, which may be the first in the world, developed by our group. Both methods complement each other, future tests are needed to validate the real-time responses from each of the methods.

Acknowledgement. This project is funded by Ministry of Education, Singapore. It is part of the effort to develop one complete general purpose vision guided modular robot.

References

[1]   R. Bajcsy, 'Active Perception', *Proc. of IEEE 76*, pp. 996-1005, 1988.
[2]   D. H. Ballard, 'Animated Vision', *Proc. of 11th International Conference on Artificial Intelligence*, 1989.
[3]   E. P. Krotkov, 'Exploratory Visual Sensing for Determining Spatial Layout with an Agile Stereo System', *Ph.D Thesis*, 1987.
[4]   J. J. Clark, N. J. Ferrier, 'Modal Control of an Attentive Vision System', *Proc. of 2nd International Conference on Computer Vision*, 1988.
[5]   E. P. Krotkov, K. Henriksen, R. Korries, 'Stereo Ranging with Verging Cameras', *Transcript of IEEE Pattern Analysis and Machine Intelligence 12*, pp. 1200-1205, 1990.
[6]   N. J. Ferrier, 'The Harvard Binocular Head', *Technical Report 91-09*, Harvard Robotics Laboratory, University of Harvard, 1991.
[7]   D. H. Ballard, A. Ozcandarli, 'Eye Fixation and Early Vision: Kinematic Depth', *Proc. of 2nd International Conference on Computer Vision*, 1988.
[8]   T. J. Olson, D. J. Coombs, 'Real-Time Vergence Control for Binocular Robots', *International Journal on Computer Vision 7*, pp. 67-89, 1991.
[9]   J. L. Crowley, P. Bobet, M. Mesrabi, 'Gaze Control for a Binocular Camera Head', *Proc. of SPIE Application of Artificial Intelligence 5: Machine Vision and Robotics*, pp. 47-61, 1992.
[10]   J. R. G. Pretlove, G. A. Parker, 'A Lightweight Camera Head for robotic-Based Binocular Stereo Vision: An Integrated Engineering Approach', *Proc. of SPIE Application of Artificial Intelligence 5: Machine Vision and Robotics*, pp. 62-67, 1992.
[11]   M. Jenkin, E. Milios, J. Tsotsos, 'TRISH: A Binocular Robot Head with Torsional Eye Movements' *Proc. of SPIE Application of Artificial Intelligence 5: Machine Vision and Robotics*, pp. 36-46, 1992.
[12]   K. Pahlavan, J. O. Eklundh, 'Heads, Eyes and Head-Eye Systems', *Proc. of SPIE Application of Artificial Intelligence 5: Machine Vision and Robotics*, pp. 14-25, 1992.
[13]   D. Schmitz, P. K. Khosla, T. Kanade, ' The CMU Reconfigurable Modular Manipulator System', *Proc. of 19th International Symposium on Experimental Robotics*, pp. 473-488, 1988.
[14]   K. H. Wurst, 'The Conceptual and Construction of a Modular Robot System', *Proc. of 16th International Symposium on Industrial Robotics*, pp. 37-44, 1986.
[15]   T. Matsumaru, 'Design and Control of the Modular Robot System: TOMMS', *Proc. of IEEE International Conference on Robotics and Automation*, pp. 2125-2131, 1995.
[16]   R. Cohen, M. G. Lipton, M. Q. Dai, B. Benhabib, 'Conceptual, Design of a Modular Robot', *ASME J. of Mechanical Design*, vol. 114, pp. 117-125, 1992.
[17]   T. Fukuda, S. Nakagawa, 'Dynamically Reconfigurable Robotic System', *Proc. of IEEE International Conference on Robotics and Automation*, pp. 1581-1586, 1988.
[18]   K. Hashimoto, 'Visual Servoing', World Scientific, 1994.
[19]   P. I. Corke, 'Visual Control of Robot Manipulator', *Visual Servoing*, Editor: K. Hashimoto, World Scientific, pp. 1-31, 1994.
[20]   M. Xie, 'Head-Eye Coordination: A Closed-Form Solution', *Proc. of 2nd World Automation Congress*, 1996.

Third Millennium Manufacturing:
Healing the Executive/Production Schism
by Computers or Conversation?

Duley K W and Maj S P

Department of Computer Science, Edith Cowan University,
Mt Lawley, Perth, Western Australia

Abstract. Since the turn of the century there has been a perceptible dichotomy between management and workshop in discrete manufacturing. Production-line Manufacturing, developed by Henry Ford, and Scientific Management, developed by Frederick Winslow Taylor, left an enduring legacy in the division between executive and production, This dichotomy has manifested itself demographically (in a new class system), sociologically (in labour unrest), technologically (in separate automation systems) and philosophically (in political division). Australian companies are seeking to close this gap - this paper looks at the two major solutions.

Computer Integrated Manufacturing and, more recently, Manufacturing Execution Systems, seeks to establish a digital link between the Management Information and Decision Support systems at executive level and the processes, control systems and general automation on the shop floor, creating a replacement for the intimate knowledge of workshop processes and progress which were the province of the owner/ manager but which vanished with the development of the multi-divisional (arid then multi-national) companies. Japanese success with a person to person solution and with modified processes rather than automated (remote) control provides an interesting alternative.

This paper follows the development and impact of the dichotomy, looks at the two solutions and seeks an appropriate Australian answer.

Keywords : Computer Integrated Manufacturing, Continuous Improvement, Cultural Differences, Discrete Manufacturing, Just-In-Time, Manufacturing Execution Systems, Pull Manufacturing, Push Manufacturing, Third Millennium Manufacturing, Zero Defects Quality

1 Evolution of the Division between Executive and Production

Discrete manufacturing has been around in some form since before man learnt to practice agriculture and could, thereby, settle and establish a defined civilisation - club makers and flint chippers produced a manufactured product. As the agricultural civilisations of Europe and the Middle East developed, a range of arts and crafts evolved becoming increasingly codified and sophisticated, culminating in the self-regulating Guild structure of Europe's Middle Ages, Renaissance and Reformation. Steam Power, providing the means to aggregate a large workforce around many machines in one workplace and revolutionising communication and trade by fast, cheap rail and sea transport, changed the techniques of manufacturing in the Industrial Revolution.

While the establishment of large factory manufacturing, the *"dark, satanic mills"* of the mid-1800s, was a significant change in the way that items were produced, little changed in the way in which production was managed. Managers tended to be Owners and also to have an understanding of the processes of production. While the crafts and trades had advanced their abilities and technologies, the economic structure which underpinned this, essentially *cottage*, style of industry remained unchanged. However, the economic foundation of production had changed. For example, prior to the Industrial Revolution the products of shearing, spinning, weaving and finishing changed ownership between the farmer and the various craftspeople involved. Each was, therefore, able to compute profit and loss by simple comparison of outlay and income (costs being defused by market forces). Under this regime, businesses used accounts primarily to record the results of these market exchanges.

By the start of the nineteenth century, textile merchant/entrepreneurs were taking control of spinning, weaving and finishing within a single enterprise. This change necessitated an emphasis on accounting for interests within the company and on the use of accounting records in administrative control of the enterprise. Despite this subtle change, the essential point is that management still understood the processes on the shop floor and were able to maintain an overview of the progress of product through a plant by personal inspection and direct, verbal reports from supervisors on the floor. This direct, almost *hands-on*, management style remained common until after the turn of the century.

Henry Ford, in moving the production of the Model 'T' to a purpose-built factory at Highland Park (IL, USA) in 1910, initiated the most significant change in the methods of discrete manufacturing since the industrial revolution itself. While increasing his staff from 1,908 to 4,110 (+215%), Ford's innovation in implementing production-line manufacturing resulted in an increase in output from 1,708 Model 'A's to 34,528 Model 'T's (+2,021%). (Ford, 1991) This, better than nine-fold, increase in *per capita* productivity attracted widespread interest.

It should be understood that this revolution in manufacturing took place in a period of US history where population, personal wealth and, consequently, market was booming. Ford's factory was designed to produce one item as well as possible, as cheaply as possible and as fast as possible. This was genuine *push* manufacturing in which parts and raw materials are thrust in one end of a factory and forced out the other as product at the greatest possible rate. Emphasis on market research did not emerge until the late 1920s (amongst Ford's competitors) since all that could be produced would find a ready market.

"No manufacturer anywhere in the world was able to exactly repeat Henry Ford's extraordinary success with mass production. This is hardly surprising. In the history of commerce there have been few opportunities to exploit such an enormous, untapped and reasonably homogeneous market. Yet mass production - modified to accommodate

different markets and labour conditions... was... widely initiated." (Batchelor, 1994, p. 66)

Despite the success of production-line manufacturing, Ford faced considerable criticism on the grounds that his practices alienated his workers, treating them as machines enslaved to the greater machine (the production line itself) and forcing him to pay extraordinarily high wage rates in an effort to stem the flow of staff turnover. This alienation was exacerbated by the concurrent work of Frederick Winslow Taylor on the concept of *Scientific Management*, the idea that the application of scientific principles to management could significantly enhance productivity.

Taylor discerned a tendency for all members of a workforce to slow their output to the pace of the slowest.

"When a naturally energetic man works for a few days beside a lazy one, the logic of the situation is unanswerable. 'Why should I work hard when that lazy fellow gets the same pay that I do and does only half as much work?" (Taylor, 1967, p. 19)

Taylor's counteraction to this tendency which he termed 'soldiering' was the study of time and motion, a process which identifies the atomic parts of a task and then discovers the most efficient way to perform that atom of work and the minimum time that efficient atom should take a suitable person without injuring that person's health or well being. Employing carefully chosen work teams, at significantly higher than average rates of pay and on the basis that the workers followed his prescription for the job to the letter, Taylor was able to obtain exceptional improvements in productivity - it is worth noting, however, that his attitude to the people in his workforce might not be seen as politically correct today.

"The [Frederick W.] Taylor method does not recognise the hidden abilities workers possess. It ignores humanity and treats workers like machines. It is no wonder that workers resent being treated that way and show no interest in their work". (Ishikawa & Lu, 1985, p. 25)

While it would be impossible to refute Taylor's success in productivity improvement, it might be said that his more enduring legacy was his enhancement of the alienation for which Ford was being criticised. Believing planning work to be clerical in nature and that every detail of a job should be thought out in a *'planning department'* Taylor developed an extensive hierarchy of middle management under the planning department to ensure that each task was carried out precisely according to plan including *'Gang Bosses', 'Speed Bosses', 'Inspectors'* and a *'Shop Disciplinarian'*. Each of these, in turn, had his foreman over him. (Taylor, 1993) As this hierarchy became institutionalised, so too did the dichotomy between white collar and blue, between executive and production. Management would manage and workers merely had to do as they were told. This dichotomy became virtually universal, persisting for more than five decades. In terms of the communications breakdown between executive and production, however, worse was on the way.

With the wall between the office and the workshop now a significant obstacle, management were entirety dependent on the accounting records as the basis for administrative control. In this they were assisted by developments in accounting theory, notably those in concert with the American Society of Mechanical Engineers (ASME), which produced information on efficiency and throughput. Two further, subtle changes rendered this unworkable. Firstly, management themselves changed - the giant, multi-divisional companies which were appearing tended to appoint accountancy and business school graduates to executive positions. Management now lacked an intuitive, practical understanding of the processes of production and could be said to have <u>had</u> to '*manage by numbers*'. Secondly, the ownership of the company were normally '*absentee industrialists*', shareholders who might never set foot in the factory and were only interested in the return on their investment. For these people, and, of course, for the taxation department, the accountants now had to produce financial reports and these reports quickly became the central focus of the accountant's employment. Communication of strategic information between production and executive levels was now virtually impossible since managers who needed relevant 'numbers' no longer received them.

"*In [the place of procedures for computing managerially relevant product costs] appeared the costing procedures that twentieth-century accountants developed to value inventories for financial reports. While those procedures yield cost information that apparently aids financial reporting, the same information is generally misleading and irrelevant for strategic product decisions.*" (Johnson & Kaplan, 1991, p. 126)

Disconnection is a term which could from this time be applied to the relationship between executive and production. This disconnection has manifested itself demographically (in a new class system). sociologically (in labour unrest), technologically (in separate automation systems) and philosophically (in political division). Its effects in manufacturing management linger:

"*A survey of information preparers and users in an automated manufacturing environment indicated that 54 percent of preparers were dissatisfied with their costing methodologies, and 62 percent of users were similarly dissatisfied. Performance measurement systems were perceived as even worse, with 57 percent of preparers and 69 percent of users dissatisfied.*" (Linnegar, 1988, p. 39)

Next we consider how companies survived this cervical severance.

2 Surviving the Communication Breakdown

Bearing in mind the market context in which manufacturing thrived, apart from the Great Depression of the '30s, through the expansion of the US, the impetus of two World Wars and the post-WWII global boom, corporate survival was relatively simple despite executive isolation from the shop floor. Inability to predict, monitor and control the passage of components and raw materials through a plant towards the finished goods inventory could be disguised and compensated for by high levels of work in progress inventory. Conventional managerial thinking of the time required all machines and processes to be running at maximum rate whenever possible - this is *push* style manufacturing - leading to rising inventories particularly at bottleneck processes. (If the mean arrival rate of batches at a workstation is 'λ' and the mean service rate is 'μ', and if any random variations exist in either, then where $\mu = \lambda$ the length of the queue (size of the inventory) will build to infinity. Where a batch arrival may occur at a process at any random time interval then the length of the queue 'L_Q' is found by:

$$L_Q = \lambda^2 / \mu(\mu - \lambda)$$

Equation 1 : Growth of Inventory

Where $\mu = \lambda$ then $\mu(\mu - \lambda) = 0$ so $L_Q \to \infty$ (Gibson, Greenhalgh, et al, 1995, p. 75)

Raw Materials and Finished Goods Inventories (RMI and FGI) also tended to rise throughout this period, RMI as an hedge against disruption of supply and FGI as a buffer against unexpected surge in demand. Conventional cost accounting permitted this rise in inventory through recognition of inventory as an asset (Mitchell & Granof, 1981, p. 195) (Kirkwood, Ryan, et al., 1989, p. 13) (Mathews, Perera, et al., 1991, p. 160) yet the results of this artificial imbalance are obvious. Company stockholders, accepting financial reports showing inflated company assets, demanded better return on their *share* of what they perceived to be their *investment*. By the 1960s, manufacturing management were seeking ways to improve that return and two major streams of development arose - one in the Western World (dominated by the USA) and the other in Japan.

3 Western Industry Embraces Computerisation

Calculation of manufacturing schedules requires the asking of four fundamental questions:
1. What are we going to make?   Master Production Schedule (MPS)
2. What does it take to make it?   Bill of Materials
3. What do we have in stock?   Inventory Records
4. What do we need to get?   Materials Requirement Plan (MRP)
                                                          (Storey, 1994, p. 159)

Time Phased Ordering (Orlicky, 1975) sought to resolve these questions by determining ordering rates and sequences (an MRP) which would meet the MPS while minimising RMI, IBM's "Production, Inventory and Control System" (PICS),which was to support sales forecasting, requirements planning, capacity planning, engineering data control, shop floor control, operations scheduling, purchasing and inventory control, lead in the mid 1960s to the development of computerised MRP

"[in] an attempt to use the 'number-crunching 'power of the computer to develop a production plan for an entire plant in which the production of each individual item is coordinated with a master production schedule for the production of end products." (Kerr, 1991, p. 17)

Although the move to MRP failed, largely due to its inability to respond to unexpected contingencies such as delivery failure or plant breakdown, PICS and the subsequent development of the 'Communications Oriented Production Information and Control System' (COPICS) by IBM in the early 1970s set Western manufacturing firmly on the road to computerisation as the solution to the executive / production communication failure. Evolution of MRP and interaction with the COPICS concept lead to MRP II (the acronym now referring to Manufacturing Resources Planning) which covered labour, machine, capital, purchasing, marketing and shipping requirements as well as the areas covered by MRP. Computerisation, however, was moving into every facet of manufacturing.

In the office, systems from word processors and spreadsheets to complex company-specific Management Information Systems (MIS) and Decision Support Systems (DSS) are, these days, ubiquitous; Computer Aided Design and Drafting (CADD) has almost completely replaced the traditional drawing board and pencil; support software is available for purchasing, quality control, supply chain, management, training, transportation and distribution and warehousing as well as the traditional ledgers and payroll. Meanwhile, on the shop floor, computerised numerical control (CNC) of production machinery, robotics, automated storage and retrieval systems, automated guided vehicles for product transfer, coordinate measuring machines and automatic data collection for quality control are typical moves towards full computerisation. Computer Integrated Manufacturing (CIM) and Manufacturing Execution Systems (MES) are not single application programs but the drawing together of a huge range of software, hardware, programming languages, operating systems, data exchange protocols, graphics display systems, databases etc. into one organism which encompasses the whole plant from reception desk to scrapyard.

"MES systems are arguably one of the more complex Computer Information Systems in existence. This is not to say that you can't think of several other systems that you personally consider more complex, perhaps you can and it's not that the code is necessarily complex either, but rather that the whole system itself is. The interaction of all the components is critical to the net result." (McDonough, 1995)

Complexity leads to expense, and until recent moves towards Open Systems development of CIM / MES software few Australian companies would have been able to support the expenditure of millions of dollars on powerful centralised computer systems and large, expensive proprietary software to run on them. Today, scores of software companies offer management support packages which run on widely used interface systems and operating systems such as DOS, OS/2, MS Windows, Windows95, WindowsNT and UNIX. System integration can start with packages which are needed and affordable, then develop as the needs of the company and the sophistication of the users grow. Reductions in inventory and lead times, smoother flows through the plant and improvements in quality are achieved yet the systems are imposed top-down and, while improving executive monitoring and control of production, do not address the fundamental problem of executive / production dichotomy. Neither, regardless of the method of implementation, does CIM / MES change the underlying processes of production.

Where two solutions to any problem exist, it is common for them to be substantially different in emphasis. For example, it is said that the NASA (when the US first started putting men into space) spent millions of dollars devising a ballpoint pen which would work in weightless conditions while the USSR solved the weightless writing problem by issuing its cosmonauts with pencils. Similarly, the Japanese response to the pressure for improved return on investment was fundamentally different from that in the West.

4 Japan's Solution (Just-In-Time) is People-based

"*After World War II, Japanese planners developed strategies for competing with the United States and Western Europe... Growing international price competition forced continual reductions in manufacturing and marketing costs. When Japan's competitive advantage derived from low-cost labour become exhausted, purchasing and inventory management systems became a focus for cost reductions. This led to JIT, although it did not become widespread throughout Japan until the OPEC embargo in 1973.*" (Meredith, Ristroph, et al., 1991, p. 448)

Japan's economy collapsed to a state of zero growth under the effects of the 1973 oil crisis yet Toyota's earnings increased and the widening gap between it and other companies, many of whom had continued to use the conventional American-style system of mass production, generated interest in the production system developed in the 1960s and 70s by Toyota under the leadership of Vice President Taiichi Ohno. (Robinson, 199 1, p. 133) Although commonly referred to as '*Just-In-Time*' (or '*Kanban*' after the inventory control cards used) the system is not the application of a single idea but the integration of a range of concepts which continue to develop under the umbrella of Continuous Improvement ('*Kaizen*').

Ohno initiated the solution of the executive/production communication problem by means of simple, human communication. By opening dialogue between workers and management, establishing *bone fides* by acting on the suggestions of shop floor staff, Toyota's middle management (with Ohno's enthusiastic support) unleashed a *bottom-up* restructuring of the actual processes of production themselves, a situation which persists within the company - from a workforce of 60,000 Toyota received 2.6

million process improvement suggestions in the calendar year 1986. Of these 96% were implemented. (Hutchins, 1988, p. 10) In this, the Toyota system not only re-opened the communications cut off at the turn of the century but also rejected Taylor's philosophies, identifying them with low motivation, low job interest and absenteeism. (Hutchins 1990, p. 49) By the time of the oil crisis in 1973, Toyota had instituted a new system of production which enabled it to weather the economic storm which threatened to wreck conventional industry worldwide. Process changes included:

- Elimination of Waste which JIT defines as any activity which doesn't add value to the product (Hernandez, 1993, p. 10) including transport time, inspection time and waiting time in inventory buffers as well as the traditional rework and scrap.

- Reorganisation of the Shop Floor including a) movement of machines into process related groups which reduces materials transfer time and permits robotic handling (Kerr, 1991) and b) the establishment of worker movement patterns to avoid workers getting in each other's way and to equitably share the workload (Sartori, 1988)

- Zero Defects Quality utilising the mistake-proofing ('Poka-Yoke') techniques devised by Shigeo Shingo in which simple changes to processes and machines, made immediately on the production of a defect, prevent the recurrence of the defect. This may be seen to supersede Total Quality Management as it is understood in the West which merely seeks to reduce the production of defects to an acceptable level. (Shingo, 1986)

- Autonomation ('Jidoka') which precipitates immediate attention by operators and supervisors to the production of a defect by stopping the production line at a signal from workers or the automatic machines themselves. Immediate abnormality identification and process improvement compensate for the down-time with a subsequent lack of defects.
  (Robinson, 199 1) (Yoshikawa & Burbridge (eds), 1991)

- Single Minute Exchange of Dies (SMED) reducing set-up times and permitting economic production in small lot sizes. (Dyck, Varzandeh, et al., 1991)

- Reduction of Inventory to the minimum required to allow the plant to operate. This is made possible only by the achievement of Zero Defects Quality (since there is little allowance to cover for mistakes). Toyota seeks to keep inter-process inventory below one tenth of a day's production. (Graham, 1988)

  Just-In-Time Process Control in which movement of batches is controlled by authorisation cards (kanbans) which replace the 'work order' in the conventional 'push' system factory and from which they differ in three ways.

1. a kanban is issued when needed and not procedurally and at a pre-determined time, thereby eliminating traditional order states and their administrative overhead

2. kanbans are issued by the downstream process and until an empty palette or bin, with its kanban, arrives at a workstation that workstation remains idle regardless of any theoretical loss of productivity

3. each kanban represents an immediate requirement to be filled in *real time* thereby directly, connecting the adjacent processes and making superfluous the holding of buffer stocks of materials between departments. (Sartori, 1988)

Inventory level in any area is controlled by the number of kanban cards which Toyota calculates according to:

$$y = D(T_w + T_p)(1 + n)a$$

Equation 2 : Number Of Kanban Cards

where 'y' the number of kanban, 'D' is the demand per unit time, 'T_w' is the waiting time for the kanban to be returned to the supplying area, 'T_p' is the process time, 'a' is the container capacity and 'n' is the policy variable which allows some excess stock to accommodate disruptions and variations in usage rate. (Graham, 1988)

Throughout the restructuring process, Toyota strenuously resisted the idea of control by management through massive, centralised computer systems. Osamu Kimura, a General Manager of Toyota Motor Corporation's Transport Administration Office, illustrated information flows in the two types of system:

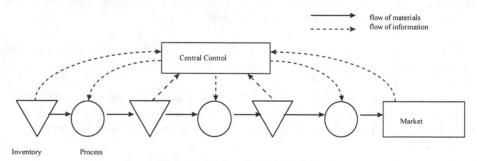

Fig 1 : Control In a Centralised System

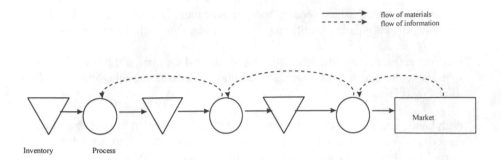

Fig 2 : Control in a Kanban System

"We should be careful not to centralise the system," he said, *"by means of mammoth computers and information networks which may only lead to death by strangulation."* (Yoshikawa & Burbridge, 1987, p. 18)

5 Solution to Schism a Question of Choice, not of Culture

Westerners sometimes correlate the success of the Toyota system, now widespread in Japan, with supposed cultural differences between Japanese workers and their Western counterparts - loyalty, long service, dedication and enthusiasm are seen to be parts of the Japanese culture and their absence in the West, therefore, to preclude the transplantation of the Japanese methodology. While some cultural differences are seen to exist, many commentators are adamant that management methodologies are exportable. (Fanuc, the Japanese CNC giant, is renowned for its high level of CIM and robotics implementation.)

Hofstede (1984) surveyed 50 countries ranking them comparatively on a variety of character traits and demonstrating quite clearly the existence of cultural differences. He concludes that the differences do not preclude the importation/exportation of systems across cultural boundaries but rather that:

"Effectiveness within a given culture, and judged according to the values of that culture, asks for management skills adapted to the local culture." (Hofstede, 1984. p. 98)

Translocation of JIT into the USA has frequently been highly successful, notably in that US bike icon Harley-Davidson. Other cases listed may include Apple, GM, General Electric, Gillette, Xerox and Hewlett Packard. (Meredith, Ristroph, et al., 1991,p.448)

"The attitude that Toyota's management and workers bring to each of these issues is not inherent to their "culture". In fact, a sizeable number of

American, firms have both the same philosophy and similar results." (Bignell, Dooner, et al., 1985, p. 154)

6 Appropriate Option may be a Mixture

Proposal 1: That clearly demonstrable benefits may accrue from the implementation of management support software.

Progressive implementation of currently available software in support of strategic management decision making, coupled with appropriate staff training and incentive, regularly produces significant improvements in efficiency, lead-time reduction, throughput and profit while reducing overhead and inventory. Despite the recent move to Open Systems implementation with reduced software costs and vendor dependence, however, the complete MES/CIM implementation remains a complex, expensive, long-term project best suited to major industrial operations.

Proposal 2: That realignment of management attitudes and restructuring of production processes may produce at least comparable benefits.

Just-In-Time manufacturing, including its associated systems and techniques, provided the impetus for the outstanding success of Japanese industry in the '80's at low cost and without massive computerisation. Management commitment to human-level communication and the consequent elimination of artificial executive / production barriers rekindled craftsmanship, pride in work, motivation and involvement at all staff levels. Resurgence of enterprise-wide focus, teamwork and willingness to make appropriate changes to fundamental processes cost little and achieved much.

Proposal 3: That Third Millennium Manufacturing may, to be successful, require a mixture of both solutions.

Manufacturing environments tend to be individually unique - JIT implementation at Nippon Steel must vary in practice from that at Sony and practice at Sony from that at Harley-Davidson; CIM implementation at Du Pont must vary from that at Fanuc and Fanuc's implementation from that at the wastewater treatment plant in Lexington, KY (USA). Australian industry is typified by small size and short production runs, consequently solutions applicable to multinational companies, directly transferred, are likely to be inappropriate. Careful blending of concepts from both mainstream solutions appears to offer the most applicable options.

Proposal 4: That reassessment of production processes should precede automation of control.

Logic demands that the best control of a flawed system can never produce the best results. Survival of Australian industry into the third millennium may depend on management accepting the challenge, and pain, of reformation far more than on management willingness to accept the cost of computerisation.

References

1. Batchelor, Ray, *Henry Ford: Mass Production, Modernism and Design.* New York, NY (USA): Manchester University Press (1994).
2. Bignell, Victor, Dooner, Mike, Hughes, John, Pym, Chris, Stone, Sheila (eds), *Manufacturing Systems: Context, Applications and Techniques.* Oxford (UK): Basil Blackwell Ltd (1985).
3. Dyck, Harold, Varzandeh, Jay, McDonnell, Jack, *Quality Impacts on JIT Performance Measures: A Factory Simulation,* Technology Management - The New International Language (Cat No 91CH3048-6) (October 27-31, 1991).
4. Ford, Henry, *Ford on Management, Harnessing, the American Spirit.* Oxford (UK): Basil Blackwell Ltd (1991)
5. Gibson, P., Greenhalgh, G., Kerr, R. *Manufacturing Management.* London, (UK): Chapman and Hall (1995)
6. Graham, Ian R, *'Just-in-Tiine' Management of Manufacturing.* Oxford (UK) : Technical Communications and Elsevier Advanced Technology Productions (1988)
7. Hernandez, Arnaldo. *Just-in-Tirne Ouality.* Englewood Cliffs, NJ (USA): PTR Prentice Hall (1993)
8. Hofstede, Geert. *Cultural Dimensions in Management and Planning.* Asia Pacific Journal of Management. 1 (2) 81-99 (Jan, 1984)
9. Hutchins, David. *Just In Time.* Aldershot, Hants. (UK) : Gower Technical Press Ltd. (1988)
10. Hutchins, David. *In Pursuit of Quality: Participative Techniques for Quality Improvement.* London (UK): Pitman Press (1990)
11. Ishikawa. Kaoru, Lu, David J. *What is Total Quality Control? The Japanese Way.* Englewood Cliffs, *NJ,* (USA): Prentice-Hall Inc. (1985)
12. Johnson, H. Thomas, Kaplan, Robert S, *Relevance, Lost : The Rise and Fall of Management Accounting.* Boston, MA (USA): Harvard Business School Press (1991)
13. Kerr, Roger, *Knowledge-Based Manufacturing Management: Applications of Artificial Intelligence to the Effective Management of Manufacturing Companies.* Sydney, NSW: Addison-Wesley Publishing Company (1991)
14. Kirkwood, L., Ryan, C., Falt, J., Stanley, T. *Accounting: An Introductory Perspective.* Melbourne, VIC: Longman Cheshire Pty Ltd (1989)
15. Linnegar, Gary J. *An Investigation into the Management of Change in Cost Accounting System Requirements During the Transition from Traditional Manufacturing to Just In Time Concepts.* Mississippi (US): Mississippi State University (1988)
16. Mathews, M. R., Perera, M.H. B., Ng, L.W., Chua,F.C., *Accounting Theory and Development.* South Melbourne, VIC. Thomas Nelson Australia (1991)
17. McDonough, Brian., *Considerations for Manufacturing Execution System Implementation.* [on -line] Available: http://www.smartdocs.com/~bmcd/accufacts9000/ art3.html (1995)
18. Meredith, Paul H, Ristroph, John H., Lee, Jim. *Implementing JIT: The Dimensions of Culture, Management, and Human Resources,* Technology Management - The New International Language (October 27-31, 1991)
19. Mitchell, Geoffrey B., Granof, Michael H., *Principles of Accounting.* Sydney,NSW.. Prentice-Hall of Australia (1981)
20. Orlicky, Joseph., *Material Requirements Planning : The New Way of Life in Production and Inventory Management.* New York, NY (USA): McGraw-Hill, Inc, (1975).

21. Robinson, Alan (ed.)., *Continuous Improvement in Operations : A Systematic Approach to Waste Reduction.* Cambridge, MT (USA), Productivity Press (1991).

22. Sartori, Luca G., *Manufacturing Information Systems.* Wokingham (UK), Addison-Wesley Publishers Ltd(1988)

23. Shingo, Shigeo, *Zero Quality Control : Source Inspection and Poka-Yoke System.* Norwalk, CT (USA), Productivity Press (1986)

24. Storey, J (ed.), *New Wave Manufacturing Strategies: Organisational and Human Resource Management Dimensions.* London (UK), Paul Chapman Publishing (1994)

25. Taylor, Frederick Winslow., *The Principles of Scientific Management.* New York, NY (USA): W. W. Norton & Company Inc. (1967)

26. Taylor, Frederick Winslow. *Principles of Scientific and Shop Management.* London (UK), Routledge/Thoemrnes Press(1993)

27. Yoshikawa, H., Burbridge, J, L. (eds). *New Technologies for Production Systems.* Amsterdam (ND): Elsevier Science Publishers B. V. (1987)

AUTHOR INDEX